GABRIEL'S
GABRIEL AERO-MARINE INSTRUMENTS LTD.
351 ST. PAUL ST. WEST
MONTREAL, P.Q.
AND HALIFAX, N.S.

REED'S
HEAT AND HEAT ENGINES
FOR
MARINE ENGINEERS

REED'S HEAT AND HEAT ENGINES FOR MARINE ENGINEERS

By
WILLIAM EMBLETON O.B.E.
EXTRA FIRST CLASS ENGINEERS CERTIFICATE
M.R.I.N.A., A.M.I.Mech.E., M.I.Mar.E., A.M.N.E.C.Inst.
*Head of the Mechanical & Marine Engineering Department
South Shields Marine and Technical College*

PUBLISHED BY THOMAS REED PUBLICATIONS LIMITED
SUNDERLAND AND LONDON

First Edition - 1963

Reprinted - 1966

Reprinted - 1969

PRINTED BY THOMAS REED AND COMPANY LIMITED
SUNDERLAND, GLASGOW AND LONDON

PREFACE

This book is the third volume of "Reed's Practical Mathematics Series', and completes the coverage of the Ministry of Transport's syllabuses for Part A of the Examinations for Second and First Class Certificates of Competency. This volume covers the Heat and Heat Engines section and, like the previous two, begins at a fairly elementary stage and progresses to the standard required for the First Class examination. Emphasis is again placed on basic principles and simple explanatory diagrammatic sketches are used to clarify them. Parts of the subject matter and the attendant test examples are marked with the prefix "f" to indicate that they are considered to be beyond the standard normally required for Second Class.

Fully worked solutions are given to all examples and problems, and detailed mathematical steps are included where such are considered to be of help to Engineers studying at sea without a tutor to turn to for assistance.

In presenting this third book of the series, the author takes the opportunity of wishing success to the student, who has carefully and diligently studied each subject from its first principles.

W. Embleton

CONTENTS

PAGE

CHAPTER 1—HEAT AND MEASUREMENT OF HEAT.
Temperature, Fahrenheit and Centigrade, conversion. Absolute temperature. Units of heat and their relationships. Specific heat. Water equivalent. Sensible and latent heat. Heat mixtures ... ... 1— 19

CHAPTER 2—EXPANSION OF METALS AND HEAT TRANSFER.
Linear, superficial and cubical expansion. Conduction, convection and radiation. Mechanical equivalent of heat. ... ... 20— 35

CHAPTER 3—LAWS OF PERFECT GASES.
Boyle's and Charles' laws. Characteristic gas equation. Specific heats of gases. Energy equation, internal energy. Dalton's law of partial pressures. 36— 53

CHAPTER 4—EXPANSION AND COMPRESSION OF GASES.
Isothermal, adiabatic and polytropic expansion and compression. Ratios of expansion and compression. Relationships between temperature, pressure and volume. Work done during expansion and compression. ... ... 54— 71

CHAPTER 5—I.C. ENGINES—ELEMENTARY PRINCIPLES.
Cycle of operations in four-stroke and two-stroke diesel engines, Doxford opposed piston engine, four-stroke and two-stroke petrol engines. Timing diagrams and indicator cards. Cooling, lubrication, starting, reversing. Mean effective pressure and indicated horse power. Brake horse power. Mechanical efficiency. Morse test. Indicated and brake thermal efficiencies. Heat balance. 72—107

CHAPTER 6—IDEAL CYCLES.   PAGE
  Constant volume, diesel, dual-combustion, and Carnot cycles. Ideal and air-standard efficiency. Reversed Carnot cycle, coefficient of performance. Engine clearance and stroke volume. ... ...   108—131

CHAPTER 7—AIR COMPRESSORS. ... ...
  Single and multi-stage compression. Effect of clearance. Work done per cycle. Time to pump up reservoirs. ... ...   132—147

CHAPTER 8—STEAM.
  Constant pressure steam generation. Steam tables. Wet and dry saturated steam, dryness fraction. Superheated steam. Specific volume. Specific heat of superheated steam. Heat mixtures. Throttling and separating calorimeters. Dalton's law with reference to air in condensers. Internal energy of steam.   148—181

CHAPTER 9—ENTROPY.
  Entropy of water, evaporation and superheated steam. Temperature-entropy diagram, dry steam and dryness fraction curves. Isothermal and adiabatic processes. Constant volume lines. ... ...   182—194

CHAPTER 10—STEAM RECIPROCATING ENGINES.
  General construction. Working cycle and action of the slide valve. Size of steam ports. Valve diagrams. Weight of steam from indicator card. Missing quantity. Compounding. Compound, triple and quadruple expansion. ...   195—224

CHAPTER 11—MEAN EFFECTIVE PRESSURE, HORSE POWER AND EFFICIENCY OF STEAM ENGINES.
  Indicated horse power. Hypothetical mean effective pressure, diagram factor and actual m.e.p. Referred mean pressure. Determination of cylinder diameters. Willans' law. Thermal efficiency. Rankine efficiency. Boiler efficiency and equivalent evaporation. ... ...   225—249

PAGE

CHAPTER 12—TURBINES.
Impulse and reaction steam turbines. Astern running. Gearing. Exhaust turbines. Regenerative condenser. Nozzles. Velocity diagrams for impulse and reaction turbines. Force on blades, horse power and blade efficiency. Gas turbines. 250—279

CHAPTER 13—COMBUSTION.
Furnace draught. Coal and oil burning. Higher and lower calorific values. Oxygen and air required for combustion. Bomb calorimeter. Composition of funnel gases, conversion from volume to weight analysis. ... ... 280—297

CHAPTER 14—REFRIGERATION.
Vapour-compression system, refrigerants, working cycle. Brine circulation, insulation, temperatures of food storage. Capacity and performance. ... ... 298—309

SOLUTIONS TO TEST EXAMPLES .... 310—420

DATA SUPPLIED AT EXAMINATIONS 421

SECOND CLASS EXAMINATION QUESTIONS 422—441

SOLUTIONS TO SECOND CLASS QUESTIONS 442—499

FIRST CLASS EXAMINATION QUESTIONS .... 500—525

SOLUTIONS TO FIRST CLASS QUESTIONS 526—619

CHAPTER 1
HEAT AND MEASUREMENT OF HEAT
THERMOMETRY

Temperature is an indication of the degree of "hotness" and is therefore a measure of the *intensity* of heat in a body.

Temperature is measured in *degrees* on either the *Fahrenheit* scale or the *Centigrade* scale, both are graduated on the basis of the natural freezing point and boiling point of fresh water at atmospheric pressure but are scaled differently. Fig. 1 illustrates two similar mercury-in-glass thermometers, one is graduated in Fahrenheit units and the other in Centigrade units. This type of mercurial thermometer is the commonest temperature measuring instrument in use. It consists of a glass tube with a bulb at the lower end, the bulb and tube are exhausted of air, partially filled with mercury and hermetically sealed at the top end. When the thermometer is placed in a substance whose temperature is to be measured, the mercury takes up the same temperature and expands (if heated) or contracts (if cooled) and the level, which rises or falls in consequence, indicates on the scale the degree of heat intensity.

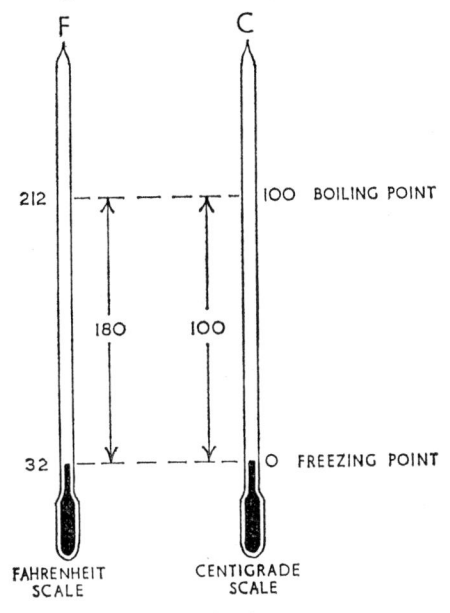

Fig. 1

At the level of temperature at which fresh water freezes, the Fahrenheit thermometer is marked 32 and the Centigrade thermometer is marked zero. At the temperature of natural boiling point of fresh water (i.e. at atmospheric pressure) the Fahrenheit thermometer is marked 212 and the Centigrade 100.

From this we can derive simple formulae to convert temperature changes and temperature readings from one scale to the other.

TEMPERATURE CHANGES

Consider a change of temperature from freezing point to natural boiling point of fresh water.

Fahrenheit temperature change = 212 — 32 = 180 degrees.
Centigrade temperature change = 100 — 0 = 100 degrees.
∴ 180 Fahrenheit degrees = 100 Centigrade degrees.

In the same proportion,
9 Fahrenheit degrees = 5 Centigrade degrees.

Hence to convert a change of temperature or a difference between the temperatures of two points, from one scale to the other:

$$\text{Change on Fahrenheit scale} = \frac{9}{5} \times \text{change on Centigrade scale}$$

$$\text{Change on Centigrade scale} = \frac{5}{9} \times \text{change on Fahrenheit scale}$$

Example. A body is heated until its temperature rises through 378 F degrees. Express this change of temperature on the Centigrade scale.

$$\text{Centigrade change of temperature} = \frac{5}{9} \times \text{Fahrenheit change}$$

$$= \frac{5}{9} \times 378$$

$$= 210 \text{ C degrees. Ans.}$$

TEMPERATURE READINGS

As zero on the Centigrade scale corresponds to 32 on the Fahrenheit scale, the two thermometer scales are not graduated from the same numerical level. Zero mark on the Fahrenheit thermometer is 32 divisions or degrees below its freezing point. This must be taken into account when comparing or converting actual temperature readings. Thus this 32 must be subtracted from the Fahrenheit reading to bring the two to a common base, and then the result multiplied by the ratio $\frac{5}{9}$ to obtain the Centigrade reading.

$$\text{C reading} = (\text{F} - 32) \times \frac{5}{9}$$

and conversely,

$$\text{F reading} = \left(\text{C} \times \frac{9}{5}\right) + 32$$

As a further explanation, the graph in Fig. 2 shows the relationship between the two temperature scales. It is a straight line graph of the form $y = a + bx$ as previously explained in Volume 1 (Mathematics), therefore,

If F = Fahrenheit temperature reading,
C = Centigrade temperature reading,

we have,
$$\text{F} = a + b\text{C}$$

where,
a = Fahrenheit temperature when the Centigrade reading is zero, this is 32,
b = Increase in Fahrenheit temperature per degree increase in Centigrade temperature, this is $\frac{9}{5}$

Hence,

$$\text{F} = 32 + \frac{9}{5}\text{C}$$

and

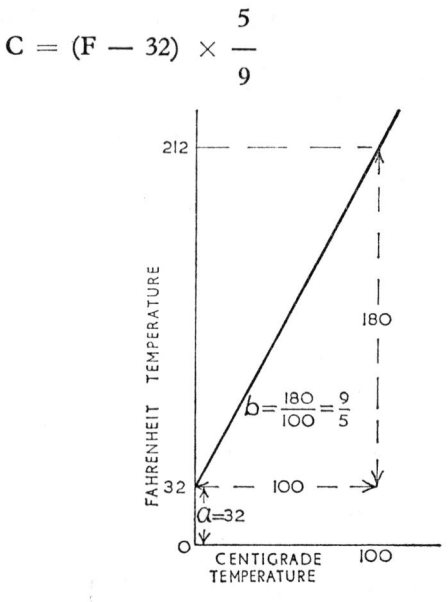

$$C = (F - 32) \times \frac{5}{9}$$

Fig. 2

Temperatures below zero on either scale are referred to as *minus* quantities. For example, 3 degrees below zero on the Centigrade scale is written — 3°C, this is 3 Centigrade degrees below freezing point and therefore sometimes expressed as "3 Centigrade degrees of frost". A temperature of 3 degrees below zero on the Fahrenheit scale is written — 3°F, this is (32 + 3) = 35 Fahrenheit degrees below freezing point and may be expressed as "35 Fahrenheit degrees of frost".

A little extra care is necessary where minus signs are concerned in converting from one scale into the other.

Example. Convert the temperatures 68°F, 5°F and — 49°F into temperature readings on the Centigrade scale.

$$C \text{ reading} = (F - 32) \times \frac{5}{9}$$

$$= (68 - 32) \times \frac{5}{9}$$

$$= 36 \times \frac{5}{9}$$

$$= 20°C \quad \text{Ans. (i)}$$

$$\text{C reading} = (5 - 32) \times \frac{5}{9}$$

$$= -27 \times \frac{5}{9}$$

$$= -15°C \quad \text{Ans. (ii)}$$

$$\text{C reading} = (-49 - 32) \times \frac{5}{9}$$

$$= -81 \times \frac{5}{9}$$

$$= -45°C \quad \text{Ans. (iii)}$$

Example. Convert the temperatures 20°C, — 10°C and — 25°C into temperature readings on the Fahrenheit scale.

$$\text{F reading} = C \times \frac{9}{5} + 32$$

$$= 20 \times \frac{9}{5} + 32$$

$$= 36 + 32$$
$$= 68°F \quad \text{Ans. (i)}$$

$$\text{F reading} = -10 \times \frac{9}{5} + 32$$

$$= -18 + 32$$
$$= 14°F \quad \text{Ans. (ii)}$$

$$\text{F reading} = -25 \times \frac{9}{5} + 32$$

$$= -45 + 32$$
$$= -13°F \quad \text{Ans. (iii)}$$

ABSOLUTE TEMPERATURE

Later we shall see that all gases expand at the same rate when heated through the same range of temperature, and contract at the same rate when cooled.

The rate of expansion or contraction of a perfect gas is $\frac{1}{273}$ of its volume at 0°C when heated or cooled at constant pressure through one Centigrade degree, or $\frac{1}{492}$ of its volume at 32°F for one degree Fahrenheit change of temperature. Hence if a gas initially at 0°C could be cooled at constant pressure until its temperature is 273 Centigrade degrees below 0°C, the volume would contract until there was nothing left and no further reduction of temperature would be possible, i.e. the gas would then have reached its *absolute zero of temperature*. Similarly on the Fahrenheit scale, if the gas initially at a temperature of 32°F was cooled at constant pressure until its temperature was 492 Fahrenheit degrees below 32 (i.e. 460 below 0°F), it would then have reached its lower limit or absolute zero of temperature (see Fig. 3).

In practice, of course, it is not possible to cool a gas down to the absolute zero and cause it to disappear. As the absolute zero of temperature is approached the gas will change into a liquid and the laws of gases are then no longer applicable.

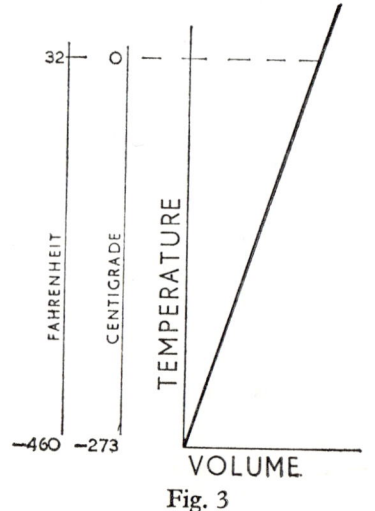

Fig. 3

We see from the above that temperatures can be expressed as *absolute* quantities, that is, stating the degrees of temperature above the level of Absolute Zero, by adding 273 to the ordinary Centigrade thermometer reading or adding 460 to the Fahrenheit thermometer reading, thus:

t°C + 273 = T°C absolute
t°F + 460 = T°F absolute

To convert an absolute temperature from one scale to the other, only the ratio 9 F degrees = 5 C degrees applies, there is no adding or subtracting 32 because both absolute scales have the common base of absolute zero.

HEAT AND UNITS OF HEAT

Heat is a form of energy which is convertible into other forms of energy and can be made available for doing work and producing mechanical and electrical power.

The quantity of heat required to raise the temperature of a substance depends upon its mass, its nature and its increase in temperature. Units of heat are based upon water. One heat unit is taken as the heat required to raise unit weight of water through one degree of temperature. As there are different units of weight and two different scales, then there are different units of heat. Those most commonly used are as follows:

The *British Thermal Unit* (Btu) is the quantity of heat required to raise the temperature of 1 lb of water through 1 degree Fahrenheit.

The *Centigrade Heat Unit* (Chu) is the quantity of heat required to raise the temperature of 1 lb of water through 1 degree Centigrade.

The *Gram-Calorie* is the quantity of heat required to raise the temperature of 1 gram of water through 1 degree Centigrade.

The above are the usual brief definitions. However, as the amount of heat to raise the temperature of water through one degree varies slightly at different temperatures, more correct definitions state that the Mean British Thermal Unit is $\frac{1}{180}$ of the heat required to raise 1 lb of water from 32 to 212°F, the Mean Centigrade Heat Unit is $\frac{1}{100}$ of the heat required to raise 1 lb of water from 0 to 100°C, and the Mean Gram-Calorie is $\frac{1}{100}$ of the heat required to raise 1 gram of water from 0 to 100°C.

One gram is the weight of one cubic centimetre of water. The gram-calorie is therefore a very small quantity of heat and it is often more convenient to express heat units in the metric system in terms of *Kilo-Calories* (kcal), one kilo-calorie being equal to 1000 gram-calories and hence equivalent to the heat required to raise 1 kilogram of water through 1 degree Centigrade.

RELATIONSHIP BETWEEN HEAT UNITS

Compare the Btu and the Chu. Both are units of heat based on 1 lb weight of water but the Btu is a rise in temperature of 1 Fahrenheit degree and the Chu a rise of 1 Centigrade degree. One Fahrenheit degree is only $\frac{5}{9}$ of a Centigrade degree, therefore:

$$\text{One Btu} = \frac{5}{9} \text{ Chu}$$

Compare the Btu and the gram-calorie.
Taking 2·2046 lb = 1 kilogram = 1000 grams

$$\text{One Btu} = 1 \text{ lb} \times 1 \text{ F}°$$
$$= \frac{1000}{2·2046} \text{ grams} \times \frac{5}{9} \text{ C}°$$

∴ **One Btu = 252 gram-calories**

Example. One pint of water is heated from 48°F to 84°F. Find the quantity of heat given to the water in (i) Btu, (ii) Chu, (iii) gram-calories, (iv) kilo-calories.

One gallon = 8 pints, weighs 10 lb

∴ One pint of water weighs 1·25 lb

Heat supplied = Weight of water × temperature rise
= 1·25 × (84 — 48)
= 1·25 × 36
= 45 Btu   Ans. (i)

$45 \times \dfrac{5}{9}$ = 25 Chu   Ans. (ii)

45 × 252 = 11340 gram-calories.   Ans. (iii)
11340 ÷ 1000 = 11·34 kilo-calories.   Ans. (iv)

HEAT VALUES PER UNIT WEIGHT

Later we shall be dealing with the quantity of heat per unit weight of a substance, such as the calorific value (i.e. heating value) of a fuel which is the quantity of heat given off when unit weight of the fuel is completely burned, also the latent heat of a substance which is the quantity of heat required to change the physical state of unit weight. These values are expressed in Btu per lb, Chu per lb, or gram-calories per gram, and we should be capable of converting from one form into the other.

Taking for an example that the latent heat of ice is 144 Btu per lb, this means that 1 lb of ice at 32°F requires 144 Btu of heat to completely melt it and change it into water at the same temperature. To express this value into other units:

$$\begin{aligned}\text{Latent heat of ice} &= 144 \text{ Btu per lb} \\ &= 144 \times \frac{5}{9} \text{ Chu per lb} \\ &= 80 \text{ Chu/lb} \quad \dots \quad \dots \quad \dots \quad \dots \quad \text{(i)}\end{aligned}$$

Thus, to convert Btu per lb into Chu per lb is simply a matter of converting the heat units of Btu to Chu by multiplying by $\frac{5}{9}$ because the same weight is involved in each case.

Now from Btu per lb into gram-calories per gram:

$$\begin{aligned}\text{L.H. of ice} &= 144 \text{ Btu per lb} \\ &= 144 \times 252 \text{ gram-calories per lb} \\ &= 144 \times 252 \times 2\cdot 2046 \text{ cal. per kilogram} \\ &= 144 \times 252 \times \frac{2\cdot 2046}{1000} \text{ cal. per gram}\end{aligned}$$

On page 8 we obtained the figure 252 from,

$$\frac{5}{9} \times \frac{1000}{2\cdot 2046}$$

Substituting this,

$$\text{L.H. of ice} = 144 \times \frac{5}{9} \times \frac{1000}{2\cdot 2046} \times \frac{2\cdot 2046}{1000}$$

$$= 144 \times \frac{5}{9}$$

$$= 80 \text{ gram-calories per gram} \quad \ldots \quad \ldots \quad (ii)$$

Hence, to convert Btu per lb into gram-calories per gram, simply multiply by $\frac{5}{9}$. Note that Chu per lb and gram-calories per gram are numerically equal.

SPECIFIC HEAT

Different substances have different heat capacities, that is, the quantity of heat required to raise different substances through the same temperature range differs and therefore the quantity of heat contained by them at the same temperature is different. The property which expresses the capability of absorbing and storing heat is known as the *specific heat*, and for purposes of comparison the specific heat of water is taken as unity.

The specific heat of a substance can therefore be defined as the ratio of the amount of heat required to raise it in temperature to the heat required to raise the same weight of water through the same range of temperature. Alternatively, the specific heat of a substance is the heat required to raise unit weight of that substance through a temperature of one degree.

From the first definition, specific heat may be regarded as a pure number. From the second definition the units of specific heat are heat units per unit weight per degree of temperature change, such as, Btu per lb per degree Fahrenheit, Chu per lb per degree Centigrade, or gram-calories per gram per degree Centigrade.

For example, to raise 1 lb of water through 1 F degree requires 1 Btu, therefore the specific heat of water is 1. To raise 1 lb of copper through 1 F degree would require 0·095 Btu, hence the specific heat of copper is 0·095. Consequently, the heat required to raise any substance in temperature is the product of its weight, its specific heat and the rise in temperature. To distinguish this heat effect from other kinds, when heat given to or taken from a substance causes a change in temperature only, it is termed sensible heat, hence,

Sensible Heat = Weight × spec. heat × change of temperature

From the definitions of units of heat:

If the weight is in lb and temperature change in F degrees, the quantity of heat will be expressed in Btu.

If the weight is in lb and temperature change in C degrees, the quantity of heat will be expressed in Chu.

If the weight is in grams and the temperature change in C degrees, the quantity of heat will be expressed in gram-calories.

Example. Calculate the quantity of heat required to raise the temperature of 5 lb of brass from 20°C to 240°C taking the specific heat of brass as 0·094.

Sensible heat = Weight × sp. ht. × temperature change
= 5 × 0·094 × (240 — 20)
= 5 × 0·094 × 220
= 103·4 Chu   Ans.

Example. Calculate the quantity of heat to be taken from 14 lb of ice at 32°F to cool it down to — 10°F, the specific heat of ice being 0·5.

Sensible heat = Wt. × sp. ht. × temperature change

Note that the change in temperature is from + 32 to — 10 which is through 42 F degrees.

∴ Sensible heat = 14 × 0·5 × [32 — (— 10)]
= 14 × 0·5 × (32 + 10)
= 14 × 0·5 × 42
= 294 Btu   Ans.

SENSIBLE HEAT AND LATENT HEAT

We have already seen that Sensible Heat is the heat given to or taken from a substance which causes a change in temperature, while its physical state remains unchanged. Thus, if a solid is given sensible heat, its temperature rises but it still remains a solid. If sensible heat is given to or taken from a liquid, its temperature rises or falls, but it still remains a liquid. Sensible heat is calculated as previously shown by the product of the weight, specific heat and the change in temperature.

Latent Heat is the heat given to or taken from a substance which *changes its physical state* while the temperature remains unchanged. The values of latent heats are found experimentally and expressed in heat units per unit weight.

The latent heat of ice is 144 Btu per lb. Thus, if ice is at its melting point of 32°F it would need to be given 144 Btu for every lb weight to melt it into water *at the same temperature*, or if water is at its freezing point of 32°F, 144 Btu would need to be extracted from each lb to freeze it into ice at the same temperature.

The latent heat of steam at atmospheric pressure is 970·6 Btu per lb. Hence, if water is at its boiling point under atmospheric pressure, i.e. 212°F, each lb would require 970·6 Btu to evaporate it into steam at the same temperature, or steam at atmospheric pressure would require to have 970·6 Btu extracted from each lb to condense it into water at the same temperature. The temperature at which water boils and the latent heat of steam, both depend upon the pressure. The higher the pressure, the higher the boiling point and the less the latent heat. This is explained in more detail later.

Example. Find the total heat required to be given to 4 lb of ice at — 20°F to change it into steam at atmospheric pressure, taking the values:

$$\text{Specific heat of ice} = 0\cdot 5,$$
$$\text{Latent heat of ice} = 144 \text{ Btu per lb,}$$
$$\text{Latent heat of steam} = 970\cdot 6 \text{ Btu per lb.}$$

As the melting point of ice is 32°F, sensible heat must first be supplied to raise the temperature of the ice from — 20°F to 32°F, which is an increase in temperature of 52 F degrees.

$$\begin{aligned}
\text{Sensible heat} &= \text{Wt.} \times \text{sp. ht.} \times \text{temperature rise} \\
&= 4 \times 0\cdot 5 \times [32 - (-20)] \\
&= 4 \times 0\cdot 5 \times (32 + 20) \\
&= 4 \times 0\cdot 5 \times 52 \\
&= 104 \text{ Btu} \quad \ldots \quad \ldots \quad \ldots \quad \ldots \quad (\text{i})
\end{aligned}$$

When the ice has received this sensible heat it is now at its melting point of 32°F. It will not rise above this temperature until it has all been changed into water by giving it latent heat:

$$\begin{aligned}
\text{Latent heat} &= 4 \times 144 \\
&= 576 \text{ Btu} \quad \ldots \quad \ldots \quad \ldots \quad \ldots \quad (\text{ii})
\end{aligned}$$

Now we have 4 lb of water at 32°F and more heat given to it will raise its temperature. As the water will boil at 212°F because it is under atmospheric pressure, the heat to be given to it next is sensible heat to raise the temperature from 32 to 212°F. The specific heat of water is unity.

HEAT AND MEASUREMENT OF HEAT

$$\text{Sensible heat} = 4 \times 1 \times (212 - 32)$$
$$= 4 \times 180$$
$$= 720 \text{ Btu} \quad \ldots \quad \ldots \quad \ldots \quad \ldots \quad \text{(iii)}$$

The water is now at its boiling point and 970·6 Btu are required to completely evaporate each lb into steam at the same temperature of 212°F.

$$\text{Latent heat} = 4 \times 970 \cdot 6$$
$$= 3882 \cdot 4 \text{ Btu} \quad \ldots \quad \ldots \quad \ldots \quad \text{(iv)}$$

$$\text{Total heat} = \text{(i)} + \text{(ii)} + \text{(iii)} + \text{(iv)}$$
$$= 104 + 576 + 720 + 3882 \cdot 4$$
$$= 5282 \cdot 4 \text{ Btu} \quad \text{Ans.}$$

The above example is set out in great detail so that the student will understand each step. Once it is fully understood, the work in such calculations as these can be very much condensed by finding the heat required to be given to 1 lb of the substance, then finally multiplying this by the total weight.

For example, the heat required to be given to 3 lb of ice at 15°F to change it into water at 45°F would be set out thus:

$$\text{Heat required} = 3 \left[0 \cdot 5 (32 - 15) + 144 + (45 - 32)\right] \text{ Btu}$$

The quantity inside the square brackets represents the heat required for each lb and this is finally multiplied by 3 to obtain the heat required for 3 lb. Note the terms inside the square brackets, the first term is the sensible heat to raise 1 lb of ice (sp. ht. = 0·5) from 15°F to 32°F, the second term is the latent heat to change the 1 lb of ice at 32°F into 1 lb of water at 32°F, and the third term is the sensible heat to raise 1 lb of water (sp. ht. = 1) from 32°F to 45°F. Such a layout as this will be used in heat mixture problems involving ice.

HEAT MIXTURES

When two substances at different temperatures are mixed together, the hotter substance will lose heat while the colder substance gains heat until both reach the same temperature. Unless otherwise stated, it would be assumed that no heat is lost to, or absorbed from, an outside source during the mixing process, and therefore the quantity of heat gained by the colder substance must be at the expense of the heat lost by the hotter. Therefore,

Heat gained by colder substance = Heat lost by hotter

When using this equation it is often advisable to state the kind of heat units employed in its construction.

Example. 50 grams of steel at a temperature of 200°C are immersed in 800 grams of oil having a temperature of 20°C. Taking the specific heat of steel as 0·116 and specific heat of oil as 0·5, find the resultant final temperature of the steel and oil.

Let t = final temperature in °C.

The steel falls in temperature from 200°C to t°C and the loss in temperature is $(200 - t)$ C degrees.

The oil rises in temperature from 20°C to t°C and the gain in temperature is $(t - 20)$ C degrees.

Working in gram-calories because weights are in grams and temperatures in degrees Centigrade:

$$\begin{aligned}
\text{Heat gained by oil} &= \text{Heat lost by steel} \\
\text{Wt.} \times \text{sp. ht.} \times \text{temp. rise} &= \text{Wt.} \times \text{sp. ht.} \times \text{temp. fall} \\
800 \times 0\cdot 5 \times (t - 20) &= 50 \times 0\cdot 116 \times (200 - t) \\
400t - 8000 &= 1160 - 5\cdot 8t \\
400t + 5\cdot 8t &= 1160 + 8000 \\
405\cdot 8t &= 9160 \\
t &= 22\cdot 57°\text{C. Ans.}
\end{aligned}$$

Example. 10 lb of lead (specific heat 0·029) at 150°F and 4 lb of copper (specific heat 0·095) at 300°F are put into 5 lb of water at 40°F. Find the resultant temperature of the mixture.

Let t = resultant temperature in °F.

Working in Btu because weights are in lb and temperatures in degrees Fahrenheit:

$$\begin{aligned}
\text{Heat gained by water} &= \text{Heat lost by lead} + \text{Heat lost by copper} \\
5 \times (t - 40) &= 10 \times 0\cdot 029 \times (150 - t) + 4 \times 0\cdot 095 \\
&\qquad\qquad\qquad\qquad\qquad\qquad\qquad\qquad \times (300 - t) \\
5t - 200 &= 43\cdot 5 - 0\cdot 29t + 114 - 0\cdot 38t \\
5\cdot 67t &= 357\cdot 5 \\
t &= 63\cdot 05°\text{F. Ans.}
\end{aligned}$$

Note that in the above example it was obvious that the water *gained* heat and the copper *lost* heat, but it was not known whether the lead would gain or lose heat, we therefore had to assume one or the other. We guessed that the final temperature would be below 150°F and therefore formed the equation on the assumption that the lead would lose heat. This however would take care of itself. If we had guessed that the final temperature would be more than 150°F and made the assumption that the lead would gain heat, the change of heat of the lead would appear on the opposite side of the equation with opposite signs, and therefore the final solution is not affected as shown below.

Heat gained by water + Heat gained by lead = Heat lost by copper

$$5 \times (t - 40) + 10 \times 0.029 \times (t - 150) = 4 \times 0.095 \times (300 - t)$$
$$5t - 200 + 0.29t - 43.5 = 114 - 0.38t$$
$$5.67t = 357.5$$
$$t = 63.05°F$$

as before.

Example. 3 lb of ice at 24°F are mixed into 2 gallons of water at 68°F. The specific heat of ice is 0·5 and its latent heat is 144 Btu per lb. Calculate the final temperature of the mixture.

One gallon of water weighs 10 lb
∴ Two gallons of water weigh 20 lb

Heat gained by ice = Heat lost by water
$$3 [0.5 (32 - 24) + 144 + (t - 32)] = 20 \times (68 - t)$$
$$3 (4 + 144 + t - 32) = 20 (68 - t)$$
$$3 (116 + t) = 20 (68 - t)$$
$$348 + 3t = 1360 - 20t$$
$$3t + 20t = 1360 - 348$$
$$23t = 1012$$
$$t = 44°F \quad \text{Ans.}$$

WATER EQUIVALENT

The *water equivalent* of a substance is the weight of water that would require the same quantity of heat as the substance to raise it through the same range of temperature. It is therefore the product of the weight of the substance and its specific heat. For example, the water equivalent of a piece of copper which weighs 8 lb and having a specific heat of 0·095 is 8 × 0·095 = 0·76 lb. That is to say 0·76 lb of water would require the same amount of heat as that 8 lb piece of copper to heat it through an equal rise of temperature.

Water equivalent = weight × specific heat

Example. 1·5 litres of water are contained in a vessel whose water equivalent is 400 grams. Find the quantity of heat, in kilo-calories, required to raise the temperature of the water and its container from 16°C to 80°C.

1 litre = 1000 cubic centimetres

As one cubic centimetre of water weighs one gram, then 1000 cubic centimetres weigh 1000 grams which is one kilogram. Hence 1·5 litres weigh 1·5 kilograms.

400 grams = 0·4 kilogram

Working in kilo-calories:

Heat required = Heat gained by water + Heat gained by vessel
= 1·5 × (80 — 16) + 0·4 × (80 — 16)
= (1·5 + 0·4) × (80 — 16)
= 1·9 × 64
= 121·6 kilo-calories.   Ans.

Note that the effect of the vessel is simply equivalent to having an extra 0·4 kilogram of water.

Example. In an experiment to determine the specific heat of lead, a quantity of lead shot weighing 2 lb is heated in a steam bath so that its temperature becomes 212°F. The lead shot is then poured into a copper calorimeter containing one pint of water at 62°F and the resultant temperature of the mixture is 68·1°F. Calculate the specific heat of lead from this data, taking the water equivalent of the calorimeter as 0·114 lb.

One pint of water weighs 1·25 lb
Let S = specific heat of lead.

Taking the original and increase of temperature of the calorimeter to be the same as the water inside it, and working in Btu:

Heat lost by lead = Heat gained by water + Heat gained by calorimeter

$$2 \times S \times (212 - 68 \cdot 1) = 1 \cdot 25 \, (68 \cdot 1 - 62) + 0 \cdot 114 \, (68 \cdot 1 - 62)$$
$$2 \times S \times 143 \cdot 9 = (1 \cdot 25 + 0 \cdot 114) \times 6 \cdot 1$$

$$S = \frac{1 \cdot 364 \times 6 \cdot 1}{2 \times 143 \cdot 9}$$

$$= 0 \cdot 02891 \quad \text{Ans.}$$

TEST EXAMPLES 1

1. Convert the following temperature readings from the Fahrenheit scale to the Centigrade scale,
    140°F   41°F   5°F   — 31°F

2. Convert the following temperature readings from the Centigrade scale to the Fahrenheit scale,
    60°C   15°C   — 10°C   — 49°C

3. (a) Boiler water is raised in temperature in the feed heater through 158 Fahrenheit degrees, express this increase in temperature in Centigrade degrees.

   (b) The difference in temperature between the inside and outside of a refrigerating chamber is 18 Centigrade degrees, express this difference in temperature in Fahrenheit degrees.

4. (a) Calculate the temperature at which the Fahrenheit and Centigrade readings are the same.

   (b) Find the temperature at which the Fahrenheit reading is twice the Centigrade reading.

5. The calorific value of a certain class of oil fuel is 18450 Btu per lb. Express this value in (i) Chu per lb, (ii) gram-calories per gram.

6. Calculate the quantity of heat required to raise the temperature of 10 lb of steel from 60°F to 1000°F, taking the specific heat of steel as 0·116. Give the answer in (i) Btu, (ii) Chu, (iii) gram-calories, (iv) kilocalories.

7. 10 lb of aluminium at 60°F are put into a crucible. Calculate the total heat required to completely melt it taking the following values for aluminium,

    specific heat = 0·211
    melting point = 1220°F
    latent heat = 173 Btu/lb.

8. A piece of brass weighing 5 lb (specific heat 0·094) at a temperature of 350°F is dropped into 2 pints of water at 56°F. Find the resultant temperature of the mixture.

HEAT AND MEASUREMENT OF HEAT

9. To ascertain the temperature of the funnel gases, a piece of copper (specific heat 0·095) weighing 4 lb is suspended in the funnel until it attains funnel temperature and then dropped into 5 lb of water at 68°F. If the resultant temperature of the copper and water is 99°F, find the temperature of the funnel gases.

10. A piece of nickel at 250°C is dropped into a quantity of oil at 25°C, the volume of the oil being twenty times the volume of the nickel. Taking the following values, find the resultant temperature.

Specific heat of nickel = 0·109
Specific gravity of nickel = 8·85
Specific heat of oil = 0·48
Specific gravity of oil = 0·88

11. 14 lb of cast iron cuttings are heated to 212°F and dropped into 1½ gallons of water at 62°F contained in a metal vessel whose water equivalent is 1·2 lb. If the resultant temperature of the mixture is 77°F, calculate the specific heat of the iron.

12. With three different quantities, A, B and C, of the same kind of liquid, of temperatures 50°F, 70°F and 90°F respectively, it is found that when A and B are mixed together the resultant temperature is 62°F, and when B and C are mixed together the resultant temperature is 82½°F. Find the resultant temperature if (i) A and C were mixed, (ii) all three were mixed.

13. 2 lb of ice were put into 22½ lb of water at 72°F and the resultant temperature of the mixture was 55°F. Calculate the initial temperature of the ice, taking its specific heat as 0·5 and its latent heat as 144 Btu/lb.

14. 4 lb of ice at 23°F are mixed into 15 lb of water at 62°F. Find the final state of the mixture and its final temperature. Specific heat of ice = 0·5, latent heat = 144 Btu/lb.

CHAPTER 2
EXPANSION OF METALS AND HEAT TRANSFER

Most metals expand when they are heated and contract when they are cooled. The amount of expansion per degree rise of temperature differs with different metals, and some alloys are manufactured for special purposes to have practically no expansion over a considerable working temperature range.

Although the expansion is in all directions so that there is an increase in all dimensions, it is sometimes convenient to consider the expansion in one direction only.

LINEAR EXPANSION. When a linear dimension is under consideration, the amount that a metal will expand lengthwise is expressed by its *coefficient of linear expansion*. This is the increase in length per unit length per degree increase in temperature. For instance, if the coefficient of linear expansion of copper is given as 0·0000095 per F degree, it means that each inch of length will expand by 0·0000095 in. when heated through one degree on the Fahrenheit scale. This coefficient may be represented by either K or $\propto$. The total increase in length for an original length of L and a temperature rise of t is therefore:

$$\text{Increase in length} = K \times L \times t$$

The new length of the metal will then be, original length + increase in length thus:

$$\text{New length} = L + KLt$$
$$= L(1 + Kt)$$

It is important to take care with the units. Change of length will be in the same units as the original length, e.g., if the original length is inserted in inches then the change of length will be expressed in inches. The coefficient of linear expansion will always be given with respect to the temperature scale, Fahrenheit or Centigrade. As one degree C is 1·8 times one degree F, the expansion per C degree will be 1·8 times the expansion per F degree, hence the temperature rise and the coefficient must be expressed in the same temperature scale.

Example. A main steam pipe is 21 ft 6 in. long when fitted at a temperature of 59°F. Find the increase in length when heated by the steam to 600°F, taking the coefficient of linear expansion of the material as 0·0000067 per F degree.

EXPANSION OF METALS AND HEAT TRANSFER 21

Increase in length = KL*t*
= 0·0000067 × 21·5 × 12 × (600 — 59)
= 0·0000067 × 21·5 × 12 × 541
= 0·9353 in. Ans.

Free expansion of steam pipes must be allowed for to avoid undue stresses in the pipe and its end connections. Two common methods of allowing for expansion in steam pipe lines are shown in Figs 4 and 5.

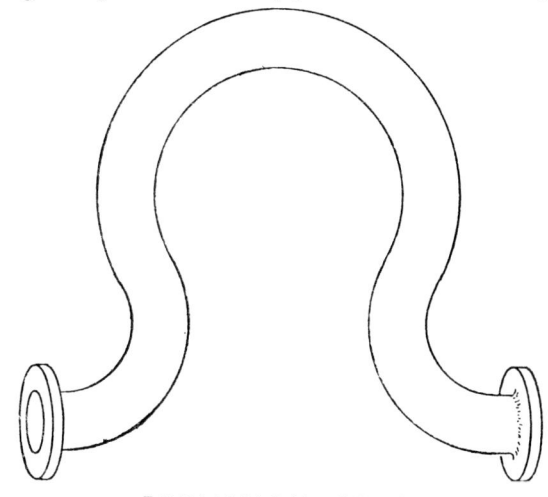

EXPANSION BEND
Fig. 4

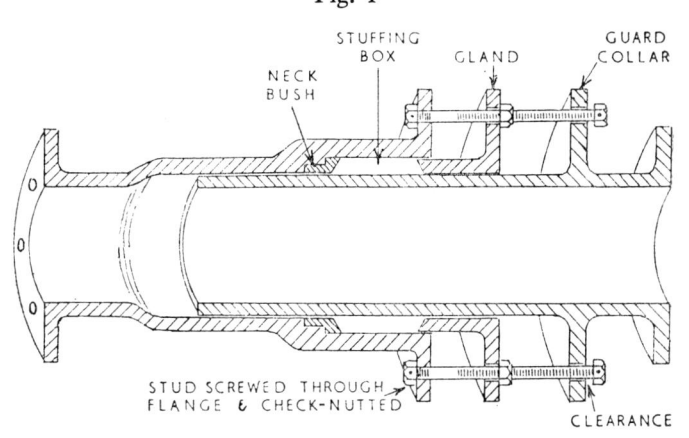

EXPANSION GLAND
Fig. 5

Example. A brass shaft liner is heated through 250 C degrees. If the original diameter was 270 mm find the increase in diameter given that the coefficient of linear expansion for brass is 0·0000105 per F degree.

Diameter is a linear dimension and therefore the same rule can be applied as for length. If the original diameter is inserted in millimetres the change in diameter will be in millimetres. Note also that since the change in temperature is given in *Centigrade* degrees and the coefficient of linear expansion as per *Fahrenheit* degree, one must therefore be converted to the same units as the other.

Coeff. of linear exp. = 0·0000105 per F degree
 = 0·0000105 × $\frac{9}{5}$ per C degree.
Change in diameter = KLt
 = 0·0000105 × $\frac{9}{5}$ × 270 × 250
 = 1·276 mm. Ans.

SUPERFICIAL EXPANSION refers to increase in area. The *coefficient of superficial expansion* is the increase in area per unit area per degree increase in temperature, therefore,

Increase in area = Coeff. of superficial exp. × orig. area
× temp. rise

Now consider an area of metal of unit length and unit breadth (Fig. 6) and let this be heated through one degree.

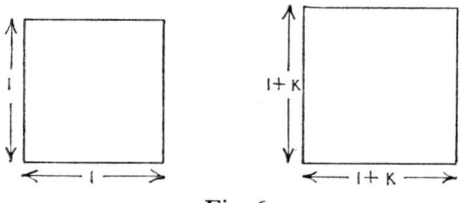

Fig. 6

The length and breadth will each increase by an amount equal to K the coefficient of linear expansion.

Original area = 1 × 1 = 1
New length (and new breadth) = 1 + K
New area = (1 + K)^2 = 1 + 2K + K^2
Increase in area = new area — original area
 = 1 + 2K + K^2 — 1
 = 2K + K^2

EXPANSION OF METALS AND HEAT TRANSFER   23

K is a very small quantity for any metal (such as 0·0000067 for steel), therefore K^2, being the second order of smallness, is a very small quantity indeed and is completely negligible as a quantity to be added for all practical purposes. We can therefore take the increase to be 2K. As this is the increase in area per unit area for one degree increase in temperature, it is the value of the coefficient of superficial expansion, hence,

Coeff. of superficial expansion $= 2 \times$ coeff. of linear expansion

and,

Increase in area $= 2KAt$
where A $=$ original area.

CUBICAL EXPANSION refers to the increase in volume. The *coefficient of cubical (or volumetric) expansion* is the increase in volume per unit volume per degree increase in temperature.

Increase in volume $=$ Coeff. of cubical exp. $\times$ orig. vol. $\times$ temp. rise

Consider a block of metal of unit length, unit breadth and unit thickness (Fig. 7) and let this be heated through one degree.

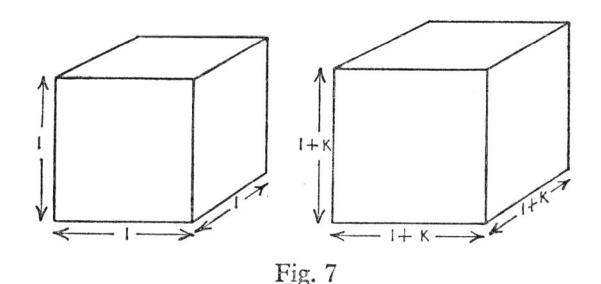

Fig. 7

The length, breadth and thickness will each increase by an amount equal to K, the coefficient of linear expansion.

Original volume $= 1 \times 1 \times 1 = 1$
New volume $= (1 + K)^3$
$= 1 + 3K + 3K^2 + K^3$

Increase in volume $=$ New volume $-$ original volume
$= 1 + 3K + 3K^2 + K^3 - 1$
$= 3K + 3K^2 + K^3$

K^2 and K^3 being the second and third orders of smallness respectively, are negligible quantities for addition, hence the increase is taken as $3K$. As this is the increase in volume per unit volume per degree increase in temperature, it is the value of the coefficient of cubical expansion.

Coeff. of cubical expansion = 3 × coeff. of linear expansion

therefore,

$$\text{Increase in volume} = 3KVt$$
$$\text{where } V = \text{original volume.}$$

It is usual to use the above value when dealing with metals but, as liquids have no linear dimensions the volumetric expansion is measured directly and therefore the coefficient of cubical expansion would be given in any problem dealing with liquids.

Sometimes the coefficient of superficial expansion is represented by K_s or β and the coefficient of cubical expansion by K_v or γ.

Example. A metal sphere is 6 in. diameter at 70°F. Find the increase in diameter, increase in surface area, and increase in volume when heated to 320°F, taking the coefficient of linear expansion of the material as 0·00001 per F degree.

Increase in temp. = 320 — 70 = 250 F degrees

Increase in diameter = KLt
= 0·00001 × 6 × 250
= 0·015 in. Ans. (i)

Surface area of sphere = πd^2

Increase in area = $2KAt$
= 2 × 0·00001 × π × 6^2 × 250
= 0·5655 in². Ans. (ii)

Volume of sphere = $\frac{\pi}{6}d^3$

Increase in volume = $3KVt$
= 3 × 0·00001 × $\frac{\pi}{6}$ × 6^3 × 250
= 0·8482 in³. Ans. (iii)

EXPANSION OF METALS AND HEAT TRANSFER

Example. Two gallons of liquid are heated through 100 F degrees. Find the increase in volume in cubic in., taking the coefficient of cubical expansion as 0·00012 per F degree.

$6\frac{1}{4}$ gallons = 1 cubic foot = 1728 cubic in.

$$\therefore \text{Original volume} = \frac{2 \times 1728}{6 \cdot 25} \text{ in}^3.$$

Increase in volume = $K_v V t$

$$= 0 \cdot 00012 \times \frac{2 \times 1728}{6 \cdot 25} \times 100$$

$$= 6 \cdot 634 \text{ in}^3. \quad \text{Ans.}$$

APPARENT CUBICAL EXPANSION. The tank or vessel which contains a liquid will also expand when heated. It is therefore usually of more practical value to know the expansion of the liquid relative to its container so that the correct allowance can be made for changes of temperature.

The apparent or relative increase in volume of a liquid is the difference between the volumetric expansion of the liquid and the volumetric expansion of its container. If both have the same initial volume and are raised through the same range of temperature, then,

Apparent increase in volume of the liquid
= vol. increase of liquid — vol. increase of container
= K_v liquid × V × t — K_v container × V × t
= (K_v liquid — K_v container) × V × t

The difference between the coefficients of cubical expansion of the liquid and its container can therefore be termed the *apparent* coefficient of cubical expansion of the liquid.

TRANSFER OF HEAT

Heat can be transferred from one place to another by three methods known as *Conduction, Convection* and *Radiation*.

CONDUCTION is the *flow* of heat through a body, or from one body to another in contact with each other, the natural flow of heat being from a region of high temperature to a region of lower temperature.

Generally speaking, metals are good conductors and liquids are bad conductors. A simple experiment to demonstrate that water is a bad conductor of heat is illustrated in Fig. 8. This represents a flask of water containing a piece of ice at the bottom (loaded to keep it down, otherwise it would float) with a flame applied near the top of the water. The water at the top begins to boil at the temperature of 212°F while the ice remains unmelted, its temperature and the water in its vicinity being at 32°F, showing a poor conduction of heat through the water.

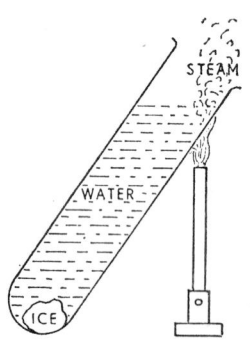

Fig. 8

The quantity of heat which flows in a given time, i.e., the rate of flow, depends upon the heat conductivity of the material, is proportional to the surface area exposed to the heat supply, is proportional to the temperature difference between the hot and cold ends, and is inversely proportional to the distance through which the heat is conducted. Hence, using the symbols:

$Q = $ quantity of heat conducted,
$k = $ coefficient of thermal conductivity (sometimes C or λ may be used)
$a = $ surface area of the material exposed to the heat supply,
$t = $ time of exposure to the heat supply,

EXPANSION OF METALS AND HEAT TRANSFER 27

θ = temperature difference between hot and cold ends,
d = distance heat is conducted,

then,
$$Q = \frac{kat\theta}{d}$$

The coefficient of thermal conductivity depends upon the nature of the material and is expressed as the quantity of heat conducted through unit area per unit time through a slab of unit thickness per degree of temperature difference. Any convenient units may be used for the coefficient provided care is taken to insert the remaining quantities of the expression, Q a t θ and d in units consistent with those of the coefficient. For instance, if the coefficient of thermal conductivity is given as the number of Btu per square foot of area per minute per Fahrenheit degree through a thickness of one inch, then Q would be in Btu, the area a in square feet, the time t in minutes, the temperature difference θ in F degrees, and the distance d in inches.

Example. Calculate the quantity of heat conducted per minute through a duralumin plate 5 inches diameter by $\frac{3}{4}$ inch thick when the temperature range across the thickness of the plate is 27 F degrees, taking the coefficient of thermal conductivity of duralumin as 0·36 gram-calorie per cm² per sec per C° through 1 cm thickness.

k = 0·36 cal per cm² per sec per C° for 1 cm thick,
a = 0·7854 × 5² × 2·54² sq. centimetres
t = 60 seconds,
θ = 27 × $\frac{5}{9}$ C degrees,
d = 0·75 × 2·54 centimetres

$$Q = \frac{kat\theta}{d}$$

$$= \frac{0·36 \times 0·7854 \times 5^2 \times 2·54^2 \times 60 \times 27 \times 5}{0·75 \times 2·54 \times 9}$$

= 21540 gram-calories. Ans.

CONVECTION is the method of transferring heat through a substance by the movement of heated particles of that substance. Liquids and gases are heated by this method.

Fig. 9 shows a vessel with an inclined tube connected at the bottom. When this contains water, and heat is applied to the tube, the heated particles of the water become less dense and rise, denser particles move to take their place and thus convection currents are set moving resulting in all the water in the vessel and tube becoming heated approximately uniformly due to the continuous circulation of the water. This is the principle of the water-tube boiler.

Fig. 10 illustrates air in a room heated by convection, the fire, radiator or other heat source being placed at the bottom of the room.

Fig. 11 illustrates the air in a room cooled by convection, the coolers (such as refrigerator pipes) being situated near the top of the room.

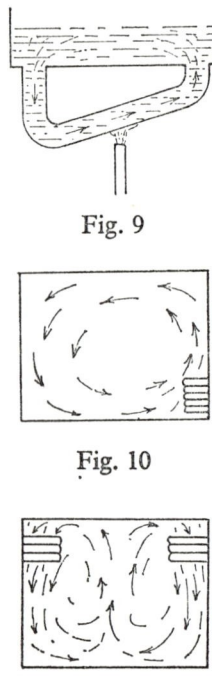

Fig. 9

Fig. 10

Fig. 11

RADIATION is the method of transferring heat from one body to another through space by heat rays in the form of waves. The rays of heat travel in straight lines at approximately the same speed as light

(186000 miles per second). Dark dull surfaces are good radiators and good absorbers of radiant heat, bright polished surfaces are bad radiators and bad absorbers of radiant heat.

The quantity of heat radiated from a radiator depends upon the condition of its surface, is proportional to the area of the surface, the time of radiation and the fourth power of its absolute temperature.

Using the symbols:

a = surface area of the radiator,
t = time of radiation,
T_1 = absolute temperature of the surface,

then, if the temperature of its surrounds was zero absolute,

Quantity of heat radiated = $atT_1^4 \times K$
where K is a constant.

If the temperature of the surrounds is T_2 degrees absolute, the heat radiated from the surrounds would be $atT_2^4 \times K$.

Assuming the radiator is the hotter and that the same constant can be applied to both radiator and its surrounds, the heat lost by the radiator and gained by its surrounds will be,

$$Kat(T_1^4 - T_2^4)$$

The constant is found experimentally and depends upon the units of heat, temperature, area and time. For example, K may be given as 16×10^{-10} Btu/ft^2 per hour, in which case the area must be measured in square feet, the time in hours, the temperature in degrees Fahrenheit absolute, and the total heat radiated will be in Btu.

Example. The temperature of the flame in the furnace of a boiler is 2340°F and the temperature of the surrounding heating surface is 500°F. Calculate the maximum theoretical quantity of heat radiated per minute to a surface area of 100 square feet, taking the radiation constant as 16×10^{-10} Btu per ft^2 per hour.

$T_1 = 2340 + 460 = 2800°F$ abs.
$T_2 = 500 + 460 = 960°F$ abs.

Heat radiated $= Kat(T_1^4 - T_2^4)$
$= 16 \times 10^{-10} \times 100 \times \frac{1}{60} \times (2800^4 - 960^4)$ Btu

Note the multiplier 10^{-10}, this is the same as dividing by 10^{10}. Also, it is much easier to factorise the term inside the brackets and so avoid very large numbers, thus:

$2800^4 - 960^4 = (2800^2 + 960^2)(2800^2 - 960^2)$
$= (7840000 + 921600)(7840000 - 921600)$
$= 8761600 \times 6918400$
$= 87 \cdot 616 \times 10^5 \times 69 \cdot 184 \times 10^5$
$= 87 \cdot 616 \times 69 \cdot 184 \times 10^{10}$

Now the multiplier of 10^{10} on the previous line will cancel with the divisor of 10^{10} in the constant.

Heat radiated $= 16 \times 10^{-10} \times 100 \times \frac{1}{60} \times 87 \cdot 616 \times 69 \cdot 184 \times 10^{10}$
$= 16 \times 100 \times \frac{1}{60} \times 87 \cdot 616 \times 69 \cdot 184$
$= 161600$ Btu   Ans.

MECHANICAL EQUIVALENT OF HEAT

Energy exists in various forms such as mechanical, heat, chemical and electrical, and any one form of energy is convertible into another. Dr. Joule determined the relationship between mechanical energy and heat energy by an apparatus consisting of a vessel containing a known weight of water, fitted with fixed vanes to minimise movement of the water, a rotating spindle with paddles between the fixed vanes, and a falling weight on the end of a cord passing over a guide pulley to rotate the paddles. The ft lb of mechanical energy expended (i.e. the work done by the rotating paddles in the water) was obtained by the product of the weight (lb) and the distance fallen (ft). The Btu of heat developed was obtained by the product of the weight of water (lb) and its rise in temperature (F degrees), the weight of water including the water equivalent of the vessel. Allowances and corrections for losses were made. After more research by later scientists the mechanical equivalent of one Btu was accepted as:

778 ft lb = 1 Btu

This is called Joule's Mechanical Equivalent of Heat and represented by J.

EXPANSION OF METALS AND HEAT TRANSFER 31

The Chu being a larger heat unit than the Btu in the ratio 9 : 5, then the mechanical equivalent of 1 Chu is $778 \times \frac{9}{5}$ ft lb:

1400 ft lb = 1 Chu

Hence, $J = 778$ ft lb/Btu $= 1400$ ft lb/Chu

Example. A shaft runs at 3000 r.p.m. in bearings 7 in. diameter. If the total load on the bearings is 600 lb and the coefficient of friction between journals and bearings is 0·04, find (i) the friction force at the surface of the journals, (ii) the ft lb of work lost due to friction per revolution, (iii) the ft lb of work lost per minute, (iv) the Btu of heat generated per minute.

Friction force $= \mu \times W$ (see Chap. 6, Vol II)
$= 0{\cdot}04 \times 600 = 24$ lb. Ans. (i)

Work lost per revolution in overcoming friction
$=$ friction force $\times$ circumference of journal
$= 24 \times \pi \times \frac{7}{12} = 44$ ft lb. Ans. (ii)

Work lost per minute
$=$ Work per rev. $\times$ rev. per minute
$= 44 \times 3000 = 132000$ ft lb. Ans. (iii)

Mechanical equivalent:
778 ft lb = 1 Btu

$\therefore$ Heat generated $= \dfrac{132000}{778} = 169{\cdot}6$ Btu/min. Ans. (iv)

HORSE POWER EQUIVALENT

One horse power = 33000 ft lb of work per minute

therefore, as 778 ft lb = 1 Btu then,

One horse power $= \dfrac{33000}{778}$

$$= 42\cdot 42 \text{ Btu per minute}$$
$$42\cdot 42 \times 60 = 2545$$
$$\therefore \text{ One horse power} = 2545 \text{ Btu per hour}$$

This is usually written:

$$\text{One horse power hour} = 2545 \text{ Btu}$$
$$\text{Since Chu} = \tfrac{5}{9} \times \text{Btu, then,}$$

$$\text{One horse power hour} = \frac{2545 \times 5}{9}$$

$$= 1414 \text{ Chu}$$

Hence, **One horse power hour** $= 2545$ **Btu** $= 1414$ **Chu**

Example. Three tons of circulating water pass through a cooler every hour, the inlet temperature of the water being 62°F and the outlet temperature 77°F, calculate the equivalent horse power carried away.

$$\text{Heat carried away} = 3 \times 2240 \times (77 - 62)$$
$$= 3 \times 2240 \times 15 \text{ Btu/hour}$$

$$\text{Equivalent horse power} = \frac{3 \times 2240 \times 15 \times 778}{60 \times 33000}$$

$$= \frac{3 \times 2240 \times 15}{2545}$$

$$= 39\cdot 6 \text{ h.p.} \quad \text{Ans.}$$

ELECTRICAL EQUIVALENT

Electrical pressure is measured in *volts* and the rate of flow of electricity is measured in *amperes*. The product of volts and amperes is the power in *watts* and 746 watts is equal to one horse power.

$$746 \text{ watts} = 1 \text{ h.p.} = 33000 \text{ ft lb/min.}$$

EXPANSION OF METALS AND HEAT TRANSFER

$$= \frac{33000 \times 60}{778} \text{ Btu/hour}$$

$$= 2545$$

$$1 \text{ watt} = \frac{2545}{746} \text{ Btu/hour}$$

$$1 \text{ kilowatt} = 1000 \text{ watts} = \frac{2545 \times 1000}{746} \text{ Btu/hour}$$

$$= 3412 \text{ Btu per hour}$$

Usually written:
1 kilowatt-hour = 3412 Btu

TEST EXAMPLES 2

1. A steam pipe is 25 ft 6 in. long when fitted at a temperature of 64°F. Find the increase in length when carrying steam at 500°F, taking the coefficient of linear expansion of the pipe material as 0·0000075 per F degree.

2. A solid cast iron sphere is 6 in. diameter. Find the increase in diameter after it has absorbed 2000 Btu of heat, taking the following values for cast iron: density 0·26 lb/in^3, specific heat 0·13, coefficient of linear expansion 0·00000618 per F°.

3. The pipe line of a hydraulic system consists of a total length of pipe of 45 ft and internal diameter 1·25 in. If the coefficient of linear expansion of steel is 0·0000067/F° and coefficient of cubical expansion of the oil in the pipe is 0·0005/F°, calculate the volumetric allowance to be made for oil overflow from the pipe when the temperature rises by 50 F degrees.

4. In an experiment to determine the coefficient of cubical expansion of an oil, a glass vessel of 40 cc capacity was filled with the oil at a temperature of 15°C. When the vessel and oil was heated to 65°C the quantity of oil spilt over from the vessel was 1·57 cc. Taking the coefficient of linear expansion of glass as 0·0000085/C°, calculate the coefficient of cubical expansion of the oil.

5. A cold storage compartment is 15 ft long by 13·5 ft wide by 8 ft high. The four walls, ceiling and floor are covered to a thickness of 6 in. with insulating material which has a coefficient of thermal conductivity of 0·00028 gram-calorie per cm^2 per second per cm per C°. Calculate the quantity of heat leaking through the insulation per minute, in Btu, when the outside temperature is 15°C and inside temperature — 5°C.

ƒ6. One wall of a refrigerated cold chamber is 20 ft long by 12 ft high and consists of 6 in. thickness of cork between outer and inner walls each of 1 in. thickness of wood. Taking the coefficients of thermal conductivity of cork and wood respectively as 0·29 and 1·45 Btu per ft^2 per hour per F degree per in. thickness, calculate the quantity of heat-leakage through the wall per day when the difference between the outside and inside temperatures of the chamber is 52 F degrees.

EXPANSION OF METALS AND HEAT TRANSFER

7. Calculate the quantity of heat radiated per sq. ft of surface area per hour from a cylinder cover when its temperature is 420°F and the temperature of the surrounding atmosphere is 110°F. Take the radiation constant as 16×10^{-10} Btu/ft² per hr.

ƒ8. The steam and water drum of a water-tube boiler has hemispherical ends, the drum is 4 ft diameter and the overall length is 20 ft. Under steaming conditions the temperature of the shell before lagging was 450°F and the surrounding atmosphere 120°F, the temperature of the cleading after lagging was 140°F and the surrounding atmosphere 90°F. Assuming three-quarters of the total area of the shell to be lagged and taking the radiation constant as 16×10^{-10} Btu/ft² per hr., estimate (i) the heat saved per hour due to lagging, (ii) the weight of fuel oil this represents per day if its calorific value is 18500 Btu/lb.

9. A water brake attached to an engine on test absorbs 95 horse power. Find the heat generated at the brake per minute and the weight of fresh water passing through if the rise of temperature of the water is 18 F degrees.

10. The journals of a shaft are 15 in. diameter, it runs at 105 r.p.m. and the coefficient of friction between the journals and bearings is 0·02. If the average total load on the bearings is 20 tons, find (i) the horse power lost due to friction, (ii) the heat generated at the bearings per minute.

11. An internal combustion engine uses 1750 lb of fuel oil per hour and each lb gives out 18700 Btu of heat. If 35% of the heat supplied is converted into useful power at the shaft, find the shaft horse power.

ƒ12. The effective radius of the pads in a single collar thrust block is 9 in. and when the shaft is running at 93 r.p.m. the total load on the block is 24 tons. Taking the coefficient of friction between thrust collar and pads as 0·025,
  (a) find (i) the horse power lost due to friction at the thrust and (ii) the heat generated per hour.
  (b) If the generated heat is all carried away by a circulation of oil, calculate the number of gallons of oil passing through the block per hour, allowing an oil temperature rise of 20 F degrees, taking the specific heat as 0·48 and specific gravity as 0·88.

CHAPTER 3

LAWS OF PERFECT GASES

When a substance has been completely evaporated it exists as a gas and one of its most important characteristics is its elastic property. For instance, if a certain volume of a liquid is put into a vessel of larger volume, the liquid will only partially fill the vessel, taking up no more nor less volume than it did before, but when a gas enters a vessel it immediately fills up every part of that vessel no matter how large it is. Practically speaking, liquids cannot be compressed nor expanded, but gases can be compressed into smaller volumes or expanded to larger volumes.

A perfect gas is a theoretically ideal gas which follows perfectly Boyle's and Charles' laws of gases.

Consider a given mass of a perfect gas enclosed in a cylinder by a gas-tight piston. When the piston is pushed inward the gas is compressed to a smaller volume, when pulled outward the gas is expanded to a larger volume. However, not only is there a change in volume but the pressure and temperature also change. These three quantities, pressure, volume and temperature, are related to each other and to determine their relationship it is usual to perform experiments with each one of these quantities in turn kept constant and observe the relationship between the other two.

In such basic laws, the pressure, volume and temperature must all be the real values, that is, measured from absolute zero, and not measured above some artificial level.

ABSOLUTE PRESSURE (P) is the pressure measured above a perfect vacuum. Ordinary pressure gauges and open-ended manometers measure pressures from the level of atmospheric pressure, for instance if a steam pressure gauge reads 200 lb/in^2 this is termed the *gauge pressure* and means that the pressure of the steam is 200 lb/in^2 above the atmospheric pressure. The absolute pressure is obtained by adding the atmospheric pressure to the gauge pressure.

F r rough calculations, it is usual to assume the atmospheric pressure to be 14·7 or 15 lb/in^2 whichever is the more convenient and applicable.

LAWS OF PERFECT GASES

For more accurate results the atmospheric pressure would probably be given in "inches of mercury" with respect to the mercurial barometer. As one cubic inch of mercury weighs 0·491 lb, a column of mercury h in. high will exert a pressure at its base of $0·491 \times h$ lb/in², thus a barometer standing at 30 in. means that 30 in. head of mercury is supported by the pressure of the atmosphere and therefore the atmospheric pressure would be $0·491 \times 30 = 14·73$ lb/in².

ABSOLUTE VOLUME (V). This is the actual volume of the gas which is equal to the full volume of the vessel containing it. For instance, when the piston of a reciprocating engine is at the beginning of its stroke there is a certain amount of volumetric clearance between the piston and the cylinder cover, therefore when the piston moves to another position, the absolute volume occupied by the gas in the cylinder is the sum of the original clearance volume and the volume swept through by the piston.

ABSOLUTE TEMPERATURE (T). This is the temperature measured above absolute zero as explained in Chapter 1, thus if $t =$ thermometer reading, and $T =$ absolute temperature, then:

On the Fahrenheit scale, $T = t + 460$
On the Centigrade scale, $T = t + 273$

BOYLE'S LAW states that *the absolute pressure of a given mass of a perfect gas varies inversely as its volume if the temperature remains unchanged.*

Expressing this mathematically,

$$P \propto \frac{1}{V}$$

$$\therefore P \times V = \text{a constant}$$
$$\text{hence, } P_1 \times V_1 = P_2 \times V_2$$

Thus, let us imagine 4 cu. ft of gas at a pressure of 15 lb/in² abs. contained in a cylinder with a gas-tight piston as illustrated in Fig. 12, when the piston is pushed inward, the pressure will increase as the gas is compressed to a smaller volume, and, if the temperature is prevented from rising or falling, the product of pressure and volume will be a constant quantity for all positions of the piston.

From the known initial conditions the constant is calculated:

$$P_1 \times V_1 = \text{constant}$$
$$15 \times 4 = 60$$

And the pressure at any other volume can be calculated:

When the volume is 3 ft^3,

$$P_2 \times V_2 = 60$$
$$P_2 \times 3 = 60$$
$$P_2 = 20 \text{ lb/in}^2 \text{ abs.}$$

When the volume is 2 ft^3,

$$P_3 \times V_3 = 60$$
$$P_3 \times 2 = 60$$
$$P_3 = 30 \text{ lb/in}^2 \text{ abs.}$$

And so on.

The variation of pressure with change of volume is shown in the graph below the cylinder in Fig. 12. Joining up the plotted points the graph produced is a rectangular hyperbola, consequently we refer to compression or expansion where PV = constant as hyperbolic compression or hyperbolic expansion. When the temperature is constant, as in the above example, the operation may also be termed isothermal compression or isothermal expansion.

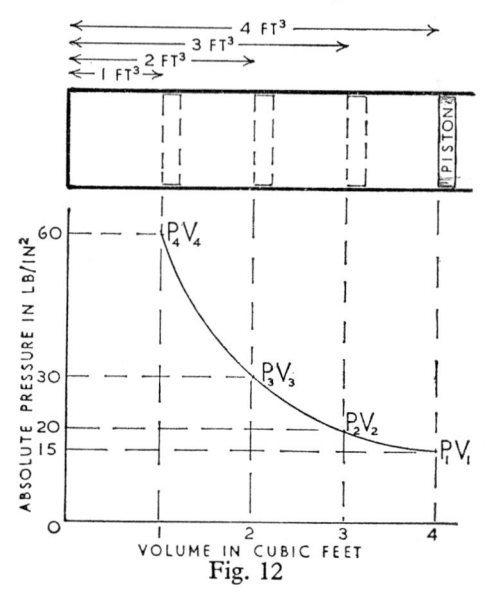

Fig. 12

LAWS OF PERFECT GASES

Note that as the ordinates (vertical measurements) represent pressure and the abscissae (horizontal measurements) represent volume, and since the product of pressure and volume is constant, then all rectangles drawn from the axes, with their corners touching the curve, will be of equal area.

Example. 8 cu. ft of air at a gauge pressure of 2·5 lb/in^2 is compressed at constant temperature to a pressure of 21·5 lb/in^2 gauge. Assuming atmospheric pressure to be 14·7 lb/in^2, find the final volume.

Initial pressure $P_1 = 2·5 + 14·7 = 17·2$ lb/in^2 abs.
Final pressure $P_2 = 21·5 + 14·7 = 36·2$ lb/in^2 abs.

$$P_1V_1 = P_2V_2$$

$$V_2 = \frac{P_1V_1}{P_2}$$

$$= \frac{17·2 \times 8}{36·2}$$

$$= 3·8 \text{ cu. ft.} \quad \text{Ans.}$$

CHARLES' LAW states that *the volume of a given mass of a perfect gas varies directly as its absolute temperature if the pressure remains unchanged*, also, *the absolute pressure varies directly as the absolute temperature if the volume remains unchanged*.

From the above statement we have:

For constant pressure, $V \propto T$

$$\therefore \frac{V}{T} = \text{constant}$$

Hence, $\dfrac{V_1}{T_1} = \dfrac{V_2}{T_2}$ or $\dfrac{V_1}{V_2} = \dfrac{T_1}{T_2}$

For constant volume, $P \propto T$

$$\therefore \frac{P}{T} = \text{constant}$$

Hence, $\dfrac{P_1}{T_1} = \dfrac{P_2}{T_2}$ or $\dfrac{P_1}{P_2} = \dfrac{T_1}{T_2}$

Example. The pressure of the air in a starting-air vessel is 600 lb/in² gauge and the temperature is 75°F. If a fire in its vicinity causes the temperature of the air to rise to 150°F, find the pressure of the air. Take the atmospheric pressure as 15 lb/in² and neglect the increase in volume of the vessel.

$$T_1 = 75 + 460 = 535°F \text{ abs.}$$
$$T_2 = 150 + 460 = 610°F \text{ abs.}$$
$$P_1 = 600 + 15 = 615 \text{ lb/in}^2 \text{ abs.}$$

For constant volume,

$$\dfrac{P_1}{T_1} = \dfrac{P_2}{T_2}$$

$$\therefore P_2 = \dfrac{P_1 \times T_2}{T_1}$$

$$= \dfrac{615 \times 610}{535}$$

$$= 701 \cdot 1 \text{ lb/in}^2 \text{ abs.}$$

or, $701 \cdot 1 - 15 = 686 \cdot 1$ lb/in² gauge. Ans.

COMBINATION OF BOYLE'S & CHARLES' LAWS

Each one of these laws states how one quantity varies with another if the third quantity remains unchanged, but if the three quantities change simultaneously, it is necessary to combine these laws in order to determine the final conditions of the gas.

Referring to Fig. 13 which again represents a cylinder with a piston, gas-tight so that the mass of gas in the cylinder is always the same. Let the gas be compressed from its initial state of pressure P_1 volume V_1 and temperature T_1 to its final state of P_2 V_2 and T_2, but to arrive at the final state let it pass through two stages, the first to satisfy Boyle's law and the second to satisfy Charles' law.

LAWS OF PERFECT GASES

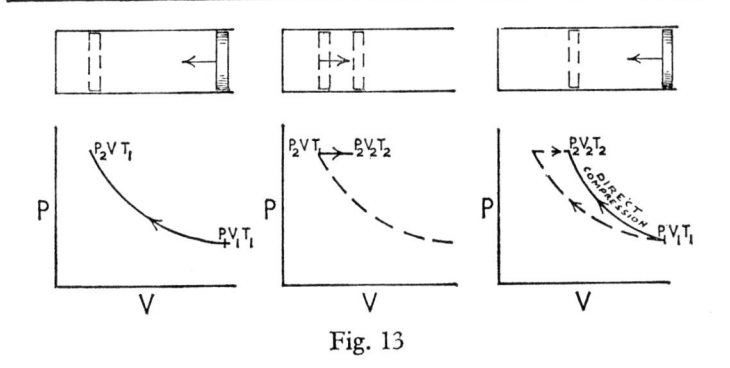

Fig. 13

Imagine the piston pushed inward to compress the gas until it reaches the final pressure of P_2 and let its volume then be represented by V. Normally the temperature would tend to increase due to the work done in compressing the gas, but any heat so generated must be taken away from it during the compression so that its temperature remains unchanged at T_1, hence following Boyle's law:

$$\therefore P_1 V_1 = P_2 V \quad \ldots \quad \ldots \quad \ldots \quad \ldots \quad \ldots \quad (i)$$

Now apply heat to raise the temperature from T_1 to T_2 and at the same time draw the piston outward to prevent a rise of pressure and keep it constant at P_2. The volume will increase in direct proportion to the increase in absolute temperature according to Charles' law:

$$\therefore \frac{V_2}{V} = \frac{T_2}{T_1} \quad \ldots \quad \ldots \quad \ldots \quad \ldots \quad (ii)$$

By substituting the value of V from (ii) into (i) this quantity will be eliminated:

From (ii), $$V = \frac{V_2 T_1}{T_2}$$

Substituting into (i),

$$P_1 V_1 = P_2 \times \frac{V_2 T_1}{T_2}$$

$$\therefore \frac{P_1V_1}{T_1} = \frac{P_2V_2}{T_2}$$

This Combined Law of Boyle's and Charles' is true for a given mass of any perfect gas subject to any form of compression or expansion.

Example. Six cubic feet of air at a pressure of 14 lb/in² abs. and temperature 65°F are compressed to a volume of 1·5 cubic feet and the final pressure is 96 lb/in² abs. Calculate the final temperature.

$$\frac{P_1V_1}{T_1} = \frac{P_2V_2}{T_2}$$

$$\frac{14 \times 6}{(65 + 460)} = \frac{96 \times 1.5}{T_2}$$

$$T_2 = \frac{525 \times 96 \times 1.5}{14 \times 6}$$

$$= 900°F \text{ abs.}$$

Expressing the final temperature as a thermometer reading:
$$t_2 = 900 - 460$$
$$= 440°F. \text{ Ans.}$$

CHARACTERISTIC EQUATION

Since the ratio PV/T is a constant, its numerical value can be determined for any quantity of any perfect gas. To form a means of comparison, the constant is calculated on the specific volume of the gas, i.e., the volume in cubic feet occupied by *one lb* weight of the gas, the pressure must therefore be expressed in lb per square *foot* absolute and the temperature in either °F abs. or °C abs. The constant so obtained is called the characteristic constant for the gas concerned, and is represented by R.

As an example, experiments show that the weight of one cubic foot of air at 14·7 lb/in² (atmospheric pressure) and at 32°F weighs 0·0807 lb, therefore:

$$P = 14.7 \times 144 \text{ lb/ft}^2$$

LAWS OF PERFECT GASES

Spec. volume $= V_s = \dfrac{1}{0.0807} = 12.39 \text{ ft}^3$ for 1 lb weight

$$T = 32 + 460 = 492°F \text{ abs.}$$

$$\frac{PV_s}{T} = R$$

$$\therefore R = \frac{14.7 \times 144 \times 12.39}{492}$$

$$= 53.3 \text{ ft lb/°F for 1 lb weight}$$

Note the units in which R is expressed, this may be more easily seen by inserting the units only of PV_s and T without any numerical values:

$$P \rightarrow \text{lb/ft}^2 = \frac{\text{lb}}{\text{ft} \times \text{ft}}$$

$$V_s \rightarrow \text{ft}^3 = \text{ft} \times \text{ft} \times \text{ft}$$

$$T \rightarrow °F$$

$$R = \frac{PV_s}{T} = \frac{\text{lb} \times \text{ft} \times \text{ft} \times \text{ft}}{\text{ft} \times \text{ft} \times °F}$$

$$= \frac{\text{lb} \times \text{ft}}{°F}$$

$$= \text{ft lb/°F for 1 lb weight}$$

Note further that the value of the constant also depends on the temperature scale in which T is expressed and therefore R must be stated as either per °F or per °C.

Taking the same example but inserting the temperature in °C,

$32°F = 0°C$

$\therefore T = 273°C \text{ abs.}$

$$R = \frac{PV_s}{T}$$

$$= \frac{14\cdot7 \times 144 \times 12\cdot39}{273}$$

$$= 96 \text{ ft lb}/°C \text{ for 1 lb weight.}$$

It must be emphasised that R is the constant for *one* lb weight because it was calculated on the volume occupied by only 1 lb weight of the gas. Therefore, if V is to represent the volume of w lb weight the equation would be:

$$\frac{PV}{T} = wR$$

which is usually written,

PV = wRT

This is known as the characteristic equation of a perfect gas, introducing the weight of the gas for the first time, it is a very useful and important expression.

Example. An air compressor delivers 6 cubic feet of air at a pressure of 370 lb/in² abs. and 120°F into an air reservoir. Taking the value of R for air as 53·3 ft/lb/°F per lb, calculate the weight of air delivered.

$$PV = wRT$$

$$w = \frac{PV}{RT}$$

$$= \frac{370 \times 144 \times 6}{53\cdot3 \times (120 + 460)}$$

$$= \frac{370 \times 144 \times 6}{53\cdot3 \times 580}$$

$$= 10\cdot35 \text{ lb.} \quad \text{Ans.}$$

SPECIFIC HEATS OF GASES

In Chapter 1 the specific heat of a substance was defined as the number of heat units required to raise unit weight of the substance through one degree of temperature. Therefore, the heat required to raise any weight of the substance through any temperature range is,

Heat supplied = Wt. × spec. ht. × temp. rise

Solids and liquids only were dealt with and it would be noted that each material or liquid had its own, but only one, value of specific heat. All gases however have *two* specific heats, one being the value when the *volume* of the gas remains constant during the heating process, the other being the value when the *pressure* of the gas remains constant during heating.

The specific heat when the gas is heated at constant volume is represented by C_v.

The specific heat when the gas is heated at constant pressure is represented by C_p. This is a higher value than C_v because to heat a gas at constant pressure it must be allowed to expand and while it increases in volume it is doing external work, extra heat must therefore be supplied equivalent to the amount of work done.

Example. 0·5 lb weight of air is heated from 70°F to 290°F in a closed vessel which prevents any increase in volume of the air. Calculate (i) the quantity of heat given to the air taking the specific heat at constant volume (C_v) as 0·169, (ii) the final pressure if the original pressure was 15 lb/in^2 abs.

Heat supplied = wt. × spec. ht. × temp. rise
= 0·5 × 0·169 × (290 — 70)
= 0·5 × 0·169 × 220
= 18·59 Btu   Ans. (i)

$$\frac{P_1 V_1}{T_1} = \frac{P_2 V_2}{T_2}$$

As the volume is constant, $V_1 = V_2$ and cancels, leaving:

$$\frac{P_1}{T_1} = \frac{P_2}{T_2} \text{ or } \frac{P_1}{P_2} = \frac{T_1}{T_2} \text{ which is Charles' law,}$$

$$\therefore P_2 = \frac{P_1 \times T_2}{T_1}$$

$$= \frac{15 \times (290 + 460)}{(70 + 460)}$$

$$= \frac{15 \times 750}{530}$$

$$= 21 \cdot 22 \text{ lb/in}^2 \text{ abs.} \quad \text{Ans. (ii)}$$

Example. 0·5 lb weight of air is heated from 70°F to 290°F at constant pressure in a cylinder with a gas-tight moveable piston to allow the air to expand. Calculate (i) the quantity of heat given to the air taking the specific heat at constant pressure (C_p) as 0·2375, (ii) the final volume if the original volume was 6·54 cubic feet.

Heat supplied = wt. × spec. ht. × temp. rise
= 0·5 × 0·2375 × 220
= 26·125 Btu   Ans. (i)

$$\frac{P_1 V_1}{T_1} = \frac{P_2 V_2}{T_2}$$

As the pressure is constant, $P_1 = P_2$ and cancels, leaving:

$$\frac{V_1}{T_1} = \frac{V_2}{T_2} \text{ or } \frac{V_1}{V_2} = \frac{T_1}{T_2} \text{ which is Charles' law,}$$

LAWS OF PERFECT GASES

$$\therefore V_2 = \frac{V_1 \times T_2}{T_1}$$

$$= \frac{6\cdot 54 \times 750}{530}$$

$$= 9\cdot 255 \text{ cubic feet.} \quad \text{Ans. (ii)}$$

Up to now, specific heats have been expressed as heat units per unit weight per degree of temperature, such as Btu/lb per °F, because that is the most convenient form for calculating the heat supplied from the expression, wt. × spec. ht. × temp. rise.

With gases however, it is sometimes convenient to express the mechanical equivalents of the specific heats, e.g., in ft lb instead of Btu, and these values are represented by the symbols K_p for the specific heat at constant pressure, and K_v for the specific heat at constant volume, i.e., $K_p = JC_p$ and $K_v = JC_v$.

For *air* the values are, approximately,

$$C_p = 0\cdot 2375 \text{ Btu/lb per °F}$$
$$K_p = 0\cdot 2375 \times 778 = 184\cdot 8 \text{ ft lb/lb per °F}$$
$$C_v = 0\cdot 169 \text{ Btu/lb per °F}$$
$$K_v = 0\cdot 169 \times 778 = 131\cdot 5 \text{ ft lb/lb per °F}$$

ENERGY EQUATION

The *internal energy* of a gas is the energy contained in it as stored up work which can be expressed in heat units or mechanical units.

When the temperature of a gas is increased, the internal energy increases because the heat absorbed by the gas to increase the temperature becomes stored up energy.

By the principle of conservation of energy, that is, that energy can neither be created nor destroyed, the total heat supplied to a gas will be the sum of the heat to increase its internal energy and the heat equivalent of any work done by the gas during the application of the heat, thus:

Heat supplied = increase in internal energy + external work done

Fig. 14

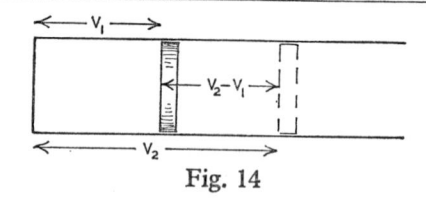

Consider a cylinder with a gas-tight piston (Fig. 14) containing w lb weight of a perfect gas at pressure P, volume V_1 and temperature T_1. If heat is supplied the temperature will rise and consequently either the pressure or volume, or both, will increase. If the piston moves outward at such a rate to allow the volume to increase without any change of pressure, the final conditions of the gas will be P, V_2 and T_2. Since the heat is supplied *at constant pressure*, then:

Total heat supplied = wt. × spec. ht. × temp. rise
$$= w \times C_p \times (T_2 - T_1)$$

As the piston is pushed forward by the gas, P is the constant pressure acting on the piston. Let this be in lb/ft², then this multiplied by the area of the piston A square feet is the total lb of force acting on the piston. If the piston moves through a distance of x feet then the work done is P × A × x ft lb. The product of the piston area in ft² and the distance moved in feet, i.e., A × x, is the *volume* moved through by the piston. This is the increase in volume of the gas which is $V_2 - V_1$ cubic feet, hence:

Work done = $P(V_2 - V_1)$ ft lb

From PV = wRT,

$$V_2 = \frac{wRT_2}{P} \text{ and } V_1 = \frac{wRT_1}{P}$$

Substituting these values,

Work done = $P(V_2 - V_1)$

$$= PV_2 - PV_1$$

$$= \frac{PwRT_2}{P} - \frac{PwRT_1}{P}$$

LAWS OF PERFECT GASES

$$= wRT_2 - wRT_1$$
$$= wR(T_2 - T_1) \text{ ft lb}$$

Using J to represent Joule's mechanical equivalent of heat (e.g. 778 ft lb = 1 Btu) the external work done can be expressed in heat units,

$$\text{Work done} = \frac{wR(T_2 - T_1)}{J} \text{ Btu}$$

The increase in internal energy of the gas will be less than the total heat supplied by an amount equal to the external work done,

Heat supplied = Increase in internal energy + External work done

$$w \times C_p \times (T_2 - T_1) = \text{Increase in internal energy} + \frac{wR(T_2 - T_1)}{J}$$

$$\therefore \text{Increase in internal energy} = w \times C_p \times (T_2 - T_1) - \frac{wR(T_2 - T_1)}{J} \quad \text{(i)}$$

Now if the gas was heated through the same range of temperature, T_1 to T_2 *at constant volume,* the heat supplied would be $w \times C_v \times (T_2 - T_1)$, but no work would be done because the piston does not move. Hence, applying the energy equation:

Heat supplied = Increase in internal energy + External work done
$w \times C_v \times (T_2 - T_1) = $ Increase in internal energy + 0
$\therefore$ Increase in internal energy $= w \times C_v \times (T_2 - T_1)$ ... ... (ii)

Therefore, from (i) and (ii),

$$w \times C_v \times (T_2 - T_1) = w \times C_p \times (T_2 - T_1) - \frac{wR(T_2 - T_1)}{J}$$

Cancelling w and $(T_2 - T_1)$ which are common to all terms,

$$C_v = C_p - \frac{R}{J}$$

$$\text{or } C_p - C_v = \frac{R}{J}$$

Multiplying throughout by J,

$$K_p - K_v = R$$

Inserting the values for air as previously given,

$$R = 184 \cdot 8 - 131 \cdot 5 = 53 \cdot 3 \text{ ft lb/lb per } °F.$$

Another very important relationship between the specific heats is the ratio C_p divided by C_v and this ratio is represented by the Greek letter gamma, thus,

$$\gamma = \frac{C_p}{C_v}$$

and for air its value is

$$\gamma = \frac{0 \cdot 2375}{0 \cdot 169} \text{ or } \frac{184 \cdot 8}{131 \cdot 5} = 1 \cdot 4$$

This will be dealt with in more detail later.

DALTON'S LAW OF PARTIAL PRESSURES

The pressure exerted in a vessel by a mixture of gases, or a mixture of a gas and a vapour, is equal to the sum of the pressures that each would exert if it alone occupied the whole volume of the vessel. The pressure exerted by each gas is termed a *partial pressure*, thus,

$$\text{Total pressure of the mixture} = \text{Partial pressure due to one gas} + \text{Partial pressure due to other gas}$$

This is particularly useful in determining the weight of air leaking into steam condensers with the steam vapour, and will be dealt with later.

LAWS OF PERFECT GASES 51

TEST EXAMPLES 3

1. Express the atmospheric pressure in lb/in² when the barometer stands at 760 millimetres of mercury. (Note, 25·4 mm = 1 inch).

2. Five cubic feet of air at 215 lb/in² abs. expand to a volume of 12·5 cubic feet at constant temperature. Find the final pressure.

3. One cubic foot of gas at 14 lb/in² abs. is compressed to a pressure of 110 lb/in² abs. If compression is according to the law PV = constant, find the final volume.

4. Calculate the volume of air at 14·7 lb/in² abs. required to compress to 5 cubic feet at 600 lb/in² abs. at the same temperature.

5. An air storage system consists of two cylindrical vessels with hemispherical ends, each being one foot diameter and 5 feet long overall. Calculate the volume of "free" air (that is, air at atmospheric pressure, say 14·7 lb/in²) to be taken from the atmosphere and pumped into the two vessels at 400 lb/in² gauge. Assume that the bottles initially contain air at 14·7 lb/in² abs. and that the temperature of the compressed air delivered is the same as the temperature of the atmosphere.

ƒ6. A single-acting air pump has an effective swept stroke volume of 3 cubic feet and the total volume of the pipe-line to which it is connected is 500 cubic feet. Commencing with a pressure of 14·7 lb/in² abs. in the line, find the number of suction strokes required to reduce the pressure to 5 lb/in² abs. assuming the temperature remains constant throughout.

7. Ten cubic feet of gas at 70°F is heated at constant pressure to a temperature of 600°F. Calculate the final volume.

8. A closed vessel contains air at a pressure of 81 lb/in² gauge and 64°F. Find the final pressure if the air is heated at constant volume to a temperature of 195°F. Take the atmospheric pressure as 15 lb/in.²

9. 0·2 ft³ of air at a pressure of 135 lb/in² abs. and temperature 80°F is expanded in a cylinder to a volume of 0·9 ft³ and pressure 25 lb/in² abs. Calculate the final temperature.

10. If one cubic foot of air at 14·7 lb/in² abs. and 32°F weighs 0·0807 lb, calculate (i) the volume of one lb weight, (ii) the characteristic gas constant for one lb weight of air in ft lb per °F.

11. Taking the characteristic gas constant for nitrogen as 99·2 ft lb per lb per °C, calculate (i) the weight of 2 cubic feet of nitrogen at 80 lb/in² abs. and 82°C, (ii) the volume of 2 lb of nitrogen at 200 lb/in² abs. and 22°C.

12. An air storage vessel has hemispherical ends. Its diameter is 3 feet and overall length 11 feet. Find the weight of air inside when the pressure is 450 lb/in² gauge and the temperature is 75°F. Take the atmospheric pressure as 15 lb/in² and the gas constant as 53·3 ft lb/lb per °F.

13. An air reservoir contains 45 lb weight of air at a temperature of 60°F and pressure 400 lb/in² gauge. Neglecting the expansion of the reservoir and taking the atmospheric pressure as 15 lb/in², calculate the pressure when the air is heated to 95°F. Calculate also the heat absorbed by the air taking the specific heat at constant volume as 0·169.

14. A ship's saloon is 42 ft by 55 ft by 10 ft. The air is completely changed once every 30 minutes and the temperature maintained at 78°F. If the temperature of the outside atmosphere is 85°F calculate the heat required to be extracted from the supply air per hour and the equivalent horse power, taking the weight of one cubic foot of air at 14·7 lb/in.² abs. and 32°F as 0·0807 lb, and the specific heat at constant pressure as 0·2375.

15. 1·5 cubic feet of air in a diesel engine cylinder is heated at a constant pressure of 500 lb/in² by the burning of the fuel until the volume is 3·3 cubic feet. Find the external work done during the fuel burning period.

16. In a diesel engine cylinder, 1·388 lb weight of air is heated at constant pressure by the fuel from 1000°F to 2752°F. Take the characteristic gas constant $R = 53·3$ ft lb/lb per °F and find the external work done during the burning period.

17. Calculate the heat given to 3 lb weight of air when heated from 40°C to 468°C (i) if the volume remains constant, (ii) if the pressure remains constant. Find also the external work done in each case. Take $C_v = 0·169$ and $C_p = 0·2375$.

ƒ18. Prove the following relationship for a perfect gas:

$$C_p - C_v = \frac{R}{J}$$

where, C_p = specific heat at constant pressure,
C_v = specific heat at constant volume,
R = characteristic gas constant,
J = Joule's mechanical equivalent of heat.

CHAPTER 4
EXPANSION AND COMPRESSION OF GASES
COMPRESSION OF A GAS

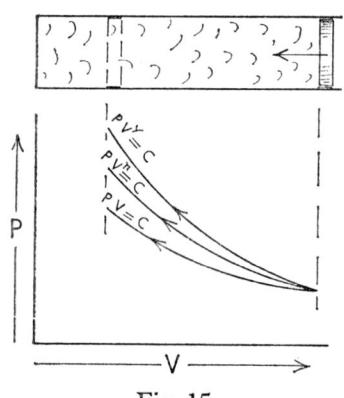

Fig. 15

When a gas is compressed in a cylinder (Fig. 15) the pressure of the gas increases as the volume decreases. The work done *on* the gas to compress it appears as heat energy in the gas and the temperature tends to rise. This effect can readily be seen with a tyre inflator; in pumping up the tyre the discharge end of the inflator gets hot due to compressing the air.

ISOTHERMAL COMPRESSION. Imagine the piston pushed inward slowly to compress the gas, and, at the same time, let heat be taken away via the cylinder walls (by a water-jacket or other means) to avoid any rise in temperature. If the gas could be compressed in this way, *at constant temperature*, the operation would be referred to as *isothermal compression* and the relationship between pressure and volume would follow Boyle's law as stated in the previous chapter, thus,

$$PV = \text{constant} \qquad \therefore P_1V_1 = P_2V_2$$

ADIABATIC COMPRESSION. Now imagine the piston pushed inward quickly so that there is insufficient time for any heat to be rejected by the gas through the cylinder walls. All the work done in compressing the gas appears as stored up heat energy. The temperature at the end of compression will, therefore, be high and, for the same ratio of compression as the first case, the pressure will consequently be higher. This form of compression, where no heat is taken from or given to the gas from an external source, is known as *adiabatic compression.* The relationship between pressure and volume for adiabatic compression is:

$$PV^\gamma = \text{constant} \qquad \therefore P_1V_1^\gamma = P_2V_2^\gamma$$

where γ (Gamma) is the ratio of the specific heat of the gas at constant pressure to the specific heat at constant volume, i.e.,

$$\gamma = \frac{C_p}{C_v}$$

POLYTROPIC COMPRESSION. In practice, neither of the above two cases can be achieved perfectly. Some heat is always lost from the gas through the cylinder walls, more especially if the cylinder is water or air cooled, but this is never as much as the whole amount of the generated heat of compression. Consequently, the compression curve representing the relationship between pressure and volume lies somewhere between the two theoretical cases of isothermal and adiabatic. Such compression, where a partial amount of heat exchange occurs between gas and cylinder walls during the operation, is termed *polytropic compression* and the compression curve follows the law:

$$PV^n = \text{constant} \qquad \therefore P_1V_1^n = P_2V_2^n$$

Thus the law $PV^n = C$ may be taken as the general case to cover all forms of compression from isothermal to adiabatic wherein the value of n for isothermal compression is unity, for adiabatic compression $n = \gamma$, and for polytropic compression n lies between 1 and γ.

EXPANSION OF A GAS

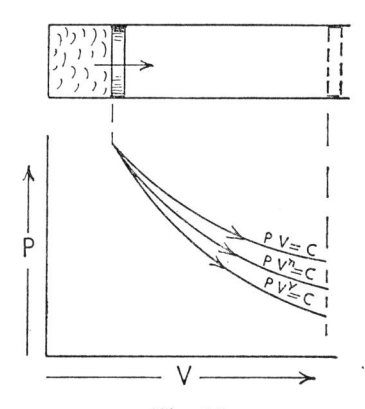

Fig. 16

When a gas is expanded in a cylinder (Fig. 16) the pressure falls and the volume increases as the piston is pushed outward by the energy in the gas.

This is exactly the opposite to compression. Work is done *by* the gas in pushing the piston outward and there is a tendency for the temperature to fall due to the heat energy in the gas being converted into mechanical energy. Therefore, to expand the gas *isothermally*, heat must be supplied to the gas from an external source during the expansion in order to maintain its temperature constant. The expansion would then follow Boyle's law, i.e. $PV = C$.

The gas would expand *adiabatically* if no heat exchange, to or from the gas, occurs during the expansion, the external work done in pushing the piston forward being entirely at the expense of the stored up heat energy. Therefore, the temperature of the gas will fall considerably during the expansion. The law for adiabatic expansion is $PV^{\gamma} = C$.

During a *polytropic* expansion, a partial amount of heat will be given to the gas but not sufficient to maintain a uniform temperature during the operation, and the law of expansion is $PV^n = C$.

With reference to Figs. 15 and 16, the student should note that the adiabatic curve is the steepest, the polytropic curve less steep and the isothermal curve the least steep. Thus, the higher the index of the law of expansion or compression, the steeper will be the curve.

It must also be noted that for any mode of expansion or compression, the combination of Boyle's and Charles' laws, and the characteristic gas equation, given in the previous chapter, are always true: viz.:

$$\frac{PV}{T} = \text{constant or } \frac{P_1 V_1}{T_1} = \frac{P_2 V_2}{T_2}$$

and $PV = wRT$

The mathematics involved in solving equations of the form $P_1 V_1^n = P_2 V_2^n$ is adequately covered in Volume 1 Chapter 4, more examples are given here.

EXPANSION AND COMPRESSION OF GASES 57

Example. Ten cubic feet of air at 12 lb/in² abs. are compressed in an engine cylinder to a volume of 2 cubic feet, the law of compression being $PV^{1.4} = C$. Calculate (i) the final pressure, (ii) the final temperature if the initial temperature was 50°F, (iii) the weight of air in the cylinder, taking the characteristic gas constant for air $R = 53.3$ ft lb/lb/°F.

$$P_1V_1^{1.4} = P_2V_2^{1.4}$$
$$12 \times 10^{1.4} = P_2 \times 2^{1.4}$$
$$P_2 = 12 \times \left\{\frac{10}{2}\right\}^{1.4}$$
$$= 12 \times 5^{1.4}$$
$$= 114.2 \text{ lb/in}^2 \text{ abs. Ans. (i)}$$

$$\frac{P_1V_1}{T_1} = \frac{P_2V_2}{T_2}$$

$$\frac{12 \times 10}{50 + 460} = \frac{114.2 \times 2}{T_2}$$

$$T_2 = \frac{510 \times 114.2 \times 2}{12 \times 10}$$

$$= 970.7°F \text{ abs.}$$
$$970.7 - 460 = 510.7°F. \text{ Ans. (ii)}$$

$$PV = wRT$$

$$w = \frac{PV}{RT}$$

$$= \frac{12 \times 144 \times 10}{53.3 \times 510}$$

$$= 0.6357 \text{ lb. Ans. (iii)}$$

Example. 2·5 cubic feet of air at 600 lb/in² abs. is expanded in an engine cylinder and the pressure at the end of expansion is 45 lb/in² abs. If expansion follows the law $PV^{1\cdot 35} = C$, find the final volume.

$$P_1 V_1^{1\cdot 35} = P_2 V_2^{1\cdot 35}$$
$$600 \times 2\cdot 5^{1\cdot 35} = 45 \times V_2^{1\cdot 35}$$

$$V_2^{1\cdot 35} = \frac{600 \times 2\cdot 5^{1\cdot 35}}{45}$$

$$V_2 = 2\cdot 5 \times \sqrt[1\cdot 35]{\frac{600}{45}}$$

$$= 17\cdot 03 \text{ cubic feet. Ans.}$$

Example. 0·5 ft³ of gas at 420 lb/in² abs. is expanded to a volume of 5·5 ft³ and the final pressure is 16 lb/in² abs. If the law of expansion is $PV^n = C$, find the value of n.

$$P_1 V_1^n = P_2 V_2^n$$
$$420 \times 0\cdot 5^n = 16 \times 5\cdot 5^n$$

$$\frac{420}{16} = \left\{\frac{5\cdot 5}{0\cdot 5}\right\}^n$$

$$26\cdot 25 = 11^n$$
$$\log 26\cdot 25 = (\log 11) \times n$$
$$1\cdot 4191 = 1\cdot 0414 \times n$$

$$n = \frac{1\cdot 4191}{1\cdot 0414}$$

$$= 1\cdot 363 \text{ Ans.}$$

DETERMINATION OF n FROM GRAPH

It will be appreciated that it would be most difficult to obtain two pairs of sufficiently accurate values of pressure and volume from a running engine to enable the law of expansion or compression to be determined. One practical method of finding the law is as follows,

EXPANSION AND COMPRESSION OF GASES

(i) measure a series of connected values of P and V from the curve of an indicator diagram, (ii) reduce the equation $PV^n = C$ to a straight line logarithmic equation, (iii) draw a straight line graph as near as possible through the plotted points of log P and log V to eliminate slight errors of measurement, (iv) determine the law of this graph to obtain the value of n. Thus:

$$P \times V^n = C$$
$$\therefore \log P + n \log V = \log C$$
$$\therefore \log P = \log C - n \log V$$

This is the same form of equation as,
$$y = a - bx$$
which represents a simple straight-line graph.

The terms log P and log V are the two variables comparable with y and x respectively, and log C and n are constants comparable with a and b respectively. The constant n (like constant b) represents the slope of the straight-line graph and, being a negative value, the line will slope downwards from left to right.

Example. The following related values of pressure P and volume V were measured from the compression curve of a diesel engine indicator diagram. Assuming that P and V are connected by the law $PV^n = C$, find the value of n.

P	500	330	250	96	38	18
V	0·3	0·4	0·5	1·0	2·0	3·5

Tabulating the values of P and V with their logarithms to obtain plotting points from the respective pairs of log P and log V:

P	log P	V	log V
500	2·6990	0·3	$\bar{1}$·4771 = $-$ 0·5229
330	2·5185	0·4	$\bar{1}$·6021 = $-$ 0·3979
250	2·3979	0·5	$\bar{1}$·6990 = $-$ 0·3010
96	1·9823	1·0	0·0000
38	1·5798	2·0	0·3010
18	1·2553	3·5	0·5441

When the value of V is less than unity, the logarithm consists of a negative characteristic. Therefore, to plot such a quantity the true numerial value must be obtained by combining the mixture of negative characteristic and positive mantissa, the result of which is an all-negative quantity.

60   REED'S HEAT AND HEAT ENGINES FOR MARINE ENGINEERS

The graph may be plotted as shown in Fig. 17. Note that it is not necessary to commence at zero origin when only the slope of the line (value of *n*) is required. A larger graph on the available squared paper can be drawn by starting and finishing to suit the minimum and maximum values to be plotted.

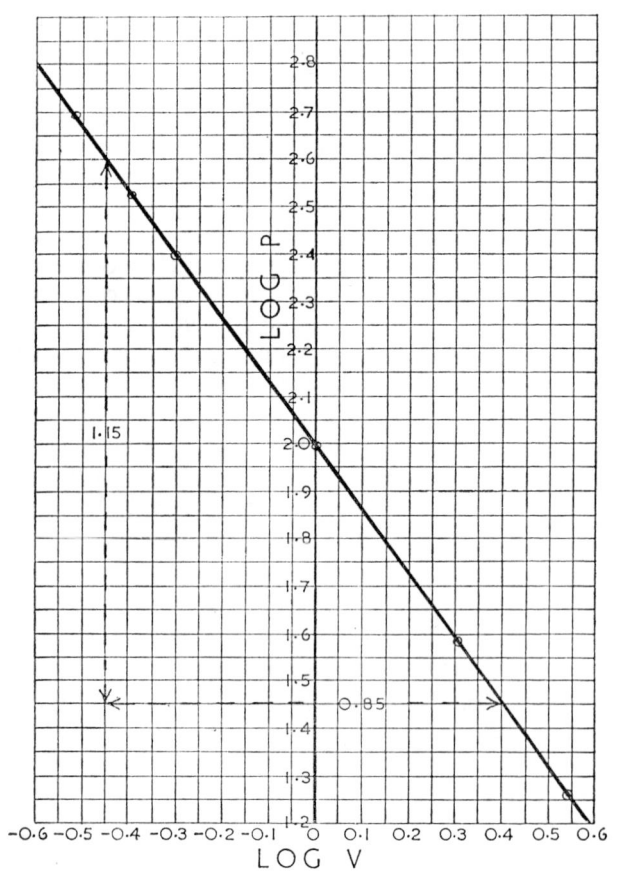

Fig. 17

Choosing two points on the line as shown:

$$n = \frac{\text{decrease of log P}}{\text{increase of log V}}$$

$$= \frac{2 \cdot 6 - 1 \cdot 45}{0 \cdot 4 - (-0 \cdot 45)}$$

$$= \frac{1 \cdot 15}{0 \cdot 85}$$

$$= 1 \cdot 353$$

∴ law of compression curve is:
$PV^{1 \cdot 353} = C$ Ans.

RATIOS OF EXPANSION AND COMPRESSION

The ratio of expansion of gas in a cylinder is the ratio of the volume at the end of expansion to the volume at the beginning of expansion. It is usually denoted by r.

$$\text{Ratio of expansion} = r = \frac{\text{Final volume}}{\text{Initial volume}}$$

The ratio of compression is the ratio of the volume of the gas at the beginning of compression to the volume at the end of compression. This is also usually denoted by r.

$$\text{Ratio of compression} = r = \frac{\text{Initial volume}}{\text{Final volume}}$$

It will be noted that in each of the above ratios, it is the larger volume divided by the smaller, therefore, the ratio of expansion and ratio of compression is always greater than unity.

RELATIONSHIPS BETWEEN TEMPERATURE AND VOLUME, AND TEMPERATURE AND PRESSURE,

when $PV^n = C$

As stated previously, the equations $P_1V_1^n = P_2V_2^n$ and $P_1V_1/T_1 = P_2V_2/T_2$ are always true for any kind of expansion or compression of a perfect gas. Some problems arise, however, where neither P_1 nor P_2 are given and the unknown temperature or volume has to be solved by substituting the value of one of the pressures from one equation into the other. Similarly, there are cases where neither volume is given and substitution has to be made for one of the volumes to obtain the unknown temperature or pressure.

Substitution can be made in one of the above equations to eliminate either pressure or volume and so derive formulae for direct solution, as follows.

$$\text{Since } P_1 V_1^n = P_2 V_2^n \qquad \text{then } P_1 = \frac{P_2 V_2^n}{V_1^n}$$

Substituting this value of P_1 into the combined law equation:

$$\therefore \frac{P_2 \times V_2^n \times V_1}{T_1 \times V_1^n} = \frac{P_2 \times V_2}{T_2}$$

$$\therefore T_1 \times V_1^n \times P_2 \times V_2 = T_2 \times P_2 \times V_2^n \times V_1$$

$$\therefore \frac{T_1}{T_2} = \frac{P_2 \times V_2^n \times V_1}{P_2 \times V_2 \times V_1^n}$$

P_2 cancels,

Dividing V_2^n by $V_2 = V_2^n \div V_2 = V_2^{n-1}$
Dividing V_1^n by $V_1 = V_1^n \div V_1 = V_1^{n-1}$

$$\therefore \frac{T_1}{T_2} = \frac{V_2^{n-1}}{V_1^{n-1}}$$

$$\frac{T_1}{T_2} = \left\{\frac{V_2}{V_1}\right\}^{n-1} \quad \ldots \quad \ldots \quad (\text{i})$$

Again from $P_1 V_1^n = P_2 V_2^n$

EXPANSION AND COMPRESSION OF GASES 63

$$V_1^n = \frac{P_2 V_2^n}{P_1} \qquad \therefore V_1 = \frac{P_2^{1/n} \times V_2}{P_1^{1/n}}$$

Substituting this value of V_1 into the combined law equation:

$$\frac{P_1 V_1}{T_1} = \frac{P_2 V_2}{T_2}$$

$$\therefore \frac{P_1 \times P_2^{1/n} \times V_2}{T_1 \times P_1^{1/n}} = \frac{P_2 \times V_2}{T_2}$$

$$\therefore T_1 \times P_1^{1/n} \times P_2 \times V_2 = T_2 \times P_1 \times P_2^{1/n} \times V_2$$

$$\therefore \frac{T_1}{T_2} = \frac{P_1 \times P_2^{1/n} \times V_2}{P_1^{1/n} \times P_2 \times V_2}$$

V_2 cancels,

Dividing P_1 by $P_1^{1/n} = P_1 \div P_1^{1/n} = P_1^{1-1/n}$
Dividing P_2 by $P_2^{1/n} = P_2 \div P_2^{1/n} = P_2^{1-1/n}$

$$\therefore \frac{T_1}{T_2} = \frac{P_1^{1-1/n}}{P_2^{1-1/n}}$$

$$\therefore \frac{T_1}{T_2} = \left\{\frac{P_1}{P_2}\right\}^{1-1/n}$$

$$\therefore \frac{T_1}{T_2} = \left\{\frac{P_1}{P_2}\right\}^{\frac{n-1}{n}} \qquad \ldots \qquad \ldots \quad \text{(ii)}$$

From (i) and (ii),

$$\therefore \frac{T_1}{T_2} = \left\{\frac{V_2}{V_1}\right\}^{n-1} = \left\{\frac{P_1}{P_2}\right\}^{\frac{n-1}{n}}$$

Since either end of the expansion or compression curve can be marked state-point 1 and the other end state-point 2, if the top end of the curve is marked $P_1V_1T_1$ and the bottom end $P_2V_2T_2$ then all the numerators in the above expression, namely $T_1 V_2$ and P_1 will be the larger respective values, and the denominators T_2 V_1 and P_2 the smaller. This is an aid to memorising the expression. However, when dealing with work done during expansion or compression, in order to determine whether work is done BY the gas (i.e. positive work done as during an expansion) or ON the gas (i.e. negative work done as during a compression), it is better to denote the *initial* conditions by the suffix 1 and the *final* conditions by the suffix 2, as will be seen later.

Example. Air is expanded adiabatically from a pressure of 100 lb/in² abs. to 16 lb/in² abs. If the final temperature is — 20°F calculate the temperature at the beginning of expansion, taking the value of γ as 1·4.

$$\frac{T_1}{T_2} = \left\{\frac{P_1}{P_2}\right\}^{\frac{\gamma-1}{\gamma}}$$

Note, $\dfrac{\gamma - 1}{\gamma} = \dfrac{1\cdot 4 - 1}{1\cdot 4} = \dfrac{0\cdot 4}{1\cdot 4} = \dfrac{2}{7}$

$$\frac{T_1}{-20 + 460} = \left\{\frac{100}{16}\right\}^{2/7}$$

$$T_1 = 440 \times 6\cdot 25^{2/7}$$
$$= 742\cdot 9°F \text{ abs.}$$
$$742\cdot 9 - 460 = 282\cdot 9°F. \text{ Ans.}$$

Example. The ratio of compression in a petrol engine is 9 to 1. Find the temperature of the gas at the end of compression if the temperature at the beginning is 75°F, assuming the compression curve to follow the law $PV^n = C$ where $n = 1\cdot 36$.

$$\frac{T_1}{T_2} = \left\{\frac{V_2}{V_1}\right\}^{n-1}$$

$$n - 1 = 1\cdot 36 - 1 = 0\cdot 36$$

EXPANSION AND COMPRESSION OF GASES

$$\frac{T_1}{75 + 460} = \left\{\frac{9}{1}\right\}^{0.36}$$

$$T_1 = 535 \times 9^{0.36}$$
$$= 1180°F \text{ abs.}$$
$$1180 - 460 = 720°F. \text{ Ans.}$$

Example. The volume and temperature of a gas at the beginning of expansion are 0·2 ft³ and 360°F, at the end of expansion the values are 0·85 ft³ and 72°F respectively. Assuming expansion follows the general law $PV^n = C$, find the value of n.

$$\frac{T_1}{T_2} = \left\{\frac{V_2}{V_1}\right\}^{n-1}$$

$$\frac{360 + 460}{72 + 460} = \left\{\frac{0.85}{0.2}\right\}^{n-1}$$

$$1.541 = 4.25^{n-1}$$

$$\log 1.541 = (\log 4.25) \times (n - 1)$$
$$0.1879 = 0.6284 (n - 1)$$
$$0.1879 = 0.6284n - 0.6284$$
$$0.6284 + 0.1879 = 0.6284n$$
$$0.8163 = 0.6284n$$

$$n = \frac{0.8163}{0.6284}$$

$$= 1.299 \text{ Ans.}$$

WORK DONE DURING EXPANSION AND COMPRESSION

First consider the case where the gas expands at constant pressure in a cylinder fitted with a movable piston.

Work done (ft lb) = Force (lb) × Distance (ft)

Let the area of the piston be A square feet and the intensity of pressure of the gas be P lb per square foot, then,

$$\text{Total force on piston} = P \times A \text{ lb}$$

If the piston moves through x feet as the gas expands,

$$\text{Work done} = P \times A \times x \text{ ft lb}$$

The area of the piston (A ft^2) multiplied by the distance it moves (x ft) is the volume swept through by the piston in cubic feet, this is also the increase in volume of the gas in the cylinder. If V_1 is the volume of the gas at the beginning of expansion and V_2 is the volume at the end of expansion, then $A \times x$ is equal to $(V_2 - V_1)$, hence,

$$\text{Work done} = P(V_2 - V_1) \text{ ft lb}$$

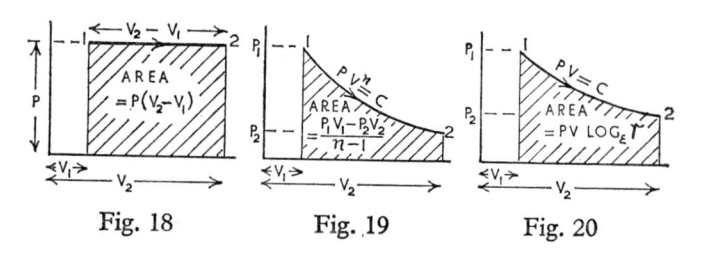

Fig. 18   Fig. 19   Fig. 20

Fig. 18 shows the pressure-volume diagram representing work done at constant pressure. The graph is a straight horizontal line and the area under it is a rectangle. The area of a rectangle is height × length which, in this case, is $P \times (V_2 - V_1)$. Hence *the area under the pressure-volume line represents work done.*

Now consider cases where the pressure falls during the expansion of the gas. The formula giving the area under the polytropic curve representing the general relationship between pressure and volume, i.e. $PV^n = C$ can only be derived satisfactorily by the use of the calculus which is beyond the scope of this book. The expression is, therefore, given here without proof and illustrated in Fig. 19.

Work done during polytropic expansion

$$= \frac{P_1 V_1 - P_2 V_2}{n - 1}$$

EXPANSION AND COMPRESSION OF GASES

The above is the general expression for work done. For pure adiabatic expansion, γ takes the place of n. For isothermal expansion, however, since the value of n is 1, and as $P_1V_1 = P_2V_2$, then substitution in this expression for work done will produce $0 \div 0$ which is indeterminate. A different expression is therefore used to obtain the work done during isothermal expansion.

Work done during isothermal expansion
$$= PV \log_\varepsilon r$$
where r is the ratio of expansion.

The above expressions give the work done BY the gas during expansion. The same expressions give the work done ON the gas during compression.

In most cases, the product of the higher pressure and smaller volume exceeds the product of the lower pressure and larger volume. Therefore, if the initial conditions of the gas are denoted by the suffix 1 and final conditions by the suffix 2, then in the case of expansion P_1V_1 will exceed P_2V_2 and the expression for work done will produce a *positive* result, indicating that work is done *by* the gas on the piston. Conversely, for compression, P_1V_1 will be less than P_2V_2 and hence a *negative* result will be obtained, indicating that work is done *on* the gas by the piston.

The student must always bear in mind that to obtain work done in ft lb, the pressure must be in lb per square foot and the volume in cubic feet.

Further, since $PV = wRT$, then the above formulae for work done can be expressed in terms of wRT instead of PV, thus,

Expansion at constant pressure:
$$\text{Work done} = P(V_2 - V_1) = wR(T_2 - T_1)$$

Polytropic expansion:
$$\text{Work done} = \frac{P_1V_1 - P_2V_2}{n-1} = \frac{wR(T_1 - T_2)}{n-1}$$

Isothermal expansion:
$$\text{Work done} = PV \log_\varepsilon r = wRT \log_\varepsilon r$$

The importance of the units of the characteristic gas constant R previously stated as ft lb/lb/°T will now be apparent, as it represents work done in ft lb per lb weight of gas per degree of temperature.

Example. Two cubic feet of gas at 215 lb/in² abs. are expanded isothermally until the volume is $4\frac{1}{2}$ cubic feet. Calculate the work done during the expansion.

Work done = $PV \log_\varepsilon r$

$$r = \frac{\text{Final volume}}{\text{Initial volume}} = \frac{4 \cdot 5}{2} = 2 \cdot 25$$

From hyperbolic logarithm tables,
$\log_\varepsilon 2 \cdot 25 = 0 \cdot 8109$

Note, if hyperbolic log tables are not to hand, the hyperbolic log of a number can be obtained by multiplying the common log of that number by 2·3026.

Work done = $PV \log_\varepsilon r$
= 215 × 144 × 2 × 0·8109
= 50210 ft lb. Ans.

ƒ Example. 0·25 ft³ of air at 200 lb/in² abs. and 635°F are expanded according to the law $PV^{1 \cdot 32} = C$ and the final pressure is 17·5 lb/in² abs. Calculate (i) the volume at the end of expansion, (ii) the work done by the air during expansion, (iii) the temperature at the end of expansion, (iv) the weight of air in the cylinder, taking R = 53·3 ft lb/lb/°F.

$$P_1 V_1^{1 \cdot 32} = P_2 V_2^{1 \cdot 32}$$
$$200 \times 0 \cdot 25^{1 \cdot 32} = 17 \cdot 5 \times V_2^{1 \cdot 32}$$

$$V_2^{1 \cdot 32} = \frac{200 \times 0 \cdot 25^{1 \cdot 32}}{17 \cdot 5}$$

$$V_2 = 0 \cdot 25 \times \sqrt[1 \cdot 32]{\frac{200}{17 \cdot 5}}$$

= 1·582 cubic feet. Ans. (i)

EXPANSION AND COMPRESSION OF GASES

$$\text{Work done} = \frac{P_1V_1 - P_2V_2}{n-1}$$

$$= \frac{200 \times 144 \times 0.25 - 17.5 \times 144 \times 1.582}{1.32 - 1}$$

$$= \frac{144}{0.32}(200 \times 0.25 - 17.5 \times 1.582)$$

$$= \frac{144}{0.32}(50 - 27.7)$$

$$= \frac{144}{0.32} \times 22.3$$

$$= 10030 \text{ ft lb. Ans (ii)}$$

$$\frac{P_1V_1}{T_1} = \frac{P_2V_2}{T_2}$$

$$\frac{200 \times 0.25}{635 + 460} = \frac{17.5 \times 1.582}{T_2}$$

$$T_2 = \frac{1095 \times 17.5 \times 1.582}{200 \times 0.25}$$

$$= 606.4°\text{F abs.}$$

$$606.4 - 460 = 146.4°\text{F. Ans. (iii)}$$

$$PV = w\text{RT}$$

$$w = \frac{PV}{RT}$$

$$= \frac{200 \times 144 \times 0.25}{53.3 \times 1095}$$

$$= 0.1234 \text{ lb. Ans. (iv)}$$

TEST EXAMPLES 4

1. A gas expands isothermally in a cylinder from a volume of 0·15 ft³ to a volume of 0·75 ft³. If the initial pressure was 100 lb/in² abs. find the final pressure.

2. 1·2 ft³ of air at 14·5 lb/in² abs. is compressed adiabatically to a volume of 0·2 ft³. Taking $\gamma = 1\cdot4$, find the pressure at the end of compression.

3. Gas is expanded in an engine cylinder following the law $PV^n = C$ where the value of n is 1·3. The initial pressure is 425 lb/in² abs. and the final pressure is 35 lb/in² abs. If the volume at the end of expansion is 7·5 cubic feet, calculate the volume at the beginning of expansion.

4. The ratio of compression in a petrol engine is 8·6 to 1. At the beginning of compression the pressure of the gas is 14 lb/in² abs. and the temperature is 82°F. Find the pressure and temperature at the end of compression assuming it follows the law $PV^{1\cdot36} = $ Constant.

5. Gas is expanded in an engine cylinder according to the law $PV^n = C$. At the beginning of expansion the pressure and volume are 250 lb/in² abs. and 2 cubic feet respectively, and at the end of expansion the values are 17·5 lb/in² abs. and 15 cubic feet. Calculate the value of n.

ƒ6. The following pairs of quantities, ordinates representing pressures (P) and abscissae representing volumes (V), were measured from the expansion curve of an indicator diagram off an I.C. engine. Assuming that expansion follows the law $PV^n = C$, find the value of n.

P	200	137	95	70	50	35
V	0·45	0·6	0·8	1·0	1·3	1·7

7. The temperature and pressure at the beginning of compression of air in a compressor are 80°F and 14·7 lb/in² abs., and the pressure at the end of compression is 220 lb/in² abs. If compression takes place according to the law $PV^n = C$ where $n = 1\cdot35$, find the temperature at the end of compression.

8. 0·5 ft³ of a gas at 150°F is expanded adiabatically in a cylinder and the temperature at the end of expansion is 35°F. Take the specific heats of the gas at constant pressure and constant volume as 0·24 and 0·17 respectively and calculate the volume at the end of expansion.

EXPANSION AND COMPRESSION OF GASES

9. One cubic foot of air at 240°F is expanded to 3·5 cubic feet and the final temperature is 30°F. If expansion follows the law $PV^n = C$, find the value of n.

10. Air is compressed in a diesel engine from 17 lb/in² abs. to 530 lb/in² abs. If the temperatures at the beginning and end of compression are 90°F and 933°F respectively, find the law of compression assuming it follows the general form $PV^n = C$.

11. 1·5 cubic feet of gas at 60 lb/in² abs. are expanded isothermally in an engine cylinder to 4·926 cubic feet. Calculate the work done during expansion.

f12. 4 cubic feet of air at 120 lb/in² abs. are expanded in a cylinder until the volume is 11·72 cubic feet. Calculate the work done if the expansion is (i) isothermal, (ii) adiabatic, taking $\gamma = 1\cdot 4$.

CHAPTER 5

I.C. ENGINES — ELEMENTARY PRINCIPLES

Internal combustion engines are so named because combustion of the fuel takes place *inside* the engine. When the fuel burns inside the engine cylinder, it gives out heat which is absorbed by the air previously taken into the cylinder, the temperature of the air is therefore increased with a consequent increase in pressure and/or volume, thus energy is imparted to the piston. The reciprocating motion of the piston is converted into a rotary motion at the crank shaft by connecting rod and crank.

The method of igniting the fuel varies. In diesel engines the air in the cylinder is compressed to a high pressure so that it attains a high temperature, and when oil fuel is injected into this high temperature air the fuel immediately ignites. When the ignition of the fuel is caused solely by the heat of compression, the engine is classed as a *compression-ignition* engine. In petrol and paraffin engines the fuel is usually taken in with the charge of air, compressed and then ignited by an electric spark.

THE FOUR-STROKE DIESEL ENGINE

In this type of engine it takes four strokes of the piston (i.e. two revolutions of the crank) to complete one working cycle of operations, hence the name *four-stroke* cycle.

Fig. 21 illustrates each of these four strokes in one cylinder. On the cylinder head is shown (i) the fuel valve which lifts to admit oil fuel (under pressure) into the cylinder, (ii) the air-induction valve through which air is drawn in, and (iii) the exhaust valve through which the exhaust gases are expelled from the cylinder. There are two more valves which are not shown here because they do not operate during the normal working cycle; one is the relief valve which opens against the compression of its spring when the pressure in the cylinder rises too high, the other is the air-starting valve which is cam-operated and opens to admit high pressure air into the cylinder to move the piston and start the engine.

I.C. ENGINES—ELEMENTARY PRINCIPLES 73

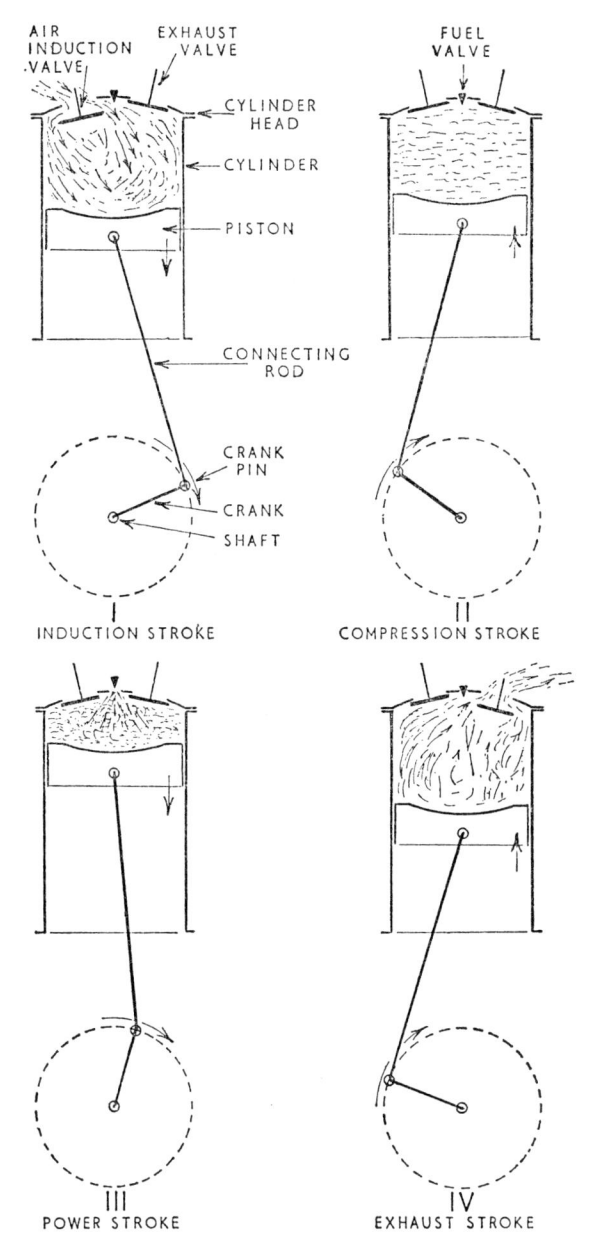

FOUR-STROKE DIESEL ENGINE Fig. 21

CYCLE OF OPERATIONS

SKETCH (i). This illustrates the *induction stroke*. The piston is moving down, the air induction valve is open and air is being drawn into the cylinder from the atmosphere by the suction effect of the piston. At the end of this stroke the cylinder is full of air and the air-induction valve closes.

SKETCH (ii). This shows the *compression stroke*. The piston is moving up, all valves are closed and therefore the air in the cylinder is being compressed. When air is compressed its temperature rises and the reason for compressing the air in a diesel engine is to obtain a sufficiently high temperature to cause the fuel oil to ignite and burn rapidly when it is injected into the cylinder at the end of this stroke. The pressure of the air at the end of compression is usually in the region of 500 lb/in^2 giving a temperature of about 1000°F.

SKETCH (iii). This is the *power stroke* and the piston is moving down. The fuel is injected into the cylinder in the form of a fine spray through the fuel valve, it mixes with the hot air and burns rapidly. The fuel is admitted a few degrees before top dead centre of the crank to give it time to reach full combustion for the beginning of the stroke and the valve remains open for about one-tenth of the downward stroke. As the oil burns it heats the air which would cause a rise in pressure or increase in volume; in the pure diesel cycle the rate of admission of the oil is controlled so that the heat evolved maintains the pressure constant while the piston is moving down during the combustion period. For this reason it is called the *constant pressure cycle* (in actual practice the pressure rises a little during combustion). When the fuel is shut off, the gases continue to push the piston down and the pressure consequently falls as the piston moves towards the end of this stroke, this is the expansion period of the power stroke. At the end of fuel combustion (and beginning of expansion) the temperature of the gases has probably risen to about 3000°F. Near the end of this stroke, when the pressure has fallen to be of little further use, the exhaust valve opens.

SKETCH (iv). This illustrates the *exhaust stroke*. The exhaust valve is open, piston moving up, and the gases are being expelled from the cylinder. At the end of this stroke the exhaust valve closes and the air-induction valve opens to begin the cycle of operations over again.

The valves and ports of an internal combustion engine do not open and close when the crank is exactly on top or bottom centre (i.e. when the piston is at the top or bottom of its stroke). For instance, the induction valve must begin to open well before the crank is on top centre so that it is fully open when the piston begins to move down on its induction stroke. As the fast-moving piston moves down, the

I.C. ENGINES—ELEMENTARY PRINCIPLES

air rushes into the cylinder at a high velocity and the momentum gained causes it to continue to 'pour' into the cylinder after the piston completes its downward stroke. Thus the induction valve is kept open until the crank has passed bottom centre and the momentum of the inrush of air is reduced to practically nil. At the end of compression the temperature of the air is sufficiently high to ignite the fuel when it is injected into the cylinder and cause it to burn rapidly. However, since it takes a fraction of a second to get the full burning effect, the oil is injected before top dead centre of the crank, and it follows that the faster the speed of the engine the earlier the beginning of the fuel injection. The fuel valve is kept open for just sufficient oil to be injected to produce the required power during the power stroke, and to allow for a reasonable drop in pressure before exhaust begins. The exhaust valve opens a little before the piston reaches the end of the power stroke, to allow any remaining pressure after expansion to fall to about atmospheric pressure, before the piston begins to move up again, and it remains open for a short while after the piston completes its upward exhaust stroke because exhaust continues for a brief period due to the momentum of the gases being entrained towards the exhaust opening.

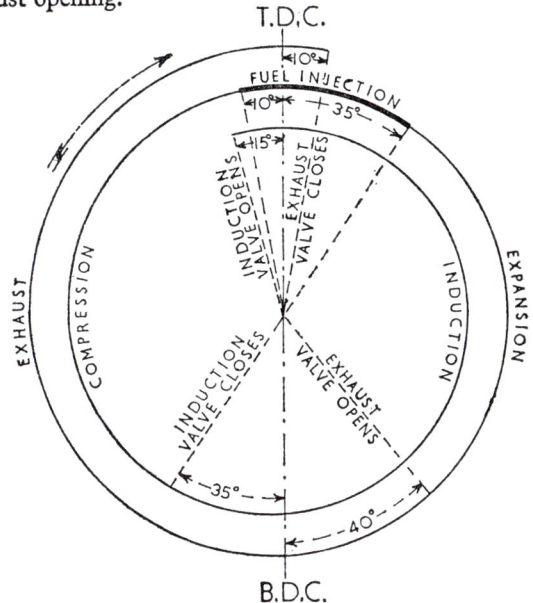

TIMING DIAGRAM FOR A FOUR-STROKE DIESEL ENGINE
Fig. 22

76   REED'S HEAT AND HEAT ENGINES FOR MARINE ENGINEERS

A diagram representing two revolutions of the crank (four strokes of the piston) showing the timing of the opening and closing of the valves with respect to the position of the crank, is shown on Fig. 22. This is called a timing diagram and the crank angles shown are typical values for a four-stroke diesel engine. These however, vary considerably for different engines and depend partly on the designed speed of the engine and the grade of fuel burned.

Note the exhaust-induction overlap, that is the short period when exhaust and induction valves are open together. The sweep of the exhaust gases towards and through the exhaust valve has the effect of pulling in air through the induction valve and so assists in scavenging the combustion space at the beginning of the air-induction stroke.

Fig. 23 is an indicator diagram that would be expected from a 4-stroke diesel engine. It is slightly exaggerated to show more distinctly the suction pressure a little below atmospheric during the induction stroke, and the pressure a little above atmospheric during the exhaust stroke.

FOUR-STROKE DIESEL
INDICATOR DIAGRAM

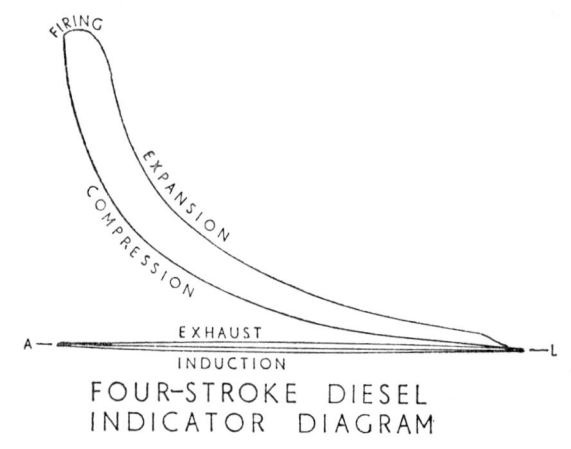

Fig. 23

VALVE MECHANISM

The air-induction, exhaust, and starting-air valves are opened by means of rocking levers fulcrummed about their centre and actuated by cams fixed to the cam shaft. Each cam has a peak, which, when it comes around to contact the roller on the end of the rocking lever, pushes this end up, and the other end is depressed to open the valve against a spring in the valve housing (see Fig. 24). Push rods may be

I.C. ENGINES—ELEMENTARY PRINCIPLES

used as distance pieces between cams and rocking levers. The cams are set in the correct position relative to the crank so that the valves open and close at the exact moment and for the required period in the working cycle. As the timing of opening and closing of the valves is relative to the crank position and direction of movement, separate cams fixed at different relative positions are required to actuate the rocking levers to run the engine in the reverse direction.

The working cycle constitutes four strokes of the piston, which is two revolutions of the crank shaft. During one cycle each valve is opened only once, hence in the four-stroke engine the cam shaft is driven at half speed of the crank shaft.

The fuel valve is usually opened by the pressure of the oil discharged by the fuel pump and closes under the action of a spring when the pressure is released. The timing of the beginning of opening, period of opening, and closing of the valve is varied by the fuel pump plunger.

VALVE MECHANISM

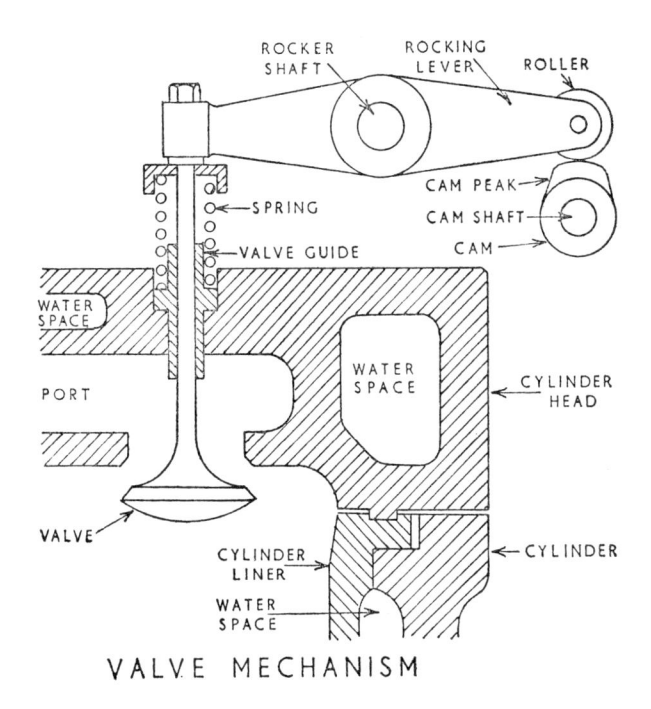

Fig. 24

THE TWO-STROKE DIESEL ENGINE

The two-stroke diesel engine is so named because it takes two strokes of the piston to complete one working cycle. Every downward stroke of the piston is a power stroke, every upward stroke is a compression stroke, the exhaust of the burned gases from the cylinder and the fresh charge of air is taken in during the late period of the downward stroke and the early part of the upward stroke. The exhaust gases pass through a set of ports in the lower part of the cylinder and the air is admitted through a similar set of ports, the ports are covered and uncovered by the piston itself which must be a long one or have a skirt attached so that the ports are covered when the piston is at the top of its stroke.

As there is no complete stroke to draw the air into the cylinder, the air must be pumped in at a low pressure from a pump, this is known as the scavenge pump, the air supplied is referred to as scavenge-air and the ports in the cylinder through which the air is admitted are termed scavenge ports. It is the function of this air to sweep around the cylinder and so 'scavenge' or clean out the cylinder by pushing the remains of the exhaust gases out, leaving a clean charge of air to be compressed.

There is therefore no air-induction valve or exhaust valve in the cylinder head as there are in a four-stroke engine and the cylinder head is consequently a much simpler and stronger casting; there are of course the fuel valve, air-starting valve and relief valve.

As the cycle of operations takes place in two-strokes of the piston, which is one revolution of the crank shaft, the cam shaft is driven at the same speed as the crank shaft.

CYCLE OF OPERATIONS IN A TWO-STROKE DIESEL ENGINE

The sketches in Fig. 25 illustrate various points in the working cycle in one cylinder.

SKETCH (i). This shows the piston moving up, the exhaust and scavenge ports are covered by the piston and the fuel valve is shut.

I.C. ENGINES—ELEMENTARY PRINCIPLES 79

I COMPRESSION

II FIRING

III EXHAUST

IV EXHAUST AND SCAVENGING

TWO-STROKE DIESEL ENGINE

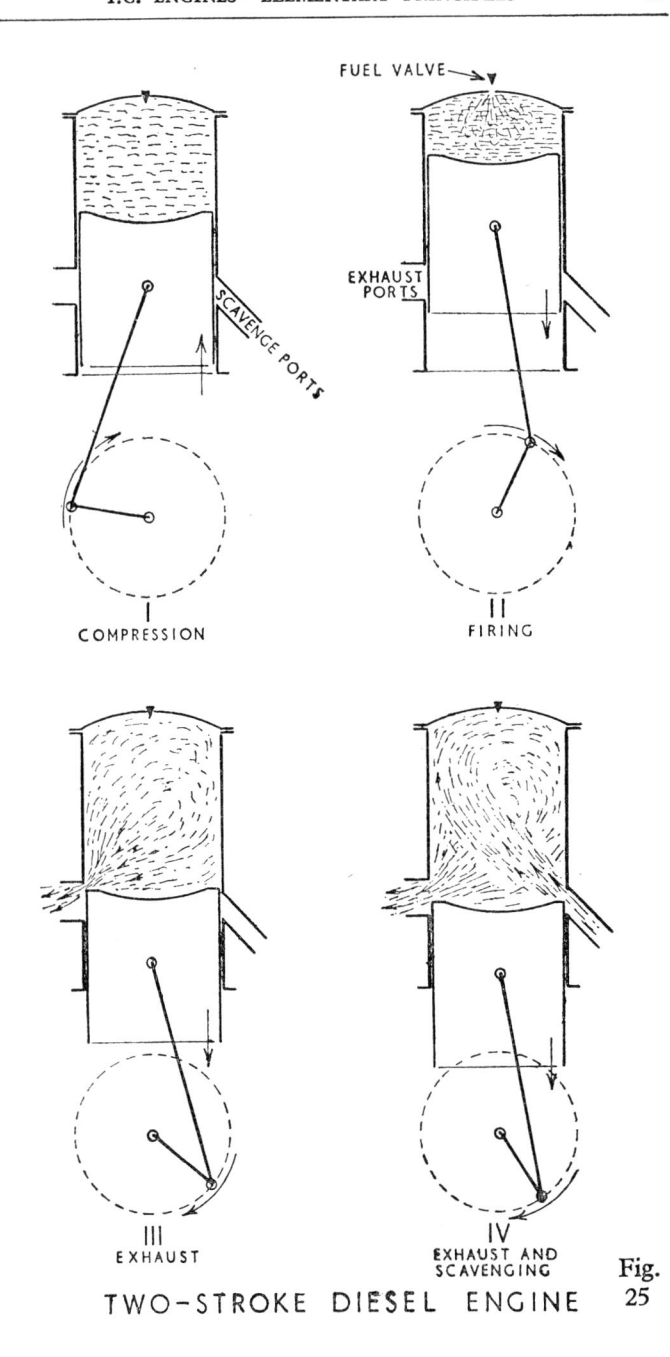

Fig. 25

Air previously taken into the cylinder is being compressed to about 500 lb/in^2 and 1000°F at the end of compression.

SKETCH (ii). This illustrates the fuel being injected into the cylinder, which, being broken up into the form of a fine spray, readily mixes with the hot air, burns and gives out heat. The fuel is injected at such a rate that the pressure of the gases inside the cylinder is kept constant (or may rise slightly in actual practice) during combustion as the piston moves forward. When the fuel is cut off at about one-tenth of the downward stroke, the hot gases contain sufficient energy to continue to do work on the piston and push it forward towards the end of the stroke, the gases consequently falling in pressure as they expand.

SKETCH (iii). The piston is still moving down and has just begun to uncover the exhaust ports (note that the top of the exhaust ports are at a slightly higher level than the top of the scavenge ports), the first rush of exhaust gases out of the cylinder is taking place, whatever pressure there was now rapidly falls to about zero.

SKETCH (iv). The piston has moved further down to uncover also the scavenge ports, the scavenge air at one or two lb per square inch pressure sweeps into the cylinder. Note the slope of the air passage which directs the air upwards into the cylinder, many engines have a hump on the top of the piston to assist this upward direction of the air; and the scavenge ports are cut through partially tangential to the cylinder to give the air a swirling movement. The cylinder is now being scavenged by expelling all the burned gases out.

When the crank passes bottom dead centre, the piston moves up and covers the scavenge and exhaust ports, the air trapped in the cylinder is then compressed to begin the cycle over again.

The timing diagram, Fig. 26, gives average values of the crank angles at the cardinal points of the cycle.

The above describes the cycle of the simplest of two-stroke engines. Most two-stroke diesel engines have scavenge control by valves or other means in addition to the sleeve action of the piston, an

I.C. ENGINES—ELEMENTARY PRINCIPLES

TIMING DIAGRAM FOR A
TWO-STROKE DIESEL ENGINE

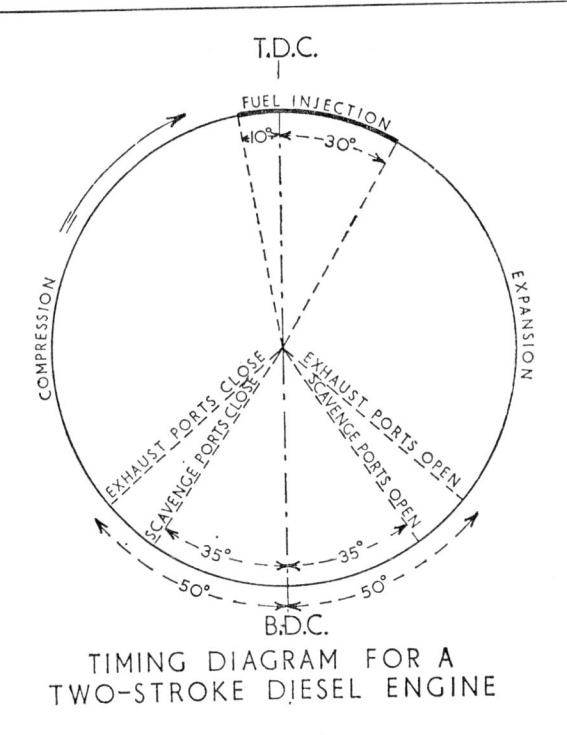

Fig. 26

extra set of scavenge ports are arranged above the main scavenge ports which are closed by the valves on the downward stroke of the piston to allow exhaust to take place first, and opened when the pressure in the cylinder falls below scavenge pressure. By this means the scavenge air remains open on the upward stroke after the exhaust ports are closed and the pressure in the cylinder at the beginning of compression can be equal to the pressure of the scavenge air. A slightly higher pressure at the beginning of compression means a greater weight of air in the cylinder and therefore more fuel can be burned to produce more power. Engines in which there is an extra pressure of air at the beginning of compression are said to be supercharged.

Fig. 27 is an indicator diagram, slightly exaggerated from a two-stroke diesel engine.

TWO-STROKE DIESEL
INDICATOR DIAGRAM

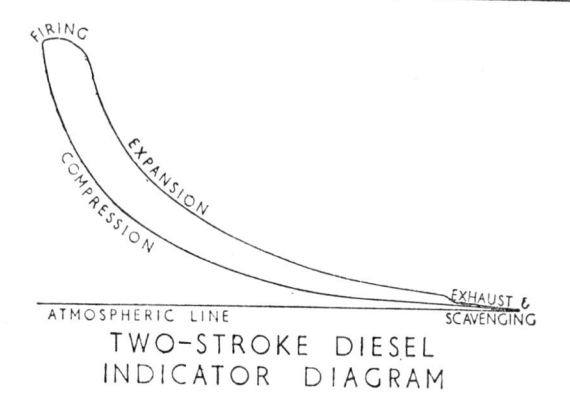

Fig. 27

THE DOXFORD OPPOSED PISTON ENGINE

This engine, which works on the two-stroke cycle, has an extra piston opposed to the main piston in each cylinder, instead of a cylinder head as in the straight diesel engines previously described. Thus each cylinder contains two pistons which are opposed to each other with regard to their direction of movement, when the bottom piston is moving up the top piston is moving down, and vice-versa. The bottom piston is connected by piston rod, crosshead and connecting rod to a centre crank; the top piston is coupled by a yoke to two side rods and connected by connecting rods to side cranks, one on either side of the centre crank. The side cranks are at 180 degrees to the centre crank which gives their respective pistons opposite directions of motion. The fuel valves are positioned at the centre of length of the cylinder, there is a complete ring of exhaust ports near the top of the cylinder which are covered and uncovered by the top piston, and a complete ring of scavenge ports near the bottom which are covered and uncovered by the bottom piston.

CYCLE OF OPERATIONS IN A DOXFORD ENGINE

Although this engine works on a two-stroke cycle, there are some important differences from an ordinary diesel engine. Firstly, the high temperature of air required to ignite and burn the fuel is not derived solely by compression, but partly by compression and partly by heat radiated from the hot crowns of the pistons. The pistons are constructed with very thick tops so that the crowns are not cooled by

I.C. ENGINES—ELEMENTARY PRINCIPLES

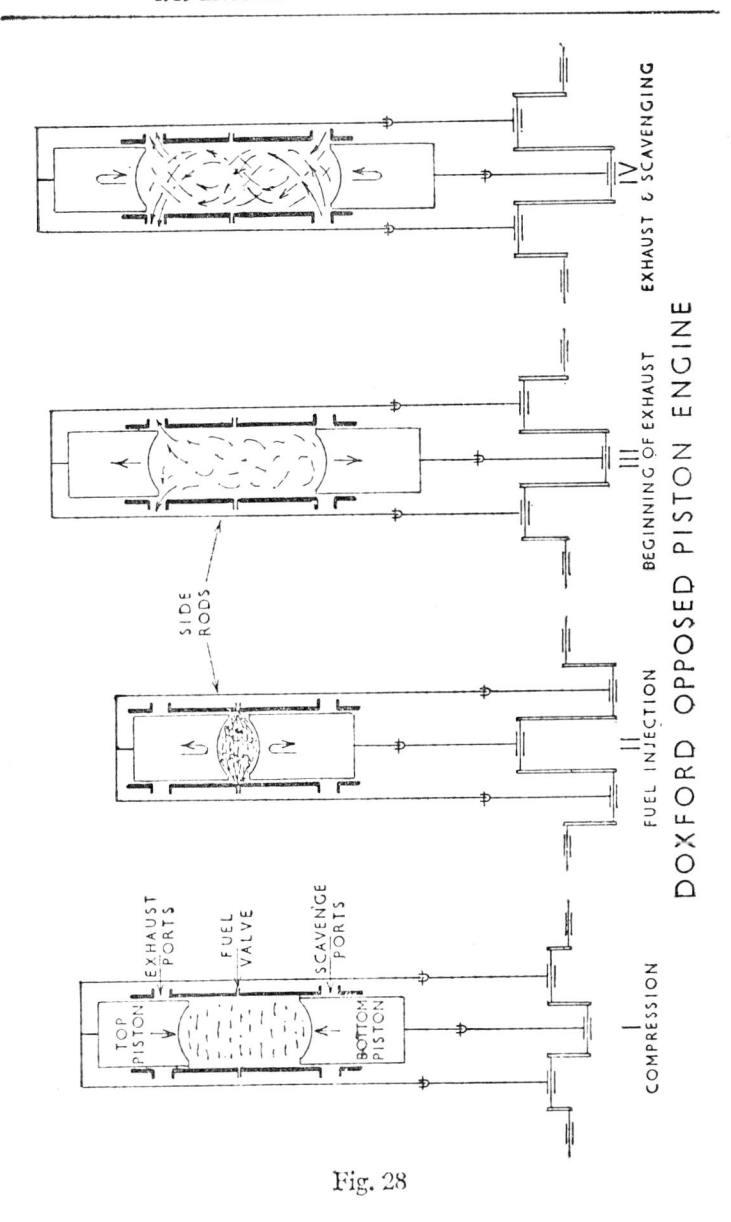

Fig. 28

the circulating water and thus form hot spots on the pistons. Secondly, combustion takes place partly at constant volume and partly at constant pressure, this is called *dual-combustion* and is attained by admitting the fuel well before the centre crank reaches top centre to obtain the constant volume combustion effect and keeping the fuel open for a while after the crank has passed top centre to obtain the constant pressure effect.

The sketches in Fig. 28 illustrate the various positions of the pistons in the working cycle.

SKETCH (i). The bottom piston is moving up, top piston moving down, exhaust and scavenge ports are covered by the top and bottom pistons respectively and the air previously admitted to the cylinder is being compressed so that near the end of the compression stroke the temperature of the air will be in the region of 1000°F due to the combined effect of compression and radiation from the hot spot on each piston.

SKETCH (ii). Fuel is injected through two fuel valves situated diametrically opposite each other and slightly tangentially offset to give the oil spray a swirling motion. The fuel ignites on coming into contact with the high temperature air, burns and gives out heat, increasing the temperature and pressure of the air to impart its energy to the pistons which are pushed apart, this is the beginning of the power stroke. The fuel was first admitted when the centre crank was about 20 degrees before the top centre and cut off when the crank was about the same angle past top centre.

SKETCH (iii). The pistons are moving outward, that is, bottom piston moving down and top piston moving up, being pushed under the action of expansion of the hot gases. Towards the end of this stroke, the top piston uncovers the exhaust ports and the gases commence exhausting from the cylinder. A little later, when the pressure in the cylinder has fallen to about atmospheric pressure, the bottom piston uncovers the scavenge ports to allow the scavenge air (under slight pressure) to rush into the cylinder. The scavenge ports are cut tangentially so that the air swirls around the cylinder as it sweeps upward, driving the burned gases before it.

SKETCH (iv). The pistons have completed their outward stroke and are beginning to move inward during the scavenging process. The scavenging effect is practically one hundred per cent under the ideal arrangement of admitting the scavenge-air in the bottom of the cylinder to drive the exhaust gases out through the top.

COOLING OF DIESEL ENGINES

The working temperatures of the gases inside the cylinders of diesel engines are very high, therefore the parts in close proximity to the combustion space must be cooled to prevent the metal from overheating. These parts are the cylinder head, cylinder and piston, and sometimes the exhaust valve. Fresh water is the most common cooling medium employed.

The cylinder is composed of a liner within a jacket, the liner is fitted at the top, with a water seal at the bottom and leaving an annular space between liner and jacket for the circulation of the water. The cylinder head is cast in box form with water passages running through it between the valve housings. The piston is hollow and telescopic or trombone pipes lead the water to and from the piston; in many heavy oil engines oil is used as the cooling medium for the pistons.

The same fresh water is used over and over again, being circulated by a pump, from a sump, through the engine, passed through the tubes of a cooler (sometimes called a heat exchanger), and back to the sump. Sea water is pumped from the sea, through the cooler on the outside of the tubes, and overboard, to keep the tubes cool.

LUBRICATION

A forced lubrication system is always employed. The lubricating oil is fed under pressure by means of a pump or gravity tank to the main lubricating oil supply line, from a sump, passing through filters on the way. Pipes lead the oil from the main line to each main bearing and cam shaft bearing. A hole is drilled through the centre of the crank shaft, through the crank webs and crank pins which allows the oil to flow from the main bearings to the crank pins thence up through a hole in the connecting rod (or pipes strapped to the connecting rod) to the crosshead and guides.

The lubrication of the inside wall of the cylinder liner on which the piston rings rub, may be adequately effected by the oil mist thrown up from the cranks, or the oil may be pumped directly into the cylinder through two or more points, timed carefully to inject a few drops on the piston rings at the moment the piston is passing the oil holes.

The lubricating oil carries away a great deal of the heat generated by friction at the various bearings and therefore must be cooled by passing it through a cooler (which is circulated by sea water) at some stage of the circuit between leaving the sump and returning to the sump.

STARTING

As previously explained, the ignition of the fuel in diesel engines is caused by the heat of compression of the air previously admitted into the cylinders, thus, for the engine to begin firing, air must first be drawn or pumped into the cylinder and this air must be compressed by the upward movement of the piston to obtain the high temperature necessary to burn the fuel when it is injected. Hence the engine must be driven for a few revolutions by some outside source before allowing the fuel into the cylinders. In heavy marine engines the practice is to drive the engine on compressed air which has previously been stored up (at pressures ranging from 300 to 600 lb/in^2 depending upon the type of engine) in starting-air reservoirs.

The compressed air is admitted to each cylinder through a cam-operated starting-air valve when the piston has just passed its top centre and commencing what will be its power stroke and remaining open until the piston has travelled part of that stroke. The period of opening depends upon the number of cylinders and whether it is a four-stroke or two-stroke engine. When the starting-air valve closes on one cylinder, another starting-air valve has already opened on another cylinder whose piston has just commenced its downward stroke, when this valve closes another valve on another cylinder has already opened and thus, no matter in what position the engine stops there will always be at least one of the cylinders with its starting-air valve open to admit compressed air to start the engine. When the engine attains sufficient speed, the fuel pumps and valves are brought into operation and the starting air valves put out of commission.

The reservoirs are pumped up by two-stage or three-stage compressors. In a two-stage compressor, atmospheric air is taken into the l.p. (low pressure) cylinder where it is compressed to about 80 lb/in^2 and discharged through an intercooler into the h.p. (high pressure) cylinder; it is now compressed to its final pressure and passed through an aftercooler into the air reservoirs. Sufficient air is stored to enable a minimum of about fourteen starts of the engine without replenishing from the compressor. In some diesel installations there is an air compressor driven off the main engines in addition to the independently driven air compressors.

SPEED CONTROL

The speed of the engine is controlled by varying the quantity of fuel injected into the cylinders, from zero quantity for stop to maximum quantity for full speed. This is commonly effected by varying the discharge period of the oil from the fuel pump.

I.C. ENGINES—ELEMENTARY PRINCIPLES

REVERSING

There are two cams for each valve, one is set so that its peak will lift the valve rocking lever to open the valve and keep it open for the correct period of the cycle when the engine is running in one direction; the other cam alongside is set for running in the opposite direction. The reversing mechanism is therefore arranged to bring either ahead or astern cams into line with the valve rocking lever as required.

The action of one type of reversing gear is first to lift the rocking levers clear of the cams, slide the cam shaft along so that the opposite cam comes into line and then return the rocking lever to its working position.

PETROL ENGINES

Engines which run with petrol (or paraffin) as the fuel are often termed 'light oil' engines. The main difference between the petrol engine and the diesel engine is that the petrol engine takes in a charge of air and petrol vapour, this explosive mixture is compressed and ignited by an electric spark; whereas in the diesel engine the cylinder is charged with air only so that only pure air is compressed and the fuel is injected at the moment ignition and burning of the fuel is required, ignition being caused solely by the heat of the compressed air.

When the air is compressed in a diesel engine there is no possibility of firing before the fuel is injected. In a petrol engine, an explosive mixture of petrol and air is compressed and there is danger of the mixture firing spontaneously due to the heat of compression alone and before the electric spark occurs, therefore the ratio of compression must be limited to prevent this. The ratio of compression in diesel engines can be high, such as twelve to one and upwards whereas the ratio of compression in petrol engines is much less, in the region of eight or nine to one.

Petrol engines usually work on the 'constant volume' cycle, that is, when combustion takes place there is theoretically no change in volume but a considerable increase in pressure.

CYCLE OF OPERATIONS IN THE FOUR-STROKE PETROL ENGINE

Fig. 29 illustrates the four strokes which constitute the cycle of operations in a four-stroke petrol engine.

D

INDUCTION VALVE EXHAUST VALVE   SPARKING PLUG

I
INDUCTION STROKE

II
COMPRESSION STROKE

III
POWER STROKE

IV
EXHAUST STROKE

FOUR-STROKE PETROL ENGINE

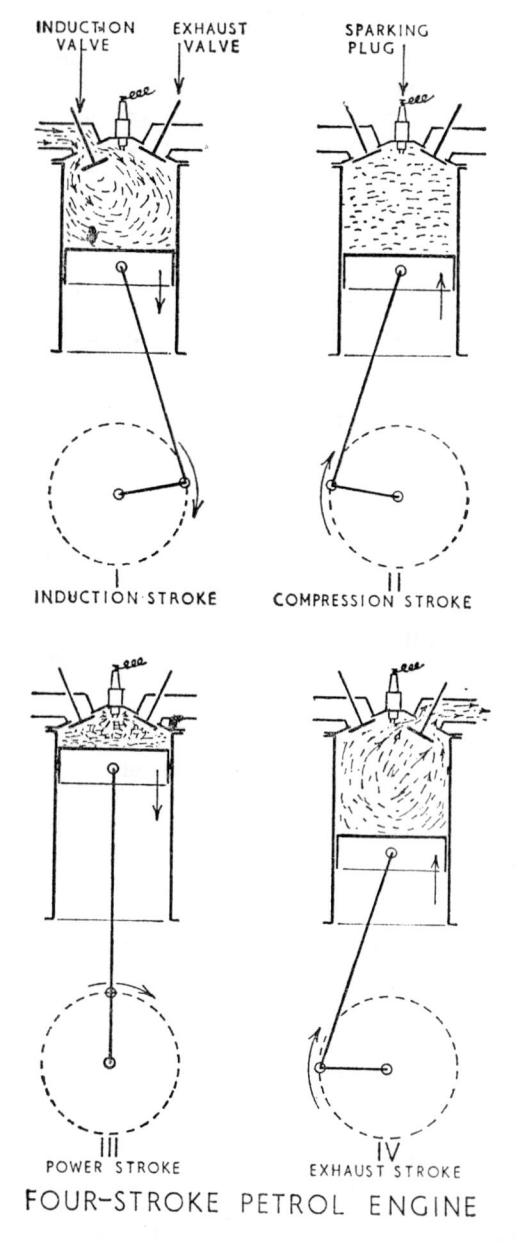

Fig. 29

I.C. ENGINES—ELEMENTARY PRINCIPLES

SKETCH (i). This shows the *Induction Stroke*. The piston is moving down, the induction valve is open and a mixture of air and petrol-vapour is being drawn into the cylinder through the induction valve, from the carburettor.

SKETCH (ii). This is the *Compression Stroke*. The induction valve has closed, no valves are open, the piston is moving up and the petrol-air mixture is being compressed. The pressure at the end of compression will be about 250 lb/in^2 and the temperature in the region of 700°F. The temperature should be as high as possible without causing spontaneous ignition so that the petrol will burn very rapidly when ignited by the electric spark towards the end of this stroke.

SKETCH (iii). This is the *Power Stroke*. An electric spark across the points of a sparking plug is timed to ignite the fuel before the crank reaches its top centre so that full burning effect takes place almost instantaneously while the piston is at the top of its stroke. The heat given out by the fuel causes a rapid rise of pressure and temperature, combustion takes place in about one three-hundredth of a second and is therefore often referred to as an 'explosion'. The hot gases in the cylinder drive the piston down, falling in pressure and expanding in volume as the piston travels towards the bottom of the cylinder.

SKETCH (iv). This shows the *Exhaust Stroke*. The exhaust valve has opened, the piston is moving up, and the burned gases are being pushed out of the cylinder leaving the cylinder empty at the end of this stroke to start the cycle of operations over again.

Timing diagrams of petrol engines vary considerably depending upon the quality of the petrol the engine is designed to run on and the range of speeds. Some average values of the opening and closing of the valves and the timing of the spark are shown in Fig. 30. As in all internal combustion engines, the valves do not open and close when the crank is exactly on top and bottom dead centres; modifications are necessary to give time for the various events to come fully into operation. For instance, the induction valve must begin to open well before the crank is on top centre so that it is full open when the piston begins to move down on its induction stroke, as the fast-moving piston moves down, the mixture rushes into the cylinder at a high velocity, the momentum gained causes the gas to continue to 'pour' into the cylinder even after the piston completes its downward stroke hence the reason for not closing the induction valve until after the crank has passed bottom centre.

The spark sets the fuel alight but it takes a fraction of a second to get the full burning effect, therefore the spark occurs before top centre, the faster the speed of the engine the earlier the spark and most engines have an automatic device to advance or retard the spark as the engine speed increases or decreases.

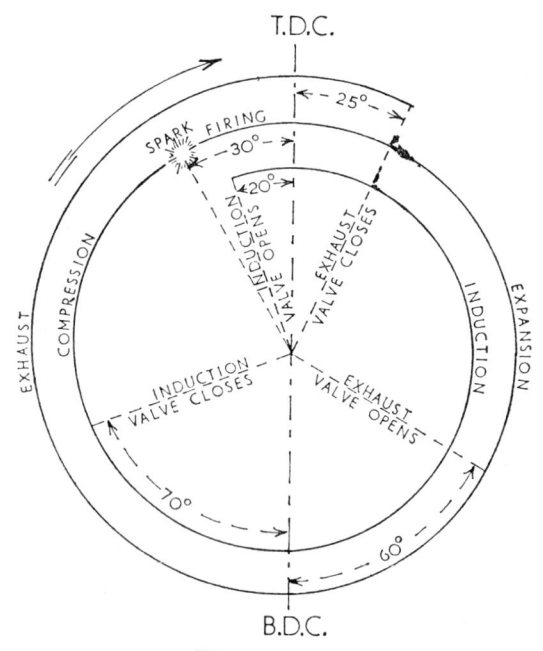

TIMING DIAGRAM FOR A
FOUR-STROKE PETROL ENGINE

Fig. 30

The exhaust valve opens before the piston gets to the end of its power stroke to allow any remaining pressure after expansion to fall to atmospheric pressure before the piston starts to move up again, and the valve remains open for a short while after the piston completes its exhaust stroke. Note the exhaust-induction valve overlap when both exhaust and induction valves are open together for a short period. The momentum of the exhaust gases rushing out through the exhaust valve induces the fresh charge to begin entering the cylinder through the induction valve, and thus scavenges the combustion space.

I.C. ENGINES—ELEMENTARY PRINCIPLES

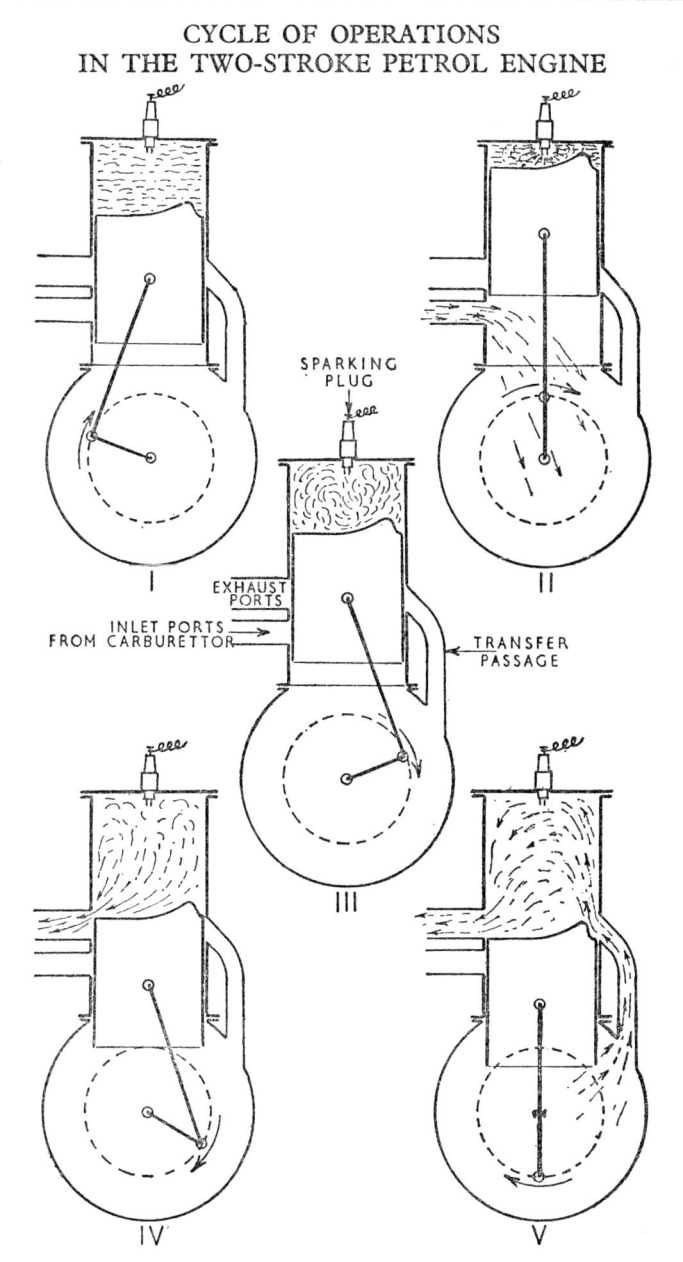

CYCLE OF OPERATIONS IN THE TWO-STROKE PETROL ENGINE

TWO-STROKE PETROL ENGINE   Fig. 31

Fig. 31 illustrates a typical two-stroke petrol engine with an enclosed crankcase to act as the scavenge pump. As in the two-stroke diesel engine, exhaust and scavenging takes place through ports in the cylinder wall and the ports are covered and uncovered by the piston.

SKETCH (i). Air and petrol vapour, previously admitted into the cylinder, is being compressed by the upward movement of the piston, at the same time a partial vacuum is formed in the crankcase.

SKETCH (ii). As the piston nears the top of its stroke, the mixture of petrol and air, now at a high temperature, is ignited by an electric spark across the gap of the points of a sparking plug, the heat of combustion causes a sudden rise of pressure which imparts its energy to the piston. The bottom of the piston has uncovered the inlet ports into the crankcase and a fresh charge of petrol-air mixture rushes in from the carburettor.

SKETCH (iii). The piston is being pushed down by the expansion of the gases in the cylinder. The mixture in the crankcase is being slightly compressed.

SKETCH (iv). End of expansion and beginning of exhaust. The piston has moved further down and uncovered the exhaust ports to allow the burned gases to leave the cylinder. The mixture in the crankcase has now been compressed to a few lb per square inch pressure.

SKETCH (v). The piston is at the bottom of its stroke, exhaust ports are full open. The scavenge ports have been uncovered by the piston and the charge of air and petrol rushes up from the crankcase, through the transfer passage and into the cylinder.

As the piston moves up, scavenge and exhaust ports are closed and the mixture in the cylinder is compressed as in sketch (i) to begin the cycle over again.

CARBURATION

The petrol is drawn into the engine cylinder by the air flow stream. Fig. 32 shows a simple float-feed carburettor. A choke-tube is fitted in the air-intake pipe and the petrol jet is situated in the centre, the float which actuates the needle valve ensures that the level of petrol is maintained just below the open top of the jet. When the air passes through the choke-tube it increases its velocity due to the restricted area of passage and consequently falls in pressure, the partial vacuum

I.C. ENGINES—ELEMENTARY PRINCIPLES

formed in the vicinity of the top of the jet causes petrol particles to lift off and mix with the air stream, the petrol vaporises and passes into the engine cylinder mixed with the air. The cross-sectional areas of the choke-tube and jet are such as to give a mixture of air and petrol in the correct ratio when running at a designed speed.

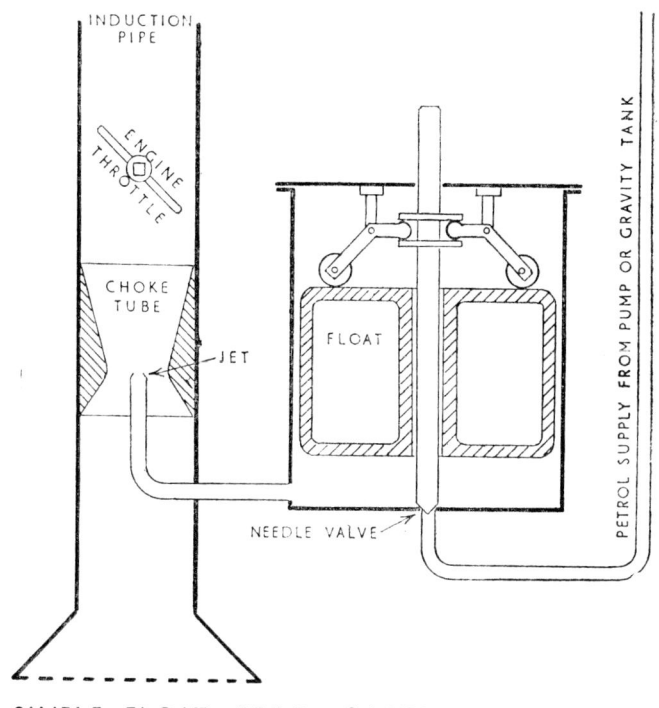

SIMPLE FLOAT-FEED CARBURETTOR

Fig. 32

When petrol engines are required to run at variable speeds the **ratio** of air to petrol is automatically controlled by special compensation, this may be done in various ways and there are many makes of carburettors which successfully achieve this purpose.

Paraffin engines work on the same principle as petrol engines, **the** only essential difference is that a vaporiser is included between **the** carburettor and the cylinder to vaporise the particles of paraffin before entering the cylinder. The vaporiser consists of a nest of tubes through which the air-paraffin mixture flows, exhaust gases are directed around the outside of the tubes to give the required heat for vaporisation. A common practice is to start the engine on petrol and run for a few minutes until the vaporiser is hot, then switch over to paraffin.

IGNITION

The usual method of producing a spark across the points of the sparking plug is by the battery and coil ignition system. A soft iron core is wound with a few turns of heavy wire to form the primary coil and connected across a battery (usually twelve volts) through a contact breaker, and on the same core is wound many turns of fine wire to form the secondary coil. When the electric circuit is broken by the contact breaker the collapse of the magnetic field induces a high voltage in the secondary coil sufficient to cause the electric current to jump across the air gap between the points of the sparking plug, the hot spark so produced ignites the explosive mixture in the cylinder. The current is conducted to the required sparking plug at the correct time by the distributor.

COOLING

Small engines may be directly cooled by air, the cylinder and cylinder head having fins of large surface area cast around the outside to dissipate the heat. Larger engines are water cooled, the water being pumped through the engine block and head and a radiator incorporated in the circuit to prevent the circulating water overheating. Sometimes there is a thermostatically controlled by-pass valve which opens when the water is cold to short-circuit the radiator until the economical working temperature is attained. Engines in ships' lifeboats are usually four-stroke engines and sea water cooled.

REVERSING

Petrol engines are designed to run in one direction only, therefore when reverse direction of rotation is required, such as in lifeboats, a clutch and gear wheel reversing mechanism is incorporated between the engine shaft and propeller shaft.

MEAN EFFECTIVE PRESSURE AND HORSE POWER

Power is the rate of doing work and the usual unit used in mechanical machines and engines is the horse-power.

One horse power (h.p.) is taken as the equivalent of doing 33000 ft lb of work per minute, or 550 ft lb per second.

The *mean effective pressure* in a reciprocating engine is the average effective gas or steam pressure causing the piston to move forward over one complete stroke.

If p = mean effective pressure, in lb/in^2,
A = area of piston, in square inches,
L = length of stroke, in feet,
N = number of power strokes per minute,

then,

$p \times A$ = average total force on piston, in lb,
$p \times A \times L$ = work done per stroke, in ft lb,
$p \times A \times L \times N$ = work done per minute, in ft lb,

therefore,

$$\text{horse power} = \frac{p \times A \times L \times N}{33000}$$

Since this is the horse power indicated by the pressure inside the cylinder, it is referred to as *indicated horse power* (i.h.p.).

$$\therefore \text{i.h.p.} = \frac{pALN}{33000}$$

The value of N, that is, the number of power strokes per minute, depends upon the working cycle of the engine, its speed, and whether it is single-acting or double-acting.

In the steam reciprocating engine cycle (dealt with in Chapter 10) there is one power stroke every revolution, and these engines are always double-acting, therefore,

$$N = \text{r.p.m.} \times 2$$

In four-stroke internal combustion engines, there is one power stroke in every two revolutions, therefore,

N = r.p.m. ÷ 2   for single-acting 4-stroke engines,
N = r.p.m.   for double-acting 4-stroke engines.

In two-stroke internal combustion engines, there is one power stroke in every revolution, therefore,

N = r.p.m.   for single-acting 2-stroke engines,
N = r.p.m. × 2   for double-acting 2-stroke engines.

Converting the above formula for i.h.p. to suit internal combustion engine dimensions expressed in the metric system:

1 metre = 100 cm, 1 inch = 2·54 cm, 1 kilogram = 2·2046 lb
One h.p. = 33000 ft lb per minute

$$= \frac{33000 \times 12 \times 2\cdot 54}{100} \text{ metre-lb per min.}$$

$$= \frac{33000 \times 12 \times 2\cdot 54}{100 \times 2\cdot 2046} \text{ metre-kilograms per min.}$$

$$= 4560 \text{ m kg/min.}$$

Hence, if p = mean effective pressure in kg/cm^2
A = area of piston in cm^2
L = length of stroke in metres
N = number of power strokes per minute,

then,

$p \times A$ = average total force on piston, in kilograms,
$p \times A \times L$ = work done per stroke, in metre-kilograms,
$p \times A \times L \times N$ = work done per minute, in metre-kilograms,

$$\therefore \text{i.h.p.} = \frac{pALN}{4560}$$

The above expressions give the i.h.p. developed in one cylinder. In multi-cylinder engines the total power developed is the sum of the

I.C. ENGINES—ELEMENTARY PRINCIPLES

powers of each cylinder. To obtain a well balanced engine, the power transmitted to the crank shaft from each cylinder should be equal.

Example. The diameter of the cylinders of a six-cylinder, single-acting, two-stroke diesel engine, is 635 mm., and the stroke is 101 cm. If the mean effective pressure in each cylinder is 5·63 kg/cm^2 when running at 132 r.p.m., calculate the total i.h.p. developed.

For a 6-cylinder engine,
$$\text{i.h.p.} = \frac{p\text{ALN}}{4560} \times 6$$

In this case, p = 5·63 kg/cm^2
A = 0·7854 × 63·5^2 cm^2
L = 1·01 metres
N = r.p.m. for a single-acting 2-stroke engine,

therefore,
$$\text{i.h.p.} = \frac{5\cdot63 \times 0\cdot7854 \times 63\cdot5^2 \times 1\cdot01 \times 132 \times 6}{4560}$$

= 3128 i.h.p. Ans.

The mean effective pressure is measured from the indicator diagram taken off the engine cylinder. One method is to measure the area of the diagram by a planimeter, divide by the length to obtain the mean height, and then multiply by the pressure scale of the indicator spring. Planimeters may be graduated to measure the area in square centimetres or square inches, the choice depending upon whether the scale of the stiffness of the indicator spring and the required mean effective pressure is in the metric or British system.

Example. The area of an indicator diagram taken off one cylinder of a 4-cylinder, 4-stroke, single-acting diesel engine is 3·78 square centimetres, the length is 7 centimetres, and the stiffness of the indicator spring is 1 cm = 10 kg/cm^2. The diameter of the cylinders is 250 mm, stroke 30 cm and speed 300 r.p.m. Find the i.h.p. developed assuming all cylinders develop equal powers.

Mean height of diagram = area ÷ length
= 3·78 ÷ 7 = 0·54 cm.
Mean effective pressure = mean height × spring scale
= 0·54 × 10 = 5·4 kg/cm^2

$$\text{i.h.p.} = \frac{p\text{ALN}}{4560}$$

$$= \frac{5 \cdot 4 \times 0 \cdot 7854 \times 25^2 \times 0 \cdot 3 \times 150}{4560} \times 4$$

$$= 104 \cdot 7 \text{ i.h.p. Ans.}$$

If a planimeter is not to hand, the mean height of the indicator diagram may be obtained by the application of the mid-ordinate rule. This method is explained fully in Volume I Chapter 11, and again in Chapter 11 of this book.

BRAKE HORSE POWER AND MECHANICAL EFFICIENCY

Some of the power developed in the cylinders is absorbed in overcoming frictional resistances at the various rubbing surfaces between moving parts, such as at the surface of the piston rings, crosshead, crank and shaft bearings. The horse power lost to friction is termed *friction horse power* (f.h.p.). The remaining useful horse power available at the shaft is termed *brake horse power* (b.h.p.) or *shaft horse power* (s.h.p.).

Brake horse power = Indicated horse power — Friction horse power

The *mechanical efficiency* expresses the relationship between the brake horse power and the indicated horse power:

$$\text{Mechanical efficiency} = \frac{\text{b.h.p.}}{\text{i.h.p.}}$$

Brake horse power is measured by applying a resisting torque as a brake on the shaft, the heat generated by the friction of the brake being carried away by circulating water.

Let F lb = resisting force applied at a radius of R ft
Work absorbed at brake = F × 2πR ft lb per revolution
= F × 2πR × r.p.m. ft lb per min.

I.C. ENGINES—ELEMENTARY PRINCIPLES

$$\therefore \text{Horse power absorbed (b.h.p.)} = \frac{F \times 2\pi R \times \text{r.p.m.}}{33000}$$

or, since Torque (T) = F × R,

$$\text{b.h.p.} = \frac{T \times 2\pi \times \text{r.p.m.}}{33000}$$

Common types of brakes for measuring brake horse power are, for small engines, a loaded rope around a flywheel on the shaft, and, for large engines, a hydraulic or electric dynamometer.

Fig. 33 illustrates a simple rope brake. One end of the rope is attached to a spring balance which is anchored to the engine base, and sufficient weights are hung on the other end of the rope, against the direction of movement of the flywheel, to control the engine at its required speed. If W = weight in lb, and S = reading of spring balance in lb, then the effective tangential braking force on the flywheel rim is (W — S) lb. If R = effective radius in feet from centre of shaft to centre of rope, then the braking torque is (W — S) × R lb ft.

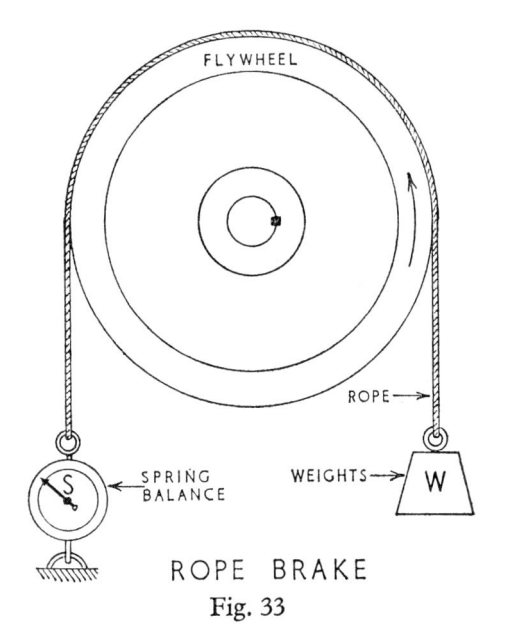

ROPE BRAKE
Fig. 33

MORSE TEST

In multi-cylinder internal combustion engines wherein all cylinders are of the same cubic capacity, a reasonable estimate of the indicated horse power developed in each cylinder can be made by the *Morse test*. This is most useful in small high speed engines where indicator diagrams cannot be taken satisfactorily by the standard mechanical indicator.

The test consists of measuring the brake horse power at the shaft when all cylinders are firing and then measuring the brake horse power of the remaining cylinders when each one is "cut out" in turn. Cutting out the power of each cylinder is done in petrol engines by shorting the sparking plug, and in diesel engines by by-passing the cylinder fuel supply. The speed of the engine and the petrol throttle or fuel pump setting is kept constant during the test so that friction and pumping losses are approximately constant.

Taking a four-cylinder engine:

With all 4 cylinders working,
 Total b.h.p. = Total i.h.p. — Total f.h.p.

$$= \frac{\text{sum of i.h.p.'s of}}{\text{the 4 cylinders}} - \frac{\text{sum of the f.h.p.'s of}}{\text{the 4 cylinders}}$$

When the power of one cylinder is cut out,

$$\text{Total b.h.p.} = \frac{\text{sum of i.h.p.'s of}}{3 \text{ cylinders}} - \frac{\text{sum of the f.h.p.'s of}}{\text{the 4 cylinders}}$$

Hence, when one cylinder is cut out, the loss of *brake* horse power at the shaft is the loss of the *indicated* horse power of that cylinder which is not firing.

Example. During a Morse test on a four-cylinder four-stroke petrol engine, the throttle was set in a fixed position and the speed maintained constant at 2000 r.p.m. by adjusting the brake, and the following readings taken:

With all cylinders working, b.h.p. developed = 57

I.C. ENGINES—ELEMENTARY PRINCIPLES

With sparking plug of No. 1 cyl. shorted, b.h.p. = 38·5
  ,,   ,,   ,,   ,, No. 2 cyl.   ,,   b.h.p. = 37
  ,,   ,,   ,,   ,, No. 3 cyl.   ,,   b.h.p. = 37·5
  ,,   ,,   ,,   ,, No. 4 cyl.   ,,   b.h.p. = 38

Estimate the i.h.p. of the engine and the mechanical efficiency.

i.h.p. of No. 1 cyl. = 57 — 38·5 = 18·5
  ,,   ,, No. 2 cyl. = 57 — 37   = 20
  ,,   ,, No. 3 cyl. = 57 — 37·5 = 19·5
  ,,   ,, No. 4 cyl. = 57 — 38   = 19

Total i.h.p. = 77·0 Ans. (i)

$$\text{Mechanical efficiency} = \frac{\text{b.h.p.}}{\text{i.h.p.}} = \frac{57}{77} = 0{\cdot}74 \text{ or } 74\% \text{ Ans. (ii)}$$

THERMAL EFFICIENCY

The thermal efficiency of an engine expresses the relationship between the quantity of heat turned into work and the quantity of heat supplied.

$$\text{Thermal efficiency} = \frac{\text{Heat turned into work}}{\text{Heat supplied}} \text{ in the same time.}$$

In internal combustion engines the heat is supplied directly into the cylinders by the burning of the injected fuel. The heat given off during complete combustion of unit weight of the fuel is known as the *calorific value*, hence the heat supplied is the product of the weight of fuel burned and its calorific value. The heat turned into work is the heat equivalent of the engine power.

In engine trials it is usual to base calculations on a running time of one hour and to reduce quantities to the amounts per horse power developed:

$$\text{Thermal efficiency} = \frac{\text{Heat equivalent of one h.p. hour}}{\text{lb fuel per h.p. hour} \times \text{calorific value}}$$

It was shown in Chapter 2 that the heat equivalent of one horse power developed continually for one hour (termed one horse-power-hour) is:

$$\frac{1 \times 33000 \times 60}{778} = 2545 \text{ Btu.}$$

or

$$\frac{1 \times 33000 \times 60}{1400} = 1414 \text{ Chu.}$$

The units of the heat equivalent of one h.p.-hour will depend upon the units in which the calorific value of the fuel is expressed, i.e., whether in Btu/lb or Chu/lb.

Thermal efficiency may be based on the heat supplied to develop one horse power in the cylinders (one i.h.p.) or the heat supplied to obtain one horse power at the shaft (one b.h.p.). In the former, the consumption of fuel is expressed as the lb of fuel per i.h.p.-hour and the efficiency is the *indicated thermal efficiency*. In the latter, the consumption of fuel is expressed as the lb of fuel per b.h.p.-hour and the efficiency is then the *brake thermal efficiency*:

$$\text{Indicated thermal efficiency} = \frac{2545 \text{ or } 1414}{\text{lb fuel/i.h.p. hour} \times \text{C.V.}}$$

$$\text{Brake thermal efficiency} = \frac{2545 \text{ or } 1414}{\text{lb fuel/b.h.p. hour} \times \text{C.V.}}$$

The brake thermal efficiency is also the product of the indicated thermal efficiency and the mechanical efficiency, and may therefore be referred to as the *overall efficiency* of the engine.

Example. The following data were taken during a one hour trial run on a single-cylinder, single-acting, four-stroke diesel engine which has a cylinder diameter of 7 inches and stroke of 9 inches, the speed being constant at 1000 r.p.m.

Mean effective pressure = 80 lb/in^2
Effective diameter of rope brake = 3 ft 6 in.
Load on brake = 90 lb
Reading of spring balance = 6 lb
Fuel consumed = 12·6 lb
C.V. of fuel = 19000 Btu/lb

I.C. ENGINES—ELEMENTARY PRINCIPLES

Calculate the i.h.p., b.h.p., specific fuel consumption per i.h.p. hour and per b.h.p. hour, mechanical efficiency, indicated thermal efficiency and brake thermal efficiency.

$$\text{i.h.p.} = \frac{p\text{ALN}}{33000}$$

$$= \frac{80 \times 0.7854 \times 7^2 \times 0.75 \times 500}{33000} = 35 \text{ i.h.p.} \quad \text{(i)}$$

$$\text{b.h.p.} = \frac{T \times 2\pi \times \text{r.p.m.}}{33000}$$

$$= \frac{(90 - 6) \times 1.75 \times 2 \times \pi \times 1000}{33000} = 28 \text{ b.h.p.} \quad \text{(ii)}$$

$$\text{Specific fuel consumption} = \frac{12.6}{35} = 0.36 \text{ lb/i.h.p. hour} \quad \text{(iii)}$$

$$\frac{12.6}{28} = 0.45 \text{ lb/b.h.p. hour} \quad \text{(iv)}$$

$$\text{Mechanical efficiency} = \frac{\text{b.h.p.}}{\text{i.h.p.}}$$

$$= \frac{28}{35} = 0.8 \text{ or } 80\% \quad \text{(v)}$$

$$\text{Indicated thermal efficiency} = \frac{2545}{\text{lb fuel/i.h.p. hour} \times \text{C.V.}}$$

$$= \frac{2545}{0.36 \times 19000}$$

$$= 0.372 \text{ or } 37.2\% \quad \text{(vi)}$$

$$\text{Brake thermal efficiency} = \frac{2545}{\text{lb fuel/b.h.p. hour} \times \text{C.V.}}$$

$$= \frac{2545}{0.45 \times 19000}$$

$$= 0.2976 \text{ or } 29.76\% \quad \ldots \quad \text{(vii)}$$

Alternatively,

$$\text{Brake thermal efficiency} = \text{indicated thermal eff.} \times \text{mech. eff.}$$

$$= 0.372 \times 0.8$$
$$= 0.2976 \text{ (as above)}$$

HEAT BALANCE

Of the total heat supplied to an engine, only a small proportion is converted into useful work. The heaviest losses are those due to the heat carried away by the cooling water and the heat in the exhaust gases. A clear picture of the distribution of heat is shown by constructing a heat balance chart, based upon taking the heat supplied in the fuel as 100%.

Fig. 34 illustrates a heat balance of a four-stroke diesel engine. In this particular case the exhaust gases were passed through a steam boiler before escaping up the funnel, and 40% of the heat in the exhaust gases was recovered by generating steam. The radiation loss is usually very small and in most cases it is included with the other losses.

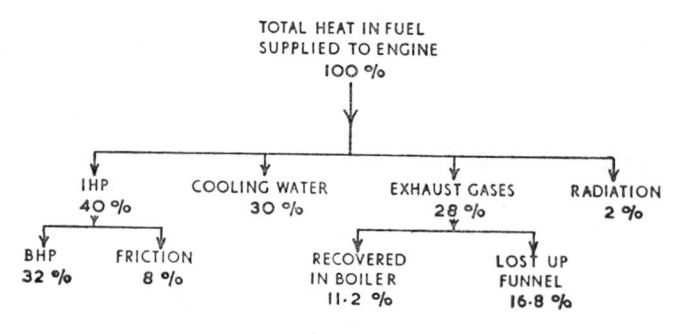

```
              TOTAL HEAT IN FUEL
              SUPPLIED TO ENGINE
                   100 %
    ┌──────────┬──────────┬──────────┐
   IHP      COOLING WATER  EXHAUST GASES  RADIATION
   40%          30%           28%           2%
  ┌──┴──┐                  ┌────┴────┐
  BHP  FRICTION         RECOVERED  LOST UP
  32%    8%             IN BOILER   FUNNEL
                         11.2%      16.8%
```

Fig. 34

TEST EXAMPLES 5

1. The area of an indicator diagram taken off a four-cylinder, single-acting, four-stroke internal combustion engine when running at 350 r.p.m. is 0·605 square inch, the length is 2·75 inches and the scale of the indicator spring is 1 inch = 300 lb/in.^{2} The diameter of the cylinders is 6 inches and the stroke is 8 inches. Find the i.h.p. of the engine assuming all cylinders develop equal power.

2. Calculate the cylinder diameters and stroke, in millimetres, of a six-cylinder, double-acting, two-stroke diesel engine to develop 3000 b.h.p. at 125 r.p.m. when the mean effective pressure in each cylinder is 4·92 kg/cm^2. Assume a mechanical efficiency of 84% and make the length of the stroke 25% greater than the cylinder diameter.

3. In a single-acting two-stroke, opposed-piston engine, the ratio of the weights of the top piston and its connected moving parts, to the bottom piston and its moving parts, is 7·5 to 6, and the strokes of the pistons are in inverse ratio to their weights. The combined stroke of top and bottom pistons is 2·43 metres. The engine has 6 cylinders of 625 mm. diameter and the mean effective pressure is 6·5 kg/cm^2 when running at 105 r.p.m. Assuming a mechanical efficiency of 90%, find the strokes of the top and bottom pistons, the indicated horse power and the brake horse power.

4. The cylinder diameters of an eight-cylinder, single-acting, four-stroke diesel engine are 750 mm. and the stroke is 1125 mm. The average m.e.p. in the cylinders when running at 110 r.p.m. is 85 lb/in.^{2} Calculate the i.h.p. and b.h.p. assuming a mechanical efficiency of 86%.

5. The flywheel of a rope brake is 4 feet diameter and the rope is one inch diameter. When the engine is running at 250 r.p.m. the load on the brake is 105 lb on one end of the rope and 16 lb on the other end. Calculate the brake horse power. If the rise in temperature of the brake cooling water is 18 C degrees, calculate the number of gallons of water passing through the brake every hour assuming that the water carries away 90% of the heat generated at the brake.

6. A compression-ignition engine under test was coupled to a hydraulic dynamometer. At a speed of 2000 r.p.m. the brake load was 42 lb, the inlet temperature of the brake circulating water was 16·4°C and outlet temperature 37·9°C. The horse-power formula for

this dynamometer is WN/3000, where W = brake load in lb and N = speed in r.p.m. If 179 gallons of water passed through the brake in one hour, find what percentage of the heat generated at the brake is carried away by the cooling water.

7. During a Morse test on a four-cylinder petrol engine the speed was kept constant at 1470 r.p.m. by adjusting the brake and the following readings taken:

With all cylinders firing, torque at brake = 143 lb ft
,, No. 1 cyl. cut out ,, ,, ,, = 96·5 ,,
,, No. 2 cyl. cut out ,, ,, ,, = 96·0 ,,
,, No. 3 cyl. cut out ,, ,, ,, = 95·8 ,,
,, No. 4 cyl. cut out ,, ,, ,, = 96·7 ,,

Calculate the b.h.p., i.h.p. and mechanical efficiency.

8. A heavy oil engine uses 0·42 lb of fuel per b.h.p. hour. The mechanical efficiency is 86% and calorific value of the fuel 18700 Btu/lb, find (i) the indicated thermal efficiency, and (ii) the brake thermal efficiency. If 35 lb of air are supplied per lb of fuel, the air inlet being 80°F and exhaust 740°F, find (iii) the heat carried away in the exhaust gases as a percentage of the heat supplied, taking the specific heat of the gases as 0·24.

9. A diesel engine uses 25·7 tons of fuel per day when developing 6650 i.h.p. and 5450 b.h.p. Of the total heat supplied to the engine, 31·7% is carried away by the cooling water and 30·8% in the exhaust gases. Calculate the indicated thermal efficiency, mechanical efficiency, overall efficiency, and the calorific value of the fuel.

10. The calorific value of the fuel used in a diesel engine is 19050 Btu/lb. The heat losses are 28·2% to the cooling water, 29·3% in the exhaust gases, and 1·5% to radiation. Taking the mechanical efficiency as 85%, find (i) the indicated thermal efficiency, (ii) the brake thermal efficiency, and (iii) the specific fuel consumption in lb per b.h.p. hour.

11. The specific fuel consumption of an engine is 0·4 lb/b.h.p.-hour, when developing 3500 b.h.p., the calorific value of the fuel being 19000 Btu/lb. 40 tons of lubricating oil circulate through the engine per hour, the inlet temperature to the oil cooler is 82°F and outlet 127°F. Taking the specific heat of the lubricating oil as 0·5, calculate the heat carried away by it as a percentage of the heat supplied to the engine. Calculate also the quantity of cooling water passing through the oil cooler in tons per hour if the inlet and outlet temperatures of the water are 71 and 89°F respectively. Take the specific heat of the water as unity.

I.C. ENGINES—ELEMENTARY PRINCIPLES

12. The following data were taken during a test on a compression-ignition engine: i.h.p. = 1750; b.h.p. = 1470; fuel consumed = 665 lb per hour; calorific value of fuel = 18850 Btu/lb; weight of circulating water through engine = 1045 lb per minute; inlet and outlet temperatures of circulating water = 82°F and 140°F respectively. Calculate the indicated and brake thermal efficiencies and draw up a heat balance chart.

13. The mean effective pressure measured from the indicator diagram taken off a single cylinder 4-stroke gas engine was 57 lb/in^2 when running at 300 r.p.m. and developing 5·8 b.h.p. The number of explosions per minute was 123 and the gas consumption 110 cubic feet per hour. The diameter of the cylinder is 7 in., stroke 12 in. and calorific value of the gas 472 Btu per cubic foot. Calculate the i.h.p., mechanical efficiency, indicated thermal efficiency and brake thermal efficiency.

14. In a six-cylinder double-acting four-stroke compression-ignition engine, the diameter of the cylinders is 28 in., stroke 54 in. and piston rod diameter 10 in. When running at 105 r.p.m. the mean effective pressures above and below the pistons are 84 and 71 lb/in^2 respectively. Calculate the i.h.p. developed above and below the pistons in each cylinder, the total engine i.h.p. and b.h.p. assuming a mechanical efficiency of 80%.

CHAPTER 6

IDEAL CYCLES

Having seen the practical working cycles of the more common types of internal combustion engines, we shall now proceed to study the ideal or theoretical cycles of these engines.

In order to make comparisons of the efficiencies of the various cycles we make the assumption that they all work as closed-circuit hot-air engines. Thus it is imagined that the cylinder contains air only as the "working fluid" which never leaves the cylinder but is made to do work on the piston by, firstly, absorbing heat from an external hot supply, converting as much of this heat as possible into mechanical work, and finally rejecting the unused heat to an external cold "sink". Any compression or expansion of the air during the cycle is assumed to be purely adiabatic. The efficiency of an ideal cycle is, therefore, termed the *ideal thermal efficiency*, or, since air is assumed to be the working fluid, it is also termed the *air-standard efficiency*.

In any cycle:

Heat converted into work = Heat supplied — Heat rejected

and,

$$\text{Thermal efficiency} = \frac{\text{Heat converted into work}}{\text{Heat supplied}} \text{ in the same time}$$

$$= \frac{\text{Heat supplied} - \text{Heat rejected}}{\text{Heat supplied}}$$

$$= \frac{\text{Heat supplied}}{\text{Heat supplied}} - \frac{\text{Heat rejected}}{\text{Heat supplied}}$$

$$= 1 - \frac{\text{Heat rejected}}{\text{Heat supplied}}$$

(Note that the heat converted into work, i.e. work done, is represented by the area of the PV diagram.)

IDEAL CYCLES

CONSTANT VOLUME CYCLE

This is also known as the OTTO cycle and is the basis on which petrol, paraffin, and gas engines usually work.

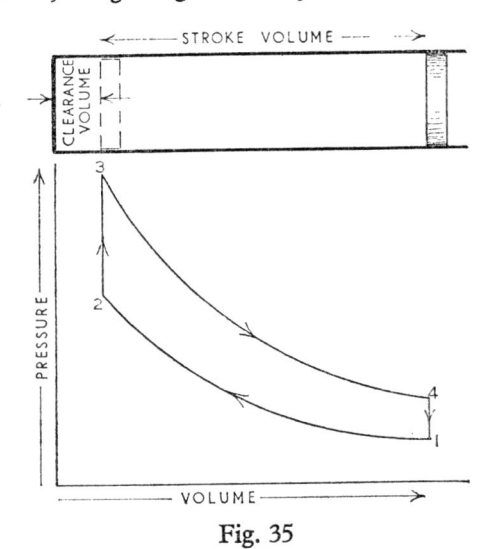

Fig. 35

Designating in sequence, the four cardinal state points of the cycle as 1, 2, 3 and 4 respectively, the cycle of operations commence with a volume of air V_1 at pressure P_1 and temperature T_1. The piston moves inward and the air is compressed adiabatically to a volume V_2 and the pressure and temperature rise to P_2 and T_2. Heat is now given from some outside source and it is assumed that the air receives this heat instantaneously so that there is no time for any change of volume to occur. The pressure and temperature consequently rise to P_3 and T_3 while the volume remains unchanged and, therefore, V_3 is equal to V_2. Adiabatic expansion of the air now takes place while the piston is pushed outward on its power stroke, the volume increasing to V_4 which is the same as the initial volume V_1 and the pressure and temperature during expansion falling to P_4 and T_4. Finally, the cycle is completed by the air rejecting heat (theoretically instantaneously) at constant volume, to an outside source, which causes the pressure and temperature to fall to their initial values of P_1 and T_1.

It should be noted that, since the compression and expansion of the air is adiabatic, then there is no exchange of heat during these operations. This means that all the heat supplied takes place at constant volume between the state points 2 and 3, and all the heat rejected takes place at constant volume between the state points 4 and 1.

Heat supplied or rejected = wt. × spec. ht. × temp. change

hence,

$$\text{Heat supplied} = w \times C_v \times (T_3 - T_2)$$
$$\text{Heat rejected} = w \times C_v \times (T_4 - T_1)$$

therefore:

$$\text{Ideal Thermal Efficiency} = 1 - \frac{\text{Heat rejected}}{\text{Heat supplied}}$$

$$= 1 - \frac{w \times C_v \times (T_4 - T_1)}{w \times C_v \times (T_3 - T_2)}$$

$$= 1 - \frac{T_4 - T_1}{T_3 - T_2}$$

This is the general expression for the ideal thermal efficiency of an engine working on the constant volume cycle.

ƒThis efficiency can be expressed in terms of the ratio of compression, as follows,

$$\text{Ratio of compression} = r = \frac{V_1}{V_2}$$

$$\text{and, ratio of expansion} = \frac{V_4}{V_3}$$

In this case the ratios of compression and expansion are the same because:

$$V_1 = V_4, \text{ and } V_2 = V_3$$

IDEAL CYCLES

Also, since $\dfrac{T_2}{T_1} = \left\{\dfrac{V_1}{V_2}\right\}^{\gamma-1} = r^{\gamma-1}$

and, $\dfrac{T_3}{T_4} = \left\{\dfrac{V_4}{V_3}\right\}^{\gamma-1} = r^{\gamma-1}$

then, $\dfrac{T_2}{T_1} = \dfrac{T_3}{T_4} = r^{\gamma-1}$

Hence, $T_3 = T_4 r^{\gamma-1}$ and $T_2 = T_1 r^{\gamma-1}$

Therefore, $T_3 - T_2 = r^{\gamma-1}(T_4 - T_1)$

Substituting this value of $(T_3 - T_2)$ into the general expression for the thermal efficiency, $(T_4 - T_1)$ cancels, leaving:

$$\text{Ideal Thermal Efficiency} = 1 - \dfrac{1}{r^{\gamma-1}}$$

If γ is taken as 1·4 (for air) then this is also the **Air Standard Efficiency**.

Also, since $r^{\gamma-1} = \dfrac{T_2}{T_1} = \dfrac{T_3}{T_4}$

then, Ideal Thermal Efficiency $= 1 - \dfrac{T_1}{T_2}$

$= 1 - \dfrac{T_4}{T_3}$

On examination of the above expression it will be seen that the greater the value of r, the greater will be the efficiency, hence the trend for higher ratios of compression in modern petrol engines. In actual practice however, petrol engines take in a mixture of petrol vapour and air during the induction stroke and this is compressed during the compression stroke. Being an explosive mixture it will burst into flame without the assistance of an electric spark or other means if it reaches its temperature of spontaneous ignition. Therefore, if the ratio of compression is too high for the grade of petrol used, pre-ignition can take place.

Fig. 36 shows the relationship between the ideal efficiency and the ratio of compression in a constant volume cycle.

[Graph: Ideal thermal efficiency vs compression ratio, with curve labelled $EFF = 1 - \frac{1}{r^{\gamma-1}}$]

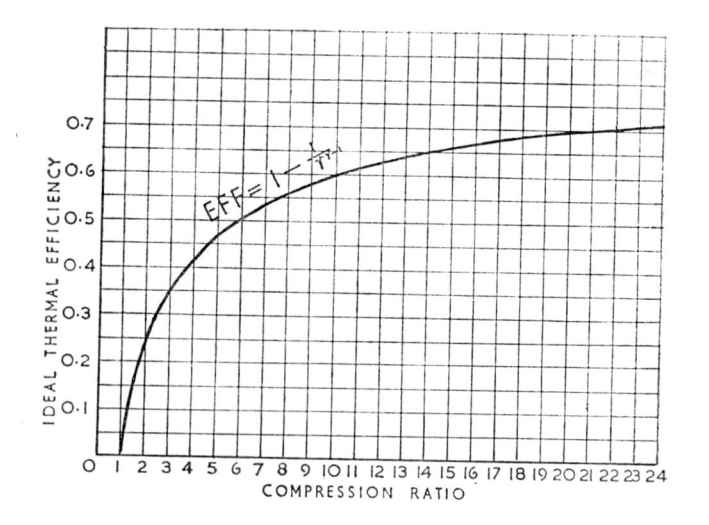

Fig. 36

From the graph we see that, although the efficiency increases with higher compression ratios, the rate of increase in efficiency becomes less as the compression ratio is increased, and there is no appreciable gain by increasing the compression ratio above about 16 to 1.

Example. The compression ratio of an engine working on the constant volume cycle is 9·3 to 1. At the beginning of compression the temperature is 87°F and at the end of combustion the temperature is 2200°F. Taking compression and expansion to be adiabatic and the value of γ as 1·4, calculate (i) the temperature at the end of compression, (ii) the temperature at the end of expansion, (iii) the theoretical thermal efficiency.

Referring to Fig. 35:

$$V_1 = 9.3 \text{ and } V_4 = 9.3$$
$$V_2 = 1 \text{ and } V_3 = 1$$

$$T_1 = 87 + 460 = 547°F \text{ abs.}$$
$$T_3 = 2200 + 460 = 2660°F \text{ abs.}$$

COMPRESSION PERIOD:

$$\frac{T_2}{T_1} = \left\{\frac{V_1}{V_2}\right\}^{\gamma-1}$$

$$\therefore T_2 = 547 \times 9.3^{0.4} = 1335°F \text{ abs.}$$

$\therefore$ Temperature at end of compression

$$= 1335 - 460 = 875°F. \text{ Ans. (i)}$$

EXPANSION PERIOD:

$$\frac{T_4}{T_3} = \left\{\frac{V_3}{V_4}\right\}^{\gamma-1}$$

from which T_4 can be calculated, but, as ratios of compression and expansion are the same, and follow the same law, then an easier method is:

$$\frac{T_2}{T_1} = \frac{T_3}{T_4}$$

$$T_4 = \frac{547 \times 2660}{1335} = 1090°F \text{ abs.}$$

$\therefore$ Temperature at end of expansion

$$= 1090 - 460 = 630°F. \text{ Ans. (ii)}$$

The theoretical efficiency can now be calculated from any of the expressions given above, thus,

$$\text{Efficiency} = 1 - \frac{T_4 - T_1}{T_3 - T_2} \quad \text{or} \quad 1 - \frac{1}{r^{\gamma-1}}$$

$$\text{or} \quad 1 - \frac{T_1}{T_2} \quad \text{or} \quad 1 - \frac{T_4}{T_3}$$

Taking the last expression,

$$\text{Efficiency} = 1 - \frac{1090}{2660} = 1 - 0{\cdot}4098$$

$$= 0{\cdot}5902 \text{ or } 59{\cdot}02\% \text{ Ans. (iii)}$$

DIESEL CYCLE

The term *constant pressure cycle* refers to one wherein the pressure remains constant during the two periods when heat is supplied and rejected. In the diesel cycle however, heat is supplied at constant pressure but rejection of heat takes place at constant volume. Thus, the diesel cycle, that upon which slow-speed diesel engines operate, is usually referred to as the *modified constant pressure cycle*.

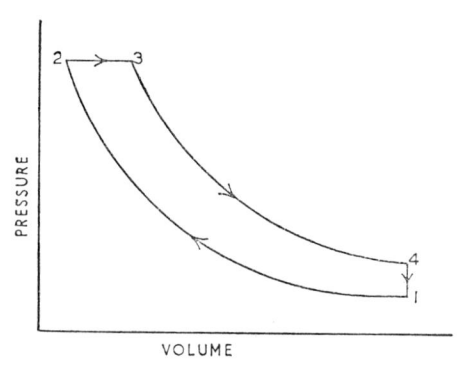

Fig. 37

Referring to Fig. 37, the cycle of operations commence with a volume of air V_1 at a pressure P_1 and temperature T_1. The air is compressed adiabatically to a volume V_2 and the pressure and temperature rise to P_2 and T_2. The piston is now at the top (inward end) of its stroke and heat is supplied at such a rate to maintain the pressure constant as the piston moves down the cylinder for a fraction of the power stroke. At the end of the heat supply period the volume is V_3, the temperature has been further increased to T_3, and the pressure represented by P_3 is the same as P_2. The air now expands adiabatically for the remainder of the power stroke until the final volume V_4 is the same as the initial volume V_1, the pressure and temperature falling during expansion to P_4 and T_4. Finally, the cycle is completed by the rejection of heat at constant volume to the initial conditions.

$$\text{Thermal Efficiency} = 1 - \frac{\text{Heat rejected}}{\text{Heat supplied}}$$

$$= 1 - \frac{w \times C_v (T_4 - T_1)}{w \times C_p (T_3 - T_2)}$$

$$= 1 - \frac{1}{\gamma} \left\{ \frac{T_4 - T_1}{T_3 - T_2} \right\}$$

This is the general expression for the ideal thermal efficiency of a diesel engine.

*f*Example. The compression ratio in a diesel engine is 13 to 1 and the ratio of expansion is 6·5 to 1. At the beginning of compression the pressure is 15 lb/in² abs. and the temperature is 90°F. Assuming adiabatic compression and expansion, calculate the temperatures at the three remaining cardinal points of the cycle, and the ideal thermal efficiency, taking the specific heats at constant pressure and constant volume as 0·238 and 0·17 respectively.

Referring to Fig. 37,

$$V_1 = 13 \qquad \text{and } V_4 = 13 \qquad V_2 = 1$$

$$\frac{V_4}{V_3} = \text{ratio of expansion} = 6 \cdot 5$$

$$\therefore V_3 = \frac{V_4}{6 \cdot 5} = \frac{13}{6 \cdot 5} = 2$$

$T_1 = 90 + 460 = 550°F$ abs.

$$\gamma = \frac{C_p}{C_v} = \frac{0.238}{0.17} = 1.4$$

FIRST STAGE, ADIABATIC COMPRESSION:

$$\frac{T_2}{T_1} = \left\{\frac{V_1}{V_2}\right\}^{\gamma-1}$$

$\therefore T_2 = 550 \times 13^{0.4} = 1535°F$ abs.

$\therefore$ Temperature at end of compression

$= 1535 - 460 = 1075°F$. Ans. (i)

SECOND STAGE, HEATING AT CONSTANT PRESSURE:

$$\frac{T_3}{T_2} = \frac{V_3}{V_2} \quad \text{(Charles' law)}$$

$T_3 = 1535 \times 2 = 3070°F$ abs.

$\therefore$ Temperature at end of combustion

$= 3070 - 460 = 2610°F$. Ans. (ii)

THIRD STAGE, ADIABATIC EXPANSION:

$$\frac{T_4}{T_3} = \left\{\frac{V_3}{V_4}\right\}^{\gamma-1}$$

$$T_4 = 3070 \times \left\{\frac{2}{13}\right\}^{0.4}$$

$$= \frac{3070}{6.5^{0.4}} = 1452°F \text{ abs.}$$

∴ Temperature at end of expansion

$$= 1452 - 460 = 992°F. \text{ Ans. (iii)}$$

Ideal thermal efficiency

$$= 1 - \frac{\text{Heat rejected}}{\text{Heat supplied}}$$

$$= 1 - \frac{1}{\gamma}\left\{\frac{T_4 - T_1}{T_3 - T_2}\right\}$$

$$= 1 - \frac{1}{1\cdot 4}\left\{\frac{1452 - 550}{3070 - 1535}\right\}$$

$$= 1 - \frac{1}{1\cdot 4} \times \frac{902}{1535}$$

$$= 1 - 0\cdot 4198$$
$$= 0\cdot 5802 \text{ or } 58\cdot 02\% \text{ Ans. (iv)}$$

ƒIn the ideal diesel cycle, where the compression and expansion are both adiabatic, the thermal efficiency can be expressed in terms of the ratio of compression and a comparison can then be made with the efficiency of a constant volume cycle of the same ratio of compression.

Expressing all temperatures in terms of T_1, substituting and simplifying:

$$\frac{T_2}{T_1} = \left\{\frac{V_1}{V_2}\right\}^{\gamma-1} = r^{\gamma-1}$$

$$\therefore T_2 = T_1 r^{\gamma-1}$$

$$\frac{T_3}{T_2} = \frac{V_3}{V_2} \quad \text{let this ratio of burning period volumes be represented by } \rho \text{ (the Greek letter rho), then,}$$

$$\frac{T_3}{T_2} = \rho$$

$$\therefore T_3 = T_2\rho = T_1 r^{\gamma-1}\rho$$

$$\frac{T_4}{T_3} = \left\{\frac{V_3}{V_4}\right\}^{\gamma-1}$$

Since $\dfrac{V_3}{V_2} = \rho$ and $\dfrac{V_4}{V_2} = r$, then $\dfrac{V_3}{V_4} = \dfrac{\rho}{r}$

$$\therefore \frac{T_4}{T_3} = \left\{\frac{\rho}{r}\right\}^{\gamma-1}$$

$$T_4 = T_3 \times \left\{\frac{\rho}{r}\right\}^{\gamma-1}$$

$$= T_1 r^{\gamma-1}\rho \times \left\{\frac{\rho}{r}\right\}^{\gamma-1}$$

$$= T_1 \rho^{\gamma}$$

Ideal thermal efficiency

$$= 1 - \frac{1}{\gamma}\left\{\frac{T_4 - T_1}{T_3 - T_2}\right\}$$

$$= 1 - \frac{1}{\gamma}\left\{\frac{T_1\rho^{\gamma} - T_1}{T_1 r^{\gamma-1}\rho - T_1 r^{\gamma-1}}\right\}$$

$$= 1 - \frac{1}{\gamma} \times \frac{1}{r^{\gamma-1}}\left\{\frac{\rho^{\gamma} - 1}{\rho - 1}\right\}$$

If $\gamma = 1\cdot 4$, this is also the Air Standard Efficiency.

Comparing this expression with the ideal efficiency of the constant volume cycle in terms of r, it will be seen that, for the same ratio of compression, the constant volume cycle has the higher thermal efficiency. This does not mean, however, that a petrol engine working on the constant volume cycle is more efficient than a diesel engine working on the modified constant pressure cycle, because, in the former, an explosive mixture is compressed and there is a limit to the ratio of compression, whereas air only is compressed in a diesel engine and the ratio of compression can be as high as required.

DUAL COMBUSTION CYCLE

In most high-speed compression-ignition engines, combustion takes place partly at constant volume and partly at constant pressure, and therefore the cycle is referred to as *dual-combustion*.

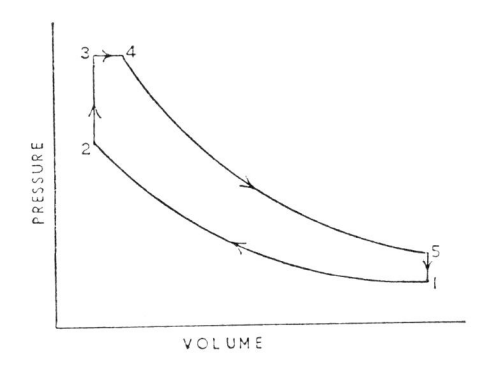

Fig. 38

Fig. 38 shows the ideal dual-combustion cycle. Commencing with a volume of air, V_1 at pressure P_1 and temperature T_1, the air is compressed adiabatically to a volume V_2 and the pressure and temperature rise to P_2 and T_2. Heat is now supplied at constant volume, the pressure and temperature is increased to P_3 and T_3 while the volume remains unchanged so that V_3 is equal to V_2. The supply of heat is continued at such a rate as to maintain the pressure constant while the piston moves outward until the volume is V_4, the temperature is further increased to T_4 and the pressure P_4 remains the same as P_3. Now adiabatic expansion takes place until the volume V_5 is the same as the initial volume V_1, the pressure and temperature falling due to expansion to P_5 and T_5. Finally, heat is rejected at constant volume and the pressure and temperature fall to the initial conditions of P_1 and T_1.

E

Ideal thermal efficiency

$$= 1 - \frac{\text{Heat rejected}}{\text{Heat supplied}}$$

$$= 1 - \frac{wC_v(T_5 - T_1)}{wC_v(T_3 - T_2) + wC_p(T_4 - T_3)}$$

$$= 1 - \frac{(T_5 - T_1)}{(T_3 - T_2) + \gamma(T_4 - T_3)}$$

f The above expression can be converted into terms of the ratio of compression in a similar manner as the previous cycles, thus,

If r = ratio of compression = V_1/V_2
 γ = ratio of specific heats = C_p/C_v
 ρ = ratio of burning period, or "cut-off" ratio = V_4/V_3
 α = ratio of pressure increase at constant volume = P_3/P_2

Ideal thermal efficiency

$$= 1 - \frac{1}{r^{\gamma-1}} \left\{ \frac{\alpha\rho^\gamma - 1}{(\alpha - 1) + \gamma\alpha(\rho - 1)} \right\}$$

Note (i) if $\alpha = 1$, the above becomes a pure diesel cycle.
 (ii) if $\rho = 1$, it becomes a pure constant-volume cycle.

Since compression-ignition oil engines depend upon the temperature of the air at the end of compression to ignite the fuel injected into the cylinder, the compression-ratio must be fairly high, usually not less than about 12 to give the necessary temperature rise during compression.

The higher compression pressures developed in this type of engine limit the use of constant volume combustion, since the maximum pressure in the cycle is limited by the consideration of strength.

As the maximum pressure is limited, increasing the compression ratio reduces the amount of fuel burned at constant volume, so that more must be burned at constant pressure and thus the gain due to increased compression ratio is partly nullified.

ƒACTUAL THERMAL EFFICIENCY

In deriving the expressions for the ideal thermal efficiency of the various cycles, compression and expansion were assumed to be purely adiabatic. In practice, however, compression and expansion will be polytropic, that is, following the law $PV^n = C$ where n is greater than unity and less than γ, therefore some heat will be rejected during compression and received during expansion. This must be taken into account when determining the actual thermal efficiency. The amount of heat received or rejected during polytropic expansion or compression may be calculated as follows:

Work done during a polytropic process

$$= \frac{P_1V_1 - P_2V_2}{J(n-1)} \text{ or } \frac{wR(T_1 - T_2)}{J(n-1)} \text{ heat units}$$

Heat exchanged = Change in I.E. + Work done

$$= wC_v(T_2 - T_1) + \frac{wR(T_1 - T_2)}{J(n-1)}$$

Since $C_p - C_v = R/J$ and $\gamma = C_p/C_v$

then, $C_v = \dfrac{R}{J(\gamma - 1)}$

$$\therefore \text{Heat exchanged} = \frac{wR(T_2 - T_1)}{J(\gamma - 1)} + \frac{wR(T_1 - T_2)}{J(n-1)}$$

$$= \frac{wR(T_1 - T_2)}{J(n-1)} - \frac{wR(T_1 - T_2)}{J(\gamma - 1)}$$

$$= \frac{wR(T_1 - T_2)}{J(n-1)} \left\{ 1 - \frac{n-1}{\gamma - 1} \right\}$$

$$= \frac{wR(T_1 - T_2)}{J(n-1)} \left\{ \frac{\gamma - 1 - n + 1}{\gamma - 1} \right\}$$

$$= \frac{wR(T_1 - T_2)}{J(n-1)} \left\{ \frac{\gamma - n}{\gamma - 1} \right\}$$

$$= \frac{\gamma - n}{\gamma - 1} \times \text{Heat equiv. of work done}$$

ƒCARNOT CYCLE

This is a purely theoretical cycle devised by the French scientist Sadi Carnot. Although it is not possible from practical considerations for an engine to work on this cycle, it has a higher theoretical thermal efficiency than any other working between the same temperature limits and, therefore, provides a useful standard for comparing the performance of other heat engines.

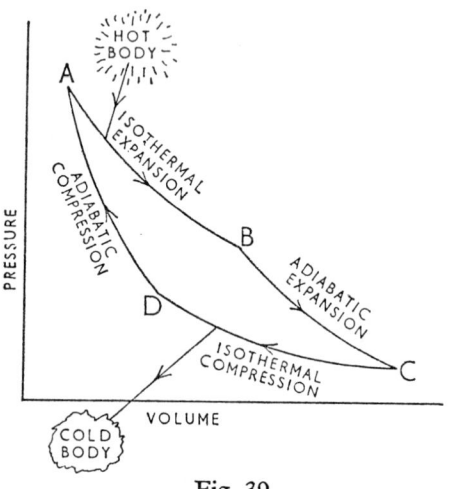

Fig. 39

Referring to the PV diagram, Fig. 39, it is usual to explain this cycle by commencing at state point A. Gas has been previously compressed in the cylinder by the piston moving inward and, at A,

IDEAL CYCLES

the piston is at the "top" of its stroke. The pressure and temperature are high, the value of the latter being represented by T_1. As the piston is pushed outward, doing work, heat is supplied to the gas from an external hot source at such a rate as to maintain its temperature constant, and during this period the gas therefore expands isothermally until point B is reached. At this point the heat supply is cut off and no heat is given to or rejected from the gas as the piston moves on to the end of the stroke at C. Hence during this period, the gas expands adiabatically as it does work and therefore the temperature falls. The temperature of the gas at point C is represented by T_2. The piston now moves inward to compress the gas from C to D and during this period it is assumed that any generated heat due to compression can flow out of the gas into a cold "sink". That is, the gas rejects heat to a cold external source at such a rate to maintain the temperature constant at T_2. This is isothermal compression. At point D, the flow of heat out of the gas is stopped and from D to A the gas is compressed adiabatically while the piston completes its stroke, the temperature of the gas rising to the initial temperature T_1.

Hence, the four stages of the Carnot cycle are briefly as follows:

A to B Isothermal expansion of the gas during which the amount of heat supplied is equal to the work done. Letting r = ratio of isothermal expansion,

$$\text{Heat supplied} = \frac{P_A V_A \log_\varepsilon r}{J} = \frac{w\, RT_1 \log_\varepsilon r}{J}$$

B to C Adiabatic expansion of the gas during which no heat is supplied or rejected.

C to D Isothermal compression. During this period heat is rejected from the gas, the quantity of heat being the equivalent of the work done on the gas and, since the ratio of isothermal compression must be the same as the ratio of isothermal expansion to form a closed cycle, then,

$$\text{Heat rejected} = \frac{P_C V_C \log_\varepsilon r}{J} = \frac{w\, RT_2 \log_\varepsilon r}{J}$$

D to A Adiabatic compression during which no heat is supplied or rejected.

Therefore,

$$\text{Thermal Efficiency} = \frac{\text{Heat supplied} - \text{Heat rejected}}{\text{Heat supplied}}$$

$$= 1 - \frac{\text{Heat rejected}}{\text{Heat supplied}}$$

$$= 1 - \frac{wRT_2 \log_\varepsilon r / J}{wRT_1 \log_\varepsilon r / J}$$

$$= 1 - \frac{T_2}{T_1}$$

$$= \frac{T_1 - T_2}{T_1}$$

This expression for the Carnot Efficiency shows that, to obtain the highest efficiency, heat should be taken in at the highest possible temperature (T_1) and rejected at the lowest possible temperature (T_2). This conclusion is applicable in the design of any heat engine.

ƒREVERSED CARNOT CYCLE

The Carnot cycle is theoretically reversible and if applied in reverse manner would act as a heat pump by taking heat from the cold source and delivering it to the hot source, as follows:

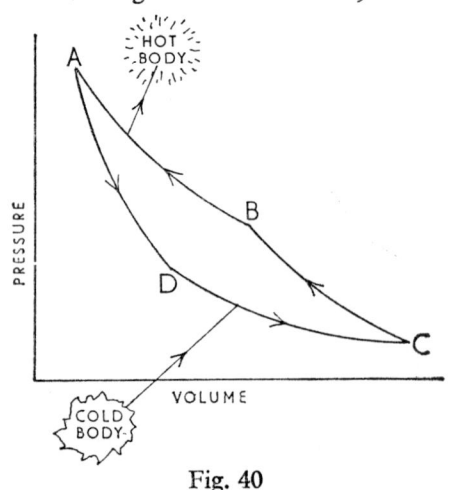

Fig. 40

Referring to Fig. 40 and commencing at state point A, the four stages of the reversed Carnot cycle consist of:

(i) Work done by the gas while it expands adiabatically from A to D and the temperature falls from T_1 to T_2. No heat is given to or taken from the gas during this process.

(ii) Further work done by the gas as it expands isothermally from D to C, a quantity of heat is taken in by the gas (from the cold body) equal to the work done, to maintain the temperature constant at T_2.

(iii) Adiabatic compression of the gas from C to B, no heat being given to or taken from the gas, therefore the temperature increases from T_2 to T_1.

(iv) Isothermal compression from B to A during which heat is rejected from the gas (to the hot body) to maintain the temperature constant at T_1

Thus an engine working on the reversed Carnot cycle would require to be driven and, as heat would be continually taken from a cold region and sent out to a hotter region, it would therefore act as a refrigerating machine. The measure of the "efficiency" of refrigeration is known as the *coefficient of performance*, and its theoretical value is:

$$= \frac{\text{Quantity of heat extracted}}{\text{Heat equivalent of work done to extract the heat}}$$

$$= \frac{w R T_2 \log_\varepsilon r / J}{w R T_1 \log_\varepsilon r / J - w R T_2 \log_\varepsilon r / J}$$

$$= \frac{T_2}{T_1 - T_2}$$

CLEARANCE AND STROKE VOLUME

Mechanical clearance is necessary between the inner face of the cylinder cover and the top of the piston when the piston is at the top of its stroke to avoid contact, and this is measured as the minimum distance between those two parts.

The *clearance volume* is the volume of the enclosed space above the piston when at the top of its stroke, including all cavities up to the valve faces when the valves are closed. Thus, in a steam reciprocating engine, it includes the ports or passages as far as the slide valve face. In internal combustion engines the clearance volume is the combustion space to accommodate sufficient air for the complete combustion of the fuel and to limit the rise of temperature during burning. It is designed as near a spherical space as practicable, in many cases by concave piston tops and concave cylinder covers.

The "stroke volume", sometimes termed the "swept volume", is the volume swept out by the piston as it moves through one complete stroke. It is, therefore, equal to the product of the cross-sectional area of the cylinder and the length of the stroke.

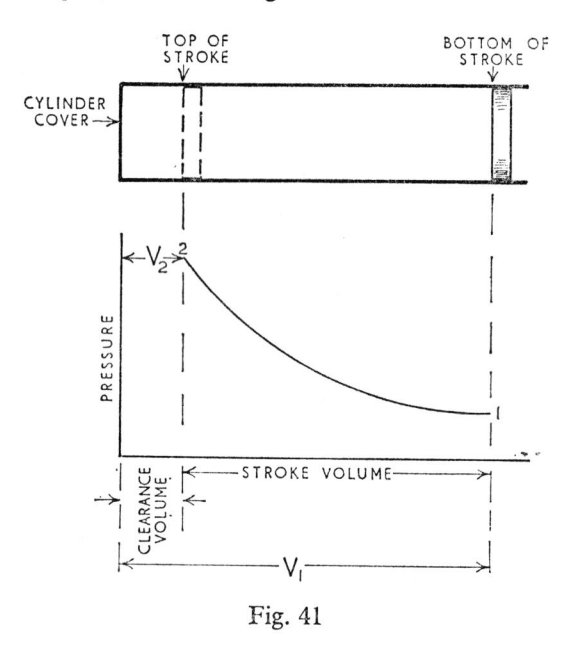

Fig. 41

Since the ratio of compression is the ratio of the volume at the beginning of compression to the volume at the end of compression, then, referring to Fig. 41,

$$r = \frac{\text{Initial volume}}{\text{Final volume}}$$

IDEAL CYCLES

$$\frac{V_1}{V_2} = \frac{\text{clearance vol.} + \text{stroke vol.}}{\text{clearance volume}}$$

Therefore, the magnitude of the clearance volume affects the ratio of compression (and expansion) and can be adjusted by shims under the foot of the connecting rod or plates in the clearance space.

In calculations involving an equation with V_1 on one side and V_2 on the other, any convenient units can be employed, such as cubic inches, cubic feet, cubic centimetres, etc., provided both are in similar units. Dividing stroke-volume by the cross-sectional area of the cylinder gives the length of the stroke. Similarly, dividing clearance-volume by the cross-sectional area of the cylinder gives the clearance in terms of length. Hence, for convenience, if the stroke is expressed in inches, the clearance may be expressed as a length in inches. Alternatively, the clearance may be expressed as "a fraction of the stroke" or "a percentage of the stroke", for instance, if the stroke is 20 inches and the clearance length is 2 inches, the clearance could be expressed as "one-tenth of the stroke", or, "10% of the stroke".

Example. The stroke of a gas engine which works on the constant volume cycle, is 18 inches. The pressure at the beginning of compression is 14·5 lb/in^2 abs. and at the end of compression it is 160 lb/in^2 abs. Assuming compression follows the law $PV^{1\cdot 36} = C$, calculate the clearance between the piston and cylinder cover at the end of compression in inches of length.

Let clearance = c inches
V_1 = stroke + clearance = $18 + c$
V_2 = clearance = c

$$P_1 V_1^{1\cdot 36} = P_2 V_2^{1\cdot 36}$$

$$\left\{\frac{V_1}{V_2}\right\}^{1\cdot 36} = \frac{P_2}{P_1}$$

$$\frac{V_1}{V_2} = \sqrt[1\cdot 36]{\frac{P_2}{P_1}}$$

$$\frac{18+c}{c} = {}^{1\cdot36}\sqrt{\frac{160}{14\cdot5}}$$

$$\frac{18+c}{c} = 5\cdot843$$

$$18 + c = 5\cdot843c$$
$$18 = 4\cdot843c$$
$$c = 3\cdot717 \text{ inches. Ans.}$$

TEST EXAMPLES 6

1. A petrol engine working on the constant volume cycle has a compression ratio of 9 to 1. If the pressure and temperature of the petrol-vapour-air mixture at the beginning of compression are 14·2 lb/in² abs. and 100°F respectively, calculate the pressure and temperature at the end of compression assuming it follows the law $PV^{1.34} = C$.

2. The stroke of an internal combustion engine is 75 mm, the diameter of the cylinder is 70 mm, and the clearance volume at the end of compression is 36 cc. Assuming compression follows the law $PV^{1.37} = C$, calculate the pressure at the end of compression if the initial pressure is 14 lb/in² abs.

3. The pressure and temperature of the air at the beginning of compression in a diesel engine are 1·1 kg/cm² and 35°C respectively, and the clearance volume is equal to 7·5% of the piston swept volume. Calculate the pressure and temperature at the end of compression assuming the law of compression is $PV^{1.36} = C$.

4. The stroke of a petrol engine is 87·5 mm and the clearance is equal to 12·5 mm. A compression plate is now fitted which has the effect of reducing the clearance to the equivalent of 10 mm. Assuming the compression period to be the whole stroke, the pressure at the beginning of compression as 14 lb/in² abs., and the law of compression $PV^{1.35} = C$, calculate the pressure at the end of compression before and after the compression plate is fitted.

5. The stroke of the piston in an internal combustion engine is 880 mm and the clearance is equal to 80 mm. The law of compression is $PV^{1.38} = C$ and the pressure at the end of compression is 462·8 lb/in² abs. Find the increase of the final pressure caused by reducing the clearance by 5 mm.

6. The ratio of compression in a diesel engine is 16 to 1 and the temperature of the air at the beginning of compression is 120°F. Calculate the temperature at the end of compression assuming it follows the law $PV^{1.34} = C$.

7. When the piston is moving up in a two-stroke diesel engine, the scavenge ports are closed when the piston is 675 mm from the top of its stroke, the pressure and temperature of the air in the cylinder then being 2 lb/in² gauge and 110°F. The clearance is

equal to 65 mm and the diameter of the cylinder is 650 mm. Calculate the weight of air compressed in the cylinder taking the weight of one cubic foot of air at 14·7 lb/in^2 and 32°F as 0·0807 lb.

8. Gas is compressed in an internal combustion engine according to the law $PV^{1\cdot 36} = C$. If the initial and final temperatures of the gas are 85°F and 650°F respectively, calculate the compression ratio.

9. The compression ratio of a petrol engine working on the constant volume cycle is 8·5. The pressure and temperature at the beginning of compression are 14·5 lb/in^2 abs. and 110°F and the maximum pressure of the cycle is 450 lb/in^2 abs. Taking compression to follow the law $PV^{1\cdot 35} = C$, calculate (i) the pressure at the end of compression, (ii) temperature at end of compression, (iii) temperature at end of combustion.

ƒ10. In an internal combustion engine working on the constant volume cycle, the pressure, volume and temperature at the beginning of compression are 14·4 lb/in^2 abs., 4 cubic feet, and 120°F respectively, and the ratio of compression is 10. During combustion at constant volume the gas receives 90 Btu. Taking compression according to the law $PV^{1\cdot 37} = C$, specific heat at constant volume 0·17 and R (the characteristic gas constant) as 54 ft lb/lb/°F, calculate:
(i) weight of gas compressed,
(ii) pressure and temperature at end of compression,
(iii) pressure and temperature at end of combustion.

ƒ11. The compression ratio of a diesel engine is 15 to 1. Fuel is admitted for one-tenth of the power stroke and combustion takes place at constant pressure. Exhaust commences when the piston has travelled nine-tenths of the stroke. At the beginning of compression the temperature of the air is 105°F. Assuming compression and expansion to follow the law $PV^n = C$ where n is 1·34, calculate the temperatures at the end of compression, end of combustion, and beginning of exhaust.

ƒ12. An engine operates on the constant volume cycle and has a ratio of compression of 7 to 1. Another engine, working on the diesel cycle, has a compression ratio of 14 to 1, and fuel is admitted at constant pressure for 6% of the stroke. Compare the Air Standard Efficiencies of the two engines by use of the following formulae:

$$\text{A.S.E. of cons. vol. cycle} = 1 - \frac{1}{r^{\gamma-1}}$$

$$\text{A.S.E. of diesel cycle} = 1 - \frac{1}{\gamma} \times \frac{1}{r^{\gamma-1}} \left\{ \frac{\rho^\gamma - 1}{\rho - 1} \right\}$$

where r = ratio of compression
γ = 1·4
ρ = fuel cut-off ratio

*f*13 The compression ratio of a compression-ignition engine is 14 to 1, the diameter of the cylinder is 20 inches and the stroke/bore ratio is 1·2 to 1. At the beginning of compression the pressure and temperature of the air in the cylinder is 15 lb/in² abs. and 125°F, and compression follows the law PV^n = constant, where n = 1·35. Assuming compression takes place over the whole stroke, calculate (i) the pressure at the end of compression, (ii) temperature at end of compression, (iii) weight of air compressed, (iv) work done on the air during compression, (v) change of internal energy, and (vi) heat exchange during compression. Take the values: R for air = 53·3 ft lb/lb/°F, and C_v for air = 0·169.

14. The compression ratio of an engine working on the dual-combustion cycle is 10·7. The pressure and temperature of the air at the beginning of compression is 14·5 lb/in² abs. and 90°F. The maximum pressure and temperature during the cycle is 600 lb/in² abs. and 2900°F. Assuming adiabatic compression and expansion, calculate the pressures and temperatures at the remaining cardinal points of the cycle and the ideal thermal efficiency. Take the values, C_v = 0·17, and C_p = 0·238.

CHAPTER 7

AIR COMPRESSORS

Compressed air, at various pressures, is used on board ship for many purposes such as scavenging, supercharging, and starting diesel engines, and as the operating fluid for many automatic control systems. Air compressors to produce medium and high pressures are usually of the reciprocating type and may be single or multi-stage. Rotary types are common for large quantities of air at low pressures.

Fig. 42 shows diagrammatically a single-stage, single-acting reciprocating compressor, and its PV indicator diagram illustrating the cycle.

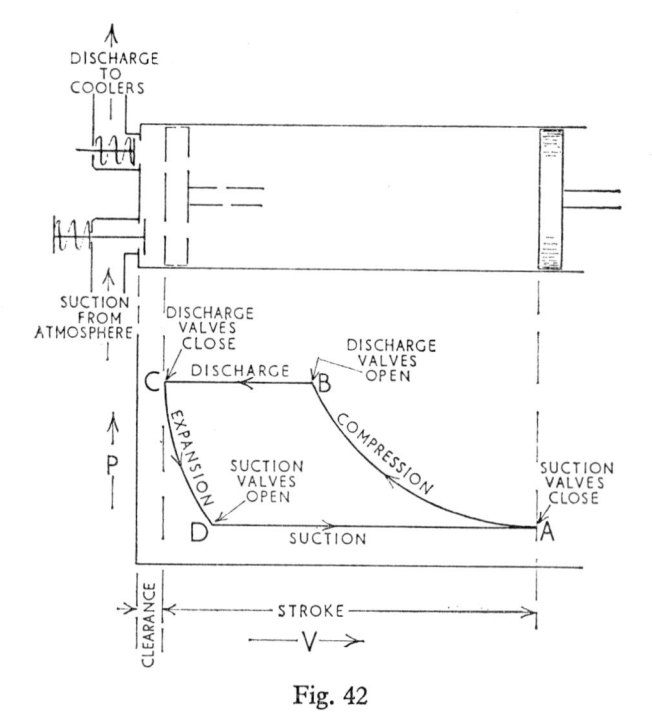

Fig. 42

Commencing at point A, the cycle of operations is as follows: A to B, compression period; with all valves closed the piston moves inward and the air which was previously drawn into the cylinder is compressed. Compression continues until the air pressure is sufficiently high to force the discharge valves open against their pre-set compression springs. Thus, at point B, the discharge valves open, and the compressed air is discharged at constant pressure for the remainder of the inward stroke, ie, from B to C. At point C the piston has completed its inward stroke and changes direction to move outward. Immediately the piston begins to move back there is a drop in pressure of the compressed air left in the clearance space, the discharge valves close, and from C to D this air expands. At point D the pressure has fallen to less than the atmospheric pressure and the lightly-sprung suction valves are opened by the greater pressure of the atmospheric air. Air is drawn into the cylinder for the remainder of the outward stroke, ie, from D to A.

Example. The stroke of the piston of an air compressor is 10 inches and the clearance volume is equal to 6% of the stroke volume. The pressure of the air at the beginning of compression is 14·2 lb/in² abs. and it is discharged at 55 lb/in² abs. Assuming compression to follow the law PV^n = constant, where n = 1·25, calculate the distance moved by the piston from the beginning of its pressure stroke before the discharge valves open and express this as a percentage of the stroke.

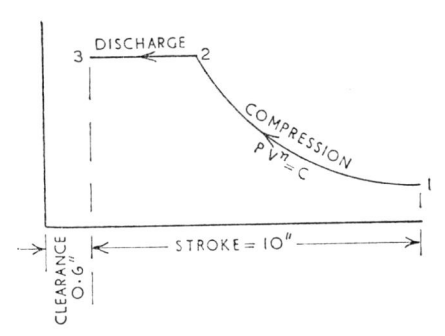

Fig. 43

Clearance length = 6% of 10 in. = 0·6 in.
P_1 = 14·2, P_2 = 55, V_1 = 10 + 0·6 = 10·6 in.
$$P_1 V_1^n = P_2 V_2^n$$
$$14·2 \times 10·6^{1·25} = 55 \times V_2^{1·25}$$

$$V_2^{1·25} = \frac{14·2 \times 10·6^{1·25}}{55}$$

$$V_2 = 10.6 \times \sqrt[1.25]{\frac{14.2}{55}} = 3.588 \text{ in.}$$

Distance moved by piston from beginning of stroke to point where discharge valves open is represented by $V_1 - V_2$,

$$V_1 - V_2 = 10.6 - 3.588$$
$$= 7.012 \text{ inches. Ans. (i)}$$

Expressed as a percentage of the stroke of 10 in.,

$$= \frac{7.012}{10} \times 100 = 70.12\% \text{ Ans. (ii)}$$

Note. If the student finds difficulty in solving the above equation, he should refer to Volume 1 (Mathematics), Chapter 4, for revision. For instance, some students may prefer to solve the above, and similar types to follow, by expressing them in the form of logarithmic equations, as shown below:

$$P_1 V_1 = P_2 V_2^n$$
$$14.2 \times 10.6^{1.25} = 55 \times V_2^{1.25}$$
$$\log 14.2 + 1.25 \times \log 10.6 = \log 55 + 1.25 \times \log V_2$$
$$1.1523 + 1.25 \times 1.0253 = 1.7404 + 1.25 \log V_2$$
$$1.1523 + 1.2816 = 1.7404 + 1.25 \log V_2$$
$$1.25 \log V_2 = 1.1523 + 1.2816 - 1.7404$$
$$1.25 \log V_2 = 0.6935$$

$$\log V_2 = \frac{0.6935}{1.25} = 0.5548$$

$$\therefore V_2 = 3.588 \text{ in. (as before)}$$

Example. The diameter of an air compressor cylinder is 14 inches, the stroke of the piston is 18 inches, and the clearance volume is 77 cubic inches. The pressure and temperature of the air in the cylinder at the end of the suction stroke and beginning of compression is 14 lb/in² abs. and 55°F. The delivery pressure is constant at 60 lb/in² abs. Taking the law of compression as $PV^{1.3}$ = constant, calculate (i) for what length of the stroke air is delivered, (ii) the volume of air delivered per stroke, in cubic feet, and (iii) the temperature of the compressed air.

AIR COMPRESSORS

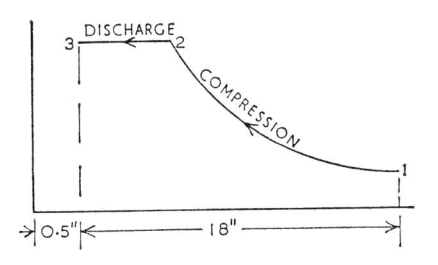

Fig. 44

$$\text{Clearance length} = \frac{\text{clearance volume}}{\text{area of cylinder}}$$

$$= \frac{77}{0.7854 \times 14^2} = 0.5 \text{ inch}$$

$$P_1 V_1^{1.3} = P_2 V_2^{1.3}$$
$$14 \times 18.5^{1.3} = 60 \times V_2^{1.3}$$

$$V_2^{1.3} = \frac{14 \times 18.5^{1.3}}{60}$$

$$V_2 = 18.5 \times \sqrt[1.3]{\frac{14}{60}}$$

$$= 6.039 \text{ inches}$$

$$\begin{aligned}\text{Delivery period} &= V_2 - V_3 \\ &= 6.039 - 0.5 \\ &= 5.539 \text{ in of stroke. Ans. (i)}\end{aligned}$$

$$\begin{aligned}\text{Volume delivered} &= \text{area} \times \text{length} \\ &= 0.7854 \times 14^2 \times 5.539 \\ &= 852.7 \text{ in.}^3\end{aligned}$$

or, dividing by 1728,
$$\text{volume delivered} = 0.4935 \text{ ft}^3 \text{ Ans. (ii)}$$

$$\frac{P_1V_1}{T_1} = \frac{P_2V_2}{T_2}$$

$$\frac{14 \times 18 \cdot 5}{(55 + 460)} = \frac{60 \times 6 \cdot 039}{T_2}$$

$$T_2 = \frac{515 \times 60 \times 6 \cdot 039}{14 \times 18 \cdot 5}$$

$$= 720 \cdot 6°F \text{ abs.}$$

∴ Temperature at end of compression
= 720·6 — 460 = 260·6°F. Ans. (iii)

EFFECT OF CLEARANCE

Clearance is necessary between the piston face and the cylinder head and valves to avoid contact. This should be kept to a minimum because the volume of compressed air left in the clearance space at the end of the inward stroke must be expanded on the outward stroke to below atmospheric pressure before the suction valves can open, thus affecting the volume of air drawn into the cylinder during the suction stroke. This effect is illustrated below.

An ideal compressor would have, theoretically, no clearance as in Fig. 45. The suction valves would open immediately the piston began to move on its outward stroke and air would be drawn into the cylinder for the whole stroke.

Fig. 46 shows the effect of a small clearance. The small volume of compressed air left in the clearance space is quickly expanded to sub-atmospheric pressure to allow the suction valves to open, and air is drawn into the cylinder from D to A which is a large proportion of the full stroke E to A.

Fig. 47 shows the effect of excessive clearance. There is a comparatively large volume of air left in the clearance space and it requires a considerable movement of the piston on its outward stroke to expand this air to a pressure below atmospheric. The suction period D to A is an inefficient proportion of the stroke E to A.

AIR COMPRESSORS 137

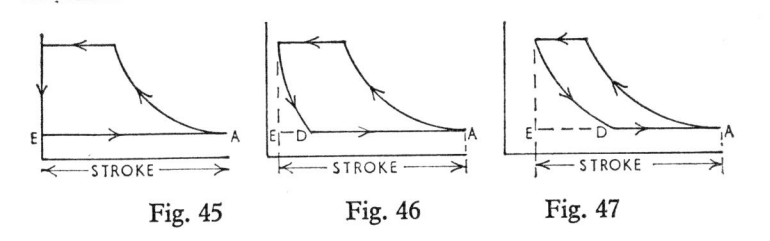

Fig. 45   Fig. 46   Fig. 47

The ratio between the volume of air drawn into the cylinder during the suction stroke and the full stroke volume swept out by the piston, is the *volumetric efficiency* of the compressor.

$$\text{Volumetric efficiency} = \frac{\text{Volume of air drawn in per stroke}}{\text{Stroke volume}}$$

and this is represented by the ratio $\dfrac{DA}{EA}$

f WORK DONE PER CYCLE

As previously shown, the area of a PV diagram represents work done, if the pressure is in lb/ft^2 and the volume in cubic feet then the area of the PV diagram and the work done per cycle is in ft lb.

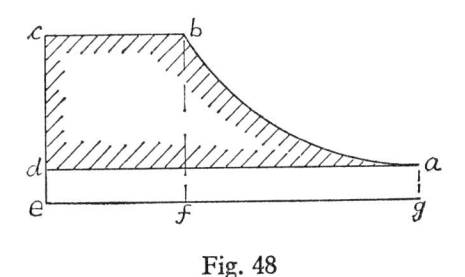

Fig. 48

Referring to Fig. 48 (neglecting clearance):

Nett area = Nett work done on the air per cycle
Area *abcd* = Area *bcef* + Area *abfg* — Area *adeg*

$$= P_2V_2 + \frac{P_2V_2 - P_1V_1}{n-1} - P_1V_1$$

$$= \frac{P_2V_2(n-1) + P_2V_2 - P_1V_1 - P_1V_1(n-1)}{n-1}$$

$$= \frac{nP_2V_2 - P_2V_2 + P_2V_2 - P_1V_1 - nP_1V_1 + P_1V_1}{n-1}$$

$$= \frac{n}{n-1}(P_2V_2 - P_1V_1) \quad \ldots \quad \ldots \quad (i)$$

This expression can be expressed in other terms as follows:

From $PV = wRT$
then, $P_1V_1 = wRT_1$ and $P_2V_2 = wRT_2$
$\therefore P_2V_2 - P_1V_1 = wR(T_2 - T_1)$

$$\therefore \text{Work done} = \frac{n}{n-1} wR(T_2 - T_1) \quad \ldots \quad \ldots \quad (ii)$$

From $\dfrac{T_2}{T_1} = \left\{\dfrac{P_2}{P_1}\right\}^{\frac{n-1}{n}}$ $T_2 = T_1 \times \left\{\dfrac{P_2}{P_1}\right\}^{\frac{n-1}{n}}$

Substituting for T_2 into (ii),

$$\text{Work done} = \frac{n}{n-1} wRT_1 \left[\left\{\frac{P_2}{P_1}\right\}^{\frac{n-1}{n}} - 1\right] \quad (iii)$$

From $\left\{\dfrac{P_2}{P_1}\right\}^{\frac{n-1}{n}} = \left\{\dfrac{V_1}{V_2}\right\}^{n-1}$ then:

$$\text{Work done} = \frac{n}{n-1} wRT_1 \left[\left\{\frac{V_1}{V_2}\right\}^{n-1} - 1\right] \quad (iv)$$

AIR COMPRESSORS

With reference to Fig. 49 it can be seen that the nearer isothermal compression can be approached, the less work will be required to compress and deliver a given weight of air. Water-cooling of the cylinder and head is necessary to prevent overheating and also has the advantage of reducing the index of compression and so reduces the work done.

f Example. The cylinder of a single-acting compressor is 9 inches diameter and the stroke of the piston is 12 inches. It takes in air at 14 lb/in² abs. and delivers it at 70 lb/in² abs. and makes 200 delivery strokes per minute. Assuming that compression follows the law PV^n = constant, and neglecting clearance, calculate the theoretical horse-power required to drive the compressor when the value of the index of the law of compression is, (i) 1·2, (ii) 1·35.

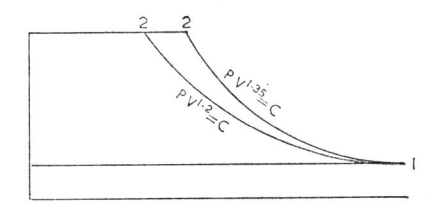

Fig. 49

$$V_1 = \frac{0.7854 \times 9^2 \times 12}{1728} = 0.4418 \text{ ft}^3$$

when $n = 1.2$:

$$P_1 V_1^{1.2} = P_2 V_2^{1.2}$$
$$14 \times 0.4418^{1.2} = 70 \times V_2^{1.2}$$

$$V_2 = \frac{0.4418}{\sqrt[1.2]{5}} = 0.1155 \text{ ft}^3$$

Work done per cycle $= \dfrac{n}{n-1} (P_2 V_2 - P_1 V_1)$

$$= \frac{1.2}{0.2} (70 \times 144 \times 0.1155 - 14 \times 144 \times 0.4418)$$

$$= 6 \times 14 \times 144\,(5 \times 0{\cdot}1155 - 0{\cdot}4418)$$
$$= 6 \times 14 \times 144 \times 0{\cdot}1357$$
$$= 1642 \text{ ft lb.}$$

Work done per minute $= 1642 \times 200$ ft lb/min.

$$\text{Equivalent horse power} = \frac{1642 \times 200}{33000}$$

$$= 9{\cdot}951 \text{ h.p. Ans. (i)}$$

when $n = 1{\cdot}35$:

$$14 \times 0{\cdot}4418^{1{\cdot}35} = 70 \times V_2^{1{\cdot}35}$$

$$V_2 = \frac{0{\cdot}4418}{\sqrt[1{\cdot}35]{5}} = 0{\cdot}1341 \text{ ft}^3$$

Work done per cycle

$$= \frac{n}{n-1}(P_2 V_2 - P_1 V_1)$$

$$= \frac{1{\cdot}35}{0{\cdot}35}(70 \times 144 \times 0{\cdot}1341 - 14 \times 144 \times 0{\cdot}4418)$$

$$= \frac{27}{7} \times 14 \times 144\,(5 \times 0{\cdot}1341 - 0{\cdot}4418)$$

$$= 27 \times 2 \times 144 \times 0{\cdot}2287$$
$$= 1778 \text{ ft lb.}$$

$$\text{Equivalent horse power} = \frac{1778 \times 200}{33000}$$

$$= 10{\cdot}77 \text{ h.p. Ans. (ii)}$$

Note that the *weight* of air delivered per stroke is the same in each case. The greater volume indicated by the value of V_2 when n is 1·35 is due purely to the air being at a higher temperature at the end of compression than it is when $n = 1{\cdot}2$.

AIR COMPRESSORS

MULTI-STAGE COMPRESSION

By compressing the air in more than one stage and intercooling between stages, the practical compression curve can approach the isothermal more closely, hence reducing the work required per lb of air compressed. Figs. 50 and 51 show the PV diagrams for two-stage and three-stage compression respectively, the shaded areas representing the work saved in each case compared with single-stage compression.

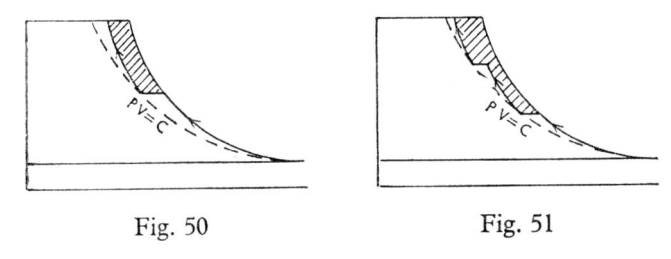

Fig. 50   Fig. 51

To obtain maximum efficiency from a multi-stage compressor, that is, to do the least work to compress and deliver a given weight of air, (i) the air should be intercooled to as near the initial temperature as possible, and (ii) the pressure ratio in each stage should be the same.

Multi-stage compressors may consist simply of separate compressor cylinders, or may be arranged in tandem. Fig. 52 shows a diagrammatic arrangement of a three-stage tandem air compressor. It is important to note that, when calculating the volume of atmospheric air drawn into the low-pressure cylinder of a tandem compressor, the *effective* area of suction is the annulus between the area of the L.P. and the area of the H.P. Thus, if D = diameter of L.P. piston, and d = diameter of H.P., then:

$$\text{Effective area} = 0.7854\, D^2 - 0.7854\, d^2$$
$$= 0.7854\, (D^2 - d^2)$$
$$= 0.7854\, (D + d)(D - d)$$

Example. In a single-acting 3-stage tandem air compressor, the piston diameters are $2\frac{3}{4}$, $13\frac{1}{4}$ and $14\frac{3}{4}$ inches diameter respectively, the stroke is 15 inches, and it is driven directly from a motor running at 250 r.p.m. The suction pressure is atmospheric (14·7 lb/in²) and the discharge pressure is 700 lb/in² gauge. Assuming that the air delivered to the reservoirs is cooled down to the initial suction

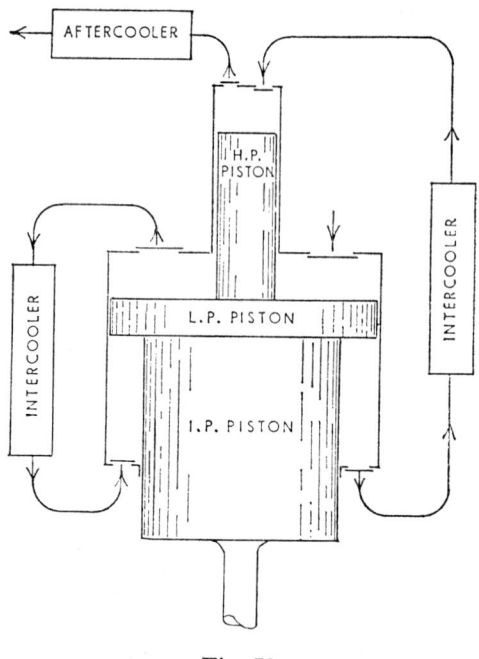

[Fig. 52: Diagram showing Aftercooler, H.P. Piston, L.P. Piston, I.P. Piston, and two Intercoolers]

Fig. 52

temperature and taking the volumetric efficiency as 90%, calculate the volume of compressed air delivered to the reservoirs per minute.

Effective area of L.P. = $0.7854 (14.75^2 - 2.75^2)$
  = $0.7854 \times 17.5 \times 12$ sq. inches.
Stroke volume = area × stroke
  = $0.7854 \times 17.5 \times 12 \times 15$ cu. inches.

Cubic feet of air drawn into L.P. per stroke

$$= \frac{0.7854 \times 17.5 \times 12 \times 15}{1728} \times \frac{90}{100}$$

  = 1.288 ft^3

Air drawn in per minute
  = 1.288×250
  = 322 ft^3

AIR COMPRESSORS 143

This is the volume of air taken into the compressor per minute at 14·7 lb/in² abs. The volume delivered per minute at 714·7 lb/in² abs. is at the same temperature, therefore, since pressure × volume = constant, volume varies inversely as the absolute pressure, then:

$$\text{Volume delivered} = 322 \times \frac{14 \cdot 7}{714 \cdot 7}$$

$$= 6 \cdot 624 \text{ ft}^3/\text{min. Ans.}$$

Example. The cylinders of a single-acting 2-stage tandem air compressor are $2\frac{1}{4}$ and $8\frac{1}{2}$ inches diameter respectively, and the stroke is 9 inches. It is connected to three air storage bottles of equal size of internal dimensions 12 in. diameter and 5 ft long overall with hemispherical ends. Taking atmospheric pressure as 15 lb/in² and assuming a volumetric efficiency of 0·88, calculate the time required to pump up the bottles to a pressure of 450 lb/in² gauge from empty, when running at 125 r.p.m.

Length of cylindrical part of bottles

$$= \text{overall length} - \text{diameter} = 5 - 1 = 4 \text{ ft}$$

$$\begin{aligned}
\text{Volume of 3 bottles} &= 3\left(\tfrac{\pi}{6} \times 1^3 + \tfrac{\pi}{4} \times 1^2 \times 4\right) \\
&= 3\pi\left(\tfrac{1}{6} + 1\right) \\
&= 3\pi \times 1\tfrac{1}{6} = 11 \text{ ft}^3
\end{aligned}$$

To produce 11 cubic feet of air at 465 lb/in² abs., the volume of atmospheric pressure air (termed "free air") required, at the same temperature, is inversely proportional to the pressure:

Volume of atmospheric air

$$= 11 \times \frac{465}{15} = 341 \text{ ft}^3$$

However, "empty" bottles contain their own volume of air at atmospheric pressure, therefore, volume of air to be taken into compressor from atmosphere

$$= 341 - 11 = 330 \text{ ft}^3 \quad \ldots \quad \ldots \quad (i)$$

Volume of air taken into compressor per minute

$$= \frac{0.7854\,(8.5^2 - 2.25^2) \times 9 \times 0.88 \times 125}{1728}$$

$$= 30.24 \text{ ft}^3/\text{min.} \quad \ldots \quad \ldots \quad \ldots \quad \text{(ii)}$$

∴ Time required $= \dfrac{330}{30.24} = 10.91$ minutes. Ans.

AIR COMPRESSORS

TEST EXAMPLES 7

1. In a single-stage air compressor the diameter of the cylinder is 25 cm, the stroke of the piston is 35 cm, and the clearance volume is 900 cc. Air is drawn in at a pressure of 14·3 lb/in² abs. and delivered at 60 lb/in² abs. Taking the law of compression to be $PV^{1·25} = C$, calculate the distance travelled by the piston from the beginning of its compression stroke when the delivery valves open.

2. A compressor cylinder is 8 cm diameter and the stroke of the piston is 16 cm. The pressures at the beginning and end of compression are 14 and 70 lb/in² abs. respectively and the delivery valves open when the piston is 12 cm from the beginning of compression. If the law of compression is $PV^{1·3} = C$, find the clearance volume.

3. The internal volume of an air-storage vessel is 20 cubic feet. Find the weight of air stored in it when the pressure is 40 atmospheres and the temperature is 26°C. Note: one cubic foot of air at atmospheric pressure and at 0°C weighs 0·0807 lb.

4. A single-stage air compressor takes in 2 lb weight of air per minute at atmospheric pressure and 75°F and compresses it to 10 atmospheres of pressure, the law of compression being $PV^{1·2} = C$. Calculate (i) the volume of air drawn in per minute, (ii) the volume of air delivered per minute, and (iii) the temperature of the compressed air. Take the weight of one cubic foot of air at atmospheric pressure and 32°F as 0·0807 lb.

5. The volume of air in a single-stage compressor at the beginning of compression is 1·5 cubic feet and its pressure is 14·5 lb/in² abs. The clearance volume in the cylinder is 75 cubic inches and the discharge pressure is 116 lb/in² abs. Calculate the volume of air delivered per stroke, in cubic inches, if it is compressed (i) isothermally, (ii) adiabatically, taking $\gamma = 1·4$.

6. The diameter of the cylinder of an air compressor is 14 cm and the stroke of the piston is 20 cm. It takes in air at 14·2 lb/in² abs. and delivers it at 100 lb/in² abs., the law of compression being $PV^{1·28} = C$. Calculate the volume of air delivered per stroke, in cubic centimetres, when the clearance volume is equal to (i) 10% of the stroke volume, (ii) 5% of the stroke volume.

7. The stroke of a compressor piston is 15 inches and the clearance volume is equal to 7% of the stroke volume. The pressure of the air at the beginning of the compression stroke is 14·5 lb/in² abs. The delivery valves open when the piston has travelled 10·5 inches from the beginning of the compression stroke and the pressure of the air is then 57 lb/in² abs. If compression follows the law $PV^n = C$, find the value of n.

ƒ8. The diameter of the cylinder of a single-acting air compressor is 10 inches and the stroke of the piston is 15 inches. It runs at 200 r.p.m. taking in air at 14·7 lb/in² abs. and delivering it at 80 lb/in² abs. Neglecting clearance, calculate the air horse-power when the law of compression is $PV^{1·2} = C$. Calculate also the percentage increase in power if the index of the law of compression increased to 1·3 without change of speed.

9. A single-acting 2-stage tandem air compressor with pistons 2 inches and 7½ inches diameter, and stroke 6 inches, runs at 140 r.p.m. Air is taken into the compressor at atmospheric pressure and discharged at 400 lb/in² gauge. The volumetric efficiency is 0·9. Assuming the temperature of the air from the after-coolers is the same as the suction temperature, and taking atmospheric pressure as 14·7 lb/in², find the volume of air delivered per minute.

10. A motor driven 3-stage single-acting tandem air compressor runs at 170 r.p.m. The H.P. and L.P. cylinder diameters are 3 inches and 14 inches diameter respectively, and the stroke is 12 inches. Find the time to pump up air reservoirs of 800 cubic feet total capacity from 275 to 425 lb/in² gauge, taking the volumetric efficiency as 0·92 and atmospheric pressure as 15 lb/in².

11. In a single-acting 3-stage tandem air compressor, the diameters of the H.P. and L.P. pistons are 3¼ and 10¾ inches respectively, the stroke is 10 inches, and it runs at 110 r.p.m. Taking atmospheric pressure as 15 lb/in² and the volumetric efficiency as 0·9, find the time taken to pump up three air-bottles from atmospheric pressure to 600 lb/in² gauge. The bottles have hemispherical ends and the internal dimensions of each are 18 inches diameter and 7 feet 6 inches long overall.

ƒ12. A single-stage single-acting air compressor, running at 200 r.p.m., takes in 0·8 ft³ of atmospheric air per stroke at 14·7 lb/in² and 75°F and compresses it to 0·2 ft³ according to the law $PV^{1·2} = C$. Calculate (i) the temperature of the air at the end of compression, (ii) the volume of air delivered per minute to the reservoirs if it is cooled at constant pressure to its initial temperature through after-coolers, (iii) the weight of air delivered per minute, (iv) the weight of sea water circulating through the cooler per minute if the difference in temperature between inlet and outlet is 20 F degrees. Take the values, R for air = 53·3 ft lb/lb/°F, C_p for air = 0·2375, spec. heat of sea water = 0·98.

ƒ13. In a two-stage compressor, 0·1 lb weight of air is taken in per stroke at 14 lb/in² abs. and 60°F and compressed in the first stage to 42 lb/in² abs. It is then passed through the intercooler where it is cooled at constant pressure to its initial temperature. In the second stage, the air is further compressed to 105 lb/in² abs. and passed through the after-cooler where it is cooled at constant pressure to the initial temperature. Calculate (i) the percentage decrease in volume due to cooling at the end of each stage, (ii) the percentage decrease in volume if the air had been compressed in a single-stage from 14 to 105 lb/in² abs. and then finally cooled to its initial temperature of 60°F. Take R for air = 53·3 ft lb/lb/°F and compression to follow the law PV^n = constant where n = 1·3 in each stage.

CHAPTER 8

STEAM

Under normal running conditions of a steam engine plant, the engines consume steam at the same rate at which it is generated in the boilers, therefore the steam is generated at constant pressure.

The temperature at which water changes into steam when heat is given to it, i.e. the boiling point of the water, depends strictly upon the pressure exerted on it. The higher the pressure the higher the temperature. A few examples are as follows:

PRESSURE lb/in^2 abs.	0·5	14·7	100	200	400
BOILING POINT °F	79·6	212	327·8	381·8	444·6
,, ,, °C	26·42	100	164·4	194·3	229·2

The temperature of the steam produced at any given pressure is the same as the temperature of the boiling water at the same pressure. Thus, if a boiler is working at a pressure of 200 lb/in^2 abs. the water begins to boil when its temperature reaches 381·8°F and the steam is generated at the same temperature.

Steam which is in physical contact with the boiling water from which it has been generated is termed *saturated steam*, its temperature is the same as the boiling water and is referred to as its *saturation temperature*. When it is steam without any water held in suspension, it is called *dry saturated steam*. If, however, the steam contains water (usually very fine particles in suspension in the form of a mist), it is called *wet saturated steam* or briefly *wet steam*, and its quality is then expressed by its *dryness fraction* (q) which is the ratio of the weight of pure steam in a given weight of the steam-plus-water-mixture.

In order to increase the temperature of steam at any specified pressure above its corresponding saturation temperature the steam must be taken away from contact with the water from which it was generated and heated externally (usually by the boiler flue gases) as it passes to the engines. Steam whose temperature is higher than its saturation temperature corresponding to its pressure is termed *superheated steam*.

CONSTANT PRESSURE STEAM GENERATION

The water which is fed into the boilers by the feed pumps first receives heat to increase its temperature up to the boiling point. As it increases in temperature it increases slightly in volume. The heat received by the water during this stage is *sensible heat*. In Chapter 1, this value was given as:

Sensible heat = weight × spec. ht. × temperature rise.

However, the specific heat of water is not constant. If it is assumed to be so and equal to unity, it is sufficiently accurate for approximate calculations only, but a true value of sensible heat can be obtained from Steam Tables which have been compiled from experimental results.

When the water reaches boiling point it begins to change its physical state from liquid (water) into vapour (steam), *at the same temperature*. The heat to convert one lb of water into one lb of steam at constant temperature is the *latent heat of steam*. Its value depends upon the temperature of evaporation and *decreases* in value as the temperature increases. During this change of state considerable expansion takes place, the volume of steam being many times greater than the volume occupied by the water from which it was generated.

In most boilers the saturated steam is subsequently passed through nests of superheater tubes where it receives more heat from the flue gases before passing on to the engines. Any water moisture present in the steam is first evaporated then the temperature of the dry steam increases. It is now receiving sensible heat and the steam is becoming *superheated*. The specific heat of superheated steam varies from about 0·48 upwards, depending upon the pressure and temperature. Approximate calculations can be made to determine the sensible heat required to superheat the steam by assuming an average value of specific heat through the temperature range of increase, or more correct values can be obtained from the steam tables. The behaviour of steam in a superheated state approximates more closely to that of a gas, and, since superheating of steam occurs at constant pressure, its volume increases with increase of temperature.

STEAM TABLES

Much experimental work has been done on the properties of steam and the results published in various forms under the title of *steam tables*. Those used in this book are "Abridged Callendar Steam Tables"—Fifth Edition. These are available separately in either the Btu version or the Chu version and the student should have a copy of each. This fifth edition of Callendar's steam tables has only recently (September 1962) been published. Compared with the earlier editions the main differences are in the headings and the symbols used, the numerical values being unaltered. Students familiar with the Fourth Edition symbols, such as *h* for sensible heat, L for latent heat, H_s for total heat, V_s for specific volume of dry saturated steam, and so on, should now aquaint themselves with those of the Fifth Edition because it is expected that these will shortly be generally adopted.

There are three tables in each set. The first table gives the properties of water and saturated steam, the second table gives the total heat (or enthalpy) values of superheated steam, and the third gives entropy values of superheated steam.

To begin with, steam tables in Fahrenheit units will be explained and used in examples, and an extract from the first table is given below for reference.

All the quantities given are not required at this stage, therefore only those to be used in this Chapter are explained here, the remaining quantities will be introduced as and when required.

Liquid-Vapour Saturation   Btu

P	*temperature*;R		$-g$	h_f	h_{fg}	h_g	s_f	s_g	v_f	v_g
lbf/in²	t_F	T_R	Btu/lb	Btu/lb	Btu/lb	Btu/lb	Btu/lbR	Btu/lbR	ft³/lb	ft³/lb
200	381·8	841·5	102·1	355·5	*844·0*	1199·5	0·5437	1·5466	0·0184	2·290
205	383·9	843·6	103·2	357·8	*842·0*	1199·8	0·5463	1·5445	0·0185	2·237
210	385·9	845·6	104·3	360·0	*840·1*	1200·1	0·5488	1·5424	0·0185	2·185
215	387·9	847·6	105·4	362·1	*838·3*	1200·4	0·5513	1·5404	0·0185	2·136
220	389·9	849·6	106·5	364·2	*836·5*	1200·7	0·5537	1·5384	0·0185	2·089
225	391·8	851·5	107·5	366·2	*834·8*	1201·0	0·5562	1·5366	0·0186	2·042
230	393·7	853·4	108·5	368·3	*833·0*	1201·3	0·5587	1·5347	0·0186	1·999
235	395·5	855·2	109·5	370·3	*831·2*	1201·5	0·5611	1·5329	0·0186	1·958
240	397·4	857·1	110·5	372·3	*829·4*	1201·7	0·5634	1·5311	0·0186	1·918
245	399·2	858·9	111·5	374·2	*827·7*	1201·9	0·5656	1·5293	0·0187	1·880

The first column lists the absolute pressure (P) of the steam in lb/in². The second column gives the saturation temperature (t_F) of the steam corresponding to the pressure P, and the third column records the corresponding absolute temperature (T_R) in °R.

Note °R (degrees Rankine) is another way of writing °F abs. just as °K (degrees Kelvin) means °C abs. Note also that the absolute temperatures recorded (T_R) are equal to ($t_F + 459 \cdot 7$) and, although this is considered to be a more accurate figure, the value of ($t_F + 460$) is generally used except when reading direct from steam tables.

The fifth column lists the sensible heat (h_f) values in Btu/lb measured above water at 32°F, i.e. it is equal to the heat required to raise one lb of water from 32°F to boiling point. Taking the example of a boiler working at 200 lb/in^2 abs., the temperature at which the water will boil under this pressure is 381·8°F, and the sensible heat required to raise one lb of water from 32°F to 381·8°F is given as 355·5 Btu. Comparing this with the assumption that the specific heat of water is unity, we would obtain:

Sensible heat = wt. × spec. ht. × temperature rise
= 1 × 1 × (381·8 − 32)
= 349·8 Btu.

Thus there is the difference of 5·7 Btu/lb between the correct figure of 355·5 and the calculated figure 349·8, which emphasises the importance of using steam tables whenever possible and especially at the higher pressures, because the discrepancy increases as the pressures increase.

The latent heat (h_{fg}) in Btu to completely evaporate one lb of water into one lb of saturated steam at constant temperature is given in column 6.

The total heat content (h_g) in one lb of dry saturated steam, measured above water at 32°F, is listed in column 7 and it will be appreciated that this is simply the sum of the two previous columns, i.e.

$$h_g = h_f + h_{fg}$$

It will be noted from the above statements that all the heat values listed in the steam tables are measured relative to water at 32°F or 0°C, i.e. the heat in water at 32°F or 0°C is taken as ZERO.

Total heat is sometimes termed enthalpy.

The last two columns record the specific volumes of water (v_f) and dry saturated steam (v_g) respectively at their specified pressures and temperatures. Specific volume refers to the volume, in cubic feet, occupied by one lb weight. In the case of steam it is obvious that this value becomes less as the pressure is increased since the steam is compressed to smaller volumes under higher pressure. Water however, is virtually incompressible, and therefore the change in specific volume depends primarily on the temperature change. Thus, as the temperature (and pressure) rises, so does the specific volume of water.

Example. Find the total heat required to produce one lb of saturated steam at a pressure of 220 lb/in² abs. from feed water at 170°F, (i) if the steam is dry, (ii) if the steam is wet with a dryness fraction of 0·9.

(i) DRY STEAM:

Reading the first steam table horizontally along the line representing 220 lb/in² abs. of pressure, h_g = 1200·7 Btu/lb. This is the total heat contained in one lb of dry saturated steam measured above water at 32°F. However, not all of this heat is required to be given to each lb of feed water because, since its temperature is 170°F it contains some sensible heat to begin with. For the usual range of feed water temperature it is sufficiently accurate to assume the specific heat of water as unity, and, as the heat for one lb only is required, then the sensible heat in the feed water above 32°F is simply the difference between the feed temperature and 32°F, thus,

$$\begin{aligned}\text{Sensible heat} &= \text{wt.} \times \text{spec. ht.} \times \text{temperature rise} \\ &= 1 \times 1 \times (\text{feed temperature} - 32) \\ &= 170 - 32 = 138 \text{ Btu/lb}\end{aligned}$$

Hence, the heat required to produce one lb of dry saturated steam from one lb of water at 170°F

$$\begin{aligned}&= h_g - (\text{feed temperature} - 32) \\ &= 1200 \cdot 7 - 138 \\ &= 1062 \cdot 7 \text{ Btu/lb. Ans. (i)}\end{aligned}$$

(ii) WET STEAM:

The dryness fraction of the wet steam is quoted as 0·9, this means that, by weight, 0·9 of it is dry steam and the remainder is water. Thus, in one lb of the wet steam, 0·9 lb is dry steam which has received latent heat to convert it from water, and the remaining 0·1 lb is water held in suspension and has therefore not received latent heat. Hence, the whole one lb of feed water receives sensible heat to raise its temperature up to its boiling point, but only 0·9 lb receives latent heat to convert it into steam.

Sensible heat given to each lb.
$$= h_f - (\text{feed temperature} - 32)$$
$$= 364 \cdot 2 - (170 - 32)$$
$$= 364 \cdot 2 - 138 = 226 \cdot 2 \text{ Btu}$$

Latent heat to evaporate 0·9 lb
$$= q \times h_{fg}$$
$$= 0 \cdot 9 \times 836 \cdot 5 = 752 \cdot 9 \text{ Btu}$$

Total heat = sensible heat + latent heat
$$= 226 \cdot 2 + 752 \cdot 9$$
$$= 979 \cdot 1 \text{ Btu/lb. Ans. (ii)}$$

From the above observations, the total heat content in one lb of wet steam of dryness fraction q, measured above water at 32°F, at any specified pressure is:

$$h_g = h_f + qh_{fg}$$

also, $q = \dfrac{\text{wt. of dry steam in sample of wet steam}}{\text{wt. of sample of wet steam}}$

Example. Dry saturated steam leaves a boiler at a pressure of 240 lb/in² abs., passes along the main steam pipe and enters the engine at a pressure of 235 lb/in² abs. when it is 5% wet. The engine develops 5000 i.h.p. and consumes 13·5 lb of steam per i.h.p. per hour. Find (i) the heat loss per lb of steam on its passage along the pipe, (ii) the weight of fuel oil this loss represents per day if the calorific value (heating value) of the oil is 18500 Btu/lb.

From steam tables,

P = 240 lb/in², h_g = 1201·7 Btu/lb
P = 235 lb/in², h_f = 370·3, h_{fg} = 831·2 Btu/lb

5% wetness means 95% dryness, i.e. q = 0·95
Total heat in steam leaving boiler = 1201·7 Btu/lb
Total heat in steam entering engine = $h_f + qh_{fg}$

$$= 370 \cdot 3 + 0 \cdot 95 \times 831 \cdot 2 = 1159 \cdot 9 \text{ Btu/lb}$$

Heat loss through pipe = 1201·7 − 1159·9
$$= 41 \cdot 8 \text{ Btu/lb Ans. (i)}$$

Total weight of steam used by engine
= 13·5 × 5000 lb/hour

Total heat loss per day
= 13·5 × 5000 × 41·8 × 24 Btu/day

Equivalent weight of oil

$$= \frac{13\cdot5 \times 5000 \times 41\cdot8 \times 24}{18500 \times 2240}$$

= 1·634 tons per day. Ans. (ii)

Example. A vessel contains 100 lb of water at its boiling point of 312°F under a pressure of 80 lb/in² abs. An escape valve at the top is opened and the pressure falls to atmospheric. Calculate the weight of water which flashes off as steam due to the drop in pressure, and the weight of water remaining in the vessel.

Consider one lb of water orginally in the vessel.

When the pressure falls from 80 lb/in² to 14·7 lb/in², the temperature falls from 312°F to 212°F and the heat released is the difference between the sensible heats of water at the two temperatures.

P = 80 lb/in², h_f = 282·1
P = 14·7 lb/in², h_f = 180·1
∴ Heat released = 102·0 Btu/lb

Assuming no external loss, this heat is absorbed by some of the water which causes it to evaporate.

Since the latent heat to evaporate one lb at atmospheric pressure is 970·6 Btu, then weight of water evaporated

$$= \frac{102}{970\cdot6} = 0\cdot105 \text{ lb}$$

Hence, for a total weight of 100 lb of water originally in vessel,

Water evaporated = 0·105 × 100 = 10·5 lb
Water remaining = 100 — 10·5 = 89·5 lb   } Ans.

Once the above principle is understood, this and similar problems could be set down in the form of an equation:

Let x = lb of water evaporated.

$$\begin{Bmatrix}\text{Initial heat in}\\ \text{water in vessel}\end{Bmatrix} = \begin{Bmatrix}\text{Heat carried away by}\\ \text{flash off into steam}\end{Bmatrix} + \begin{Bmatrix}\text{Heat in remaining}\\ \text{water in vessel}\end{Bmatrix}$$

$$100 \times 282 \cdot 1 = x(180 \cdot 1 + 970 \cdot 6) + (100 - x) \times 180 \cdot 1$$
$$28210 = 180 \cdot 1x + 970 \cdot 6x + 18010 - 180 \cdot 1x$$
$$28210 - 18010 = 970 \cdot 6x$$
$$x = 10 \cdot 5 \text{ lb evaporated (as before).}$$

SPECIFIC VOLUME OF WET STEAM

The specific volume of dry saturated steam is given in the steam tables under the heading v_g. This is the volume occupied, in cubic feet, by one lb weight of dry saturated steam at the given pressure.

In one lb of *wet* steam with a dryness fraction of q, the weight of dry steam present in the mixture is q lb and the remainder $(1 - q)$ lb is water. Hence the total volume of one lb of wet steam is the sum of the volumes occupied by q lb of dry steam and $(1 - q)$ lb of water.

If v_f represents the specific volume of water, then,

$$\text{Spec. vol. of wet steam} = qv_g + (1 - q)v_f$$

However, the volume of one lb of water is comparatively small (about 0·0167 ft³ at atmospheric pressure), and the value of $(1 - q)$ is seldom more than 0·2 (corresponding to a dryness fraction of 0·8), hence the volume occupied by the water particles in wet steam is negligible for most practical purposes. Therefore we assume:

$$\text{Spec. vol. of wet steam} = qv_g \text{ ft}^3/\text{lb}$$

SUPERHEATED STEAM

As previously stated, **steam** is said to be *superheated* when its temperature is higher than the saturation temperature corresponding to its pressure. In practice, steam is superheated at *constant pressure*, the saturated steam being taken from the boiler steam space and passed through superheater tubes where it receives additional heat to dry the steam and raise its temperature. The properties of superheated steam are approximately similar to those of a gas, and, since superheating occurs under constant pressure conditions, then for approximate calculations we may assume that the volume of the steam varies directly as its absolute temperature.

Thus, if v_g = spec. vol. of dry saturated steam,
T_R = abs. temp. of saturated steam,
T = abs. temp. of superheated steam,

then, Specific Volume of superheated steam =

$$v = v_g \times \frac{T}{T_R}$$

A more accurate value of the specific volume of superheated steam can be obtained by using Callendar's formula:

$$v = \frac{1 \cdot 248 \, (h - 835 \cdot 2)}{P} + 0 \cdot 0123 \text{ ft}^3/\text{lb}$$

where, h = total heat in Btu/lb of the superheated steam,

P = absolute pressure in lb/in^2

The second table of Callendar's steam tables gives the total heat contents of one lb of superheated steam, again measured above water at 32°F or 0°C, for various pressures and different degrees of superheat.

The numerical values of the column headings, 20, 40, 60, etc., under *superheat*/rankine-fahrenheit degree, are the "degrees of superheat" of the steam. These are the number of degrees of temperature *above the saturation temperature*. For instance, reading from the steam tables, the temperature of saturated steam at a pressure of 250 lb/in² abs. is 401°F, therefore superheated steam at the same pressure but having a temperature of 501°F has 100 F degrees of superheat, and the total heat per lb (specific enthalpy) is 1263·4 Btu/lb.

Example. Find the quantity of heat required to produce one lb of superheated steam at 300 lb/in² abs. with 100 F degrees of superheat, from feed water at 240°F. Calculate also the specific volume of this steam, (i) assuming that the volume varies directly as the absolute temperature, (ii) by Callendar's formula.

Referring to the second table, reading along the horizontal line corresponding to the pressure of 300 lb/in² abs., in the column headed 100°, we read the total heat (h) in one lb of the superheated steam measured above water at 32°F.

$$h = 1267\cdot 2 \text{ Btu/lb}$$

Sensible heat in feed water above 32°F

$$= 240 - 32 = 208 \text{ Btu/lb}$$
∴ Heat required $= 1267\cdot 2 - 208$
$= 1059\cdot 2$ Btu/lb of feed. Ans. (i)

Referring to the first table, we read along the line for 300 lb/in² abs. of pressure, that the temperature of saturated steam at this pressure is 877°F absolute, and the specific volume is 1·543 cubic feet per lb.

∴ Temperature of superheated steam

$$= 877 + 100 = 977°\text{F abs.}$$

By direct proportion, spec. vol. of superheated steam

$$= 1\cdot 543 \times \frac{977}{877} = 1\cdot 719 \text{ ft}^3/\text{lb. Ans. (iia)}$$

By Callendar's formula,

$$v = \frac{1\cdot 248\,(h - 835\cdot 2)}{P} + 0\cdot 0123 \text{ ft}^3/\text{lb}$$

$$= \frac{1\cdot 248\,(1267\cdot 2 - 835\cdot 2)}{300} + 0\cdot 0123$$

$$= \frac{1\cdot 248 \times 432}{300} + 0\cdot 0123$$

$$= 1\cdot 8093 \text{ ft}^3/\text{lb. Ans. (iib)}$$

Example. A marine turbine installation consists of an H.P. and an L.P. turbine. The H.P. is supplied with steam at 400 lb/in² abs. having 200 F degrees of superheat, and leaves as dry saturated steam at 60 lb/in² abs. 10% of the steam is now bled off to the feed heaters and the remainder passes to the L.P. The pressure of the exhaust steam from the L.P. is 0·5 lb/in² abs. and its dryness fraction is 0·92. Calculate (i) the heat drop through each turbine per lb of steam supplied by the boilers, (ii) the percentage greater or less power developed by the L.P. compared with the H.P., assuming that the power is proportional to the heat drop.

Referring to superheat tables,
$P = 400$ lb/in², superheat 200°, $h = 1331\cdot 6$ Btu/lb

Referring to saturation tables,
$P = 60$ lb/in², $h_g = 1178\cdot 4$ Btu/lb
$P = 0\cdot 5$ lb/in², $h_f = 47\cdot 6$ $h_{fg} = 1048\cdot 5$ Btu/lb

Heat drop per lb of steam through H.P.

$$= 1331\cdot 6 - 1178\cdot 4 = 153\cdot 2 \text{ Btu. Ans. (ia)}$$

10% of 1 lb = 0·1 lb of steam is now bled off to the heaters, leaving 0·9 lb of steam to pass through the L.P.

STEAM

Heat drop through L.P. per lb of boiler steam
$$= 0.9 [1178.4 - (47.6 + 0.92 \times 1048.5)]$$
$$= 0.9 [1178.4 - 1012.2]$$
$$= 0.9 \times 166.2 = 149.6 \text{ Btu. Ans. (ib)}$$

Percentage less heat drop (= power developed) in L.P. compared with H.P.
$$= \frac{153.2 - 149.6}{153.2} \times 100$$
$$= 2.35\% \text{ less. Ans. (ii)}$$

SPECIFIC HEAT OF SUPERHEATED STEAM

If the specific heat of superheated steam is given in a problem, it infers that the sensible heat of superheat is to be calculated instead of reading the total heat per lb of the superheated steam direct from the superheat tables, thus,

Sensible heat required to superheat one lb of dry saturated steam
$$= C_P \times (t - t_F)$$

where C_P = mean spec. ht. of the superheated steam,
t = temperature of the superheated steam,
t_F = saturation temperature at the same pressure.

and therefore, Total Heat content of one lb of superheated steam, measured above water at 32°F =

$$\mathbf{h} = \mathbf{h}_g + \mathbf{C}_P (\mathbf{t} - \mathbf{t}_F)$$

$(t - t_F)$ being the degree of superheat.

Example. Find the heat required to raise one lb of feed water at 190°F to steam at a pressure of 225 lb/in² abs. with 200 F degrees of superheat, taking the mean specific heat of superheated steam over this range as 0.57.

From steam tables, $P = 225$ lb/in², $h = 1201$ Btu/lb

To produce 1 lb of dry saturated steam:

heat required $= h_g -$ (feed temp. $-$ 32)
$= 1201 - (190 - 32) = 1043$ Btu.

Additional heat to superheat the steam:

$$\text{sensible heat} = \text{wt.} \times \text{spec. ht.} \times \text{temp. rise}$$
$$= 1 \times 0.57 \times 200 = 114 \text{ Btu.}$$

∴ Total heat required
$$= 1043 + 114$$
$$= 1157 \text{ Btu/lb. Ans.}$$

Example. Determine the mean specific heat of superheated steam at a pressure of 200 lb/in² abs. when the temperature of the steam is (i) 421·8°F, and (ii) 661·8°F.

(i) When steam temperature = 421·8°F:

From steam tables, P = 200 lb/in², sat. temp. = 381·8°F

∴ degree of superheat $= (t - t_F)$

$$= 421 \cdot 8 - 381 \cdot 8 = 40 \text{ F degrees}$$

From superheat tables,
P = 200 lb/in² superheat 40°, $h_g = 1199 \cdot 5$, $h = 1224$ Btu/lb

$$h = h_g + C_P (t - t_F)$$

$$\therefore C_P = \frac{h - h_g}{t - t_F}$$

$$= \frac{1224 - 1199 \cdot 5}{40} = 0 \cdot 6125 \text{ Ans. (i)}$$

(ii) When steam temperature = 661·8°F:
Degree of superheat = 661·8 − 381·8 = 280 F degrees

From superheat tables, P = 200 lb/in² supht. 280°,
$h = 1353 \cdot 7$ Btu/lb

$$\therefore C_P = \frac{1353 \cdot 7 - 1199 \cdot 5}{280} = 0 \cdot 5507 \text{ Ans. (ii)}$$

INTERPOLATION

Cases may arise when the total heat per lb of superheated steam is required at a pressure or temperature not included in the abridged steam tables and the mean specific heat of superheated steam is not given. An estimate can be made by taking proportionate values from the nearest quoted pressures or temperatures above and below the given quantity. The following examples demonstrate the usual methods.

Example. Find the total heat per lb of steam at a pressure of 100 lb/in² abs. with 85 F degrees of superheat.

From superheat tables,

h @ 100 lb/in² with 100° supht = 1241·8
h @ 100 lb/in² with 80° supht = 1231·4

Difference for 20° supht = 10·4 Btu

Difference for 5° = $\frac{5}{20}$ × 10·4 = 2·6

∴ h @ 100 lb/in² with 85° supht
= 2·6 Btu more than for 80°.
∴ h = 1231·4 + 2·6 = 1234 Btu/lb. Ans.

Note that the mean specific heat at the range of 100 lb/in², 80 to 100° of superheat is:

$$\frac{\text{Difference in heat/lb}}{\text{Difference in temp.}} = \frac{10\cdot 4}{20} = 0\cdot 52$$

This could be used to obtain the additional heat required to give the extra 5° above 80° of superheat:

Spec. ht. × temp. rise = 0·52 × 5 = 2·6 Btu

Example. Find the total heat per lb of steam at 220 lb/in² abs. when it is superheated 120 F degrees.

From superheat tables,
h @ 250 lb/in^2 with 120° supht = 1274·9
h @ 200 lb/in^2 with 120° supht = 1269·3

Difference for 50 lb/in^2 pressure = 5·6 Btu
Difference for 20 lb/in^2 = $\frac{20}{50}$ × 5·6 = 2·24
h @ 220 lb/in^2 with 120° supht = 2·24 Btu more than for 200 lb/in^2
= 1269·3 + 2·24 = 1271·54 Btu

As total heat is usually expressed to the first decimal place,
h = 1271·5 Btu. Ans.

Example. Find the total heat per lb of steam at a pressure of 325 lb/in^2 abs. with 260 F degrees of superheat.

From superheat tables,
h @ 400 lb/in^2 with 280° supht = 1375·9
h @ 300 lb/in^2 with 280° supht = 1366·5

Difference for 100 lb/in^2 pressure = 9·4 Btu

Difference for 25 lb/in^2 = $\frac{25}{100}$ × 9·4 = 2·35
h @ 325 lb/in^2 with 280° supht = 1366·5 + 2·35
= 1368·85 Btu ... (i)

h @ 400 lb/in^2 with 240° supht = 1353·9
h @ 300 lb/in^2 with 240° supht = 1345·1

Difference for 100 lb/in^2 pressure = 8·8 Btu
Difference for 25 lb/in^2 = $\frac{25}{100}$ × 8·8 = 2·2
h @ 325 lb/in^2 with 240° supht = 1345·1 + 2·2
= 1347·3 Btu ... (ii)

From (i) and (ii):
h @ 325 lb/in^2 with 280° supht = 1368·85
h @ 325 lb/in^2 with 240° supht = 1347·3

Difference for 40° supht = 21·55 Btu
Difference for 20° supht = 10·775

∴ h @ 325 lb/in^2 with 260° supht = 1347·3 + 10·775
= 1358·075
h = 1358·1 Btu/lb. Ans.

STEAM TABLES—Chu VERSION

Steam tables in Centigrade units give the saturation temperature (t_C) of the steam in degrees Celsius (°C) and the corresponding absolute temperature (T_K) in degrees Kelvin (°K).

Note the absolute temperatures recorded (T_K) are equal to ($t_C + 273 \cdot 15$), though the value ($t_C + 273$) is usually used except when reading direct from steam tables.

The sensible (h_f), latent (h_{fg}), and total heat (h_g) values are given in Centigrade heat units (Chu) measured above water at 0°C (corresponding to the level of 32°F).

At the present time, most heat calculations on steam are performed in Fahrenheit units but there is a definite tendency to change over to the Centigrade system.

One example is given here and some in the test examples at the end of the chapter.

Example. Calculate the total heat required to produce one lb of steam at a pressure of 250 lb/in² abs. from feed water at 105°C if the steam is (i) wet, having a dryness fraction of 0·95, (ii) dry and saturated, (iii) superheated 100 C degrees.

From first table of Chu version,

$P = 250$ lb/in², $\quad h_f = 209, \quad h_{fg} = 459 \cdot 2, \quad h_g = 668 \cdot 2$ Chu/lb

From superheat tables,
$P = 250$ lb/in², supht 100 C°, $\quad h = 726 \cdot 6$ Chu/lb

To produce wet steam:
  Heat supplied $= (h_f + qh_{fg})$ — feed temp.
  $= 209 + 0 \cdot 95 \times 459 \cdot 2 - 105$
  $= 540 \cdot 2$ Chu/lb. Ans. (i)

To produce dry steam:
  Heat supplied $= h_g$ — feed temp.
  $= 668 \cdot 2 - 105$
  $= 563 \cdot 2$ Chu/lb. Ans. (ii)

To produce superheated steam:
Heat supplied = h — feed temp.
     = 726·6 — 105
     = 621·6 Chu/lb. Ans. (iii)

HEAT MIXTURES

In Chapter 1, heat mixtures involving ice, liquids and metals were explained, and now further notes are given to include steam. The same principles apply, namely that unless otherwise stated, it is assumed that no heat is lost to, or received from, an outside source during the mixing process. Therefore the heat gained by the colder substance is equal to the heat lost by the hotter substance. Another very useful approach, especially when steam is involved, is to apply the principle that the sum of the total heats contained in the various substances before mixing is equal to the total heat in the resultant mixture.

In the example to follow, the two methods will be shown. One may appear clearer and more straightforward to the student than the other. Although either method can be used in any mixture problem, in general, the total heat method is more likely to be easier to visualise in more complicated problems to follow.

Example. 3 lb of wet steam at 205 lb/in² abs. and 0·95 dry, are blown into 100 lb of water at 40°F. Find the resultant temperature of the water.

From steam tables,
P = 205 lb/in², h_f = 357·8, h_{fg} = 842 Btu/lb

Method (i)

Heat lost by steam = Heat gained by water

When the steam begins to lose heat, it first condenses at its saturation temperature into water at the same temperature, the latent heat lost being qh_{fg}. The water then loses sensible heat as it falls in temperature to the final temperature of the mixture. If t represents the final temperature of the mixture, the heat lost by the steam is the difference between the initial heat contained in it and the sensible heat in the condensate (water) at $t°$.

STEAM

The heat gained by the cold water is sensible heat only since it rises in temperature from 40° to $t°$.

Heat lost by steam = Heat gained by water

$$3[0.95 \times 842 + 357.8 - (t - 32)] = 100(t - 40)$$
$$3569 - 3t = 100t - 4000$$
$$7569 = 103t$$
$$t = 73.48°F. \text{ Ans.}$$

Method (ii)

Heat in 3 lb steam before mixing = $3(h_f + qh_{fg})$
$= 3(357.8 + 0.95 \times 842) = 3 \times 1157.7$ Btu

Heat in 100 lb water before mixing = $100(40 - 32) = 100 \times 8$ Btu

After mixing there are $3 + 100 = 103$ lb of water at a final temperature of $t°F$, therefore, heat in this water = $103(t - 32)$ Btu

Total heat before mixing = Total heat after mixing
$$3 \times 1157.7 + 100 \times 8 = 103(t - 32)$$
$$3473 + 800 = 103t - 3296$$
$$t = 73.48°F \text{ (as before)}$$

Example. Steam is tapped from an intermediate stage of an engine at a pressure of 36 lb/in² abs. when its dryness fraction is 0·94, and passed to a contact feed heater to heat the feed water. If the hotwell temperature is 115°F, find the temperature of the boiler feed when the amount tapped off is 9% of the steam supplied to the engine.

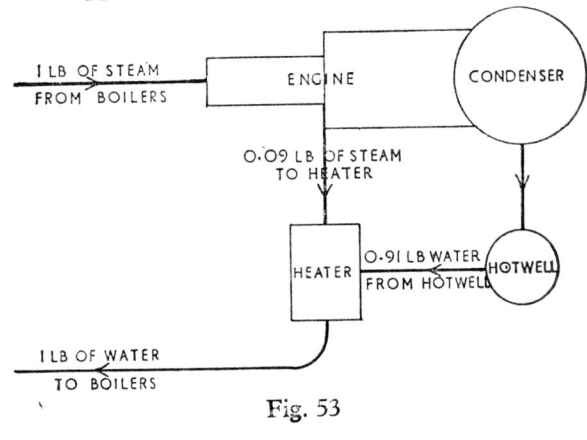

Fig. 53

Let one lb of steam be supplied from boilers to engine, then 0·09 lb is tapped off to the heater. This leaves (1 — 0·09) = 0·91 lb of steam to continue through the engine, into the condenser and hotwell, and pumped as water into the feed heater. Being a *contact* heater, the 0·09 lb of heating steam makes contact and mixes with the 0·91 lb of condensate, making one lb of feed to be pumped back into the boilers. This is illustrated in Fig. 53.

From steam tables,
P = 36 lb/in², h_f = 229·7, h_{fg} = 938·5 Btu/lb

Let t = temperature of feed in °F
Total heat entering heater = Total heat leaving
Heat in steam + Heat in condensate
= Heat in feed
0·09[229·7 + 0·94 × 938·5] + 0·91 (115 — 32)
= 1 × (t — 32)
100·07 + 75·53 = t — 32
t = 207·6°F. Ans.

Example. Steam is bled from the main steam pipe line at a pressure of 250 lb/in² abs. and dryness fraction 0·98, to a surface feed heater, and the remainder passes through the engine. The condensate at 105°F from the engine condenser and the drain from the feed heater passes into the hotwell. Calculate the percentage of main steam to bleed off to the heater so that the feed temperature to the boilers will be 220°F.

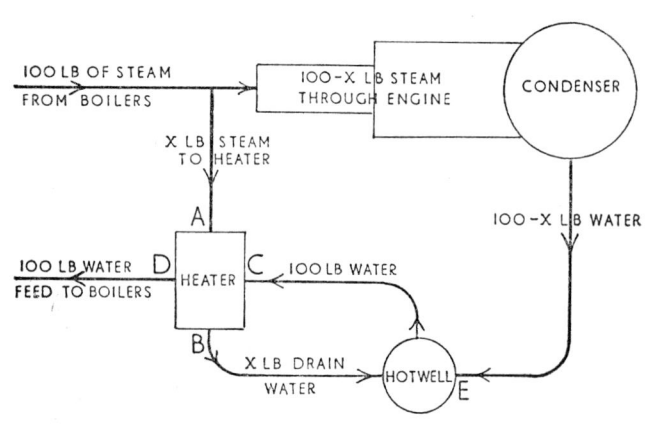

Fig. 54

Let 100 lb of steam be supplied from boilers,
x lb of steam bled off to heater,
$(100 - x)$ lb of steam passed through engine.

Considering diagrammatic sketch of the circuit in Fig. 54. To avoid lengthy expressions:

Let A = heat in x lb steam to heater,
 B = heat in x lb drain water from heater to hotwell,
 C = heat in 100 lb feed water from hotwell to heater,
 D = heat in 100 lb feed water leaving heater,
 E = heat in $(100 - x)$ lb water entering hotwell from condenser.

Total heat entering hotwell = Total heat leaving hotwell
$$B + E = C \quad \ldots \quad \ldots \quad \ldots \quad (i)$$
Total heat entering heater = Total heat leaving heater
$$A + C = B + D \ldots \quad \ldots \quad \ldots \quad (ii)$$

Substituting value of C in (i) into (ii):
$$A + B + E = B + D$$
$$\therefore A + E = D$$

This means that, since no heat is lost, the heat in the steam to the heater plus the heat in the condensed water from the engine, is equal to the heat in the feed water leaving the heater. The two pipes connecting heater and hotwell can therefore be disregarded, and heater and hotwell considered as one combined unit.

From steam tables,
$P = 250$ lb/in^2 $h_f = 376 \cdot 1$ $h_{fg} = 826$ Btu/lb

Total heat entering heater and hotwell = Total heat leaving heater
$$x(376 \cdot 1 + 0 \cdot 98 \times 826) + (100 - x)(105 - 32)$$
$$= 100(220 - 32)$$
$$1185 \cdot 6x + 7300 - 73x = 18800$$
$$1112 \cdot 6x = 11500$$
$$x = 10 \cdot 34$$
$\therefore$ Percentage of main steam to heater = 10·34% Ans.

THROTTLING OF STEAM

When steam is reduced in pressure by wire-drawing or throttling, such as when it passes through a reducing valve, no external work is done by the steam. If no heat is converted into work or into any other form of energy, and no heat is received or rejected, then the total heat contained in each lb of steam remains the same. That is, the total heat per lb of steam at the reduced pressure is equal to the total heat per lb at the initial pressure.

Example. Steam is passed through a reducing valve and reduced in pressure from 250 lb/in^2 abs. to 120 lb/in^2 abs. Find the condition of the reduced pressure steam if the high pressure steam was (i) wet, having a dryness fraction of 0·95, (ii) dry saturated, (iii) superheated by 60 F degrees. Take the mean specific heat of superheated steam at 120 lb/in^2 as 0·55.

Firstly, note the total heat per lb in dry saturated steam at 120 lb/in^2, from the tables this is 1191·4 Btu. Therefore, if the total heat in the supply steam is less than this, the final steam must be wet and we will be required to find its dryness fraction. If the total heat is more than 1191·4 Btu the final steam must be superheated and we will be required to find its degree of superheat.

(i) When the high pressure steam is wet, dryness 0·95:

Total heat in supply steam
$= 376·1 + 0·95 \times 826 = 1160·8$ Btu/lb

This is equal to the total heat per lb of the reduced pressure steam. Since it is less than h_g for 120 lb/in^2 then it must be wet.

$$\therefore 312·5 + q \times 878·9 = 1160·8$$
$$q \times 878·9 = 848·3$$
$$q = 0·9654$$

The reduced pressure steam is wet
its dryness fraction being 0·9654. Ans. (i)

(ii) When the high pressure steam is dry saturated:
Total heat h_g @ 250 lb/in^2 = 1202·1 Btu
This is more than h_g for 120 lb/in^2, therefore the reduced pressure steam must be superheated.

STEAM

$$\text{Total heat} = h_g + C_P \times \text{degree of superheat}$$
$$\therefore 1202\cdot1 = 1191\cdot4 + 0\cdot55 \times t° \text{ supht}$$
$$10\cdot7 = 0\cdot55 \times t$$
$$t = 19\cdot5 \text{ F degrees of superheat. Ans. (ii)}$$

Note also that, as the saturation temperature at 120 lb/in² is 341·3°F, the temperature of the reduced pressure steam is 341·3 + 19·5 = 360·8°F.

(iii) When the high pressure steam is superheated 60°F:

From tables, $P = 250 \text{ lb/in}^2$, supht 60°, $h = 1239\cdot7 \text{ Btu/lb}$

The reduced pressure steam is obviously superheated.

$$\text{Total heat} = h_g + C_P \times \text{degree of superheat}$$
$$1239\cdot7 = 1191\cdot4 + 0\cdot55 \times t° \text{ supht.}$$
$$48\cdot3 = 0\cdot55 \times t$$
$$t = 87\cdot8 \text{ F degrees of superheat. Ans. (iii)}$$

Note also that the temperature of this steam is 341·3 + 87·8 = 429·1°F.

Examination of the above results show that the effect of passing steam through a reducing valve is to (a) reduce the pressure, (b) reduce the temperature, and (c) increase the quality of the steam, i.e. to either increase the dryness fraction or increase the degree of superheat.

THROTTLING CALORIMETER

The *throttling calorimeter* is an instrument for measuring the dryness fraction of steam. It consists of a calorimeter admitting steam through a very small orifice near the top and an exit open to atmosphere at the bottom. It is fitted with a thermometer pocket so that the temperature of the steam in the calorimeter can be observed, and a mercurial manometer to measure the pressure. The arrangement is illustrated diagrammatically in Fig. 55. In the steam pipe the tendency is for the drier steam to flow along the upper part of the pipe bore and the wetter (and heavier) steam along the lower half. An attempt is therefore made to obtain a reasonably true representative sample by fitting the sampling tube vertically in the pipe, the perforations facing the steam flow.

The principle of operation depends upon the fact that if steam is reduced in pressure by a throttling process without doing work, the heat per lb after throttling is equal to the heat per lb before throttling, as previously explained.

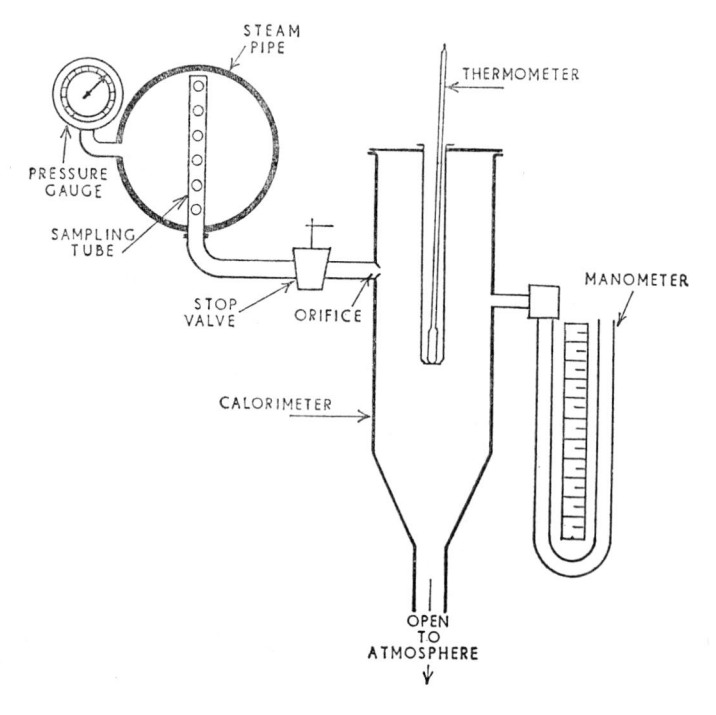

Fig. 55

Example. Steam at a pressure of 215 lb/in² abs. from the main steam pipe is passed through a throttling calorimeter. The pressure in the calorimeter is 18 lb/in² abs. and the temperature is 247°F. Taking the specific heat of low pressure superheated steam as 0·5, calculate the dryness fraction of the main steam.

From the steam tables we see that the saturation temperature of steam at 18 lb/in² is 222·4°F, therefore the steam in the calorimeter is superheated by (247 — 222·4) = 24·6 F degrees.

Heat/lb of steam before throttling = Heat/lb after throttling
$$(h_f + qh_{fg}) @ 215 \text{ lb/in}^2 = (h_g + \text{supht}) @ 18 \text{ lb/in}^2$$
$$362 \cdot 1 + q \times 838 \cdot 3 = 1154 \cdot 6 + 0 \cdot 5 \times 24 \cdot 6$$
$$q \times 838 \cdot 3 = 804 \cdot 8$$
$$q = 0 \cdot 96 \text{ Ans.}$$

LIMITATIONS OF THE THROTTLING CALORIMETER

It will be noted that the throttling calorimeter has its limitations for measuring the dryness fraction. The dryness fraction can only be determined by it when the temperature of the throttled steam in the calorimeter is higher than its saturation temperature corresponding to its pressure; that is, when it is superheated. If the throttled steam was not superheated, the thermometer would record the saturation temperature whether it was dry or wet and its condition would not be known.

For instance, referring to the previous example of main steam at 215 lb/in², if the throttled steam at 18 lb/in² had no superheat but was just dry and saturated, the equation would be:

$$362 \cdot 1 + q \times 838 \cdot 3 = 1154 \cdot 6$$
$$q \times 838 \cdot 3 = 792 \cdot 5$$
$$q = 0 \cdot 945$$

This would be the limiting dryness of the main steam, below which the steam in the calorimeter would still be wet and therefore the dryness fraction would be indeterminable. Thus, if the throttled steam at 18 lb/in² was wet, then the equation would become
$$362 \cdot 1 + q \times 838 \cdot 3 = 190 \cdot 6 + q_1 \times 964$$

Here there are two unknown quantities and therefore the equation cannot be solved.

SEPARATING CALORIMETER

When very wet steam samples are to be tested, an approximate value for the dryness fraction may be determined by passing the steam through a *separating calorimeter*.

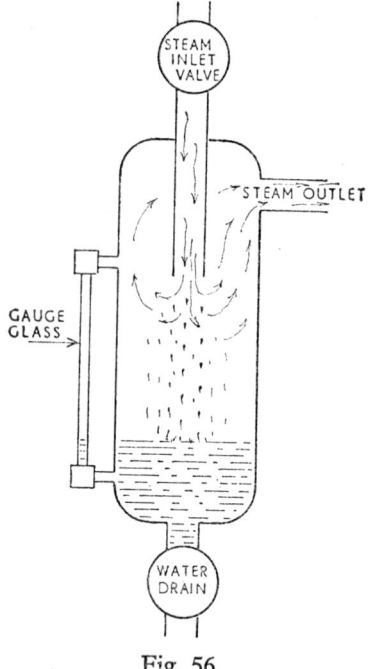

Fig. 56

This consists of a vessel with an inlet at the top leading into an open-ended internal tube, an outlet near the top, and a drain valve at the bottom. When steam is passed through it changes its direction of flow from the internal tube to the outlet which causes the heavier water particles to be thrown out of suspension by centrifugal action. The separated water collects in the bottom of the separator which is later drained off and weighed, let this be w. The remaining steam passes out through the outlet and is led to a small condenser where it is condensed into water, collected and weighed, let this be W. Thus:

W = weight of steam (assumed dry),
w = weight of water separated,
then,

$$\text{dryness fraction } q = \frac{W}{W + w}$$

However, the dryness fraction determined in this manner is only approximate because, although the steam leaving the separator is assumed to be dry saturated, it will actually be slightly wet since perfect separation is not achieved.

A more accurate value for the dryness fraction may be determined by passing the wet steam sample first through a separating calorimeter and then through a throttling calorimeter by connecting the two in series. Such an arrangement is called a:

ƒCOMBINED SEPARATING AND THROTTLING CALORIMETER

In this arrangement, the outlet from the separator is connected directly to the inlet of the throttling calorimeter, and the condenser is connected to the exit of the throttling calorimeter. When the sample of wet steam passes through the separator, some (but not all) of the water is separated out of the steam and this water is collected in the bottom of the separator to be measured (w). The partially dried steam is now throttled through the orifice in the throttling calorimeter where its temperature and pressure is noted. It then passes through the exit of the throttling calorimeter and into the condenser where it is condensed into water and collected for measuring (W).

Using the throttling calorimeter as previously explained, the dryness fraction of the steam leaving the separator and entering the throttling calorimeter is calculated, let this be q_2. Then each lb of steam passed into the throttling calorimeter contains q_2 lb of dry steam, and in W lb there are $q_2 W$ lb of dry steam.

The total weight of the sample is $W + w$ and, since the dryness fraction of the sample is the weight of dry steam divided by the total weight of the wet steam sample, then,

Dryness fraction of sample =

$$q = \frac{q_2 W}{W + w}$$

Note that $\dfrac{W}{W + w}$ would be the apparent dryness fraction if the separating calorimeter only was used, if this is represented by q_1 then,

$$\mathbf{q = q_1 \times q_2}$$

In words this is:

True dryness fraction = $\begin{Bmatrix} \text{dryness fraction by} \\ \text{separating calorimeter} \\ \text{used alone} \end{Bmatrix} \times \begin{Bmatrix} \text{dryness fraction by} \\ \text{throttling calorimeter} \\ \text{used alone} \end{Bmatrix}$

Example. Steam at a pressure of 115 lb/in^2 abs. in a steam pipe was tested by passing a sample through a combined separating and throttling calorimeter. The weight of water collected in the separator was 0·5 lb and the weight of condensate collected from the condenser after throttling was 5 lb. The pressure of the steam in the throttling calorimeter was 17 lb/in^2 abs. and its temperature was 233·5°F. Taking the specific heat of the superheated throttled steam as 0·5, find the dryness fraction of the sample of steam from the steam pipe.

From steam tables,

P = 115 lb/in^2, h_f = 309·2, h_{fg} = 881·5 Btu/lb
P = 17 lb/in^2, t_F = 219·5°F, h_g = 1153·5 Btu/lb

Dryness fraction from separating calorimeter,

$$q_1 = \frac{W}{W+w} = \frac{5}{5\cdot 5}$$

Dryness fraction from throttling calorimeter:
Superheat of throttled steam = 233·5 − 219·5 = 14 F°
Heat before throttling = Heat after throttling
$309\cdot 2 + q_2 \times 881\cdot 5 = 1153\cdot 5 + 0\cdot 5 \times 14$
$q_2 \times 881\cdot 5 = 851\cdot 3$
$q_2 = 0\cdot 9658$

$$q = q_1 \times q_2 = \frac{5}{5\cdot 5} \times 0\cdot 9658$$

$$= 0\cdot 878 \text{ Ans.}$$

ƒMIXTURES OF STEAM AND AIR IN CONDENSERS

It was stated in Chapter 3 under Dalton's law of partial pressures that the pressure exerted in a vessel occupied by a mixture of gases, or a mixture of gases and vapours, is equal to the sum of the pressures that each would exert if it alone occupied the whole volume of the vessel. The pressure exerted by each gas is termed a partial pressure. Thus, in the case of steam condensers which contain a mixture of leakage air and steam:

$$\text{Total pressure of mixture} = \text{Partial pressure due to air} + \text{Partial pressure due to steam}$$

Therefore an estimate can be made on the amount of air leakage into steam condensers.

ƒ Example. The pressure in a steam condenser is 1·8 lb/in² abs. and the temperature is 112°F. If the total internal volume of the condenser is 300 cubic feet, estimate the weight of air in the condenser, taking R for air as 53·3 ft lb/lb/°F.

From the steam tables we see that a steam temperature of 112°F corresponds to a pressure of 1·35 lb/in² abs. As the total pressure in the condenser is 1·8 lb/in² abs. then the difference is due to the presence of air.

$$\text{Partial press. due to air} = \text{Total press.} - \text{Steam press.}$$
$$= 1\cdot 8 - 1\cdot 35 = 0\cdot 45 \text{ lb/in}^2$$

From $PV = wRT$, $\quad w = \dfrac{PV}{RT}$

where $P = 0\cdot 45 \times 144 \text{ lb/ft}^2$
$V = 300$ cubic feet
$R = 53\cdot 3$ ft lb/lb per °F
$T = 112 + 460 = 572°\text{F abs.}$

$$\therefore w = \frac{0\cdot 45 \times 144 \times 300}{53\cdot 3 \times 572}$$

$$= 0\cdot 6377 \text{ lb. Ans.}$$

*f*INTERNAL ENERGY OF STEAM

During the evaporation of water into steam at constant pressure, the volume of each lb changes from that of water v_f to the volume of steam v_g. External work must therefore be done during evaporation against the pressure P lb/ft² exerted on it.

External work done during evaporation
$$= P(v_g - v_f) \text{ ft lb} = P(v_g - v_f)/J \text{ heat units}$$
where J = 778 Btu/lb, or 1400 Chu/lb

v_f is comparatively very small, it can be neglected for all normal practical purposes, hence,

$$\text{Work done} = \frac{Pv_g}{J} \text{ heat units}$$

Since the above amount of heat is expended to provide the work for expansion, the true latent heat now possessed by the steam will be less than the latent heat supplied (h_{fg}), i.e. $h_{fg} - Pv_g/J$. This is referred to as the *internal latent heat* possessed by the steam. The total internal heat energy in the steam, measured above 32°F (0°C) is the sum of the internal latent heat and the sensible heat of the water.

$$\text{Internal Energy} = h_f + h_{fg} - \frac{Pv_g}{J}, \text{ or } h_g - \frac{Pv_g}{J}$$

If the steam is wet with a dryness fraction of q, the latent heat supplied is qh_{fg} and the increase in volume is qv_g then,

$$\text{Internal Energy of wet steam} = h_f + qh_{fg} - \frac{Pqv_g}{J}$$

and when the steam is superheated:

$$\text{Internal Energy of superheated steam} = h - \frac{Pv}{J}$$

Example. Calculate the internal energy in one lb of steam at a pressure of 200 lb/in² abs. if the steam is (i) wet, with a dryness fraction of 0·9, (ii) dry and saturated, (iii) superheated 200 F degrees.

From saturation and superheat steam tables,

$P = 200 \text{ lb/in}^2$, $h_f = 355.5$, $h_{fg} = 844$, $h_g = 1199.5$, $v_g = 2.29$
$P = 200 \text{ lb/in}^2$ supht $200°$, $h = 1312.1$

(i) Internal Energy $= h_f + qh_{fg} - Pqv_g/J$

$$= 355.5 + 0.9 \times 844 - \frac{200 \times 144 \times 0.9 \times 2.29}{778}$$

$$= 355.5 + 759.6 - 76.3$$
$$= 1038.8 \text{ Btu/lb. Ans. (i)}$$

(ii) Internal Energy $= h_g - Pv_g/J$

$$= 1199.5 - \frac{200 \times 144 \times 2.29}{778}$$

$$= 1199.5 - 84.8$$
$$= 1114.7 \text{ Btu/lb. Ans. (ii)}$$

(iii) Volume of superheated steam by Callendar's formula:

$$v = \frac{1.248 \, (h - 835.2)}{P} + 0.0123$$

$$= \frac{1.248 \, (1312.1 - 835.2)}{200} + 0.0123$$

$$= 2.988 \text{ ft}^3/\text{lb}$$

Internal Energy $= h - Pv/J$

$$= 1312.1 - \frac{200 \times 144 \times 2.988}{778}$$

$$= 1312.1 - 110.6$$
$$= 1201.5 \text{ Btu/lb. Ans. (iii)}$$

TEST EXAMPLES 8

1. Steam at a pressure of 135 lb/in² abs. is generated in an exhaust gas boiler from feed water at 180°F. If the dryness fraction of the steam is 0·96, find the heat supplied per lb of steam.

2. If wet saturated steam at a pressure of 115 lb/in² abs. requires 35·3 Btu per lb to completely dry it, what is the dryness fraction of the steam?

3. Wet saturated steam at a pressure of 245 lb/in² abs., dryness fraction 0·97, is produced from feed water at 180°F. Find (i) the heat required per lb, (ii) the percentage less heat required per lb of steam at the same pressure and quality if the feed temperature is increased to 230°F.

4. A turbo-generator is supplied with superheated steam at a pressure of 400 lb/in² abs. and 200 F degrees of superheat. The pressure of the exhaust steam from the turbine is 0·9 lb/in² abs. with a dryness fraction of 0·88. (i) Calculate the heat drop per lb of steam through the turbine. (ii) If the turbine uses 2000 lb of steam per hour, calculate the horse-power equivalent of the total heat drop.

5. Steam enters the superheaters of a boiler at a pressure of 300 lb/in² abs. and 0·98 dry, and leaves at the same pressure with 180 F degrees of superheat. Find (i) the heat supplied per lb of steam in the superheaters, and (ii) the volume per lb before and after the superheaters. Use Callendar's equation for the specific volume of superheated steam:

$$v = 1·248\,(h - 835·2)/P + 0·0123 \text{ ft}^3/\text{lb}$$
where h = total heat in Btu/lb
P = absolute pressure in lb/in².

6. Wet steam at a pressure of 250 lb/in² abs., 0·95 dry, is produced in a boiler from feed water at 205°F. If the temperature of the feed remains constant, find the percentage extra heat to produce (i) dry saturated steam, (ii) superheated steam with 200 F degrees of superheat, at the same pressure.

7. One lb of wet saturated steam at a pressure of 120 lb/in² abs., dryness fraction 0·94, is expanded until the pressure is 60 lb/in² abs. If expansion follows the law $PV^n = C$ where $n = 1·12$, find the dryness fraction of the steam at the lower pressure.

8. Two boilers of equal evaporative capacities generate steam at the same pressure of 200 lb/in² abs. to a common pipe line. One boiler produces steam with 80 F degrees of superheat, and the other produces wet steam. If the mixture is just dry and saturated, find the dryness fraction of the wet steam.

9. Eight cubic feet of steam at 115 lb/in² abs. and 0·95 dry is expanded to a pressure of 80 lb/in² abs. Calculate the dryness fraction of the expanded steam if the expansion follows the law (i) PV = constant, (ii) $PV^{1 \cdot 14}$ = constant.

10. Superheated steam at a pressure of 300 lb/in² abs. with 160 C degrees of superheat, is cooled at constant volume, and the pressure falls to 180 lb/in² abs. Find (i) the condition of the steam at the lower pressure, and (ii) the Chu extracted per lb of steam. Use Callendar's equation for the volume of superheated steam:

$$v = 2 \cdot 246 \, (h - 464)/P + 0 \cdot 0123 \text{ ft}^3/\text{lb}$$

where h = total heat in Chu/lb
P = absolute pressure in lb/in².

11. Five cubic feet of superheated steam at a pressure of 320 lb/in² abs. and superheated 250 F degrees, is expanded in an engine to a pressure of 1·5 lb/in² abs. when its dryness fraction is 0·9. Find the final volume of the steam. Assume that the volume of superheated steam varies as the absolute temperature.

12. 3 lb of wet steam at a pressure of 74 lb/in² abs. and dryness fraction 0·95, is blown into 15 gallons of water at 55°F. Find the final temperature of the mixture.

13. Exhaust steam from an engine passes into a condenser at a pressure of 2 lb/in² abs. and 12% wet. The temperature of the condensate from the condenser is 105°F. The circulating water enters the condenser at 53°F and leaves at 84°F. Find (i) the weight of circulating water per lb of steam condensed. If 500 tons of circulating water pass through the condenser per hour, find (ii) the horse-power equivalent of the heat pumped overboard. Take spec. ht. of the cooling water as 1.

14. In an experiment to determine the dryness fraction of steam, a sample at a pressure of 15 lb/in² abs. was blown into a vessel containing 20 lb of water at 60°F. The final weight of water in the vessel was 21·5 lb and the final temperature 133°F. Find the dryness fraction of the steam, taking the water equivalent of the vessel as 0·5 lb.

*f*15. Find the total heat per lb of steam at a pressure of 125 lb/in² abs. having 130 F degrees of superheat. If 2 lb of this steam is blown into 50 lb of water at a temperature of 57°F, find the resultant temperature of the mixture.

16. $2\frac{1}{2}$ lb of crushed ice at 20°F are dropped into 10 lb of water at 60°F contained in a vessel whose water equivalent is 2 lb, and one lb of dry saturated steam at atmospheric pressure (14·696 lb/in²) is blown into the mixture. Find the resultant temperature, taking the specific heat of ice as 0·5, and latent heat 144 Btu/lb.

17. Dry saturated steam is tapped off the inlet branch of an L.P. turbine at a pressure of 35 lb/in² abs. to supply heating steam in a contact feed heater. The temperature of the feed water inlet to the heater is 105°F and the outlet is 210°F. Find the percentage weight of steam taken from the engine.

18. 10% of the weight of steam passing through an engine is tapped off at an intermediate stage and led to the surface feed heater. The pressure of the steam to the heater is 44 lb/in² abs. and the drain from the heater passes as water to the hotwell. The temperature of the condensate from the condenser is 109°F and the temperature of the feed water to the boilers is 214°F. Calculate the dryness fraction of the steam supplied to the heater.

19. Wet saturated steam at 215 lb/in² abs. and dryness fraction 0·97 enters a reducing valve, and leaves at 115 lb/in² abs. Find the dryness fraction of the reduced pressure steam.

20. A throttling calorimeter was fitted to a pipe carrying steam at a pressure of 180 lb/in² abs. in order to measure the dryness fraction. The pressure in the calorimeter was 17 lb/in² abs. and its temperature was 241·5°F. Taking the specific heat of superheat of the steam in the calorimeter as 0·5, find the dryness fraction of the main steam.

*f*21. A combined separating and throttling calorimeter was connected to a main steam pipe carrying steam at 210 lb/in² abs. and the following data recorded:

Weight of water collected in separator = 1·1 lb
Weight of condensate after throttling = 20 lb
Press. of steam in throttling calorimeter = 16 lb/in² abs.
Temp. of steam in throttling calorimeter = 230·7°F.

Taking the specific heat of the throttled superheated steam as 0·5, find the dryness fraction of the steam taken from the pipe.

*f*22. The pressure and temperature in a steam condenser is 1·15 lb/in² abs. and 100°F respectively, and the dryness fraction of the exhaust steam is 0·86. Taking R for air as 53·3 ft lb/lb per °F, find the ratio of the weight of steam to weight of air present in the condenser.

*f*23. During the process of raising steam in a boiler, when the reading of the pressure gauge was 25 lb/in² the temperature inside the boiler was 128°C, and when the pressure gauge reading was 105 lb/in² the temperature was 170°C. If the volume of the steam space is constant at 300 ft³, calculate the weights of steam and air present in each case. Take R for air = 96 ft lb/lb per ° C, atmospheric pressure = 14·7 lb/in², and assume the steam is dry in each case.

CHAPTER 9

ƒ ENTROPY
(FIRST CLASS STUDENTS ONLY)

We have seen that mechanical energy, being the product of two quantities, can be represented as an area. Fig. 57 is a diagram in which the ordinates represent pressure in lb/ft^2 and the abscissae represent volume in ft^3, the area enclosed represents mechanical work in ft lb. The availability to do work depends on the magnitude of the pressure.

In a similar manner, heat energy can be represented as an area. The availability to transfer heat depends upon the temperature and therefore the ordinates of a heat energy diagram represent absolute temperature. The abscissae are termed *entropy* and denoted by the letter *s*. Thus a diagram whose area represents heat units, ordinates representing absolute temperature (T) and abscissae representing entropy (*s*), is referred to as a *temperature-entropy* (T–*s*) diagram, as shown in Fig. 58.

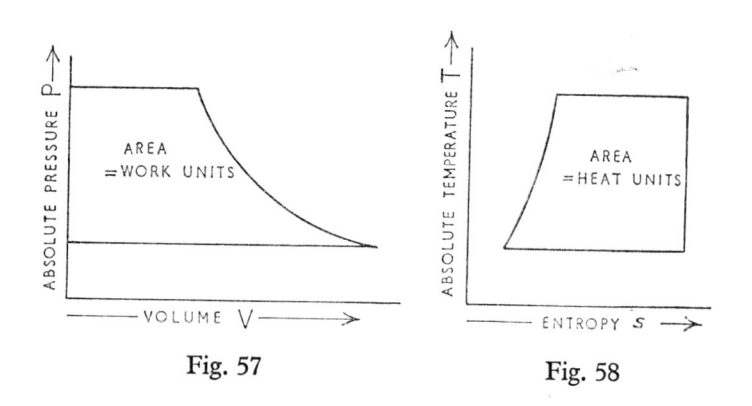

Fig. 57

Fig. 58

ENTROPY OF WATER

Consider the heating of one lb of water from a temperature of 32°F (i.e. 491·7°F abs. from steam tables). Assuming that the specific heat of water remains constant at unity during the heating process, then,

ENTROPY 183

if the water is heated through 50 F degrees, i.e., from 491·7°F abs. to 541·7°F abs.:

Heat added = Area of T–s diagram (Fig. 59) = 50 Btu.

The mean height of the diagram is the average absolute temperature = $\frac{1}{2}$ (491·7 + 541·7) = 516·7°F abs.

Width of diagram = area ÷ mean height
The width of the diagram is the increase in entropy,

$$\therefore s = 50 \div 516 \cdot 7 = 0 \cdot 09679$$

Now let the water be heated through a further 50 F degrees, i.e., from 541·7°F abs. to 591·7°F abs.

Heat added = area of T–s diagram (Fig. 60) = **50 Btu**

Mean height of diagram = $\frac{1}{2}$ (541·7 + 591·7)
= 566·7°F abs.
Width of diagram = increase in entropy
$$\therefore s = 50 \div 566 \cdot 7 = 0 \cdot 08824$$

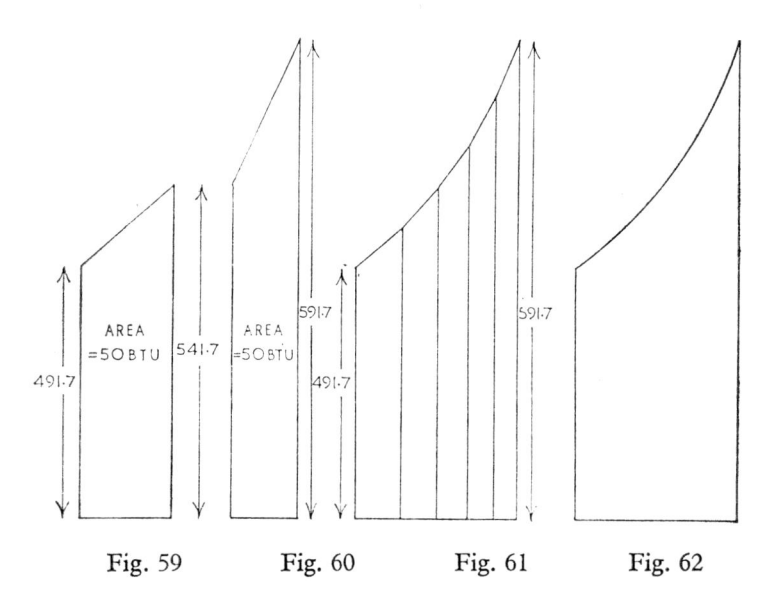

Fig. 59   Fig. 60   Fig. 61   Fig. 62

G

Thus, comparing Figs. 59 and 60, the area (50 Btu) is the same in each case, but, since their mean heights are different then their widths must be different. As the increase in entropy in Fig. 60 is less than in Fig. 59, then the slope at the top of Fig. 60 is steeper.

If the water was heated through the total 100 F degrees in five stages instead of two, there would be five boundary lines each of increasing slope. (Fig. 61).

If an infinite number of steps were taken, the boundary line would be a curve as shown in Fig. 62.

The expression to give the change in entropy of water when the temperature changes from T_1 to T_2, the specific heat of water being represented by k, is given by:

$$k \log_\varepsilon \frac{T_2}{T_1}$$

the proof of which is beyond the scope of the present work.

Since the total heat (or enthalpy) of water and steam is measured above the level of 32°F (0°C), then entropy is measured from the same level. Hence the total heat and entropy of water at 32°F are taken as zero.

It is also usual to take the specific heat of water as unity and the absolute temperatures corresponding to 32°F and 0°C as 492 and 273 respectively. Therefore the expressions to calculate the entropy in water (s_f) measured above 32°F or 0°C are:

$$s_f = \log_\varepsilon \frac{T_R}{492} \text{ where } T_R = \text{abs. temp. in °F} \quad \ldots \quad \text{(i)}$$

$$\text{or } s_f = \log_\varepsilon \frac{T_K}{273} \text{ where } T_K = \text{abs. temp. in °C}$$

Example. Calculate the entropy in water at a temperature of 228°F, measured above water at 32°F, and compare result with figure given in the steam tables.

From steam tables, when the temperature is 228°F, the absolute temperature is given as 687·7°F. Steam table values of absolute temperature will be used here but, if preferred, the slight approximation $t_F + 460$ may be taken. The difference in the final result is negligible.

$$s_f = \log_\varepsilon \frac{T_R}{492} = \log_\varepsilon \frac{687\cdot 7}{492}$$

For convenience of reading hyperbolic log tables, this may be taken as $\log_\varepsilon \dfrac{6\cdot 877}{4\cdot 92}$

From hyperbolic log tables:

$$\log_\varepsilon 6\cdot 877 = 1\cdot 9282$$
$$\log_\varepsilon 4\cdot 92 = 1\cdot 5933$$
$$\text{subtracting:} = 0\cdot 3349$$

∴ $s_f = 0\cdot 3349$. Ans. by calculation.

Reading from steam tables,

When $t = 228°F$, $s_f = 0\cdot 3358$ Ans. from tables.

The difference in the answers is mainly due to making the approximation that the specific heat of water is unity.

ENTROPY OF EVAPORATION

When water boils it changes from water to steam at constant temperature and the heat supplied to produce this change of state is the latent heat (h_{fg}) of the steam. Therefore, referring to Fig. 63,

Heat added = area of T–s diagram = latent heat.
Height of diagram is constant at the absolute temperature at which the water boils (T_R), hence,

Entropy of evaporation = area ÷ height

$$= \frac{h_{fg}}{T_R}$$

Note, however, that if the steam produced was wet with a dryness fraction of q, the latent heat supplied would be qh_{fg}. For instance, if the dryness fraction was 0·9 the latent heat absorbed during the evaporating process would be 0·9 of h_{fg} and the increase in entropy would be 0·9 of the entropy of complete evaporation for dry steam, the dryness fraction point being denoted by x in Fig. 63 where $Ax/AB = 0·9 = q$. The general expression for the entropy of evaporation is therefore:

$$s_{fg} = \frac{qh_{fg}}{T_R} \qquad \ldots \qquad \ldots \qquad \ldots \qquad (ii)$$

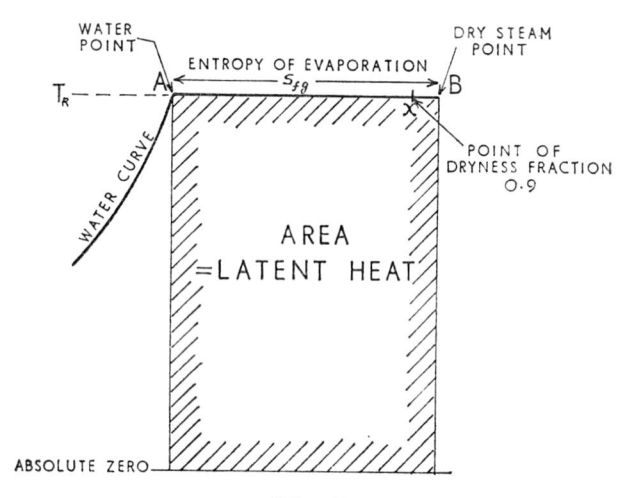

Fig. 63

Example. Calculate the increase in entropy when one lb of water is completely evaporated at a temperature of 228°F from water at 228°F.

From steam tables,
$t_F = 228°F,\qquad T_R = 687·7°F$ abs.,$\qquad h_{fg} = 960·4$ Btu/lb

$$s_{fg} = \frac{h_{fg}}{T_R} = \frac{960·4}{687·7} = 1·396 \text{ Ans.}$$

Note that in the steam tables, s_g is the entropy in steam measured above water at 32°F and is equal to the entropy in water above 32°F (s_f) plus the entropy of evaporation (s_{fg}).

ENTROPY

Therefore, taking entropy values from steam tables for a temperature of 228°F,

$s_{fg} = s_g - s_f = 1\cdot7327 - 0\cdot3358 = 1\cdot3969.$

ENTROPY OF SUPERHEATED STEAM

During the process of superheating steam, the absolute temperature is increasing and the entropy line of superheated steam is therefore similar to the water curve, but steeper, because the specific heat of superheated steam is much less than unity.

If C_P represents the specific heat of superheated steam, and the absolute temperature is raised from T_R to T, then increase in entropy is:

$$C_P \log_\varepsilon \frac{T}{T_R} \quad \ldots \quad \ldots \quad \ldots \quad \text{(iii)}$$

Hence, the total entropy in superheated steam at an absolute temperature of T, measured above water at 32°F is, from (i), (ii) and (iii):

$$s = \log_\varepsilon \frac{T_R}{492} + \frac{qh_{fg}}{T_R} + C_P \log_\varepsilon \frac{T}{T_R}$$

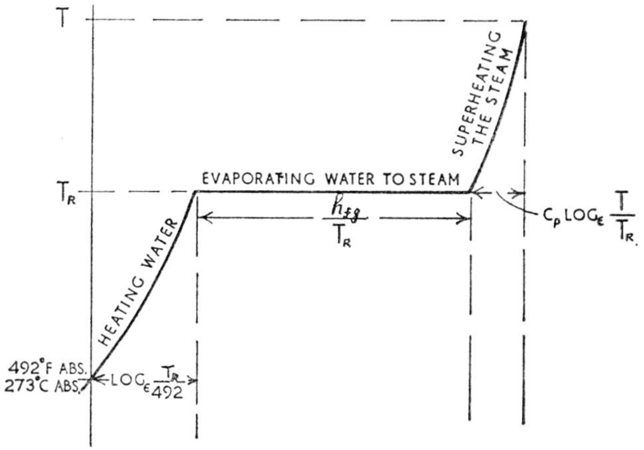

Fig. 64

Working in Chu, the absolute temperatures will be in °C abs. (T_K) and the figure 492 will be replaced by 273. See Fig. 64. The first two terms only in the general formula are required to calculate the entropy of saturated steam. If the steam is dry, the value of q is unity.

Example. Calculate the total entropy in one lb of superheated steam at a pressure of 250 lb/in² abs. with 200 F degrees of superheat, taking the mean specific heat of the superheated steam as 0·58, and compare result with the value of entropy given in the third table of the steam tables.

From first table of steam tables,

$P = 250$ lb/in², $T_R = 860·7$°F abs., $h_{fg} = 826$ Btu/lb
∴ $T = 860·7 + 200 = 1060·7$°F abs.

$$s = \log_\varepsilon \frac{T_R}{492} + \frac{h_{fg}}{T_R} + C_P \log_\varepsilon \frac{T}{T_R}$$

$$= \log_\varepsilon \frac{860·7}{492} + \frac{826}{860·7} + 0·58 \times \log_\varepsilon \frac{1060·7}{860·7}$$

$= 0·5593 + 0·9596 + 0·58 \times 0·209$
$= 0·5593 + 0·9596 + 0·1212$
$= 1·6401$ Ans. (i) by calculation.

Reading third table of steam tables,
$P = 250$ lb/in² 200F° supht., $s = 1·6485$ Ans. (ii)

DRY STEAM AND DRYNESS FRACTION CURVES

If the increase in entropy for complete evaporation of one lb of water for a series of temperatures were scaled off to the right of the water curve, and a line drawn through these plotted points, the 'dry steam curve' would be produced. Any point on this line represents dry steam conditions at the given level of temperature.

Dryness fraction curves are plotted by scaling off the values of qh_{fg}/T_R from the water curve. For instance, if 0·9 of the values of entropy of complete evaporation for a series of temperatures were scaled off from the water curve, the line drawn through these points

ENTROPY

is the curve for a dryness fraction of 0·9. Curves for dryness fractions of 0·8, 0·7, etc. down to 0·1 are obtained in a similar manner and appear as shown in Fig. 65.

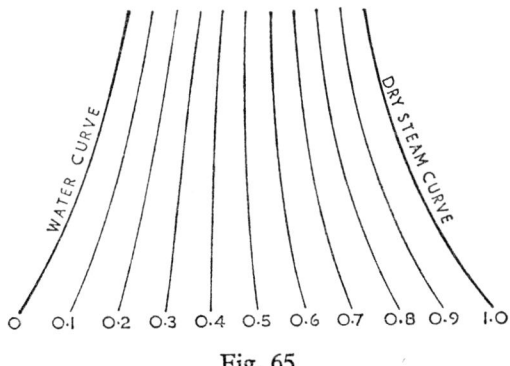

Fig. 65

ISOTHERMAL AND ADIABATIC PROCESSES

During an isothermal process the temperature remains constant, therefore an isothermal operation is represented on the T–s diagram by a straight horizontal line.

During an adiabatic process, no external heat is added or rejected, hence there is no increase or decrease in entropy and an adiabatic operation is therefore represented on the T–s diagram by a straight vertical line. Since there is no change in entropy, an adiabatic operation may be said to be 'isentropic', i.e., at constant entropy.

As an example, Fig. 66 shows the pressure-volume diagram of the Carnot cycle (explained in Chapter 6), and the corresponding temperature-entropy diagram is shown in Fig. 67.

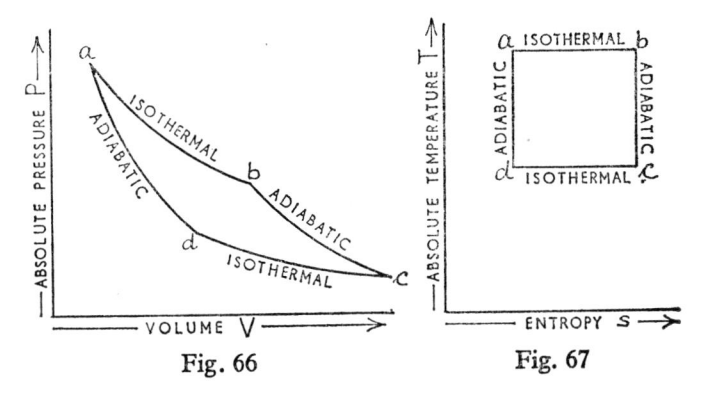

Fig. 66

Fig. 67

ADIABATIC EXPANSION OF STEAM

Since entropy remains constant during an adiabatic expansion, it is shown by a vertical line on the T–s diagram. Therefore, if initially dry steam expands adiabatically, the final condition will be wet because the vertical adiabatic line falls inside the dry steam curve. Steam of initially normal wetness becomes wetter when it expands adiabatically.

The final dryness fraction can be found by calculation making use of the formulae previously given, or taking values of entropy from the steam tables, or, if a T–s chart is at hand the final dryness can be found graphically by drawing the vertical adiabatic line and scaling off.

Example. Dry steam at a pressure of 100 lb/in² abs. expands adiabatically to a pressure of 3 lb/in² abs. Find the final dryness of the steam.

From steam tables,

$P = 100$ lb/in², $T_R = 787 \cdot 5°$F abs., $h_{fg} = 889 \cdot 7$ Btu/lb
$P = 3$ lb/in², $T_R = 601 \cdot 2°$F abs., $h_{fg} = 1013 \cdot 2$ Btu/lb

Method (i) by calculation using formulae:

As the expansion is isentropic,
Final entropy of steam = Initial entropy of steam

$$\log_\varepsilon \frac{T}{492} + \frac{qh_{fg}}{T_R} @ \ 3 \ \text{lb/in}^2 = \log_\varepsilon \frac{T_R}{492} + \frac{h_{fg}}{T_R} @ \ 100 \ \text{lb/in}^2$$

$$\log_\varepsilon \frac{601 \cdot 2}{492} + \frac{q \times 1013 \cdot 2}{601 \cdot 2} = \log_\varepsilon \frac{787 \cdot 5}{492} + \frac{889 \cdot 7}{787 \cdot 5}$$

$$0 \cdot 2004 + 1 \cdot 685q = 0 \cdot 4704 + 1 \cdot 129$$
$$1 \cdot 685q = 1 \cdot 399$$
$$q = 0 \cdot 8305 \ \text{Ans.}$$

Note, instead of working out separately,

$$\log_\varepsilon \frac{787 \cdot 5}{492} = 0 \cdot 4704 \quad \text{and} \quad \log_\varepsilon \frac{601 \cdot 2}{492} = 0 \cdot 2004$$

ENTROPY 191

which, when the latter term is brought over to the other side of the equation is to be subtracted, viz. $0.4704 - 0.2004 = 0.27$, the student should know that:

$$\log_e \frac{787.5}{492} - \log_e \frac{601.2}{492} \text{ is equal to } \log_e \frac{787.5}{601.2}$$

This gives the same result of 0·27 to this part and hence reduces the amount of arithmetic.

Method (ii) taking values of entropy from steam tables:

$$\text{Final entropy of steam} = \text{Initial entropy of steam}$$
$$0.2008 + q(1.8869 - 0.2008) = 1.6038$$
$$0.2008 + 1.6861q = 1.6038$$
$$1.6861q = 1.403$$
$$q = 0.832$$

Method (iii), graphically, by measurement on the T–s chart.

On the chart draw a vertical line downwards from the dry steam curve at the level for 100 lb/in² to the level corresponding to 3 lb/in². It will be seen that the final point lies between the dryness fraction curves of 0·8 and 0·9. By careful measurement the dryness can be estimated to the second decimal place to give the value of 0·83.

T–s charts are obtainable from most students' book shops and the student is advised to obtain a copy.

Example. Steam at a pressure of 400 lb/in² abs. and superheated 320 F degrees expands isentropically to a pressure of 10 lb/in² abs. Taking the specific heat of superheated steam as 0·6, calculate the dryness of the steam at the end of expansion.

From steam tables,
P = 400 lb/in², T_R = 904·3°F abs., h_{fg} = 781·3 Btu/lb
P = 10 lb/in², T_R = 652·9°F abs., h_{fg} = 982·5 Btu/lb

Method (i) employing the formulae,

Entropy of steam at 10 lb/in² = Entropy of steam at 400 lb/in²

$$\log_e \frac{T_R}{492} + \frac{qh_{fg}}{T_R} = \log_e \frac{T_R}{492} + \frac{h_{fg}}{T_R} + C_P \log_e \frac{T}{T_R}$$

$$\frac{q \times 982\cdot5}{652\cdot9} = \log_\varepsilon \frac{904\cdot3}{652\cdot9} + \frac{781\cdot3}{904\cdot3} + 0\cdot6 \log_\varepsilon \frac{1224\cdot3}{904\cdot3}$$

$$1\cdot505q = 0\cdot3257 + 0\cdot8642 + 0\cdot1818$$
$$1\cdot505q = 1\cdot3717$$
$$q = 0\cdot9116 \text{ Ans.}$$

Method (ii) taking values of entropy from steam tables I and III,

$$\text{Entropy at 10 lb/in}^2 = \text{Entropy at 400 lb/in}^2$$
$$0\cdot2836 + q\,(1\cdot7884 - 0\cdot2836) = 1\cdot6679$$
$$1\cdot5048q = 1\cdot3843$$
$$q = 0\cdot92$$

CONSTANT VOLUME LINES

Each constant volume curve on a T–s diagram represents one particular volume occupied by one lb of saturated steam under varying conditions from dry to very wet.

Taking a simple volume as an example, we read from the steam tables that 5 cubic feet is the volume occupied by one lb of dry saturated steam when its temperature is 318·7°F (159·4°C), its corresponding pressure being 88 lb/in² abs. The curve for a volume of 5 cubic feet therefore begins at the point on the dry steam curve at the level of 318·7°F of temperature.

Now taking some other temperature, say 281°F (138·3°C) which corresponds to a pressure of 50 lb/in², we note that the specific volume of dry saturated steam is given as 8·516 cubic feet. Hence, for one lb of steam at 281°F to occupy a volume of 5 cubic feet, it must be wet, and its dryness fraction will be $5 \div 8\cdot516 = 0\cdot5872$.

At another temperature, say 212°F (100°C), v_g is 26·8 ft³/lb. For one lb of steam at 212°F to occupy a volume of 5 cubic feet the dryness fraction will be $5 \div 26\cdot8 = 0\cdot1866$.

Plotting these three points on the T–s diagram and drawing a line through them, the constant volume curve for 5 cubic feet is produced as shown in Fig. 68.

ENTROPY

The above neglects the volume occupied by the water in the wet steam, and obviously more than three plotted points would be needed to obtain a true curve, but the principle is the same. Constant volume curves for volumes ranging from 0·5 to 500 cubic feet are usually given on T–s charts.

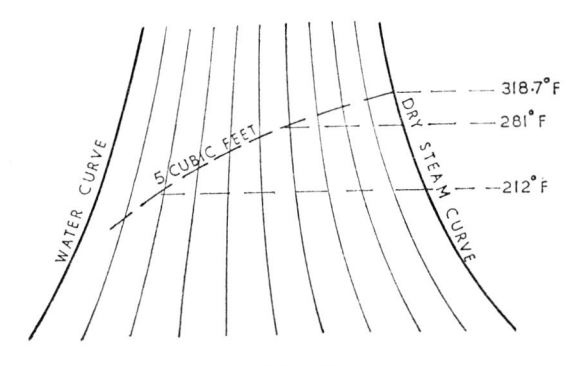

Fig. 68

ƒTEST EXAMPLES 9

Note. In this set of problems the values of entropy are to be calculated from the following formula when working in Fahrenheit degrees of temperature and Btu of heat. T_R is replaced by T_K and the figure 492 is replaced by 273 when working in Centigrade degrees and Chu. In all cases the student is advised to further his practice by comparing calculated results with those obtained by reading the entropy values from steam tables and by use of the T–s chart when applicable.

$$s = \log_e \frac{T_R}{492} + \frac{qh_{fg}}{T_R} + C_p \log_e \frac{T}{T_R}$$

1. Calculate the entropy of one lb of wet steam at a pressure of 250 lb/in² abs. measured above water at 32°F, when the dryness fraction of the steam is 0·95.

2. Calculate the entropy per lb of wet steam which has a temperature of 192°C and a dryness fraction of 0·9.

3. Find the entropy per lb of steam at a pressure of 180 lb/in² abs. which is superheated 200 F degrees, taking the mean specific heat of the superheated steam as 0·58.

4. An engine is supplied with dry saturated steam at 80 lb/in² abs. and exhausts at 3 lb/in² abs. If the expansion of the steam takes place at constant entropy, calculate the dryness fraction of the exhaust steam.

5. Superheated steam at a pressure of 250 lb/in² abs. with 100 F degrees of superheat is expanded adiabatically to a pressure of 25 lb/in² abs. Taking the specific heat of the superheated steam as 0·613, find the dryness fraction of the steam at the end of expansion.

6. Dry saturated steam at a pressure of 195 lb/in² abs. is throttled to a pressure of 100 lb/in² abs. Calculate the degree of superheat (Fahrenheit) in the reduced pressure steam and the gain in entropy. Assume the specific heat of the superheated steam to be 0·55.

CHAPTER 10

STEAM RECIPROCATING ENGINES

A steam engine is a machine which converts the heat energy in steam into mechanical energy. There are two distinct types of steam engine, namely the reciprocating steam engine and the steam turbine.

The general construction of a single cylinder steam reciprocating engine is shown in Fig. 69. Steam from the boiler is admitted through the engine stop valve into the valve chest which houses the slide valve, which may be a flat type, or of circular section like a bobbin and termed a piston valve. Ports in the valve chest connect to passages leading to the top and bottom of the cylinder and a central port communicates with the exhaust range. The function of the slide valve is to open and close the ports to steam and exhaust so that the piston is pushed up and down in the cylinder. The reciprocating motion of the piston is converted into a rotary motion at the crank shaft by connecting mechanism consisting of piston rod, connecting rod, and crank. The junction of the piston rod and connecting rod is known as the crosshead, to which is fitted a guide-shoe which slides on the face of the guide. The bottom end of the connecting rod carries a bearing which envelops the crank-pin carried between a pair of crank-webs shrunk and keyed on to the crank-shaft. The crank-shaft is supported in main bearings seated in the engine bedplate. The cylinder is supported by columns fixed to the bedplate. The slide valve is actuated by an eccentric-sheave and strap, eccentric-rod and slide valve rod, this mechanism gives the slide valve a reciprocating motion from the rotary motion of the eccentric-sheave which is keyed to the crank shaft.

The piston is fitted with piston rings to prevent steam leakage from one side of the piston to the other. The piston rod passes through a neck bush, stuffing box and gland in the bottom of the cylinder to prevent leakage of steam as the rod reciprocates through the bottom of the cylinder. Likewise with the valve rod through the bottom of the valve chest. In high pressure engines, specially designed metallic packing is fitted in the stuffing boxes. In low pressure engines, soft packing is inserted which is squeezed up by the gland until steam tightness is obtained.

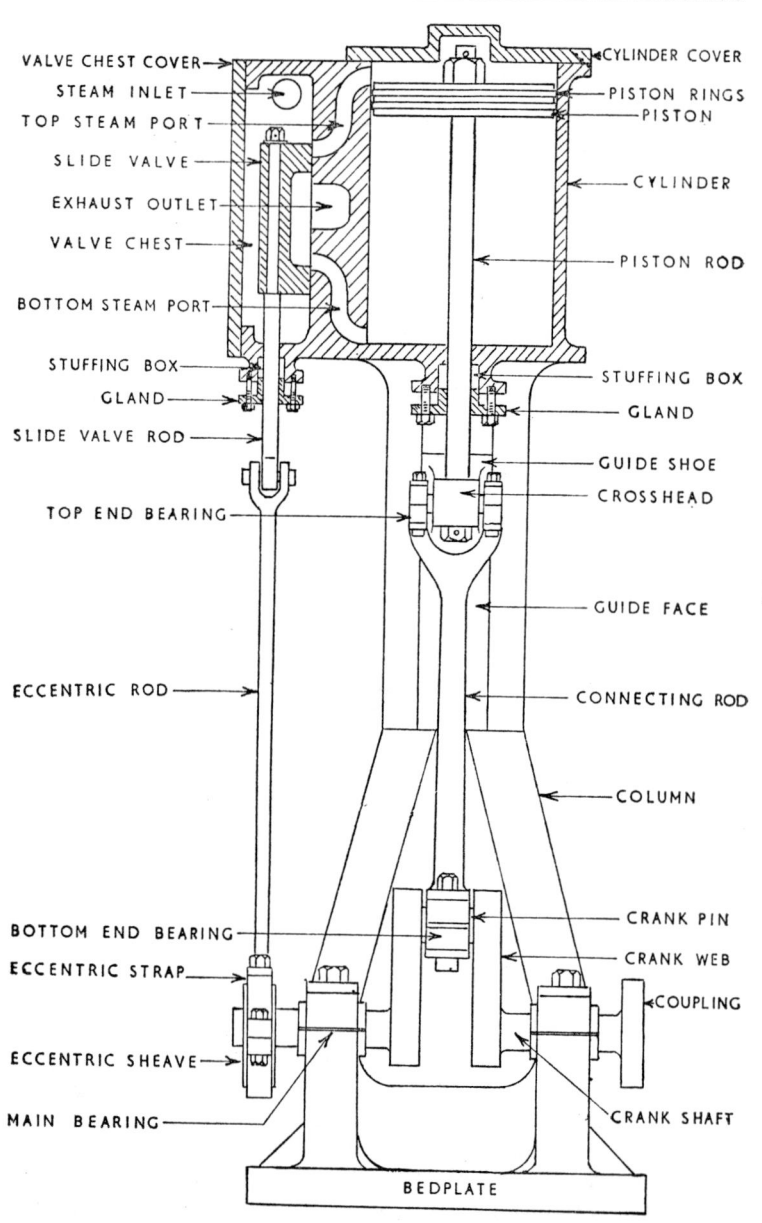

SINGLE CYLINDER STEAM ENGINE
Fig. 69

ACTION OF THE SLIDE VALVE

A simple slide valve is made of cast iron, rectangular in shape with a recess in its face and a hollow boss on the back through which is fitted the slide valve rod. A section through the axis of the valve resembles the letter D and is often referred to as the simple D slide valve. See Fig. 70.

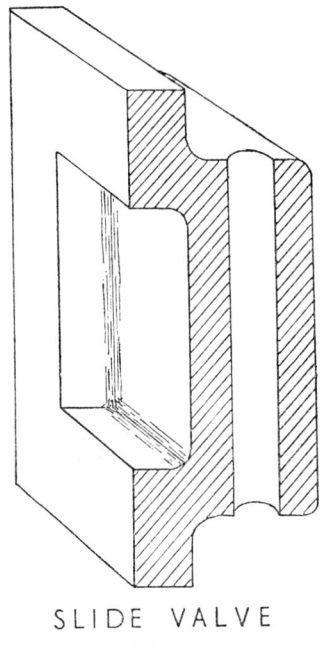

SLIDE VALVE
Fig. 70

Diagrammatic sketches of a simple slide valve showing its position relative to the piston and crank at various points in the cycle are given in Fig. 71 to which the following notes refer.

(i) Steam begins to be admitted to the top of the cylinder to start the piston on its outward stroke.

(ii) The port is full open to steam at the top of the cylinder, piston moving outward.

(iii) The steam supply to the top of the cylinder is now cut off by the slide valve, the piston having travelled about half stroke, and the piston continues to be pushed forward by the expansion of the steam as it falls in pressure.

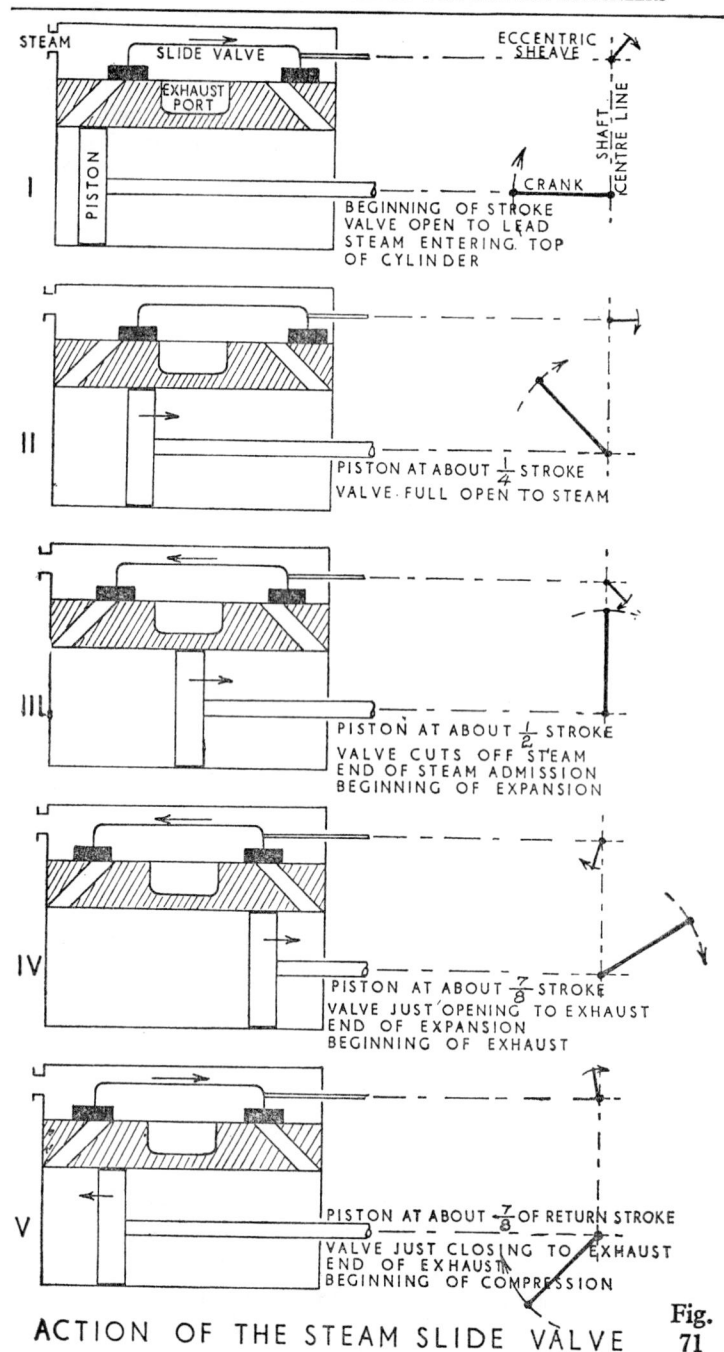

198 REED'S HEAT AND HEAT ENGINES FOR MARINE ENGINEERS

Fig. 71

ACTION OF THE STEAM SLIDE VALVE

(iv) The steam is now at a low pressure and beginning to be released from the cylinder into the exhaust range. It will now exhaust for about three-quarters of the return stroke.

(v) The valve now closes the port to stop exhaust, the remainder of the steam trapped in the cylinder is compressed as the piston completes its return stroke and acts as a cushion to bring the reciprocating parts quietly to rest at the end of the stroke.

A similar cycle of events comprising steam admission, expansion, exhaust and compression takes place on the bottom side of the piston to make the engine "double-acting".

In high pressure steam engines a piston valve is fitted instead of a flat slide valve. Its function and operation is the same but, being cylindrical, has not the disadvantage of a flat valve which is subjected to a heavy steam load pressing it against the valve face and necessitating considerable force to move it against friction. Another advantage is that steam can be admitted 'on the inside' of a piston valve, i.e. in the place of the exhaust of a slide valve, with exhaust over the outside edges; thus, by making the top of the valve slightly larger in diameter than the bottom, the extra upward steam thrust can compensate for the weight of the valve and gear to reduce wear on its bearings. Further, with steam admission on the inside, the valve rod gland is subjected only to the pressure of the exhaust steam and is therefore easier to keep steam tight.

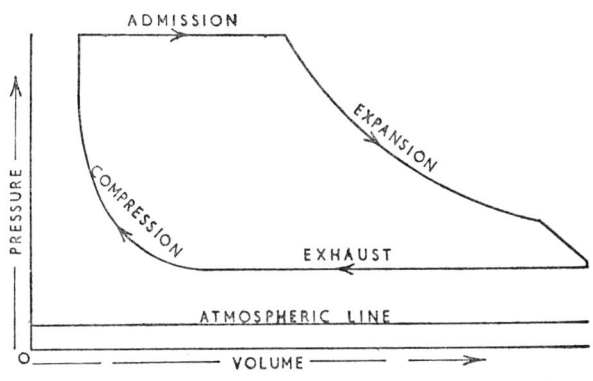

THEORETICAL INDICATOR DIAGRAM

Fig. 72

A PV diagram, showing the variation of steam pressure on one side of the piston as it moves the two strokes which constitute the working cycle, is shown in Fig. 72.

By means of indicator mechanisms, these diagrams are drawn to scale automatically on to cards when the engine is running. As all steam engines are double-acting, two diagrams, one from each end of the cylinder (top and bottom of the piston) are drawn on one card and appear as shown in Fig. 73.

PAIR OF INDICATOR DIAGRAMS

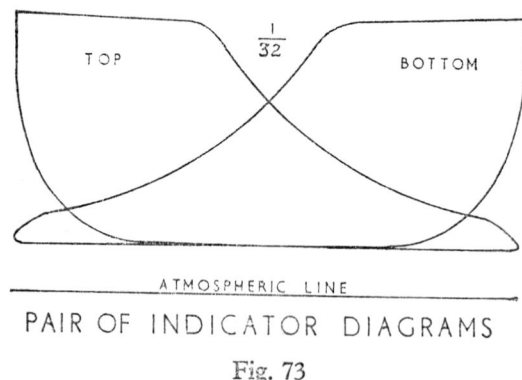

Fig. 73

The simplest type of slide valve is one in which the depths of the top and bottom bars of the valve are equal to the depths of the top and bottom steam ports in the valve chest face. In this case the valve is in mid-position (at half its travel) when the piston is at the beginning of its stroke. The eccentric sheave setting is therefore 90 degrees ahead of the crank in direction of rotation (90 degrees *behind* the crank for a piston valve with inside steam admission) so that the valve will open to steam and close again while the piston moves one complete stroke, and steam is admitted for the whole stroke of the piston. This is uneconomical as no use is made of the expansive properties of steam. One example of an engine with steam admission for the full stroke is the steam steering engine, where reliability of running and instant starting is more important than economy.

Except in such special cases, the steam is cut off when the piston has travelled a fraction of its stroke (as explained under 'action of the slide valve'). Work continues to be done by the steam after cut off while it expands from its initial supply pressure to a pressure near exhaust, and thus, although the work done per stroke is less, more work is obtained per lb of steam used. To enable this to be done, the bars of the slide valve are made deeper than the steam ports. When the valve is in mid-position, the bars of the valve will therefore overlap the steam ports and the amount of overlap on the steam edge is referred to as *steam lap*. Similarly, to cut off the exhaust from the cylinder before the end of the exhaust stroke, the valve is given *exhaust lap*.

With the crank on dead centre and the piston at the beginning of its stroke, the eccentric sheave setting must now be more than the normal 90 degrees ahead of the crank. It must be advanced sufficiently to move the valve through the distance equal to the steam lap, and a little further so that the port is actually open to steam to ensure a good head of steam in the cylinder to start the piston on its working stroke. This latter quantity is referred to as the *lead* of the valve and it necessarily follows therefore, that for a valve with lead, the steam admission will actually commence before the piston reaches its dead centre.

The following usual brief definitions should now be readily understood.

STEAM LAP is the amount that the steam edge of the valve overlaps the steam port when the valve is in mid-position. (See Fig. 74).

EXHAUST LAP is the amount that the exhaust edge of the valve overlaps the steam port when the valve is in mid-position. (See Fig. 74).

LEAD is the amount the port is open to steam when the piston is at the beginning of its stroke. (See Fig. 75).

ANGLE OF ADVANCE is the angle through which the eccentric sheave is advanced beyond its normal position of 90 degrees to the crank to allow for steam lap and lead.

THROW OF ECCENTRIC SHEAVE is the distance from the shaft centre to the geometrical centre of the sheave, and this is equal to half the travel of the slide valve. It will also be seen by reference to Fig. 77 that the full travel of the valve is equal to the difference between the thickest part and thinnest part of the sheave, i.e. $ab - cd$.

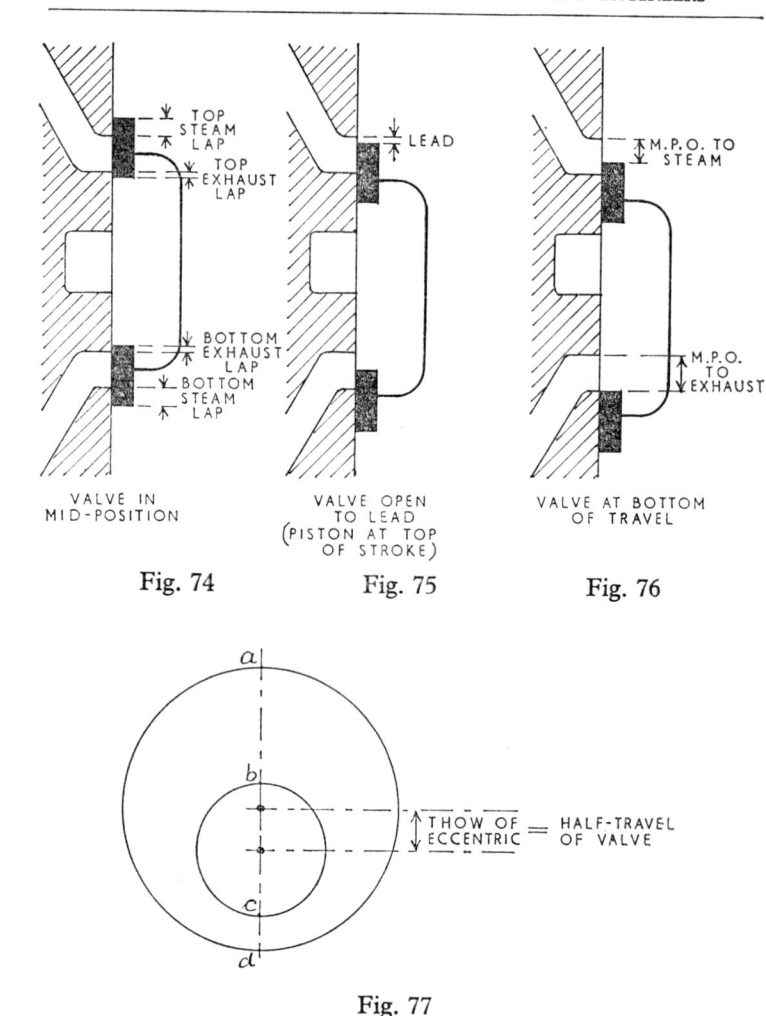

Fig. 74 — VALVE IN MID-POSITION (TOP STEAM LAP, TOP EXHAUST LAP, BOTTOM EXHAUST LAP, BOTTOM STEAM LAP)

Fig. 75 — VALVE OPEN TO LEAD (PISTON AT TOP OF STROKE) — LEAD

Fig. 76 — VALVE AT BOTTOM OF TRAVEL — M.P.O. TO STEAM, M.P.O. TO EXHAUST

Fig. 77 — THROW OF ECCENTRIC = HALF-TRAVEL OF VALVE

It will be noted that when we speak of the full movement of the piston from one end of the cylinder to the other, it is referred to as the *stroke* of the piston, whereas the full movement of the valve from one extreme end to the other is referred to as the *travel* of the slide valve.

When the valve is at the bottom of its travel, the port opening to steam is at its maximum. This amount is termed the *maximum port opening to steam* (M.P.O.) and is shown in Fig. 76. At this position the valve has moved down from its mid-position (mid-travel) a distance equal to the steam lap plus maximum port opening to steam, thus:

$$\text{Half travel} = \text{Steam lap} + \text{M.P.O. to steam}$$

The port in the valve chest is deeper than the maximum port opening required for the steam admission into the cylinder because, later in the cycle, the exhaust steam which has been expanded to a larger volume is returned through the same port into the exhaust range and the valve will move sufficiently to allow the port to be fully open during exhaust.

In vertical engines it is usual to allow for a greater maximum port opening to steam at the bottom, by having a smaller steam lap, than at the top. This allows more steam into the cylinder during the up-stroke of the piston to overcome the weight of the moving parts.

In some cases, the effect of the angle of advance of the eccentric is sufficient to cause cut-off to exhaust and attain the required amount of compression without the need of any exhaust lap. In fact, to avoid excessive compression the exhaust edge of the valve may be cut away so that when the valve is in mid-position the port is actually open a little to exhaust. Such an amount of exhaust opening is referred to as 'negative exhaust lap', or 'minus exhaust lap' or 'exhaust lead'.

Example. The following measurements were taken on a D-type slide valve:

Top steam lap = 2 in.,   Bottom steam lap = $1\frac{7}{8}$ in.
Top exhaust lap = $\frac{1}{4}$ in.,   Bottom exhaust lap = $\frac{1}{8}$ in.
Top lead = $\frac{1}{16}$ in.,   Bottom lead = $\frac{3}{16}$ in.

Calculate the port opening to exhaust when the piston is (a) at the top of its stroke, (b) at the bottom of its stroke.

Distance valve has moved down from its mid-position when piston is at the top of its stroke,

$$= \text{Top steam lap} + \text{Top lead}$$
$$= 2 + \tfrac{1}{16} = 2\tfrac{1}{16} \text{ in.}$$

(This is clearly seen by reference to Figs. 74 and 75).

The bottom of the valve has moved down the same distance, therefore the bottom port opening to exhaust when the piston is at the top of its stroke is this distance of $2\frac{1}{16}$ in. less the amount of bottom exhaust lap.

$$\text{P.O. to exhaust} = 2\tfrac{1}{16} - \tfrac{1}{8} = 1\tfrac{15}{16} \text{ in. Ans. (a)}$$

Similarly, when the piston is at the bottom of its stroke, the valve has moved $2\frac{1}{16}$ in. up from its mid-position. Note that when the crank is on *either* dead centre, the movement of the valve from its mid-position is the same, in other words:

(Steam lap + lead) @ the top = (steam lap + lead) @ the bottom

Therefore, the port opening to exhaust when the piston is at the bottom of its stroke,
$$= 2\tfrac{1}{16} - \tfrac{1}{4} = 1\tfrac{13}{16} \text{ in. Ans. (b)}$$

SIZE OF STEAM PORTS

As previously seen, the quantity of fluid flowing through a pipe is:

Volume (ft³/sec) = Area (ft²) × Velocity (ft/sec)

and, since the pressure of the steam should be constant at any part of the steam passage, the product of area and velocity should be constant. Therefore,

Area of steam pipe × steam velocity through pipe
= Area of steam port × steam velocity through port
= Area of cylinder × steam velocity through cylinder.

With regard to the steam velocity through the cylinder, this depends upon the speed of the piston and, since the piston speed varies throughout its stroke, the maximum piston speed should be taken in order to avoid wiredrawing of the steam.

For all practical purposes it can be taken that the maximum speed of the piston is equal to the linear speed of the crank and occurs when the crank and connecting rod are at right angles to each other.

STEAM RECIPROCATING ENGINES

If r = radius of crank pin circle (= ½ stroke)
then, maximum piston speed = speed of crank pin
$$= 2\pi \times r \times \text{r.p.m.}$$
and, mean piston speed $= 2 \times \text{stroke} \times \text{r.p.m.}$
$$= 2 \times 2r \times \text{r.p.m.}$$

$$\therefore \text{Ratio of } \frac{\text{max. piston speed}}{\text{mean piston speed}}$$

$$= \frac{2\pi \times r \times \text{r.p.m.}}{2 \times 2r \times \text{r.p.m.}} = \frac{\pi}{2}$$

That is,
$$\text{Maximum piston speed} = \frac{\pi}{2} \times \text{mean piston speed}$$

Example. The h.p. cylinder of a steam engine is 20 inches diameter, the stroke is 3 feet and the engine runs at 105 r.p.m. Assuming a steam velocity of 8000 ft/min and a steam port width of 15 inches, calculate (i) the port opening to steam, and (ii) the diameter of the steam pipe.

Max. piston speed = linear speed of crank pin
$$= 2\pi \times 1.5 \times 105$$
$$= 990 \text{ ft/min.}$$

$$\begin{array}{c}\text{Area of}\\ \text{cylinder}\end{array} \times \begin{array}{c}\text{maximum}\\ \text{piston speed}\end{array} = \begin{array}{c}\text{Area of}\\ \text{port}\end{array} \times \begin{array}{c}\text{Steam speed}\\ \text{through port}\end{array}$$

$$\frac{0.7854 \times 20^2}{144} \times 990 = \frac{15 \times \text{port opening}}{144} \times 8000$$

$$\text{Port opening} = \frac{0.7854 \times 20^2 \times 990}{15 \times 8000}$$

$$= 2.592 \text{ in. Ans. (i)}$$

$$\begin{array}{c}\text{Area of}\\ \text{port}\end{array} \times \begin{array}{c}\text{steam speed}\\ \text{through port}\end{array} = \begin{array}{c}\text{Area of}\\ \text{pipe}\end{array} \times \begin{array}{c}\text{Steam speed}\\ \text{through pipe}\end{array}$$

$$\frac{2 \cdot 592 \times 15}{144} \times 8000 = \frac{0 \cdot 7854 d^2}{144} \times 8000$$

$$d = \sqrt{\frac{2 \cdot 592 \times 15}{0 \cdot 7854}}$$

$$= 7 \cdot 036 \text{ in. Ans. (ii)}$$

VALVE DIAGRAMS

The construction of a valve diagram provides a means of solution, by calculation or measurement, of problems which involve the eccentric sheave setting relative to the crank.

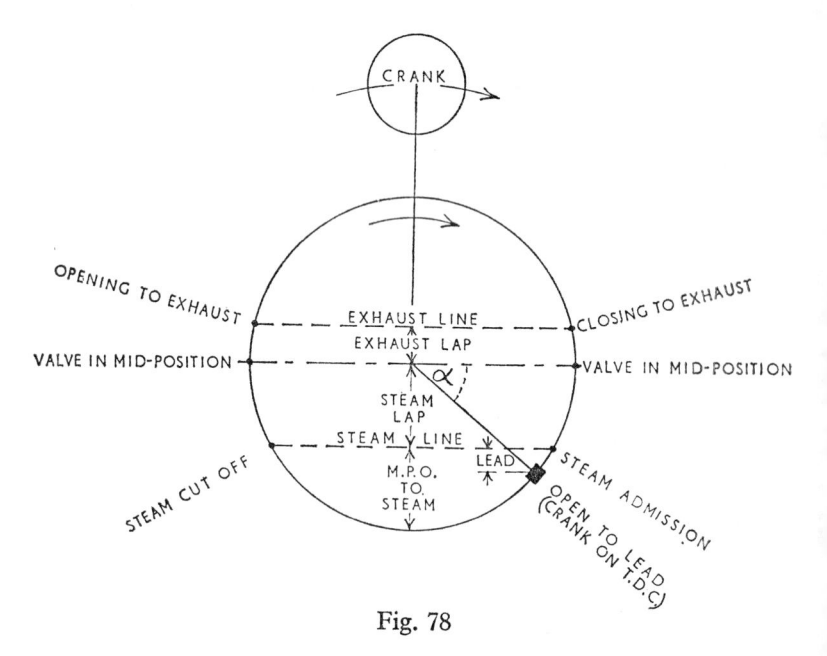

Fig. 78

A simple form of diagram is given in Fig. 78, the diameter of the circle represents the travel of the valve. It shows the position of the eccentric sheave when the crank is on top dead centre and the piston is at the beginning of its stroke, the valve is open to lead and corresponds with sketch I of Fig. 71. This diagram should be studied with reference to all the sketches of Fig. 71 which illustrates the relative positions of crank and eccentric for the cardinal points of the cycle.

Assuming clockwise rotation and the crank on T.D.C., for a valve with outside steam admission the position of the eccentric is 90 degrees ahead of the crank plus the angle through which it is advanced to allow for steam lap and lead. Thus, in Fig. 78, α is the angle of advance and, since the radius of the circle is equal to half-travel of the slide valve, it can be seen that:

$$\text{Sine of angle of advance} = \frac{\text{steam lap} + \text{lead}}{\text{half-travel}}$$

also,   **Half-travel = Steam lap + M.P.O. to steam**

For a piston valve with inside steam admission, the eccentric lags behind the crank by 90 degrees minus the angle of advance, hence the same valve diagram will serve if the crank be assumed to be on bottom dead centre instead of at the top.

Example. The throw of an eccentric sheave is 3 inches, the angle of advance is 35° 41′, and the steam lap of the valve is $1\frac{5}{8}$ inches. Calculate (i) the lead, (ii) maximum port opening to steam, (iii) crank angle from T.D.C. at cut off, (iv) crank angle before T.D.C. at admission, and (v) fraction of stroke completed by the piston when cut off takes place.

$$\sin \alpha = \frac{\text{steam lap} + \text{lead}}{\text{half-travel}}$$

$\therefore$ Lead = $\frac{1}{2}$ travel $\times$ sin α — steam lap
     = 3 $\times$ sin 35° 41′ — $1\frac{5}{8}$
     = $\frac{1}{8}$ inch. Ans. (i)

  $\frac{1}{2}$ travel = steam lap + M.P.O. to steam
$\therefore$ M.P.O. = 3 — $1\frac{5}{8}$
     = $1\frac{3}{8}$ inches. Ans. (ii)

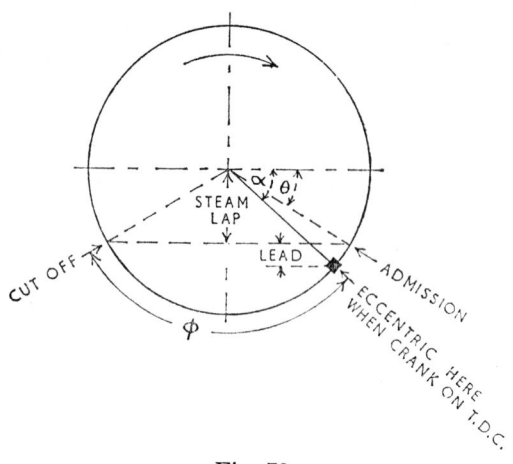

Referring to Fig. 79, since radius of circle represents half-travel,

$$\sin \theta = \frac{\text{steam lap}}{\frac{1}{2} \text{ travel}}$$

$$= \frac{1 \cdot 625}{3} = 0 \cdot 5417$$

$$\therefore \theta = 32° \ 48'$$

$$\phi = 180 - 35° \ 41' - 32° \ 48' = 111° \ 31'$$

∴ at cut off point, crank has moved
111° 31′ past T.D.C. Ans. (iii)

35° 41′ — 32° 48′ = 2° 53′

∴ at admission point, crank is
2° 53′ before T.D.C. Ans. (iv)

At cut off, angle of crank from bottom dead centre
= 180 — 111° 31′ = 68° 29′

Let stroke = 1, then radius of crank = 0·5

STEAM RECIPROCATING ENGINES

Neglecting angularity of connecting rod, distance of piston past mid-stroke when cut off occurs

$$= 0{\cdot}5 \times \cos 68° 29' = 0{\cdot}1834$$

∴ Fraction of whole stroke when cut off occurs

$$= 0{\cdot}5 + 0{\cdot}1834 = 0{\cdot}6834 \text{ Ans. (v)}$$

Example. The travel of a piston valve with outside steam admission is $7\frac{1}{2}$ inches, the steam lap is 2 inches and the lead is $\frac{1}{4}$ inch. Calculate (i) the angle of advance of the eccentric, (ii) the maximum port opening to steam, and (iii) the port opening to steam when the crank is 80 degrees past top dead centre.

$$\sin \alpha = \frac{\text{steam lap} + \text{lead}}{\frac{1}{2}\text{ travel}}$$

$$= \frac{2 + \frac{1}{4}}{3{\cdot}75} = 0{\cdot}6$$

∴ Angle of advance = 36° 52′ Ans. (i)

$$\text{Steam lap} + \text{M.P.O.} = \tfrac{1}{2} \text{ travel}$$
$$\therefore \text{M.P.O.} = 3{\cdot}75 - 2$$
$$= 1{\cdot}75 \text{ in. Ans. (ii)}$$

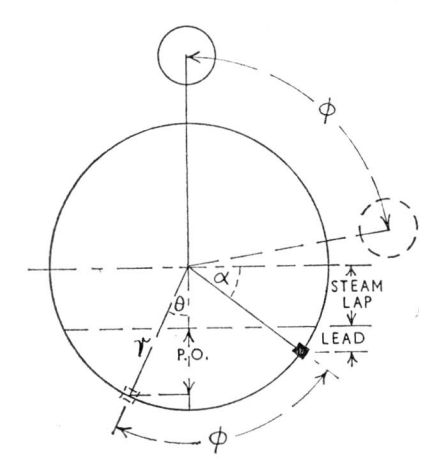

Fig. 80

When crank has moved $\phi°$ past T.D.C., eccentric has also moved $\phi°$ as shown in Fig. 80.

$$\theta = \phi + \alpha - 90$$
$$= 80 + 36° 52' - 90 = 26° 52'$$

Displacement of valve from mid-position $= r \cos \theta$

$\therefore$ Port opening $= r \cos \theta -$ steam lap
$= 3.75 \times \cos 26° 52' - 2$
$= 1.345$ in. Ans. (iii)

*f*Example. The steam lap of a slide valve is $2\frac{1}{4}$ inches, the lead is $\frac{1}{8}$ inch, and the port is one inch open to steam when the crank is 70° past T.D.C. Calculate the angle of advance of the eccentric and the travel of the valve.

Referring to Fig. 80,
$r =$ radius of diagram $=$ half-travel
$\alpha =$ angle of advance
$\theta = \phi + \alpha - 90$
$= 70 + \alpha - 90 = \alpha - 20$

$$\sin \alpha = \frac{\text{steam lap} + \text{lead}}{\frac{1}{2} \text{ travel}} = \frac{2.25 + 0.125}{r}$$

$$\therefore r = \frac{2.375}{\sin \alpha} \quad \ldots \quad \ldots \quad \ldots \quad \ldots \quad (i)$$

$r \cos \theta =$ steam lap $+$ port opening to steam
$\therefore r \cos \theta = 2.25 + 1 = 3.25$ in.

Expressing $\cos \theta$ in terms of $\sin \alpha$:
$\cos \theta = \sin (90 - \theta) = \sin [90 - (\alpha - 20)] = \sin (110 - \alpha)$

$\therefore r \sin (110 - \alpha) = 3.25$

$$r = \frac{3.25}{\sin (110 - \alpha)} \ldots \quad \ldots \quad (ii)$$

Equating r values from (i) and (ii),

$$\frac{2\cdot 375}{\sin \alpha} = \frac{3\cdot 25}{\sin (110-\alpha)}$$

$$\frac{3\cdot 25 \sin \alpha}{2\cdot 375} = \sin (110 - \alpha)$$

$$1\cdot 369 \sin \alpha = \sin (110 - \alpha)$$

Expanding $\sin (110 - \alpha)$ according to the identity:
$\sin (A - B) = \sin A \cos B - \cos A \sin B$
$\sin (110 - \alpha) = \sin 110 \cos \alpha - \cos 110 \sin \alpha$
$ = 0\cdot 9397 \cos \alpha + 0\cdot 342 \sin \alpha$

and, since $1\cdot 369 \sin \alpha = \sin (110 - \alpha)$
then, $1\cdot 369 \sin \alpha = 0\cdot 9397 \cos \alpha + 0\cdot 342 \sin \alpha$

Dividing throughout by $\cos \alpha$,
$1\cdot 369 \tan \alpha = 0\cdot 9397 + 0\cdot 342 \tan \alpha$
$1\cdot 027 \tan \alpha = 0\cdot 9397$
$\tan \alpha = 0\cdot 9149$
$\alpha = 42° 27'$ Ans. (a)

From (i), $r = \dfrac{2\cdot 375}{\sin 42° 27'} = 3\cdot 518$ in.

∴ Travel of valve $= 2 \times 3\cdot 518 = 7\cdot 036$ in. Ans. (b)

THE REULEAUX VALVE DIAGRAM

The Reuleaux valve diagram combines the valve travel, the stroke of the piston and the crank circle. The diameter of the diagram represents to scale the travel of the valve, and also the stroke of the piston to some other scale.

Referring to Fig. 81, assuming clockwise rotation of the crank and considering the top end of the cylinder, the diagram is constructed as follows.

With centre O and radius equal to the throw of the eccentric (half-travel of valve), draw the circle.

Insert the diameter TB. This represents the stroke of the piston, T being the top of the stroke (inner dead centre of the crank), and B the bottom of the stroke (outer dead centre of the crank).

With centre O and radius equal to the steam lap of the valve, describe the arc OL.

With centre T and radius equal to the lead, draw the lead circle.

Draw the line $S_1 S_2$ as a common tangent to the lead circle and the steam lap arc, this is referred to as the *steam line*.

Draw the centre-line XY through the centre O and parallel to the steam line.

Draw another centre-line MN through centre O and perpendicular to $S_1 S_2$ and XY.

Through the point l on the line ON, where the distance Ol is equal to the exhaust lap, draw a line $E_1 E_2$ parallel to $S_1 S_2$ and XY. This line $E_1 E_2$ is referred to as the *exhaust line*. If the valve has no exhaust lap, the exhaust line will coincide with XY. If the valve has negative exhaust lap, Ol is measured in the opposite direction (i.e. along OM).

Since OL is the steam lap, and OM is the half-travel, then LM is the maximum port opening to steam.

Since TZ (perpendicular to XY) is steam lap + lead, and TO is half-travel, then the angle TOZ which is marked α is the angle of advance of the eccentric.

Horizontally from T to B and back to T represents the two strokes of the piston during one complete cycle, therefore the circle represents the motion of the crank pin through one revolution. Following the crank movement in a clockwise direction:

When the crank is at S_1 the valve is just opening the port to steam. The angle of the crank before top centre is TOS_1 and this is usually referred to as the *angle of pre-admission*.

At T, the crank is on top centre, the piston is at the top of its stroke, and the valve is open to lead.

STEAM RECIPROCATING ENGINES 213

When the crank is at M the port opening to steam is maximum. Note that the valve is opening to steam while the crank is moving from S_1 to M, and closing to steam while the crank is moving from M to S_2. Thus S_1 to S_2 is the steam admission period and cut off takes place when the crank is at S_2.

While the crank moves from S_2 to E_1 the port is closed by the valve, this is the expansion period.

When the crank is at E_1 the valve is just opening the port to exhaust, this is the release point. The valve is opening to exhaust

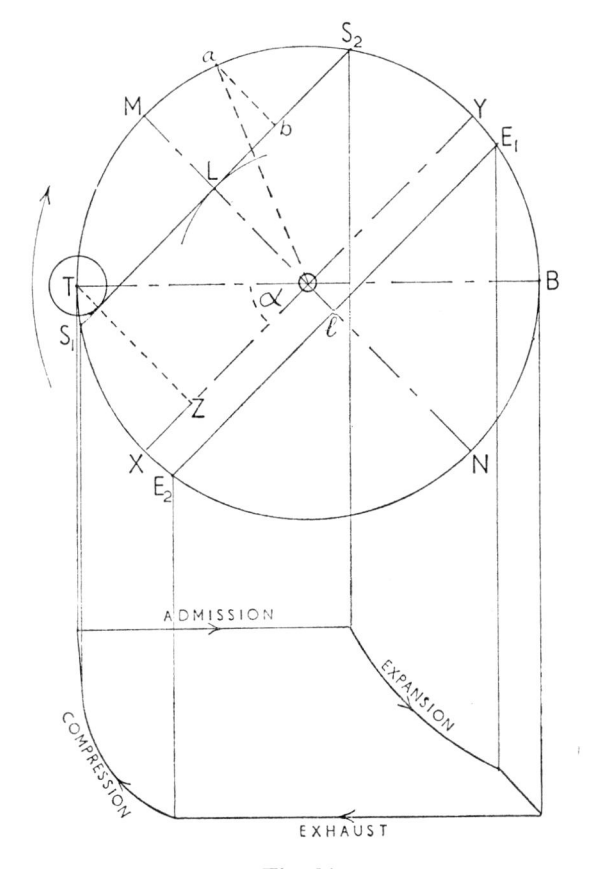

Fig. 81

while the crank moves from E_1, past bottom centre B, to position N where the exhaust opening is maximum, then it is closing as the crank moves from N to E_2.

At E_2 the valve closes to exhaust and compression begins. The compression period is from E_2 to S_1.

The projection of the theoretical indicator diagram from the valve diagram will now be readily understood.

The port opening to steam for any position of the crank is obtained by dropping a perpendicular from the crank pin position to the steam line S_1S_2, such as when the crank pin is at a, the port opening to steam is ab.

Similarly the port opening to exhaust for any position of the crank is obtained by erecting a perpendicular from the crank pin position to the exhaust line E_1E_2.

Neglecting the angularity or obliquity of the connecting rod, the fraction of the stroke completed by the piston is obtained by dropping a perpendicular from the crank pin position to the piston stroke line TB. Thus, cut off takes place when the crank is at S_2, the perpendicular from S_2 cuts TB at F in Fig. 82, therefore the fraction of the stroke when cut off occurs is represented by TF/TB.

The effect of the angularity of the connecting rod can be taken into account by striking an arc from the crank pin position to the piston stroke line TB, of radius equivalent to the ratio of connecting rod length to crank length on the valve circle scale. For instance, if the ratio connecting rod length divided by crank length is represented by n, then the radius of the arc $= R = n \times$ radius of valve circle. Thus, when cut off takes place, the piston has actually completed TG/TB of its stroke, as shown in Fig. 82. This means that, due to the connecting rod obliquity, cut-off occurs later on the down-stroke, since the effect is to pull the piston a distance of FG further down the stroke. For the cycle of events on the bottom side of the piston, the steam and exhaust laps are drawn on the opposite sides of the centre-line to those for the top side of the piston, it will therefore readily be seen that the effect of the angularity of the connecting rod is to cause cut off to occur earlier on the up-stroke.

STEAM RECIPROCATING ENGINES   215

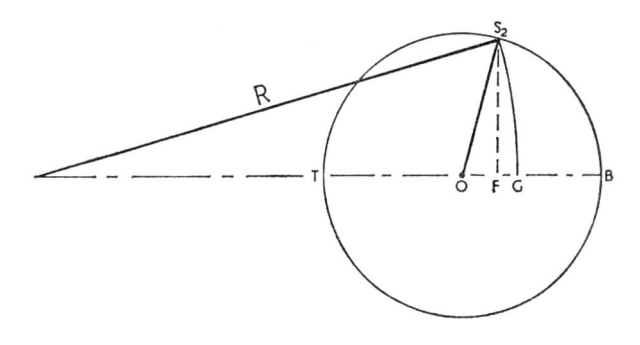

Fig. 82

WEIGHT OF STEAM FROM INDICATOR DIAGRAM

The weight of steam used by the engine can be estimated from the indicator diagrams. Referring to Fig. 83, *ad* is a line drawn parallel to the atmospheric line at such a height that it cuts both the compression curve and the expansion curve. The distance from the cylinder cover to point *c* represents to scale the volume of steam in the cylinder at point *c* on the expansion line, at the gauge pressure represented by the height $d\text{L}$. This is known as the volume of indicated steam. Similarly, the distance from the cylinder cover to point *b* represents the volume of steam remaining in the cylinder at point *b* on the compression curve, at the same pressure. This is known as the volume of cushion steam. Hence, the distance *bc* represents the volume of steam supplied to the cylinder at the gauge pressure represented by the height $d\text{L}$. Since *ad* represents the stroke or swept volume, then,

Volume of steam supplied to the cylinder

$$= \frac{bc}{ad} \times \text{stroke volume}$$

To obtain the best results, the measurements should be taken where the steam is driest, namely, in the H.P. cylinder. The level of *ad* should be near the end of expansion where a certain amount of re-evaporation of the initially condensed steam takes place due to the cylinder walls being at a higher temperature than the expanded steam.

H

Fig. 83 is an indicator diagram taken from the top end of an H.P. cylinder. The cylinder is 28 in. diameter and the stroke of the piston is 3 feet. The scale of the indicator spring is one inch = 100 lb/in². *ad*, representing the stroke volume, is 2·97 inches, and *bc* representing the volume of steam used measures 1·98 inches. Therefore:

$$\text{Pressure } d\text{L} = 0.9 \times 100 \text{ lb/in}^2 \text{ gauge}$$
$$= 105 \text{ lb/in}^2 \text{ absolute}$$

$$\text{Volume } bc = \frac{1\cdot98}{2\cdot97} \times \frac{0\cdot7854 \times 28^2}{144} \times 3 \text{ ft}^3$$

$$= 8\cdot553 \text{ ft}^3$$

From the steam tables, the specific volume of saturated steam at a pressure of 105 lb/in² abs. is 4·23 lb/ft³.

$$\therefore \text{ Weight of steam} = \frac{8\cdot553}{4\cdot23} = 2\cdot022 \text{ lb.}$$

If it be assumed that the same weight of steam is used on the upstroke, then weight per revolution is twice the above quantity. Alternatively, the same method can be applied to the indicator diagram taken from the bottom of the cylinder and the sum of the two quantities is then the weight per revolution.

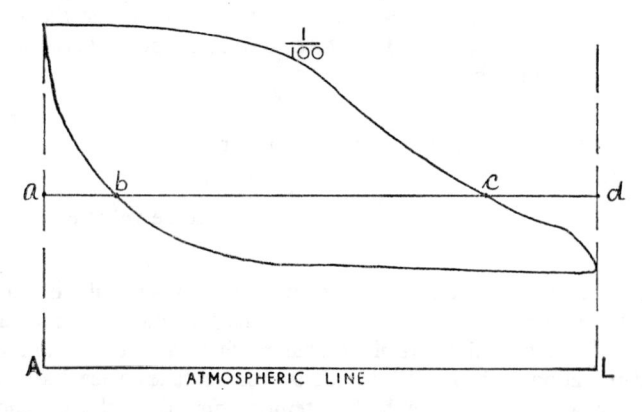

Fig. 83

STEAM RECIPROCATING ENGINES

MISSING QUANTITY

If the actual weight of steam used by the engine is measured by weighing the condensate discharged by the air pump over a period of, say, one hour, then this divided by the number of strokes made by the engine per hour will be the weight of steam actually used per stroke. This quantity is often referred to as the *cylinder feed*. Adding to this the weight of cushion steam in the cylinder gives the actual weight of steam in the cylinder after cut off and during the expansion period.

With the aid of steam tables, the dry saturation curve or expansion curve of constant steam weight can be plotted, and will appear as shown in Fig. 84.

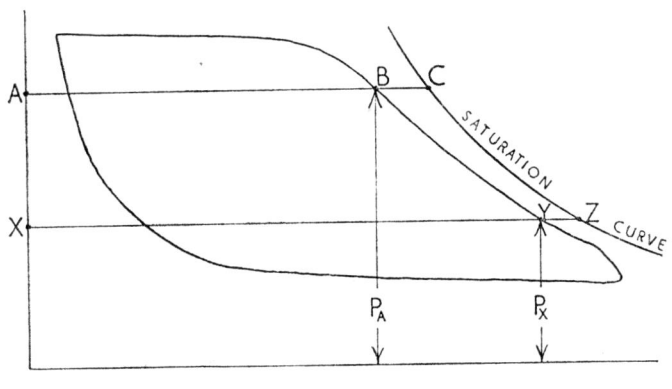

Fig. 84

AC represents the volume of steam in the cylinder if the steam was dry and saturated, at absolute pressure P_A.

AB represents the actual volume of the expanding steam in the cylinder when the pressure is P_A.

Therefore the dryness fraction of the steam at this point of expansion is $\dfrac{AB}{AC}$.

Similarly, at the pressure P_X the dryness fraction at this point of expansion is $\dfrac{XY}{XZ}$.

The horizontal difference between the actual expansion curve and the dry steam curve is the loss of volume due to wetness of the steam, and this is known as the *missing quantity*. Thus, at pressure P_A the missing quantity is represented by BC, and at the pressure P_X the missing quantity is represented by YZ. The missing volume is converted into weight to express the missing quantity in terms of lb per stroke, or lb per hour.

*f*Example. The cylinder diameter of an engine is 21 in., stroke 30 in. and clearance volume equal to 8% of the stroke volume. When the engine was running at 120 r.p.m. the condensate was weighed and found to be 15120 lb per hour. Measuring the indicator card from the engine, the pressure was 90 lb/in² abs. at 0·7 of the outward stroke during the expansion period, and 70 lb/in² abs. at 0·9 of the return stroke during the compression period. Find the missing quantity (i) in lb per stroke, (ii) in lb per hour, and (iii) as a percentage of the actual steam consumption. Find also the dryness fraction of the steam in the cylinder at 0·7 of the outward stroke.

$$\text{Stroke volume} = \frac{0\cdot7854 \times 21^2 \times 30}{1728} = 6\cdot013 \text{ ft}^3$$

Clearance volume $= 0\cdot08 \times 6\cdot013 = 0\cdot481$ ft³
0·7 of stroke volume $= 0\cdot7 \times 6\cdot013 = 4\cdot2091$ ft³
0·1 of stroke volume $= 0\cdot1 \times 6\cdot013 = 0\cdot6013$ ft³

From measurements of indicator card, volume of steam in cylinder when piston has travelled 0·7 of outward stroke and pressure is 90 lb/in²

$$= 4\cdot2091 + 0\cdot481 = 4\cdot6901 \text{ ft}^3$$

Specific volume @ 90 lb/in² $= 4\cdot896$ ft³/lb

∴ Weight of steam in cylinder during expansion as indicated by card $= \dfrac{4\cdot6901}{4\cdot896} = 0\cdot9581$ lb

Weight of steam used as measured by weighing condensate

$$= \frac{15120}{60 \times 120 \times 2} = 1\cdot05 \text{ lb per stroke.}$$

Volume of cushion steam (left in cylinder after exhaust is cut off) when piston has travelled 0·9 of return stroke, that is, when piston is 0·1 of its stroke before top dead centre, and the pressure is 70 lb/in²

$$= 0{\cdot}6013 + 0{\cdot}481 = 1{\cdot}0823 \text{ ft}^3$$

Specific volume at 70 lb/in² = 6·206 ft³/lb

$$\therefore \text{ Weight of cushion steam} = \frac{1{\cdot}0823}{6{\cdot}206} = 0{\cdot}1744 \text{ lb}$$

Hence, actual weight of steam in cylinder during expansion
 = wt. of steam supplied + wt. of cushion steam
 = 1·05 + 0·1744 = 1·2244 lb

$$\text{Missing Quantity} = \begin{array}{c}\text{actual wt. of}\\ \text{steam in cylinder}\end{array} - \begin{array}{c}\text{wt. as indicated}\\ \text{by card}\end{array}$$

$$= 1{\cdot}2244 - 0{\cdot}9581$$
$$= 0{\cdot}2663 \text{ lb per stroke. Ans. (i)}$$
$$0{\cdot}2663 \times 2 \times 120 \times 60 = 3835 \text{ lb per hour. Ans. (ii)}$$

As a percentage of the actual steam consumption

$$= \frac{3835}{15120} \times 100 = 25{\cdot}36\% \text{ Ans. (iii)}$$

Dryness fraction at 0·7 of outward stroke

$$= \frac{0{\cdot}9581}{1{\cdot}2244} = 0{\cdot}7825 \text{ Ans. (iv)}$$

CAUSES OF MISSING QUANTITY:

(i) Condensation, initially and during expansion.
(ii) Steam leakage past piston rings, and also past piston (or slide) valve direct from supply to exhaust without passing through cylinder.

MEANS OF REDUCING CYLINDER CONDENSATION:

(i) By superheating the supply steam. This allows superheat (sensible heat) to be lost by the steam when it comes into contact with the colder cylinder before its saturation temperature is reached when condensation begins.

(ii) By compressing some of the exhaust steam (i.e. cushion steam) the temperature of the compressed steam is raised as its pressure is increased as near as practicable to the admission steam temperature, and the temperature gradient over which the cylinder is working is thereby reduced. This is achieved by giving the valve exhaust lap.

(iii) By steam jacketing the cylinder walls.

(iv) By 'compounding'. That is, by carrying out the expansion of the steam in two or more cylinders, the effect of this being to reduce the temperature difference between admission and exhaust over each cylinder.

COMPOUNDING

Steam engines are usually constructed with two, three, or four cylinders in series, their pistons being connected to a common crank shaft, so that the full expansion of the steam from boiler supply pressure to final exhaust pressure can be spread throughout those cylinders.

Taking as an example saturated steam supplied to the engine at a temperature of 380°F (about 195 lb/in^2) and exhaust at a temperature of 140°F (about 24 in. of vacuum). This represents a temperature difference of 240 F degrees. If only one cylinder was used for the expansion, the temperature of the cylinder walls and steam ports would be somewhere between 380 and 140°F but nearer the lower figure due to the longer period of exhaust, say about 250°F. When the supply steam at 380°F comes into contact with the colder metal, it causes a considerable proportion of the steam to condense before it can do work in the cylinder. This initial condensation results in a loss of efficiency.

In a two-cylinder engine, the total pressure drop can take place in two stages. The difference of 240 F degrees between initial supply and final exhaust is divided into two separate temperature drops of 120 F degrees each. The high-pressure (H.P.) cylinder takes in the steam at 380°F and exhausts it at 260°F, the working temperature of the cylinder and ports being about 310°F. The low-pressure (L.P.) cylinder takes in the steam exhausted from the H.P. at 260°F and exhausts it at 140°F, the working temperature of this cylinder being about 190°F. Thus there is less difference between the temperatures of the supply steam and the cylinder with consequent less initial condensation and higher efficiency.

A three-cylinder engine, in which the steam expansion or pressure drop occurs in three stages, is termed a *triple expansion engine*, and the temperature drop of the steam through each cylinder would be one-third of 240 = 80 F degrees.

A four-cylinder engine in which the steam expansion takes place in four stages, is termed a *quadruple expansion engine*, and the steam temperature drop per cylinder would be one-quarter of 240 = 60 F degrees.

Thus, the greater number of cylinders used to expand the steam from initial supply to final exhaust, the less the cooling effect and initial condensation of the steam. However, more cylinders mean a longer engine, more heavy parts to move and more frictional resistances to overcome, and therefore compounding the steam is usually limited to three cylinders in average sized marine engines, and four cylinders in the larger powered engines.

It must be understood that the terms 'triple' and 'quadruple' expansion refer to the number of expansion stages and not necessarily to the number of cylinders. For instance, a triple expansion engine could have four cylinders, the arrangement being, one H.P., one I.P., and two L.P. cylinders; the two L.P.'s take the steam simultaneously from the exhaust of the I.P., they can be made of moderate size and may be preferred over one large cumbersome L.P. cylinder. Such an engine is termed a 'four-cylinder-triple', or, 'four-crank-triple'.

FURTHER ADVANTAGES OF COMPOUNDING:

In addition to the advantage of reducing cylinder condensation, as explained above, an engine with more than one cylinder gives a better balance and a more even turning moment at the crank shaft. Re-heating of the steam can be arranged between the exhaust from one cylinder and the supply to the next, this improves the quality of the steam for expansion in succeeding cylinders. The design of the engine is more satisfactory, the H.P. cylinder is made small and strong to take the high pressure steam, whereas the L.P. cylinder is made sufficiently large to accommodate the large volume of the steam at the end of expansion, but need not be so strong because the steam pressures at this stage are very low. Leakage steam past the pistons and valves is not entirely lost (except in the L.P.) since it will be available for work in the succeeding cylinders.

A PV diagram showing the expansion of steam through the three cylinders of a normal triple expansion engine is shown in Fig. 85, and a typical set of indicator cards is given in Fig. 86.

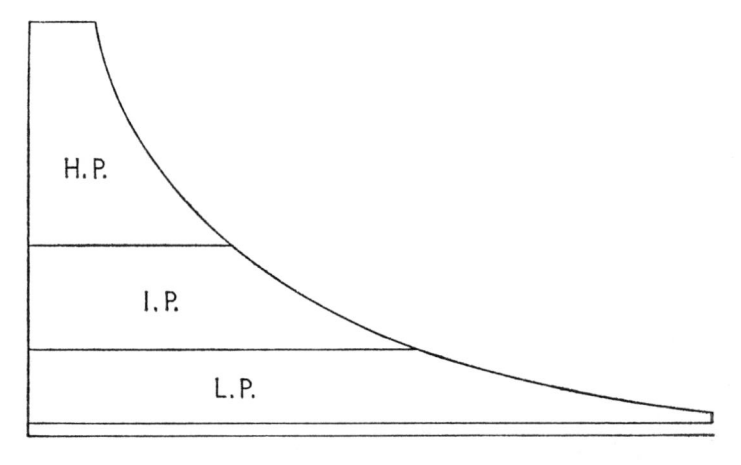

Fig. 85

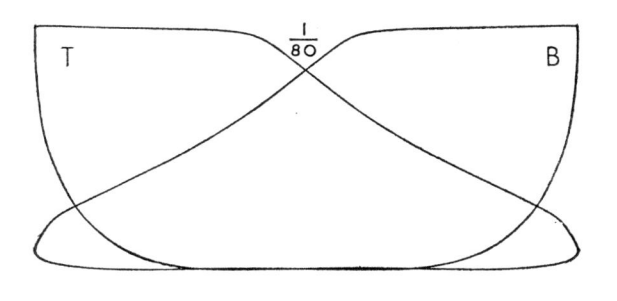

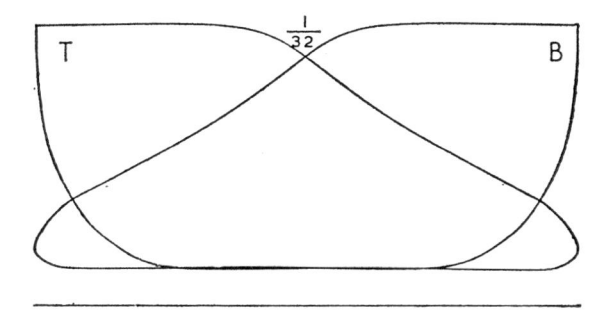

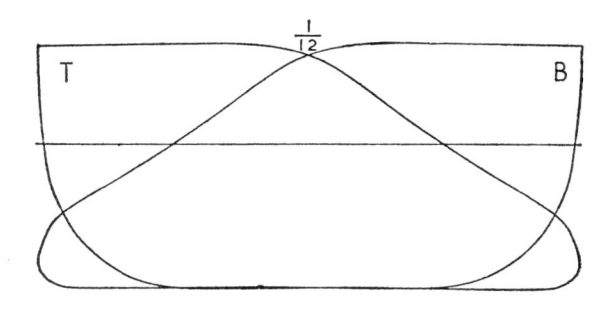

Fig. 86

TEST EXAMPLES 10

1. The top lead of a slide valve is $\frac{1}{8}$ inch and the bottom exhaust lap is $\frac{3}{16}$ inch. When the crank is on top dead centre the bottom exhaust opening is $1\frac{7}{16}$ in., and when the crank is on bottom dead centre the top exhaust opening is $1\frac{1}{2}$ in. Find the top steam and exhaust laps.

2. The travel of a slide valve is $6\frac{1}{2}$ in., the steam lap is $1\frac{3}{4}$ in. and the lead is $\frac{1}{4}$ in. Find the angle of advance of the eccentric and the maximum port opening to steam.

3. A slide valve has $1\frac{1}{2}$ in. steam lap, $\frac{1}{8}$ in. lead, and 1 in. maximum port opening to steam. Find the valve travel and the angle of advance of the eccentric sheave.

4. The throw of an eccentric sheave is 3 in. and the angle of advance is $38°\ 19'$. The maximum port opening to steam is $1\cdot3$ in. Find the steam lap of the valve and the lead.

5. The travel of a slide valve is 8 in., the top steam lap is 2 in. and the angle of advance of the eccentric sheave is $33°\ 22'$. Calculate (a) the opening to steam when the crank is on top dead centre, (b) the maximum port opening to steam, (c) the crank angle from top dead centre at points of (i) cut off, and (ii) admission to steam.

6. Calculate the travel of a slide valve which has $0\cdot18$ inch lead and $1\cdot25$ in. maximum port opening to steam, when the angle of advance of the eccentric is $37°\ 39'$.

7. The travel of a slide valve is 7 in., top steam lap $1\cdot75$ in., and lead $0\cdot125$ in. Calculate (i) the port opening to steam when the crank is $77°$ past T.D.C., (ii) the crank angle from T.D.C. when the port opening to steam is maximum.

8. A slide valve has a travel of 6 in. The lead at the top is $0\cdot1$ in., top steam lap $1\cdot7$ in., bottom steam lap $1\cdot6$ in., top exhaust lap $0\cdot15$ in., and bottom exhaust lap $0\cdot2$ in. Calculate the port openings to steam and exhaust when the crank is at $90°$ to the line of stroke.

ƒ9. The steam lap of a slide valve is $1\cdot75$ in., the lead is $0\cdot17$ in. and the maximum port opening to steam is 1 inch. Calculate (i) the valve travel, (ii) the angle of advance of the eccentric, and (iii) the fraction of the stroke completed by the piston when the valve opens to exhaust, if the exhaust lap is $0\cdot375$ in. (a) neglecting obliquity of the connecting rod, (b) taking into account that the connecting rod is $3\cdot5$ times the crank length.

CHAPTER 11

M.E.P., HORSE POWER & EFFICIENCY OF STEAM ENGINES

The formulae for calculating the indicated horse power of reciprocating engines were derived in Chapter 5, one being applicable to the metric system and the other when British units of measurement are used, thus:

		$\dfrac{pALN}{4560}$	$\dfrac{pALN}{33000}$
i.h.p.	= indicated horse power		
p	= mean effective pressure	kg/cm²	lb/in²
A	= area of cylinder	sq. cm.	sq. in.
L	= length of stroke	metres	feet

N = number of power strokes per minute

In the steam reciprocating engine cycle there is one power stroke in every revolution and, as all these engines are double-acting, then N = 2 × r.p.m.

Methods of finding the mean effective pressure from the indicator diagram were also explained in Chapter 5. The mean height of the diagram may be obtained either by dividing the area (as found by a planimeter) by the length, or, by using the mid-ordinate rule.

Fig. 87 represents an indicator card taken from the top side of the cylinder of a steam engine. If the area of this is measured by a planimeter it will be found to be 3·15 square inches. The length is 3·3 inches, therefore the mean height is 3·15 ÷ 3·3 = 0·955 inch. Since the scale of the indicator spring is marked on the card as $\frac{1}{80}$, this means that one inch of height represents 80 lb/in² of pressure, therefore the mean effective pressure is 0·955 × 80 = 76·4 lb/in².

When employing the mid-ordinate rule it is convenient to use ten mid-ordinates. A simple procedure is as follows and illustrated in Fig. 87.

226 REED'S HEAT AND HEAT ENGINES FOR MARINE ENGINEERS

0.84" 1.32" 1.45" 1.46" 1.44" 1.13" 0.81" 0.55" 0.35" 0.20"

1/80

ATMOSPHERIC LINE

SUM = 9.55
DIVIDING BY 10
MEAN HEIGHT = 0.955"

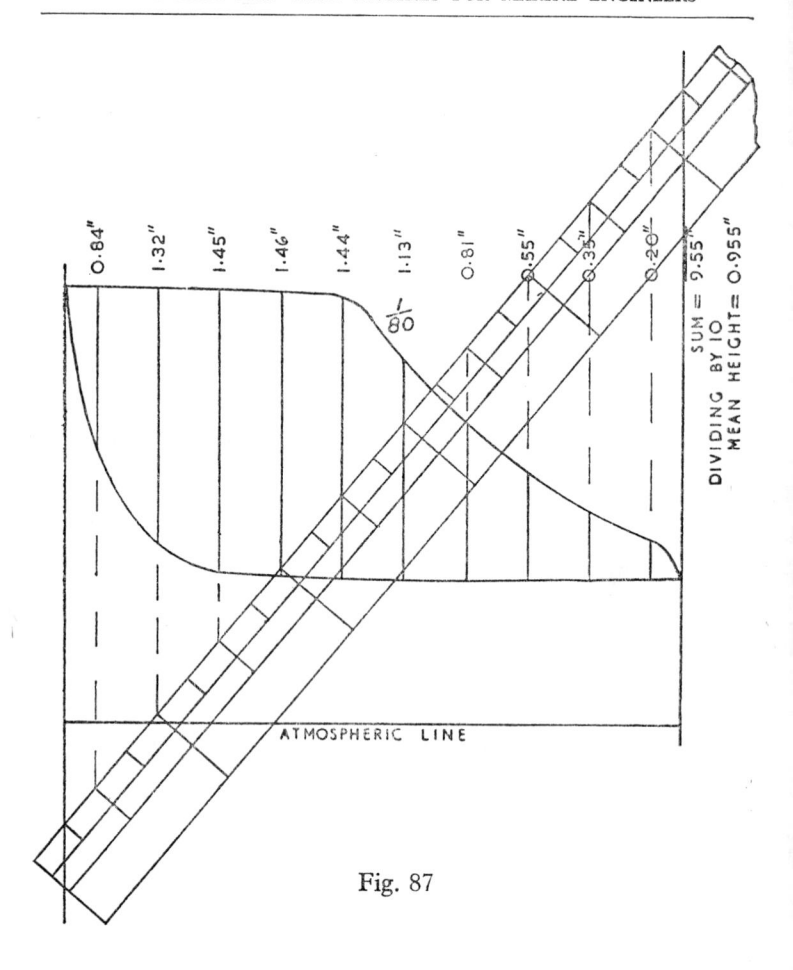

Fig. 87

Erect a vertical line at each of the two extreme ends of the diagram, perpendicular to the atmospheric line.

Place a rule so that it measures 5 inches between these perpendiculars. If the length of the diagram is exactly 5 inches, the rule will lie parallel to the atmospheric line, but, since almost all diagrams are less than this, then the rule must be inclined until it registers 5 inches between the lines.

Instead of marking the ten spaces and then the middle of these spaces to get the mid-ordinates, it is quicker to set the rule measuring

¼ inch at the first perpendicular, and 5¼ inches at the other, then mark off every half-inch point to indicate the position of the mid-ordinates directly.

Erect perpendicular lines through these half-inch marks across the diagram, these are the mid-ordinates.

Measure the mid-ordinates, add the measurements together and divide the sum by the number of mid-ordinates, in this case 10, to obtain the mean height of the diagram.

$$\begin{aligned} \text{m.e.p.} &= \text{mean height} \times \text{spring scale} \\ &= 0.955 \times 80 \\ &= 76.4 \text{ lb/in}^2 \text{ (as shown)} \end{aligned}$$

HYPOTHETICAL MEAN EFFECTIVE PRESSURE

The hypothetical or theoretical mean effective pressure can be obtained in the manner explained above, from the hypothetical indicator diagram. However, as the theoretical diagrams are regular shaped figures, the simplest method is to first calculate the area and then divide by the length.

In steam engines it is usual to assume that the steam expands hyperbolically, i.e. according to the law $PV = C$ unless stated otherwise, and in I.C. engines the gas is assumed to expand according to the law $PV^n = C$ where n is between 1·34 and 1·4.

We have previously seen that the area of a PV diagram during expansion is given by:

$$PV \log_e r \quad \text{when } PV = C$$

$$\text{and} \quad \frac{P_1V_1 - P_2V_2}{n - 1} \quad \text{when } PV^n = C$$

The areas of rectangular portions are simply, height × length, which is P × V, therefore the following examples should readily explain the procedure.

Example. Steam at 185 lb/in² gauge is admitted to an engine cylinder and cut off when the piston has travelled 0·4 of its stroke. The back pressure is 70 lb/in² gauge. The clearance volume is equal to 10 per cent of the stroke volume. Calculate the hypothetical mean effective pressure.

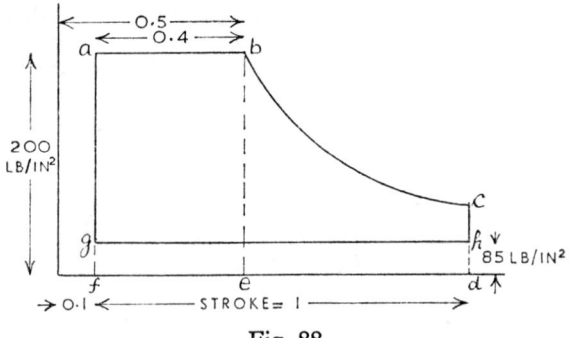

Fig. 88

Let stroke volume = 1
then, clearance volume = 10% of 1 = 0·1
Volume at cut off = 0·1 + 0·4 = 0·5
Initial steam pressure = 185 + 15 = 200 lb/in² abs.
Back pressure = 70 + 15 = 85 lb/in² abs.

$$\text{Ratio of expansion} = \frac{\text{volume at end of expansion}}{\text{volume at beginning of expansion}}$$

$$= \frac{1\cdot 1}{0\cdot 5} = 2\cdot 2$$

Area of diagram during admission period = *abef*
= 200 × 0·4 = 80

Area of diagram during expansion period = *bcde*
= PV log$_\varepsilon$ *r*
= 200 × 0·5 × log$_\varepsilon$ 2·2
= 200 × 0·5 × 0·7885
= 78·85

∴ Gross area = 80 + 78·85 = 158·85

Back pressure area = *ghdf*
= 85 × 1 = 85

∴ Effective area = Gross area — back pressure area
= 158·85 — 85
= 73·85

M.E.P., HORSE POWER AND EFFICIENCY OF STEAM ENGINES

Since mean height = area ÷ length, then the mean effective pressure is the effective area of the diagram divided by the length.

∴ Hypothetical m.e.p.= 73·85 ÷ 1
= 73·85 lb/in². Ans.

Note that the *gross* area divided by the length would give the mean *gross* pressure. That is, the average absolute pressure acting on that side of the piston. In the above case the mean gross pressure is 158·85 lb/in² absolute.

When the back pressure is constant as it is in the above example at 85 lb/in² abs., then the mean effective pressure can be obtained simply by subtracting the back pressure from the mean gross pressure:

Mean effective pressure = mean gross pressure — back pressure
= 158·85 lb/in² abs. — 85 lb/in² abs.
= 73·85 lb/in².

Note also that the mean *effective* pressure is not an absolute pressure nor a gauge pressure. It is merely an effective difference between two pressures.

DIAGRAM FACTOR

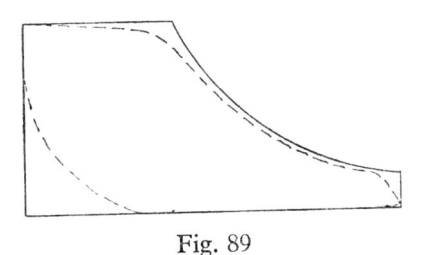

Fig. 89

The difference between the actual indicator diagram and the theoretical diagram is shown in Fig. 89. Wire-drawing of the steam through the ports during the admission period accounts for the sloping admission line. Near cut off, the valve closes the port gradually and causes a further pressure drop before complete cut off takes place. The volume at cut off and during expansion is less than the theoretical saturation curve due to initial condensation of the steam (explained in previous Chapter under 'missing quantity').

However, the actual expansion curve usually approaches more closely to the theoretical curve near the end of expansion due to re-evaporation. In practice, exhaust commences before the end of the power stroke and ceases before the end of the return stroke, as previously explained. Hence the result of these differences is to reduce the area of the diagram. Since the length remains the same, then the mean height, which represents the mean effective pressure, is reduced in the same ratio.

The ratio of the area of the actual diagram to the area of the hypothetical diagram, or the ratio of the actual mean effective pressure to the theoretical mean effective pressure, is known as the *diagram factor*.

This value is usually between 0·6 and 0·75.

Assuming a diagram factor of 0·7, the actual mean effective pressure in the last example would be 73·85 × 0·7 = 51·7 lb/in².

*f*Example. An engine is supplied with steam at a pressure of 205 lb/in² gauge and cut off takes place when the piston has travelled 0·45 of its stroke. The clearance volume is equal to 0·1 of the piston swept volume. The exhaust pressure is 80 lb/in² gauge and the exhaust is cut off when the piston has completed 0·88 of the exhaust stroke. Calculate the mean effective pressure, assuming the steam expands to the end of the stroke and that expansion and compression follows the law PV = constant.

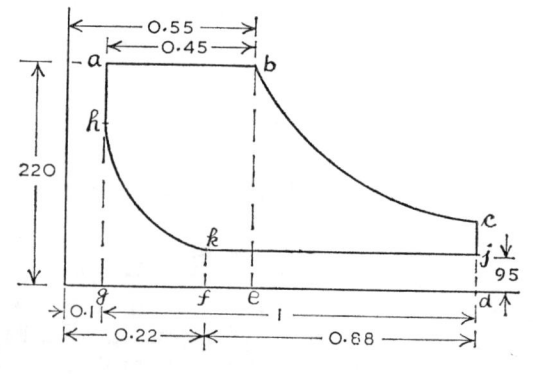

Fig. 90

Let piston swept volume = 1
Volume at beginning of expansion = 0·55
Volume at end of expansion = 1·1
∴ Ratio of expansion = 1·1 ÷ 0·55 = 2·0
Volume at beginning of compression = 1·1 — 0·88 = 0·22
Volume at end of compression = 0·1
∴ Ratio of compression = 0·22 ÷ 0·1 = 2·2

$$\begin{aligned}
\text{Admission area} &= abeg \\
&= 220 \times 0\cdot45 = 99 \\
\text{Expansion area} &= bcde \\
&= P_b V_b \log_\varepsilon r \\
&= 220 \times 0\cdot55 \times \log_\varepsilon 2\cdot0 \\
&= 83\cdot87
\end{aligned}$$

Gross area = 99 + 83·87 = 182·87 ... ... (i)

$$\begin{aligned}
\text{Exhaust area} &= jkfd \\
&= 95 \times 0\cdot88 = 83\cdot6 \\
\text{Compression area} &= hkfg \\
&= P_k V_k \log_\varepsilon r \\
&= 95 \times 0\cdot22 \times \log_\varepsilon 2\cdot2 \\
&= 16\cdot47
\end{aligned}$$

Total back pressure area = 83·6 + 16·47
= 100·07 ... ... (ii)

Effective area = Gross area — Back pressure area
= 182·87 — 100·07
= 82·8

Mean height = area ÷ length = 82·8 ÷ 1 = 82·8
∴ Mean effective pressure = 82·8 lb/in². Ans.

REFERRED MEAN PRESSURE

If it is to be assumed for convenience of design, that the whole power of a compound, triple or quadruple expansion steam engine could be developed in one of its cylinders, the only cylinder which could accommodate the volume of steam after full expansion is the largest cylinder in the engine, namely the L.P. cylinder.

Taking as an example a triple expansion engine of cylinder diameters 20·5, 34·5 and 58 in., and cut off in the H.P. at 0·5 stroke, neglecting clearances, the overall ratio of expansion of the steam throughout the engine is:

$$R = \frac{\text{Full volume of L.P. cylinder}}{\text{Volume of H.P. cyl. up to cut off}}$$

$$= \frac{0.7854 \times 58^2 \times \text{stroke}}{0.5 \times 0.7854 \times 20.5^2 \times \text{stroke}}$$

$$= 16$$

Thus, if we imagine the L.P. cylinder to be of sufficient strength to take steam direct from the boiler, cut off would be at one-sixteenth of the stroke so that the steam would be expanded to 16 times its initial volume when the piston completed its power stroke.

Suppose the length of the stroke of this engine is 3 feet and the required i.h.p. is 2500 when running at 150 r.p.m. The mean effective pressure necessary in the L.P. for the whole power to be developed in this cylinder would be:

$$p = \frac{33000 \times \text{i.h.p.}}{\text{ALN}}$$

$$= \frac{33000 \times 2500}{0.7854 \times 58^2 \times 3 \times 150 \times 2}$$

$$= 34.7 \text{ lb/in}^2.$$

This is termed the 'mean effective pressure all referred to the L.P. cylinder', or usually more briefly, the *referred mean pressure*, and is defined thus: The referred mean pressure is the mean effective pressure that would be required in the L.P. cylinder if the whole engine power be assumed to be developed in that cylinder.

M.E.P., HORSE POWER AND EFFICIENCY OF STEAM ENGINES 233

If equal powers are developed in each cylinder, then, for a triple expansion engine, one-third of the total engine power is developed in each cylinder. The actual mean effective pressure in the L.P. will then be one-third of the referred mean pressure.

$$\text{m.e.p. in L.P.} = \tfrac{1}{3} \times 34\cdot 7 = 11\cdot 6 \text{ lb/in}^2.$$

Since the power in the other two cylinders is to be the same as that in the L.P., then:

$$\text{i.h.p. (H.P.)} = \text{i.h.p. (I.P.)} = \text{i.h.p. (L.P.)}$$

$$\frac{pALN}{33000}\text{ (H.P.)} = \frac{pALN}{33000}\text{ (I.P.)} = \frac{pALN}{33000}\text{ (L.P.)}$$

Cancelling quantities common to all terms:

$$pd^2 \text{ (H.P.)} = pd^2 \text{ (I.P.)} = pd^2 \text{ (L.P.)}$$

Hence, to develop power in the H.P. and I.P. equal to the L.P. power:

$$\text{m.e.p. in H.P.} = \frac{\text{m.e.p. (L.P.)} \times d^2 \text{ (L.P.)}}{d^2 \text{ (H.P.)}}$$

$$= \frac{11\cdot 6 \times 58^2}{20\cdot 5^2} = 92\cdot 8 \text{ lb/in}^2.$$

$$\text{m.e.p. in I.P.} = \frac{11\cdot 6 \times 58^2}{34\cdot 5^2} = 32\cdot 8 \text{ lb/in}^2.$$

Example. The cylinder diameters of a quadruple expansion steam engine are 25, 37, 54 and 80 in. respectively. Find the mean effective pressure in each cylinder when the mean pressure all referred to the L.P. is 32 lb/in², assuming all cylinders develop equal power. If the length of the stroke is 4 feet and the engine runs at 90 r.p.m., find the total i.h.p. of the engine.

$$\text{m.e.p. in L.P.} = 32 \div 4 = 8 \text{ lb/in}^2.$$

$$\text{m.e.p. in 1st I.P.} = \frac{8 \times 80^2}{54^2} = 17 \cdot 6 \text{ lb/in}^2.$$

$$\text{m.e.p. in 2nd I.P.} = \frac{8 \times 80^2}{37^2} = 37 \cdot 4 \text{ lb/in}^2.$$

$$\text{m.e.p. in H.P.} = \frac{8 \times 80^2}{25^2} = 81 \cdot 9 \text{ lb/in}^2.$$

$$\text{i.h.p.} = \frac{p\text{ALN}}{33000}$$

$$= \frac{32 \times 0 \cdot 7854 \times 80^2 \times 4 \times 90 \times 2}{33000}$$

$$= 3510 \text{ i.h.p. Ans.}$$

Example. The cylinder diameters of a triple expansion engine are 26, 44 and 74 in. and the mean effective pressures are 85, 30 and 10·5 lb/in² respectively when the mean piston speed is 760 ft/min. Calculate the mean effective pressure referred to the L.P. cylinder and the total i.h.p. developed.

m.e.p. of H.P. referred to L.P.

$$= \frac{85 \times 26^2}{74^2} = 10 \cdot 49 \text{ lb/in}^2.$$

m.e.p. of I.P. referred to L.P.

$$= \frac{30 \times 44^2}{74^2} = 10 \cdot 6 \text{ lb/in}^2.$$

Total mean pressure referred to L.P.
$$= 10 \cdot 49 + 10 \cdot 6 + 10 \cdot 5 = 31 \cdot 59 \text{ lb/in}^2. \text{ Ans. (i)}$$

M.E.P., HORSE POWER AND EFFICIENCY OF STEAM ENGINES 235

Note that the mean piston speed in feet per minute is the feet of distance travelled by the piston during one minute, and is equal to stroke length (ft) $\times$ 2 $\times$ r.p.m., which is L $\times$ N in the i.h.p. formula.

$$\text{i.h.p.} = \frac{p\text{ALN}}{33000}$$

$$= \frac{31{\cdot}59 \times 0{\cdot}7854 \times 74^2 \times 760}{33000}$$

$$= 3129 \text{ Ans. (ii)}$$

HYPOTHETICAL REFERRED MEAN PRESSURE

Referring to Fig. 85, neglecting clearances:

Let P = initial absolute pressure of steam supplied to engine.
P_b = final absolute back pressure.
v = volume of steam before expansion, i.e. volume admitted to the H.P. cylinder, this is H.P. cyl. volume $\times$ cut off fraction.
V = volume of steam after expansion, i.e. full volume of L.P. cylinder.
R = overall ratio of expansion throughout engine
  = V $\div$ v

Then,
Admission area = Pv
Expansion area = Pv $\log_e$R
Gross area = Pv + Pv $\log_e$R
= Pv (1 + $\log_e$R)
Mean gross pressure = area $\div$ length

$$= \frac{Pv\,(1 + \log_e R)}{V}$$

Since $\dfrac{v}{V} = \dfrac{1}{R}$

Mean gross pressure $= \dfrac{P}{R}(1 + \log_\varepsilon R)$

Mean effective press. = mean gross press. — back press.

$$= \dfrac{P}{R}(1 + \log_\varepsilon R) - P_b$$

This is the hypothetical mean effective pressure referred to the L.P. cylinder. The actual referred mean pressure is the hypothetical value multiplied by the overall diagram factor:

Actual R.M.P. $= \left\{\dfrac{P}{R}(1 + \log_\varepsilon R) - P_b\right\} \times f_o$

ƒ DETERMINATION OF CYLINDER DIAMETERS

When calculating the sizes of the cylinders for a compound, triple or quadruple expansion engine, to develop a given horse power, certain assumptions are made based upon experience, such as the most suitable boiler pressure, condenser pressure, ratio of expansion, and piston speed.

The dimensions of intermediate cylinders between the H.P. and L.P. are usually such that the ratio of the cubic capacity of any two successive cylinders are equal.

Since the stroke is common for all cylinders, the volume depends upon the area, and this is proportional to the square of the diameter.

Thus, for a triple expansion engine, if d = diameter of H.P., D = diameter of L.P., and x = diameter of I.P., then,

$$\dfrac{d^2}{x^2} = \dfrac{x^2}{D^2}$$

Taking square root of both sides,

$$\dfrac{d}{x} = \dfrac{x}{D}$$

from which, $x = \sqrt{dD}$

M.E.P., HORSE POWER AND EFFICIENCY OF STEAM ENGINES

Similarly, for a quadruple expansion engine, if x = diameter of 1st I.P., and y = diameter of 2nd I.P.,

$$\frac{d}{x} = \frac{x}{y} = \frac{y}{D}$$

Equating the first and last terms,

$$xy = dD, \quad \therefore y = \frac{dD}{x}$$

Equating the first and second terms,

$$x^2 = dy \quad \therefore x^2 = d \times \frac{dD}{x}$$

from which,

$$x = \sqrt[3]{d^2 D}$$
$$\text{and } y = \sqrt[3]{dD^2}$$

Example. Determine the cylinder diameters of a triple expansion engine to develop 3000 i.h.p. at a mean piston speed of 800 ft/min and suggest a suitable stroke and engine speed, given the following data:

$$\begin{aligned}
\text{Steam supply pressure} &= 195 \text{ lb/in}^2 \text{ gauge} \\
\text{Back pressure} &= 4 \text{ lb/in}^2 \text{ abs.} \\
\text{Cut off in H.P. cyl.} &= 0 \cdot 5 \text{ stroke} \\
\text{Overall ratio of expansion} &= 14 \\
\text{Overall diagram factor} &= 0 \cdot 7
\end{aligned}$$

$$\begin{aligned}
\text{Referred mean press.} &= \left\{\frac{P}{R}(1 + \log_e R) - P_b\right\} \times f_o \\
&= \left\{\frac{210}{14}(1 + \log_e 14) - 4\right\} \times 0 \cdot 7 \\
&= \{15(1 + 2 \cdot 639) - 4\} \times 0 \cdot 7 \\
&= \{15 \times 3 \cdot 639 - 4\} \times 0 \cdot 7 \\
&= 50 \cdot 59 \times 0 \cdot 7 \\
&= 35 \cdot 41 \text{ lb/in}^2.
\end{aligned}$$

From, i.h.p. $= \dfrac{p\text{ALN}}{33000}$

$3000 \times 33000 = 35{\cdot}41 \times 0{\cdot}7854 \times D^2 \times 800$

$$D = \sqrt{\dfrac{3000 \times 33000}{35{\cdot}41 \times 0{\cdot}7854 \times 800}}$$

$= 66{\cdot}7$ in.

$$R = \dfrac{\text{volume at end of expansion}}{\text{volume at beginning of expansion}}$$

$$= \dfrac{\text{full volume of L.P. cyl.}}{\text{volume of H.P. up to cut off}} = \dfrac{D^2}{0{\cdot}5 d^2}$$

$\therefore d = \sqrt{\dfrac{66{\cdot}7^2}{0{\cdot}5 \times 14}} = 25{\cdot}22$ in.

$x = \sqrt{dD} = \sqrt{25{\cdot}22 \times 66{\cdot}7} = 41{\cdot}01$ in.

Practical diameters would be:
H.P. $= 25\frac{1}{4}$ in., I.P. $= 41$ in., L.P. $= 67$ in. Ans. (i)

A suitable stroke would be about equal to, or a little more than, the diameter of the I.P.,

$\therefore$ Stroke $= 42$ in. $= 3\frac{1}{2}$ ft. Ans. (ii)

$2 \times$ stroke $\times$ r.p.m. $=$ mean piston speed

$\therefore$ Engine speed $= \dfrac{800}{2 \times 3{\cdot}5} = 114$ r.p.m. Ans. (iii)

WILLANS' LAW

The usual methods of varying the power of a steam engine are (i) by throttling the steam at the engine stop valve, (ii) by altering the cut off point of the H.P. piston valve, or (iii) by a combination of (i) and (ii).

M.E.P., HORSE POWER AND EFFICIENCY OF STEAM ENGINES 239

When the power is varied by throttling only, the weight of steam used by the engine varies proportionally as the indicated horse power. This is known as Willans' law and can be expressed by a linear equation:

$$W = a + bP$$

where, W = weight of steam consumed (usually lb/hr)
P = i.h.p. developed
a and b = constants

The specific steam consumption is usually expressed in lb per i.h.p. per hour, and is an indication of the efficiency of the engine.

Example. The following data were taken during a trial run on a quadruple expansion steam engine. Plot graphs of steam consumption in lb per hour, and lb per i.h.p. per hour, on a common base of indicated horse power and find Willans' law for this engine.

i.h.p.	20	40	60	80	100	120
lb steam/hr	510	730	970	1220	1430	1650

Dividing lb of steam per hour by i.h.p. to obtain the specific steam consumption at each stage, weights of steam per i.h.p. per hour = 25·5, 18·25, 16·17, 15·25, 14·3 and 13·75 respectively.

The plotted graphs appear as in Fig. 91.

Reading from the graph of Willans' line:

a = steam consumption in lb per hour when the i.h.p. is theoretically zero, this reads 280.

b = increase in steam consumption in lb per hour for an increase of one i.h.p. This is the slope of the line and equal to $y \div x = 920 \div 80 = 11·5$
∴ Willans' law is $W = 280 + 11·5P$ Ans.

Example. The weight of steam used by an engine is 69500 lb per hour when developing 5000 i.h.p., and 50000 lb per hour when the i.h.p. is 3500. Assuming the consumption of steam follows Willans' law, find (i) the weight of steam used per hour when the i.h.p. is 4000, and (ii) the specific consumption in lb per i.h.p. per hour at this power.

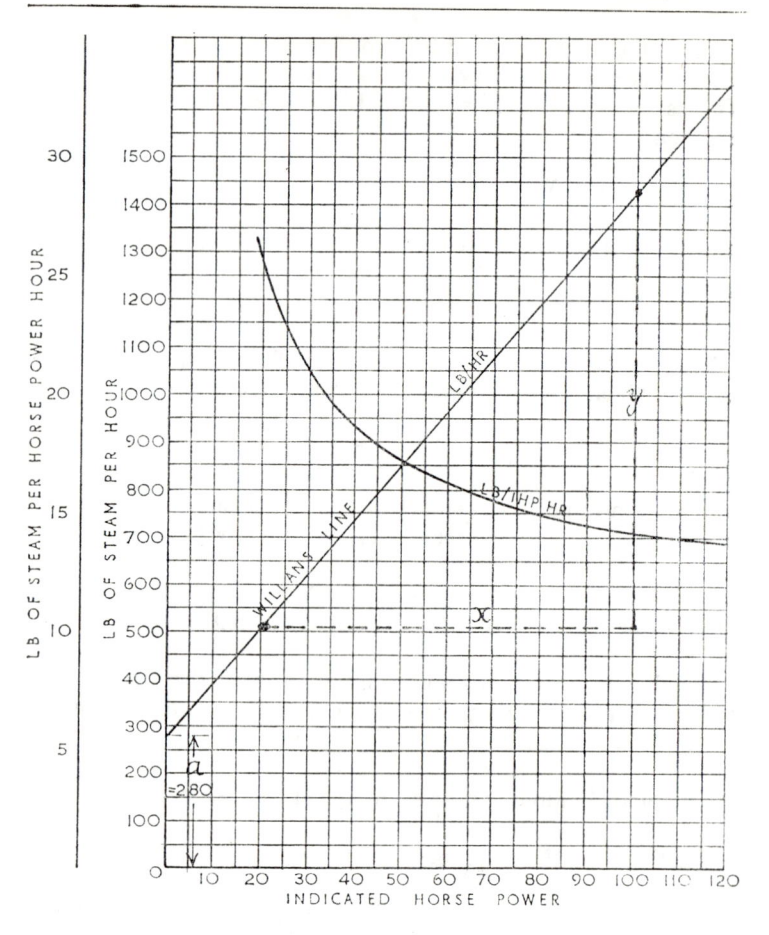

240 REED'S HEAT AND HEAT ENGINES FOR MARINE ENGINEERS

Fig. 91

Putting in the pair of given quantities in the form of a simultaneous equation, solving the constants a and b, and expressing Willans' law for this engine:

$$W = a + bP$$
$$69500 = a + b \times 5000 \quad \ldots \text{(i)}$$
$$50000 = a + b \times 3500 \quad \ldots \text{(ii)}$$

$$19500 = \quad b \times 1500 \text{ by subtraction}$$

$$\therefore b = 13$$

M.E.P., HORSE POWER AND EFFICIENCY OF STEAM ENGINES 241

From (i),
$$69500 = a + 13 \times 5000$$
$$\therefore a = 4500$$
$$\therefore \text{Willans' law is } W = 4500 + 13P$$
When i.h.p. = 4000:
$$W = 4500 + 13 \times 4000$$
$$= 56500 \text{ lb per hour. Ans. (i)}$$
$$\text{Specific consumption} = 56500 \div 4000$$
$$= 14 \cdot 125 \text{ lb/i.h.p. hr. Ans. (ii)}$$

THERMAL EFFICIENCY

The indicated thermal efficiency of a steam engine is the ratio of the heat equivalent of the work done in the cylinders to the heat supplied to the engine by the steam. The heat supplied by the steam is the heat given to it in the boilers. It is usual to assume that the boilers are fed with water at the temperature at which the steam is condensed in the condenser, any overcooling of the condensate to a temperature below the exhaust steam temperature being disregarded since this is a condenser loss and not an engine loss of heat.

Example. A steam engine is supplied with dry saturated steam at a pressure of 185 lb/in² abs. and the pressure of the exhaust to the condenser is 3 lb/in² abs. When developing 500 i.h.p., the weight of steam used by the engine is 7250 lb per hour. Calculate the indicated thermal efficiency.

Total heat in supply steam (from tables)
$$= 1198 \cdot 4 \text{ Btu/lb}$$

Total heat in condensate (heat in water when it is condensed from steam at 3 lb/in²) = 109·4 Btu/lb

Heat given to feed in boilers to produce this steam
$$= 1198 \cdot 4 - 109 \cdot 4 = 1089 \text{ Btu/lb}$$

Weight of steam used = 7250 lb/hour

$$\frac{7250}{500} = 14 \cdot 5 \text{ lb/i.h.p. hour}$$

Indicated thermal effic. = $\dfrac{\text{Heat equivalent of work done in cyl.}}{\text{Heat supplied}}$

= $\dfrac{\text{Heat equivalent of one h.p. hour}}{\text{Heat supplied to produce one h.p. hour}}$

= $\dfrac{2545}{14{\cdot}5 \times 1089}$

= 0·1612 or 16·12% Ans.

ƒRANKINE EFFICIENCY

The ideal cycle adopted as a basis of comparison for the actual performance of a steam engine is known as the *Rankine cycle*, and the thermal efficiency of an engine working on the Rankine cycle is the *Rankine efficiency*. The ratio of the indicated thermal efficiency of an engine to the Rankine efficiency is termed the *relative efficiency* or *efficiency ratio*.

The Rankine cycle is illustrated in Fig. 92 on a PV diagram, and in Fig. 93 on a T-s diagram, and consists of:

(i) *a* to *b*, feed water pumped into the boiler and receiving sensible heat as it is increased in pressure from P_2 to P_1 and in temperature from T_2 to T_1.

(ii) *b* to *c*. The water is completely evaporated into steam in the boiler at constant pressure P_1 and constant temperature T_1 and the steam is supplied to the engine as it is generated.

(iii) *c* to *d*. The steam supply to the engine is cut off from the boiler and expands adiabatically in the engine cylinder from the highest to the lowest limits of pressure and temperature.

(iv) *d* to *a*. The steam is exhausted from the engine into the condenser and condensed into water at constant pressure P_2 and constant temperature T_2.

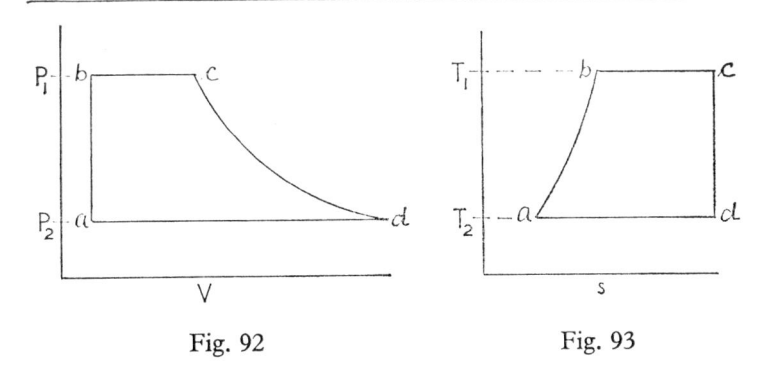

M.E.P., HORSE POWER AND EFFICIENCY OF STEAM ENGINES 243

Fig. 92

Fig. 93

The work which would be done in the ideal engine is equal to the heat energy given up by the steam on its passage through the engine. This is the heat drop, i.e. the difference between the total heat in the supply steam (h_{g1}) and the total heat in the exhaust steam (h_{g2}).

The heat supplied to the feed water in the boiler to produce the steam is the difference between the total heat in the supply steam (h_{g1}) and the total heat in the feed water (h_{f2}). Thus:

$$\text{Rankine efficiency} = \frac{h_{g1} - h_{g2}}{h_{g1} - h_{f2}}$$

The value of h_{g1} depends upon whether the supply steam is wet, dry, or superheated. Since the exhaust steam is always wet, the value of h_{g2} will be obtained from $h_{f2} + qh_{fg2}$.

Taking the figures in the previous example, i.e. dry saturated steam at 185 lb/in² abs. supplied to the engine, and exhausted at 3 lb/in² abs.:

Heat supplied to steam in boiler
 = 1198·4 — 109·4 = 1089 Btu/lb

To calculate the heat in the steam after adiabatic expansion we must first find the dryness fraction at point d, and we will assume that the steam expands isentropically.

Reading entropy from steam tables,

$$\text{Entropy at } c = 1\cdot 5531$$

s_g at d is equal to s_g at c if the expansion is isentropic.

$$\begin{aligned}
s_g \text{ at } c &= s_f + qs_{fg} \text{ for 3 lb/in}^2. \\
1\cdot 5531 &= 0\cdot 2008 + q\,(1\cdot 8869 - 0\cdot 2008) \\
1\cdot 5531 &= 0\cdot 2008 + q \times 1\cdot 6861 \\
1\cdot 6861 q &= 1\cdot 3523 \\
q &= 0\cdot 802 \\
\text{Heat rejected} &= \text{Heat in steam at } d \\
&= 109\cdot 4 + 0\cdot 802 \times 1013\cdot 2 \\
&= 921\cdot 8
\end{aligned}$$

$$\text{Rankine efficiency} = \frac{1198\cdot 4 - 921\cdot 8}{1198\cdot 4 - 109\cdot 4}$$

$$= \frac{276\cdot 6}{1089}$$

$$= 0\cdot 254 \text{ or } 25\cdot 4\%$$

$$\text{Efficiency ratio} = \frac{\text{Indicated thermal efficiency}}{\text{Rankine efficiency}}$$

$$= \frac{0\cdot 1612}{0\cdot 254}$$

$$= 0\cdot 6346$$

BOILER EFFICIENCY

The thermal efficiency of a boiler is the ratio of the heat absorbed in converting the feed water into steam to the heat supplied by the burning of the fuel, over the same period of time.

M.E.P., HORSE POWER AND EFFICIENCY OF STEAM ENGINES

The heat given to the water and steam is the difference between the total heat in the steam leaving the boiler and the total heat in the feed water entering the boiler.

The heat supplied to the boiler is the product of the weight of fuel burned and its calorific value.

Taking a basis of one lb of fuel burned, let W = weight of steam produced from each lb of fuel:

$$\text{Boiler efficiency} = \frac{W \text{ lb of steam} \times \text{heat given to each lb}}{\text{Calorific value of the fuel}}$$

Example. A boiler working at a pressure of 220 lb/in² abs. generates 14·5 lb of steam for each lb of oil fuel burned in the furnace. The feed water temperature is 210°F and the steam leaves the boiler 0·98 dry. If the calorific value of the oil is 18240 Btu/lb, calculate the thermal efficiency of the boiler.

From steam tables,
P = 220 lb/in², h_f = 364·2, h_{fg} = 836·5 Btu/lb

Heat absorbed per lb of steam formed

$$= (364 \cdot 2 + 0 \cdot 98 \times 836 \cdot 5) - (210 - 32)$$
$$= 1006 \text{ Btu}$$

Heat given to steam per lb of fuel burned
$$= 14 \cdot 5 \times 1006 \text{ Btu}$$

Heat supplied to boiler per lb of fuel burned
$$= 18240 \text{ Btu}$$

$$\text{Boiler efficiency} = \frac{14 \cdot 5 \times 1006}{18240}$$

$$= 0 \cdot 7998 \text{ or } 79 \cdot 98\% \text{ Ans.}$$

EQUIVALENT EVAPORATION

The evaporative capacity of a boiler is usually expressed as the lb weight of steam it can produce per hour.

However, since the feed water temperature and steam pressure varies with different boilers, it is necessary, for purposes of comparison, to reduce the evaporative capacity to the common standard of weight of water evaporated per lb of fuel burned if the feed water was 212°F (100°C) and dry saturated steam was produced at the same temperature, i.e. if the boiler worked under atmospheric pressure conditions. This basis of comparison is termed the *equivalent evaporation from and at* 212°F (or 100°C).

Thus, in the previous example, when one lb of fuel is burned, the heat received by the feed water at 210°F to produce steam at 220 lb/in^2 0·98 dry is 14·5 × 1006 Btu. If the feed water had been 212°F and dry saturated steam was produced at 212°F, each lb of steam would have received latent heat only. From the steam tables this is 970·6 Btu (or 539·22 Chu), therefore, the weight of steam that would be produced by one lb of fuel is:

$$\frac{\text{Total heat given to steam per lb of fuel}}{970·6 \text{ (or } 539·22)}$$

Hence, for this boiler burning the same grade of fuel,

$$\text{Equiv. evap. from and at } 212 = \frac{14·5 \times 1006}{970·6}$$

$$= 15·03 \text{ lb of steam per lb of fuel}$$

TEST EXAMPLES 11

1. The area of an indicator diagram taken off the H.P. cylinder of a quadruple expansion engine is 2·24 in², the length is 3·2 in. and the scale of the indicator spring is $\frac{1}{100}$. The diameter of the cylinder is 24 in., stroke 42 in. and speed 98 r.p.m. Calculate the mean effective pressure and i.h.p. developed in this cylinder. If the power developed in each cylinder is the same, find the total i.h p. of the engine.

2. The mid-ordinates measured from the I.P. engine of a triple expansion engine are: 0·5, 1·1, 1·45, 1·45, 1·4, 1·2, 0·8, 0·7, 0·45 and 0·25 inches respectively. The scale of the indicator spring is $\frac{1}{40}$. Find the mean effective pressure. If the diameter of the cylinder is 35 in., stroke 36 in., and speed 110 r.p.m., find the i.h.p. developed in this cylinder.

3. Steam at a pressure of 95 lb/in² gauge is admitted to an engine cylinder and cut off at 0·34 of the stroke. The back pressure is 10 lb/in² gauge. Neglecting clearance, calculate the hypothetical mean gross pressure and mean effective pressure.

4. Steam is supplied to the cylinder of an engine at 105 lb/in² gauge and cut off takes place when the piston has travelled 0·45 of its stroke. The clearance volume at each end of the cylinder is equal to 5% of the stroke volume. If the back pressure is 20 lb/in² gauge, calculate the theoretical mean effective pressure. Taking a diagram factor of 0·72, find the actual mean effective pressure.

ƒ5. Steam at 210 lb/in² gauge is supplied to the H.P. cylinder of a steam engine and cut off at 0·4 stroke. The back pressure is 75 lb/in² gauge, clearance volume equal to 8% of the stroke volume, and compression begins at 0·9 of the exhaust stroke. Assume expansion and compression to follow the law $PV = C$ and to continue to the ends of their respective strokes and calculate (i) the pressure at the end of compression, (ii) the mean effective pressure.

6. The cylinders of a triple expansion engine are 24, 38 and 60 inches diameter, and the stroke is 42 inches. When running at 108 r.p.m. the mean effective pressures in the cylinders are 75, 30 and 12 lb/in² respectively. Calculate (i) the i.h.p. developed in each cylinder, (ii) the mean effective pressure all referred to the L.P. cylinder, and (iii) the total i.h.p. of the engine.

7. The ratio of the cylinder volumes of a triple expansion engine is 1 : 2·5 : 6·5, the diameter of the H.P. cylinder is 27 in. and the stroke is 45 in. When the engine is running at 85 r.p.m., the mean effective pressures in the cylinders are 78·8, 31·7 and 12·5 lb/in^2 respectively. Calculate the mean effective pressure all referred to the L.P. and the total i.h.p. developed by the engine.

8. The diameters of the cylinders of a quadruple expansion engine are 27, 38, 53 and 75 in. respectively, and the referred mean pressure is 45 lb/in^2 when the mean piston speed is 800 ft/min. Assuming equal powers are developed in all cylinders, calculate the mean effective pressure in each cylinder and the total i.h.p. of the engine.

9. The diameters of the cylinders of a triple expansion engine are 28, 46 and 76 in. and the mean effective pressure referred to the L.P. is 34·5 lb/in^2. Find the mean effective pressure in each cylinder if the powers developed in the H.P. and I.P. are 10% and 5% respectively more than the power in the L.P.

ƒ10. A triple expansion engine is to develop 3500 i.h.p. with a mean piston speed of 720 ft/min. Calculate suitable cylinder diameters, taking a steam supply pressure of 220 lb/in^2 abs., back pressure 3 lb/in^2 abs., cut off in H.P. at 0·5 stroke, number of expansions 13·5, indicator diagram factor 0·7.

ƒ11. One cylinder of a vertical steam engine is 21 in. diameter. The stroke is 3 ft and the length of the connecting rod is 6 ft. The weight of the reciprocating parts is 1,200 lb. When the crank is 45° past top centre and the piston is moving down, the effective steam pressure on the piston is 105 lb/in^2 and the acceleration of the piston is 116 ft/sec^2. Calculate the effective downward load on the crosshead at this position, the turning moment, and the crank effort.

ƒ12. An engine is supplied with steam at a pressure of 200 lb/in^2 abs. with 100 F degrees of superheat, and the pressure of the exhaust is 2 lb/in^2 abs. Assuming isentropic expansion, find (i) the dryness fraction of the steam after expansion, (iii) the Rankine efficiency.

M.E.P., HORSE POWER AND EFFICIENCY OF STEAM ENGINES

13. The H.P. and L.P. cylinders of a triple expansion engine are $25\frac{1}{2}$ and 68 in. dia. respectively, and the stroke is 45 in. The boiler pressure is 210 lb/in² gauge. Calculate the shaft diameter using the formula:

$$\text{Shaft diameter} = \sqrt[3]{\frac{C \times P \times D^2}{f\left(2 + \frac{D^2}{d^2}\right)}}$$

Where C = length of crank in inches
P = boiler pressure in lb/in² abs.
D = diameter of L.P. in inches
d = diameter of H.P. in inches
f = 1110

14. During a test on an oil fired water-tube boiler the following data were recorded:
Pressure of steam leaving boiler = 250 lb/in² abs.
Temperature of steam = 581°F
Wt. of feed water entering boiler = 2500 lb/hour
Temperature of feed water = 190°F
Weight of fuel burned = 193 lb/hour
Calorific value of fuel = 18200 Btu/lb

Calculate the boiler efficiency and the equivalent evaporation from and at 212°F.

15. When the fuel burned in the boiler furnace has a calorific value of 10 kilo-calories per gram, the equivalent evaporation from and at 100°C is 15 kilograms of steam per kilogram of fuel. Find the thermal efficiency of the boiler.

16. An engine consumes 29500 lb of steam per hour when developing 2000 i.h.p., and 41300 lb of steam per hour when the i.h.p. is 3000. Estimate Willans' law for this engine and find the steam consumption (i) in lb per hour, (ii) in lb per i.h.p. per hour, when developing 2800 i.h.p.

CHAPTER 12

TURBINES

A steam turbine, like the steam reciprocating engine, is a machine for converting the heat energy in the steam into mechanical energy at the shaft, but the principles upon which these two engines work are entirely different. In the reciprocating engine the steam pressure acts as a static load on the piston to cause it to move up and down and this motion is converted from a reciprocating one into a rotary motion at the shaft by connecting rod and crank mechanism. In the turbine the rotor coupled to the shaft receives its rotary motion direct from the action of high velocity steam impinging on blades fitted into grooves around the periphery of the rotor, thus the action of the steam in a turbine is 'dynamic' instead of static as it is in the reciprocating engine.

There are two types of turbine, the impulse and the reaction. In both cases the steam is allowed to expand from a high pressure to a lower pressure so that the steam acquires a high velocity at the expense of pressure, and this high velocity steam is directed on to curved section blades which absorb some of its velocity; the difference is in the methods of expanding the steam.

In impulse turbines the steam is expanded in nozzles in which the high velocity of the steam is attained before it enters the blades on the turbine rotor, the pressure drop and consequent increase in velocity therefore takes place in these nozzles. As the steam passes over the rotor blades it loses velocity but there is no fall in pressure.

In reaction turbines, expansion of the steam takes place as it passes through the moving blades on the rotor as well as through the guide blades fixed to the casing.

THE IMPULSE TURBINE

As stated above, in the impulse turbine the high pressure steam passes into nozzles wherein it expands from a high pressure to a lower pressure and thus the pressure energy in the steam is converted into velocity energy (kinetic energy). The high velocity steam is directed on to blades fitted around the turbine wheel, the blades being of curved section so that the direction of the steam is changed thereby imparting a force to the blades to push the wheel around. The

TURBINES 251

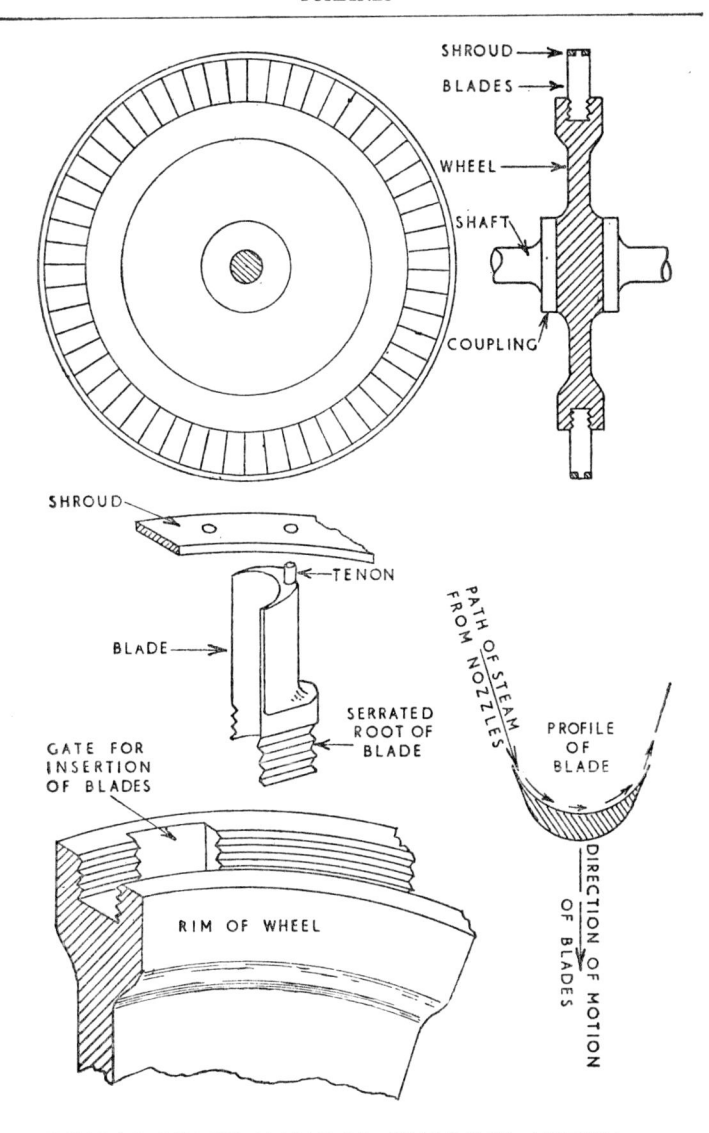

SINGLE STAGE IMPULSE TURBINE WHEEL

Fig. 94

simplest form of impulse turbine is the single-stage De Laval shown in Fig. 94. This consists of a solid wheel bolted by flanges to a shaft, blades of bronze or nickel steel are fixed into a groove around the rim of the wheel, caulked in, and a shroud or strap wrapped around the tips of the blades to strengthen them.

The best efficiency is obtained when the linear speed of the blades is half of the velocity of the steam entering the blades, thus, when one set of nozzles is used to expand the steam from its high supply pressure right down to the final low pressure, the resultant velocity of the steam leaving the nozzles is very high, say about 4000 ft/sec. To obtain a high efficiency it means therefore that the wheel should run at a very high speed so that the linear velocity of the blades approaches 2000 ft/sec, for example, in the case of a turbine wheel diameter of two feet, the speed would be about 19000 rev/min. Lower speeds, which are more suitable, can be obtained by pressure-compounding, or velocity-compounding, or a combination of these termed pressure-velocity-compounding.

In the pressure-compounded impulse turbine, the drop in pressure is carried out in stages, each stage consisting of one set of nozzles and one bladed turbine wheel, the series of wheels being keyed to the one shaft with nozzle plates fixed to the casing between the wheels.

In the velocity-compounded impulse turbine, the complete drop in steam pressure takes place in one set of nozzles but the drop in velocity of the steam is carried out in stages, by absorbing only a part of the steam velocity in each row of blades on separate wheels and having guide blades fixed to the casing at each stage between the wheels to guide the steam in the proper direction onto the moving blades.

The pressure-velocity-compounded turbine is a combination of the two.

CONSTRUCTION

In marine engines where the shaft is required to run in the astern direction as well as ahead, a separate astern turbine is necessary. Fig. 95 shows the principal parts of a pressure-velocity-compounded impulse turbine in which there are four pressure stages consisting of four sets of nozzles and four wheels in the ahead turbine, and two similar pressure stages in the astern turbine. Each wheel carries two rows of blades and there is one row of guide blades fixed to the casing protruding radially inwards between each row of moving blades to drop the velocity in two steps from each set of nozzles. The wheels are of forged steel and fitted on to a mild steel stepped shaft. The nozzle plates and casing to which they are fixed, are in halves. Leakage of steam between the pressure stages is prevented by the

TURBINES 253

Fig. 95

PRESSURE-VELOCITY COMPOUNDED IMPULSE TURBINE

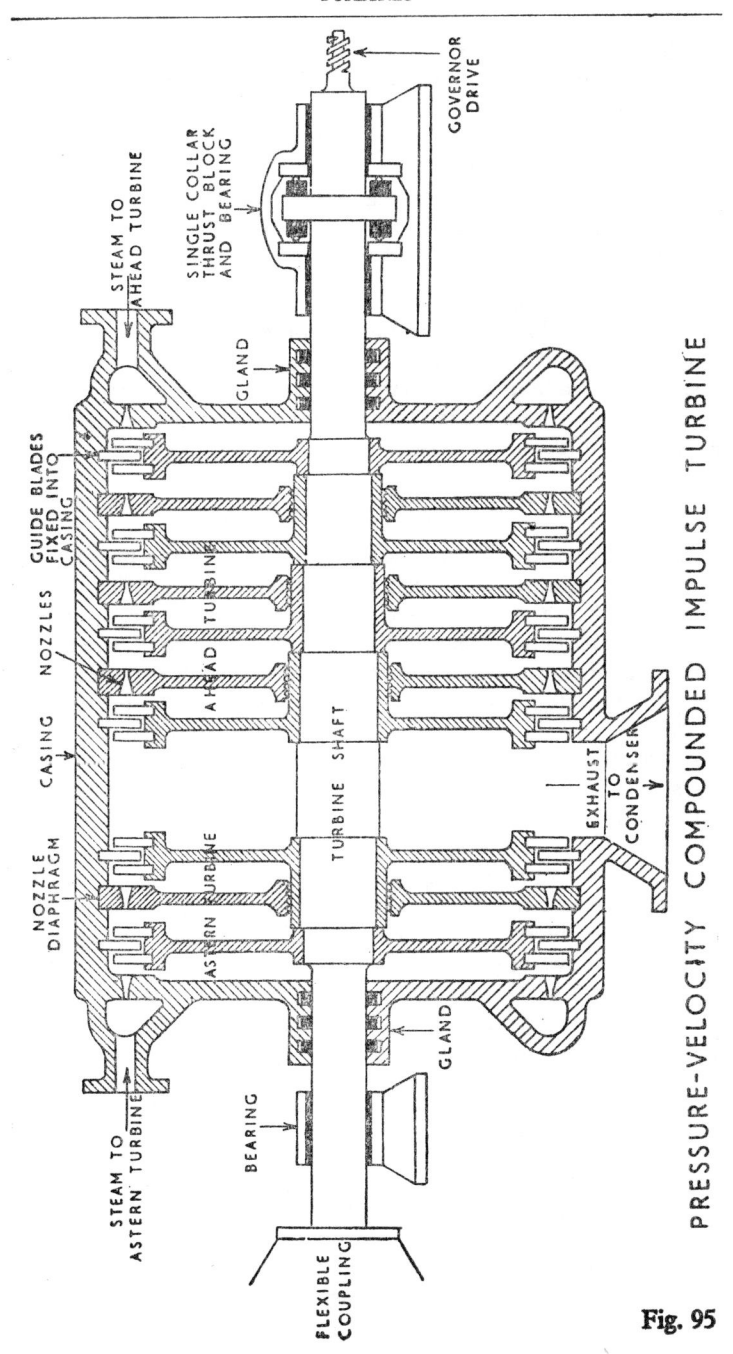

nozzle plates having a series of thin rings almost touching the bosses of the wheels, this is known as labyrinth packing. Leakage is prevented at the ends of the casing through which the shaft passes by glands containing carbon rings, each ring being composed of a number of segments with slight clearance between the butts and a garter spring is stretched around a groove in a periphery so that the carbon bears lightly on the shaft. A deflector plate is usually incorporated between the ahead and astern turbine.

To run in the ahead direction, the ahead steam stop valve is opened which allows steam to pass into the ahead turbine. To run astern, the ahead stop valve is closed and the astern stop valve is opened which admits steam to the astern turbine. The astern power, which is required mainly to brake the headway of the ship, is usually about 70% of the ahead power.

THE REACTION TURBINE

In the type usually known as the reaction turbine, the steam is expanded continuously through guide blades fixed to the casing and also as it passes through the moving blades on the rotor, on its way from the inlet end to the exhaust end of the turbine. There are no nozzles as in the impulse turbine. When the high pressure steam enters the reaction turbine, it is first passed through a row of guide blades in the casing through which the steam is expanded slightly, causing a little drop in pressure with a resulting increase in velocity, the steam being guided on to the blades in the first row of the rotor gives an impulse effect to these blades. As the steam passes through the rotor blades it is allowed to expand further so that the steam issues from them at a high relative velocity in a direction approximately opposite to the movement of the blades, thus exerting a further force due to reaction. This operation is repeated through the next pair of rows of guide blades and moving blades, then through the next and so on throughout a number of rows of guide and moving blades until the pressure has fallen to exhaust pressure. As explained, the action of the steam on the blades is partly impulse and partly reaction, and a more correct name for this type of turbine might be 'impulse-reaction' but it is generally known as 'reaction' to distinguish it from the pure impulse type.

As the steam falls in pressure it consequently increases in volume and to accommodate for the increasing volume of the steam as it passes through the turbine, the rotor and casing are made progressively larger in diameter from the H.P. end to the L.P., usually in steps, with larger area of annulus and longer blades.

TURBINES 255

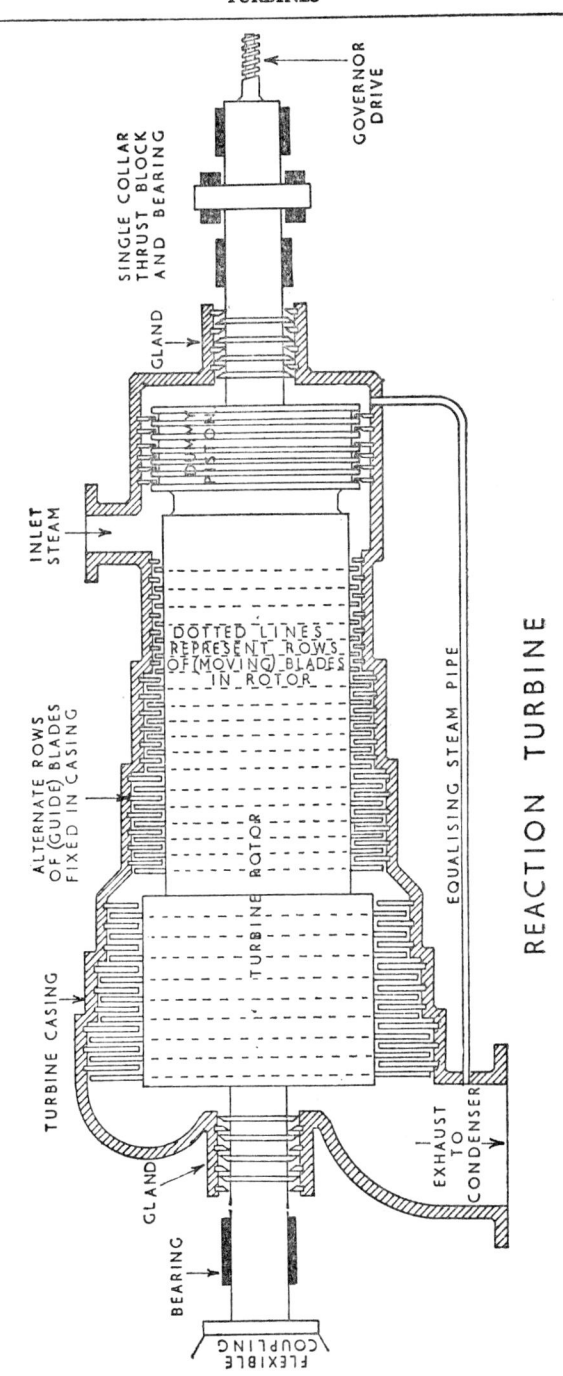

Fig. 96

CONSTRUCTION

The principal parts of a reaction turbine are shown in Fig. 96. The rotor consists of a steel drum with grooves around the outer circumference into which the blades are fitted, the drum is carried on a framework or spiders mounted on the shaft. At the steam inlet end of the rotor, a dummy piston is mounted for the steam pressure to act upon to balance the opposite axial thrust on the rotor blades; any unbalanced axial thrust being taken up by a single-collar thrust block at the forward end of the shaft, outside the casing.

The casing (or stator) is in halves with grooves around the inside circumference, into which the guide blades are fixed. The steam inlet branch is at the forward end and the exhaust branch at the after end. Carbon ring glands or glands of the labyrinth type, steam packed, seal the ends through which the shaft passes. White-metal lined brass bearings, forced lubricated, are incorporated in the casing outside each end which carry the weight of the rotor and shaft.

ASTERN RUNNING

As in all turbine installations on board ship, a separate astern turbine is required to drive the propeller in the reverse direction. This astern turbine may be mounted on the same shaft as the ahead turbine, it may be a completely separate unit geared to the main shaft, or the lay-out may be H.P. and I.P. ahead turbines on one turbine shaft and another shaft in parallel carrying the L.P. ahead and the astern turbines, each shaft being geared to the main shafting, (see Figs. 95 and 97). The ahead and astern turbines have their own separate steam stop valves, or it could be one steam chest with two valves and branches, specially designed so that it is not possible to open both ahead and astern steam valves at the same time.

GEARING

The speed at which turbines should run to obtain the best efficiency, is high, much higher than the economical speed of a ship's propeller, therefore to obtain the best performance from both turbine and propeller, reduction gearing is interposed between the turbine shaft and the propeller shaft, to enable both to run at their best speeds.

Single reduction gearing consists of one stage of speed reduction in which a pinion on the turbine shaft meshes with a gear wheel on the main shaft.

TURBINES

Double reduction gearing consists of two stages of speed reduction by means of a pinion on the turbine shaft meshing with an intermediate wheel, on the intermediate shaft a secondary pinion meshes with the main gear wheel on the propeller shafting.

The teeth in the wheels are cut at an angle to the axis of the shaft (these are known as helical teeth) for smooth running, and all gear wheels are arranged in pairs to balance any axial thrust caused by the teeth being helical. Fig. 97 shows diagrammatically a typical turbine lay-out including astern turbine and double-reduction gearing. Note the flexible couplings between turbine shafts and pinion shafts, these are to prevent thermal expansion of the turbine or misalignment of its shaft affecting the perfect meshing of the gear wheel teeth.

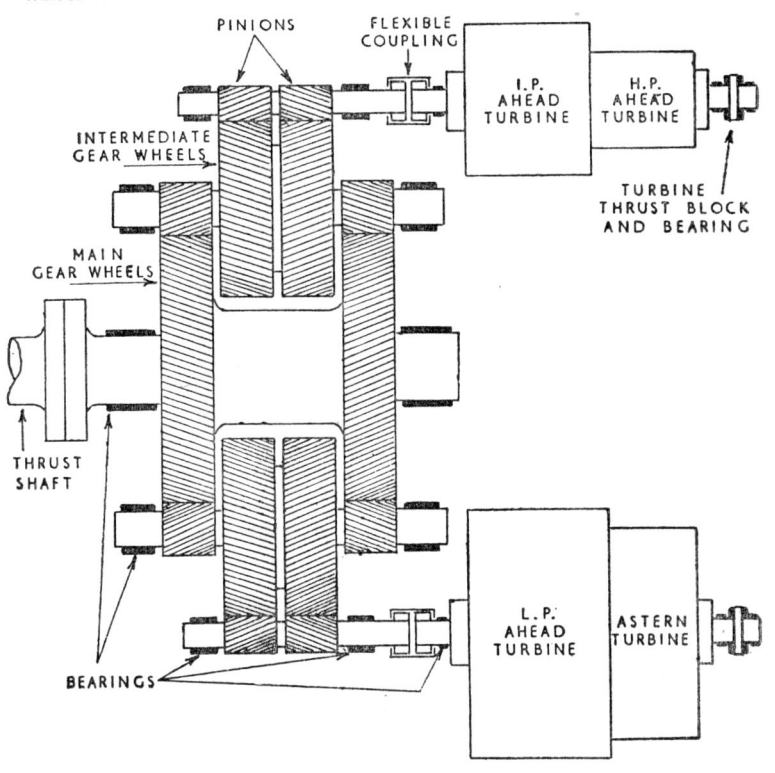

DOUBLE-REDUCTION GEARING

Fig. 97

EXHAUST TURBINES

One very great advantage of the steam turbine over the steam reciprocating engine is that the steam can be expanded to a lower pressure before exhausting it to the condenser, and hence extract more work out of the steam. When steam expands its volume increases, the lower the pressure the greater the volume, therefore to expand steam to a very low pressure in a reciprocating engine would require a very large diameter L.P. cylinder with a big heavy piston to be pushed up and down, whereas in a turbine the large volume can be accommodated simply by increasing the annular space between rotor and casing. The area of a pressure-volume diagram represents work done, and Fig. 98 shows the extra work that can be got out of a turbine by expanding the steam down to a lower pressure than that

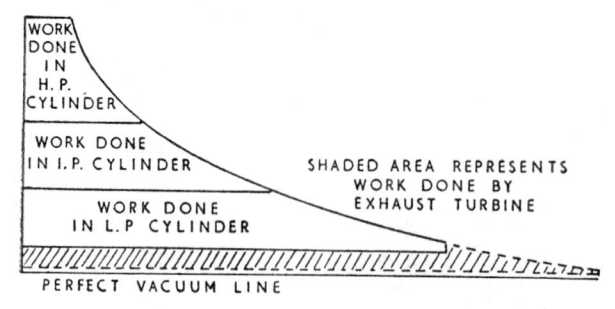

P V DIAGRAM FOR TRIPLE EXPANSION ENGINE AND EXHAUST TURBINE

Fig. 98

practicable in a reciprocating engine, this is done by carrying as high a vacuum as possible in a turbine condenser compared with a moderate vacuum of about 25 in. in a reciprocating engine condenser.

Some reciprocating engine installations include an exhaust steam turbine wherein the exhaust steam from the L.P. cylinder is directed into a low pressure turbine so that more work can be extracted from the steam before it is finally exhausted into the condenser.

Such an arrangement is illustrated in Fig. 99. As only an ahead-direction turbine is fitted, and used during normal running at sea, a change-over valve in the exhaust line from the reciprocating engine allows the exhaust steam to pass direct into the condenser, thus only the reciprocating engine is used for manoeuvring in and out of port; a hydraulic clutch allows the turbine shaft to be disengaged from the main shaft when the turbine is not in use.

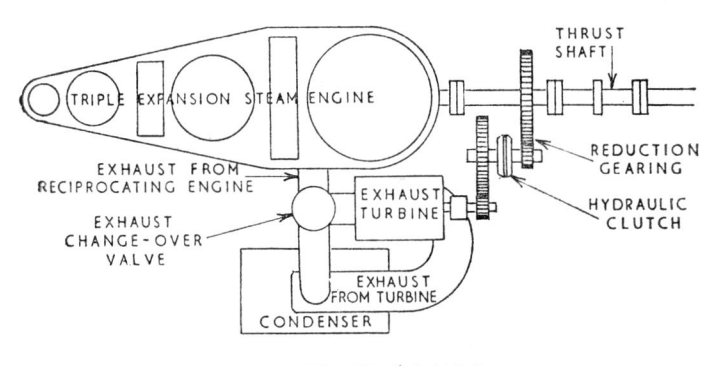

EXHAUST TURBINE

Fig. 99

ADVANTAGES OF STEAM TURBINES OVER STEAM RECIPROCATING ENGINES

Turbines are more suitable for high pressure and high temperature steam supply, and low terminal pressure; there is no initial condensation of the steam caused by cooler working parts due to the lower temperature exhaust steam returning along the same passages as the supply steam as happens in reciprocating engines; the result of the above is a higher thermal efficiency of engine with consequent less fuel required per given horse power.

Being a pure rotary motion, perfect balance of the engine can be obtained, therefore no vibration. No heavy reactionary forces on the bearings, therefore less wear. Not so many moving parts such as crossheads, cranks, eccentrics, link motion gear, etc. to be adjusted hence less maintenance required, less friction and higher mechanical efficiency.

Power for power, the size of a turbine is less than a reciprocating engine.

Against the turbine we have the disadvantage that it is non-reversible and therefore a separate astern turbine must be provided;

as the astern turbine is kept as small as practicable, the astern power available is less than the ahead power.

HORSE POWER OF TURBINES

As indicator cards cannot be taken off a turbine to obtain the indicated horse power as done in reciprocating engines, the brake horse power is measured by a special device fitted on the main shafting. This is known as a Torsionmeter and the principle upon which it works is to measure the angle of twist in the shaft over a given length, since the shaft is always slightly twisted when transmitting power and the angle of twist is proportional to the power transmitted. This information, together with the speed at which the shaft is running, is multiplied by a constant depending upon the diameter of the shaft, the length over which the measured twist takes place and the material of which the shaft is made, to obtain the shaft horse power.

There are different designs of torsionmeters. One type consists of a sheath or tube over part of the shafting, secured to the shaft at one end and supported but free at the other end. A link is connected from the free end of the sheath to a hinged mirror mounted on the shaft. A thin beam of light is directed on to the mirror which reflects on to a scale on some fixture nearby. The light therefore flashes on to the scale every time the mirror passes through the light beam as the shaft rotates. Since the shaft twists between that part of the shaft where the sheath is fixed and the part where the mirror is mounted, the link displaces the mirror slightly, the amount depending upon the angle of twist, and the reflection of the flashes of light move across the graduated scale from which the angle of twist is read.

REGENERATIVE CONDENSER

The function of a condenser is to condense the exhaust steam from the engine into water. The disadvantage of an ordinary condenser is that it goes beyond this requirement and cools the water to a lower temperature after it has been condensed, this is known as overcooling of the condensate and constitutes a loss of efficiency as the lost heat must be made up again at the expense of fuel. The regenerative condenser is designed so that the condensate leaves the condenser as near as possible to the temperature of the exhaust steam entering, and therefore in effect, only the latent heat is extracted from the steam.

Fig. 100 shows a section through a regenerative condenser. As the steam condenses, it drops on to sloping baffles and falls off like rain into a well at the bottom of the condenser. A gap is left between the nests of tubes to allow some of the exhaust steam to pass directly to the bottom part of the condenser where it mixes with the water droppings and surface of the water in the well, this heats up any water which is below the steam temperature; as this free

REGENERATIVE · CONDENSER

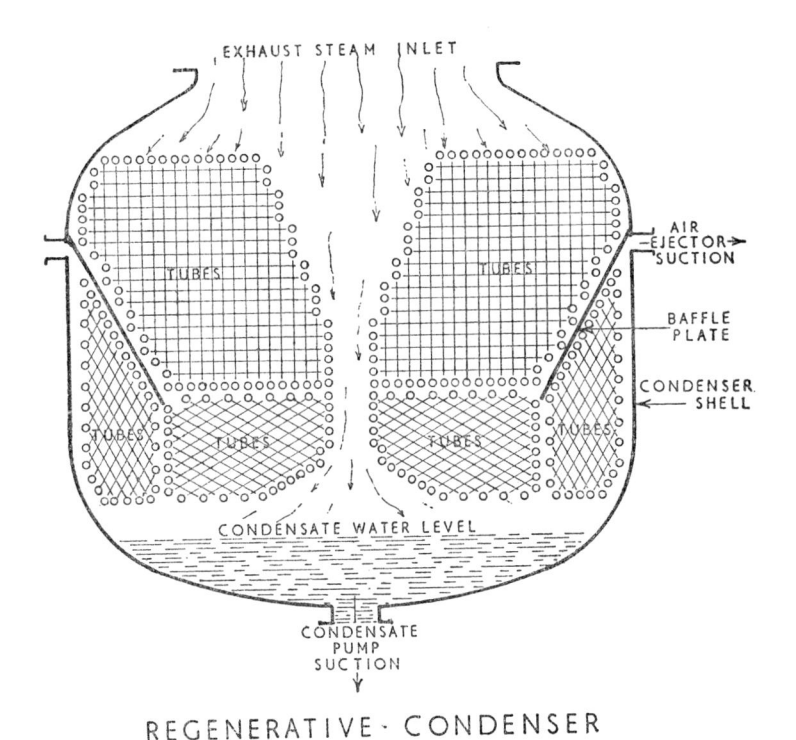

Fig. 100

passage steam condenses when it gives up its heat to the water, it falls itself as water to the bottom, any free steam not so condensed tends to rise towards the air-extraction outlet and condenses immediately on coming into contact with the lower tubes.

A condensate pump, float-controlled by the water level in the condenser well, extracts the condensed water, the air is extracted separately by a steam-jet air extractor, thus a very high vacuum is maintained which makes this type of condenser ideally suited to turbine installations; also most of the air is taken out of the water.

CLOSED FEED SYSTEM

After the condensate has been de-aerated in a regenerative condenser, the water, on its passage back to the boilers should not be allowed to come into contact with the atmosphere where it could absorb more air. This is achieved by having a completely closed feed water circuit known as the closed feed system. By feeding the boilers with water which carries a minimum quantity of air, corrosion due to air in the boilers is minimised.

NOZZLES

The function of a steam nozzle is to produce a jet of high velocity steam which can be directed on to the blades of a turbine.

The velocity is produced by allowing the steam to expand from a high pressure at the inlet to a lower pressure at the open exit. As no external work is done by the steam on its passage through the nozzle, the steam does work upon itself and the heat energy is converted into kinetic energy. The maximum heat drop between initial and exit pressures is obtained when the expansion is adiabatic, the nozzle efficiency is therefore expressed by the ratio of the actual heat drop to the adiabatic heat drop.

$$\text{Kinetic Energy} = \frac{Wv^2}{2g} \text{ ft lb} = \frac{Wv^2}{2gJ} \text{ heat units}$$

For one lb weight of steam, $\text{K.E.} = \dfrac{v^2}{2gJ}$, therefore:

$$\text{Heat drop} = \text{Gain in K.E.}$$

$$= \frac{v_2^2}{2gJ} - \frac{v_1^2}{2gJ}$$

If the initial velocity v_1 is negligible compared with v_2 then the velocity of the steam leaving the nozzle is given by:

$$v = \sqrt{2gJ \times \text{heat drop}}$$

or, as the value of the constant $\sqrt{2gJ} = \sqrt{2 \times 32 \cdot 2 \times 778} = 223 \cdot 9$
then, $\qquad v = 223 \cdot 9 \sqrt{\text{heat drop}}$

Example. Steam enters a nozzle at a pressure of 100 lb/in² abs. with 20 F degrees of superheat, and leaves at 60 lb/in² abs. 0·98 dry. Find (i) the velocity of steam at exit, and (ii) the weight of steam discharged per minute if the area of exit is 0·5 sq. in.

From steam tables,

Heat in steam at 100 lb/in², supht 20° = 1199·3 Btu/lb
Heat in steam at 60 lb/in², 0·98 dry = $h_f + qh_{fg}$
$\qquad = 262 \cdot 2 + 0 \cdot 98 \times 916 \cdot 2 = 1160 \cdot 1$ Btu/lb
Heat drop = 1199·3 — 1160·1 = 39·2 Btu/lb

$$v = \sqrt{2gJ \times \text{heat drop}}$$
$$= \sqrt{2 \times 32 \cdot 2 \times 778 \times 39 \cdot 2}$$

$$= 1402 \text{ ft/sec. Ans. (i)}$$

Rate of discharge (ft³/sec) = Area (ft²) × Velocity (ft/sec)
Spec. vol. of steam at exit = qv_g
$\qquad = 0 \cdot 98 \times 7 \cdot 175 \text{ ft}^3/\text{lb}$

$$\therefore \text{Weight discharged} = \frac{0 \cdot 5 \times 1402 \times 60}{144 \times 0 \cdot 98 \times 7 \cdot 175} \text{ lb/min}$$

$$= 41 \cdot 54 \text{ lb/min. Ans. (ii)}$$

Example. Dry saturated steam enters a convergent-divergent nozzle at a pressure of 50 lb/in² abs. The heat drop between entrance and throat is 41·8 Btu/lb and the pressure there is 29 lb/in² abs. The pressure at the exit from the nozzle is 1·5 lb/in² abs., the total possible heat drop from entrance to exit is 226·5 Btu/lb and this is reduced by 12% due to the effect of friction in the divergent part of the nozzle. Calculate the nozzle area in square inches at the throat and the mouth to pass 15 lb weight of steam per minute.

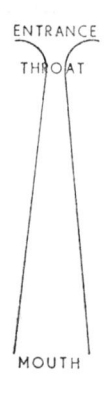

Fig. 101

From steam tables,
$P = 50$ lb/in², $\quad h_g = 1174 \cdot 8$ Btu/lb, $\quad v_g = 8 \cdot 516$ ft³/lb
$P = 29$ lb/in², $\quad h_f = 217,\quad h_{fg} = 947,\quad v_g = 14 \cdot 19$
$P = 1 \cdot 5$ lb/in², $\quad h_f = 83 \cdot 7,\quad h_{fg} = 1028 \cdot 1,\quad v_g = 228$

Heat in steam at 29 lb/in² = h_g at 50 — 41·8
$$217 + q \times 947 = 1174 \cdot 8 - 41 \cdot 8$$
$$q \times 947 = 916$$

∴ dryness fraction of the steam at the throat = 0·9674

Spec. vol. of steam at throat = qv_g
$$= 0 \cdot 9674 \times 14 \cdot 19 = 13 \cdot 73 \text{ ft}^3/\text{lb}$$

$$v = \sqrt{2gJ \times \text{heat drop}}$$

∴ Velocity through throat
$$= \sqrt{2 \times 32 \cdot 2 \times 778 \times 41 \cdot 8} = 1447 \text{ ft/sec}$$

Quantity flowing (ft³/sec) = area (ft²) × velocity (ft/sec)

$$\frac{15 \times 13 \cdot 73}{60} = \frac{\text{area (in}^2)}{144} \times 1447$$

TURBINES

$$\text{Area at throat} = \frac{15 \times 13 \cdot 73 \times 144}{60 \times 1447}$$

$$= 0 \cdot 3415 \text{ in}^2. \text{ Ans. (i)}$$

Heat drop between entrance and exit
$= 0 \cdot 88 \times 226 \cdot 5 = 199 \cdot 3$ Btu/lb

Heat in exit steam $= 1174 \cdot 8 - 199 \cdot 3 = 975 \cdot 5$ Btu/lb
$\therefore 83 \cdot 7 + q \times 1028 \cdot 1 = 975 \cdot 5$
$ q \times 1028 \cdot 1 = 891 \cdot 8$

$\therefore$ Dryness fraction of steam at exit $= 0 \cdot 8676$

spec. vol. at exit $= qv_g$
$\phantom{\text{spec. vol. at exit }} = 0 \cdot 8676 \times 228 = 197 \cdot 8 \text{ ft}^3/\text{lb}$

Velocity at exit $= \sqrt{2gJ \times \text{heat drop}}$
$\phantom{\text{Velocity at exit }} = \sqrt{2 \times 32 \cdot 2 \times 778 \times 199 \cdot 3}$
$\phantom{\text{Velocity at exit }} = 3160$ ft/sec.

Quantity discharged (ft^3/sec) = area (ft^2) $\times$ velocity (ft/sec)

$$\frac{15 \times 197 \cdot 8}{60} = \frac{\text{area (in}^2)}{144} \times 3160$$

$$\text{Area} = \frac{15 \times 197 \cdot 8 \times 144}{60 \times 3160}$$

$$= 2 \cdot 253 \text{ in}^2. \text{ Ans. (ii)}$$

VELOCITY DIAGRAMS FOR IMPULSE TURBINES

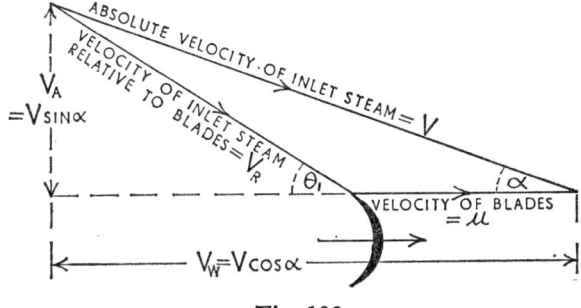

Fig. 102

Fig. 102 illustrates the vector diagram of velocities at the entrance side of the rotor or 'moving' blades. V represents the absolute velocity of the steam directed towards the blades at an angle α (as close as practicable) to the direction of their movement. The linear velocity of the blades is represented by u. Drawing the vectors of the steam velocity and blade velocity towards a common point, the vector joining these to form a closed figure is the velocity of the steam relative to the moving blades.

Velocity vector diagrams and relative velocity were explained in Volume II (Applied Mechanics), Chapter 2, to which the student may refer if revision is required.

The relative direction of the steam to the blades is θ_1 and in order that the steam should enter the blades without shock, the entrance edge of the blades must be in this direction. Therefore θ_1 is the *entrance angle* of the blades.

$V_W = V \cos \alpha$, this is the component of the velocity of the steam jet in the direction of blade movement, and is referred to as the *velocity of whirl at entrance*.

$V_A = V \sin \alpha$, this is the axial component of the steam jet, i.e. the component in the direction of the axis of the turbine.

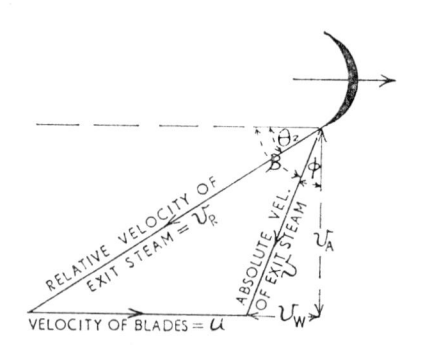

Fig. 103

Fig. 103 is the vector diagram of velocities at the exit side of the moving blades. θ_2 is the exit angle of the blades and the relative velocity of the exit steam, v_R, is in this direction. In impulse turbines, since there is no fall in steam pressure as it passes over the rotor blades, if friction is neglected the relative velocity at exit is the same magnitude as the relative velocity at entrance, that is, $v_R = V_R$. Friction between the steam and the blade surface reduces the velocity, and, to take friction into account, $v_R = kV_R$ where the velocity coefficient k is in the region of 0·8 to 0·95.

In the simple impulse turbine it is usual to make the blades symmetrical, whence $\theta_1 = \theta_2$.

u is the vector of the blade velocity. v is the vector of the absolute velocity of the exit steam. The direction of the exit steam is at angle β to the direction of blade movement, or at angle ϕ to the axis of the turbine.

$v_w = v \cos \beta$ or $v \sin \phi$, this is the component of the exit steam velocity in the direction of blade movement and is referred to as the *velocity of whirl at exit*. The axial component of the exit steam is $v_A = v \sin \beta$ or $v \cos \phi$.

Example. Steam at a velocity of 2000 ft/sec from a nozzle is directed on to the blades at 20° to the direction of blade movement. Calculate the inlet angle of the blades so that the steam will enter without shock when the linear velocity of the blades is 800 ft/sec. If the exit angle of the blades is the same as the inlet angle, find, neglecting blade friction, the magnitude and direction of the steam leaving the blades.

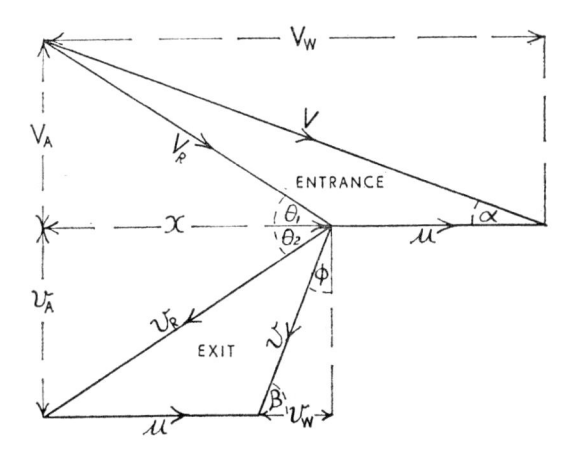

Fig. 104

$$V_A = V \sin \alpha = 2000 \times \sin 20° = 684 \text{ ft/sec}$$
$$V_W = V \cos \alpha = 2000 \times \cos 20° = 1879 \text{ ft/sec}$$
$$x = V \cos \alpha - u = 1879 - 800 = 1079$$

$$\tan \theta_1 = \frac{V_A}{x} = \frac{684}{1079} = 0{\cdot}6340$$

∴ Inlet angle of blades $\theta_1 = 32° 22'$ Ans. (i)

Neglecting friction across the blades, $v_R = V_R$, and, since $\theta_2 = \theta_1$, then x is common to both entrance and exit triangles, and $v_A = V_A$.

$$v_W = x - u = 1079 - 800 = 279 \text{ ft/sec}$$

$$\tan \phi = \frac{v_W}{v_A} = \frac{279}{684} = 0{\cdot}4079$$

∴ $\phi = 22° 11'$, and $\beta = 90 - 22° 11' = 67° 49'$

$$v = \frac{v_W}{\sin \phi} = \frac{279}{0{\cdot}3776} = 738{\cdot}9 \text{ ft/sec}$$

∴ Absolute velocity of exit steam = 738·9 ft/sec at 67° 49' to direction of blade movement, or 22° 11' to axis of turbine. } Ans.

Both entrance and exit velocity diagrams contain the vector of the blade velocity u, therefore they can be combined together by using the blade velocity as a common base. This is a more convenient diagram to solve either graphically or by calculation. Fig. 105 is a combined diagram. The student is advised to draw a diagram to scale for every problem as a check on his calculations.

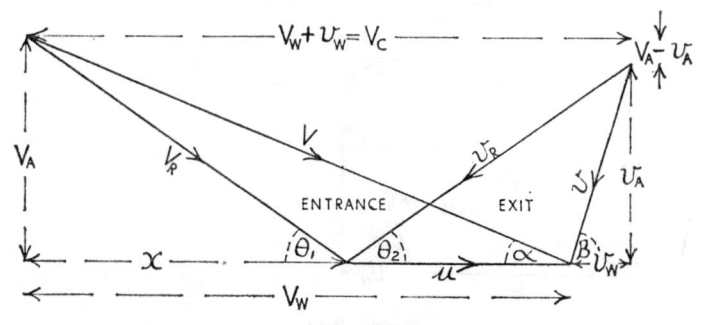

Fig. 105

ƒFORCE ON BLADES AND HORSE POWER

The effective velocity of the steam is the component in the direction of movement of the blades. This is the velocity of whirl. The velocity of whirl at entrance, V_W, is $V \cos \alpha$. The velocity of whirl at exit, v_w, is $v \cos \beta$. The effective change of velocity of the steam is the total velocity of whirl V_C, this is the algebraic difference of V_W and v_w. In the previous example, and in most cases, v_w is in the opposite direction to V_W, therefore,

Effective change of velocity, $V_C = V_W + v_w$

If v_w had been in the same direction as V_W, the effective change of velocity would be $V_W - v_w$.

Since the direction of motion of the steam is changed as it passes over the blades, then the *blades* must exert a force *on the steam* to cause this change.

From Newton's 3rd law of motion, which states that action and reaction are equal and opposite, this is also the magnitude of the force exerted by the *steam* on the *blades*.

Force = Change of momentum per second
Let W lb = weight of steam used per second, then:

$$\text{Force (lb)} = \frac{W \text{ (lb per sec)}}{g} \times \text{effective change of velocity (ft/sec)}$$

$$\therefore \text{Force on blades} = \frac{WV_C}{g} \text{ lb}$$

Work done = force × distance. As the distance through which the force acts on the blades is the linear distance moved by the blades, then:

Work done per second = Force on blades × blade velocity

$$= \frac{WV_C}{g} \times u \text{ ft lb/sec}$$

One horse power is equivalent to 33000 ft lb of work done per minute, or 550 ft lb per second, therefore:

$$\text{Horse power supplied} = \frac{WV_C u}{g \times 550}$$

The work supplied to the blades is the kinetic energy of the steam jet $= \dfrac{WV^2}{2g}$, hence,

$$\text{Blade efficiency} = \frac{\text{work done on blades}}{\text{work supplied}}$$

$$= \frac{WV_C u}{g} \div \frac{WV^2}{2g}$$

$$= \frac{2V_C u}{V^2}$$

The *axial* force of the steam on the blades is due to the difference between the axial components of the steam at entrance and exit, thus,

$$\text{Axial thrust} = \frac{W}{g}(V_A - v_A)$$

If $v_A = V_A$ as in the previous example, there is no axial thrust.

ƒExample. Steam leaves the nozzles of a single stage impulse turbine at a velocity of 2200 ft/sec at 19° to the plane of the wheel; and the steam consumption is 45 lb/min. The mean diameter of the blade ring is 3 ft 6 in. Find (i) the inlet angle of the blades to suit a rotor speed of 5,000 r.p.m. If the velocity coefficient of the steam across the blades is 0·9 and the blade exit angle is 32°, find (ii) the force on the blades, (iii) the horse power given to the wheel, and (iv) the blade efficiency.

TURBINES

Linear velocity of blades = mean circumference × revs. per sec.

$$= \frac{\pi \times 3.5 \times 5000}{60} = 916.4 \text{ ft/sec}$$

Referring to Fig. 105,
$V_A = V \sin \alpha = 2200 \times \sin 19° = 716.3$ ft/sec
$V_W = V \cos \alpha = 2200 \times \cos 19° = 2080$ ft/sec
$x = V_W - u = 2080 - 916.4 = 1163.6$ ft/sec

$$\text{Tan } \theta_1 = \frac{V_A}{x} = \frac{716.3}{1163.6} = 0.6156$$

∴ Entrance angle = 31° 37′ Ans. (i)

$$V_R = \frac{V_A}{\sin \theta_1} = \frac{716.3}{\sin 31° 37′} = 1367 \text{ ft/sec}$$

$v_R = 0.9 \, V_R = 0.9 \times 1367 = 1229$ ft/sec
$v_w = v_R \cos \theta_2 - u = 1229 \times \cos 32° - 916.4$
     $= 1042 - 916.4 = 125.6$ ft/sec

Effective change of velocity = $V_C = V_W + v_w$
     = 2080 + 125.6 = 2205.6 ft/sec

Weight of steam supplied = 45 ÷ 60 = 0.75 lb/sec

$$\text{Force on blades} = \frac{WV_C}{g}$$

$$= \frac{0.75 \times 2205.6}{32.2} = 51.37 \text{ lb. Ans. (ii)}$$

$$\text{Horse power} = \frac{\text{Force} \times u}{550}$$

$$= \frac{51.37 \times 916.4}{550} = 85.59 \text{ h.p. Ans. (iii)}$$

$$\text{Blade efficiency} = \frac{\text{work done on blades}}{\text{work supplied}} = \frac{2V_C^m u}{V^2}$$

$$= \frac{2 \times 2205 \cdot 6 \times 916 \cdot 4}{2200^2}$$

$$= 0 \cdot 8352 \text{ or } 83 \cdot 52\% \text{ Ans. (iv)}$$

VECTOR DIAGRAMS FOR REACTION TURBINES

It has been seen that, in the impulse turbine, the steam expands whilst passing through fixed nozzles and there is no expansion on its passage through the channels between the moving blades. All generation of velocity takes place in the nozzles.

In the reaction turbine, expansion of the steam takes place during its passage through the fixed (guide) blades which take the place of nozzles, and it also expands as it passes through the moving blades. Therefore the velocity of the steam is increased as it passes through the fixed blades, and the relative velocity of the steam to the moving blades is increased as it passes through the moving blades.

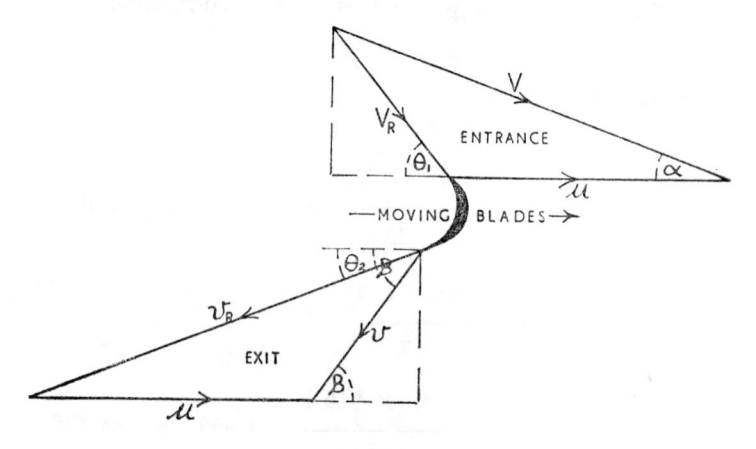

Fig. 106

In the Parson's reaction turbine, the fixed and moving blades are of the same section and reversed in direction, i.e. the entrance and exit angles of the fixed blades are the same as those of the moving blades. The velocity vector diagram at entrance is identical with the velocity vector diagram at exit, therefore the combined diagram is symmetrical. See Figs. 106 and 107. Thus the relative velocity of the steam at exit from the moving blades is equal to the absolute velocity at entrance, $v_R = V$, and the absolute velocity of the steam at exit from the moving blades is equal to the relative velocity at entrance, $v = V_R$. Hence, $\theta_2 = \alpha$, and $\theta_1 = \beta$.

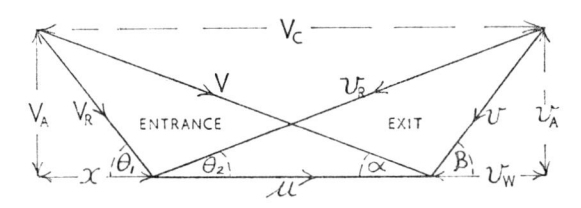

Fig. 107

Example. At one stage of a reaction turbine the velocity of the steam leaving the fixed blades is 300 ft/sec and the exit angle is 20°. The linear velocity of the moving blades is 200 ft/sec and the steam consumption is 1·2 lb/sec. Assuming the fixed and moving blades to be of identical section, calculate (i) the entrance angle of the blades, (ii) the force on the blades, and (iii) the stage horse power.

$$V_A = V \sin \alpha = 300 \times \sin 20° = 102·6 \text{ ft/sec}$$
$$V_W = V \cos \alpha = 300 \times \cos 20° = 281·9 \text{ ,,}$$
$$x = V_W - u = 281·9 - 200 = 81·9 \text{ ,,}$$

$$\text{Tan } \theta_1 = \frac{102·6}{81·9} = 1·253$$

∴ Entrance angle = 51° 24'. Ans. (i)

Effective change of velocity = $V_C = V_W + v_w$
(note that $v_w = x$)

$$V_C = 281·9 + 81·9 = 363·8 \text{ ft/sec}$$
$$\text{or, } V_C = 2V_W - u$$
$$= 2 \times 281·9 - 200 = 363·8$$

$$\text{Force on blades} = \frac{WV_C}{g}$$

$$= \frac{1\cdot 2 \times 363\cdot 8}{32\cdot 2} = 13\cdot 56 \text{ lb. Ans. (ii)}$$

$$\text{Horse power} = \frac{\text{Force} \times u}{550}$$

$$= \frac{13\cdot 56 \times 200}{550} = 4\cdot 929 \text{ h.p. Ans. (iii)}$$

GAS TURBINES

Marine type gas turbines usually work on the constant pressure (Joule) cycle. Fig. 108 is a diagrammatic sketch of a simple open cycle gas turbine which consists of the three essential parts—air compressor, combustion chamber, and turbine. Referring to Figs. 108 and 109, air is drawn in from the atmosphere and compressed from $P_1 V_1 T_1$ to the higher pressure, smaller volume and higher temperature $P_2 V_2 T_2$. The compressed air is delivered to the combustion chamber. Some of this air is used for burning the fuel which is admitted through the burner into the combustion chamber, the remainder of the air passes through the jacket surrounding the burner housing, mixes with the products of combustion, and is heated at constant pressure while the volume and temperature increases, the conditions now being $P_3 V_3 T_3$. The mixture of hot air and gases now passes through the turbine where it expands to $P_4 V_4 T_4$ as it does work in driving the rotor. Finally, the gases exhaust at constant pressure.

Of the power developed in the turbine, some is absorbed in driving the compressor, the remainder being available for external use such as driving an electric generator. A starting motor is fitted at the opposite end of the shaft to that the the external drive.

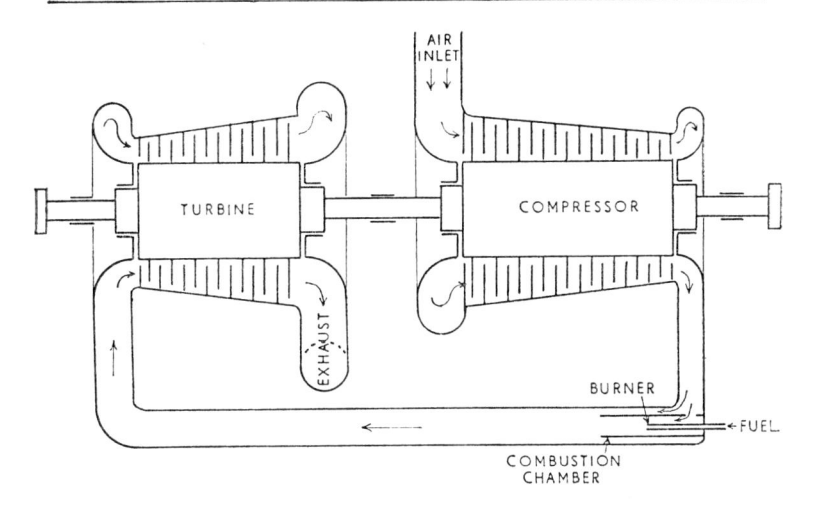

Fig. 108

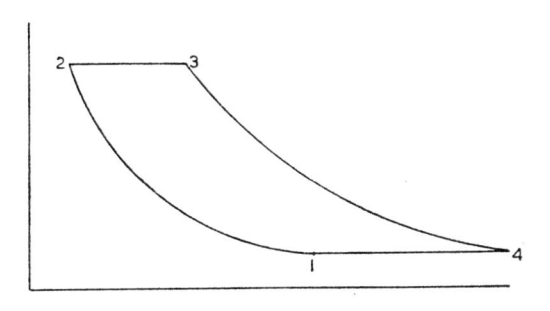

Fig. 109

In the ideal cycle, the ratio of compression V_1/V_2 is equal to the ratio of expansion V_4/V_3 and compression and expansion is adiabatic, i.e. following the law $PV^\gamma =$ constant. Therefore, referring to Fig. 109:

$$\text{Ideal thermal effic.} = \frac{\text{Heat converted into work}}{\text{Heat supplied}}$$

$$= \frac{\text{Heat supplied} - \text{Heat rejected}}{\text{Heat supplied}}$$

$$= 1 - \frac{\text{Heat rejected}}{\text{Heat supplied}}$$

$$= 1 - \frac{W \times C_P \times (T_4 - T_1)}{W \times C_P \times (T_3 - T_2)}$$

$$= 1 - \frac{T_4 - T_1}{T_3 - T_2}$$

This, it will be noted, is a similar expression as that for the constant volume cycle as explained in Chapter 6.

$$\text{Since } T_3 = T_4 \, r^{\gamma-1}, \text{ and } T_2 = T_1 \, r^{\gamma-1}$$
$$\text{then, } T_3 - T_2 = r^{\gamma-1}(T_4 - T_1)$$

$$\therefore \text{ Ideal thermal effic.} = 1 - \frac{T_4 - T_1}{T_3 - T_2}$$

$$= 1 - \frac{1}{r^{\gamma-1}}$$

$$\text{or } 1 - \frac{T_1}{T_2}$$

$$\text{or } 1 - \frac{T_4}{T_3}$$

Example. In a gas turbine working on the ideal constant pressure cycle, the ratio of compression is 10, the inlet temperature of the air taken into the compressor is 65°F and the inlet temperature to the turbine is 2050°F. Taking $\gamma = 1 \cdot 4$, calculate the temperatures at the end of compression and at exit from the turbine, and the ideal efficiency of the cycle.

Ratio of expansion = ratio of compression
$$\therefore V_4/V_3 = V_1/V_2 = 10$$

$$\frac{T_2}{T_1} = \left\{\frac{V_1}{V_2}\right\}^{\gamma-1}$$

$$\therefore T_2 = (65 + 460) \times 10^{0.4} = 1319°F \text{ abs.}$$

Temperature at end of compression
$$= 1319 - 460 = 859°F \text{ Ans. (i)}$$

$$\frac{T_4}{T_3} = \left\{\frac{V_3}{V_4}\right\}^{\gamma-1}$$

$$\therefore T_4 = \frac{(2050 + 460)}{10^{0.4}} = 999.3°F \text{ abs.}$$

Temperature at end of expansion
$$= 999.3 - 460 = 539.3°F \text{ Ans. (ii)}$$

$$\text{Ideal efficiency} = 1 - \frac{1}{r^{\gamma-1}} \quad \text{(or any of the alternatives given above)}$$

$$= 1 - \frac{1}{10^{0.4}} = 1 - 0.3981$$

$$= 0.6019 \text{ or } 60.19\% \text{ Ans. (iii)}$$

TEST EXAMPLES 12

1. Steam is expanded through a nozzle and the heat drop per lb of steam from the initial pressure to the final pressure is 57 Btu. Neglecting friction, find the velocity of discharge.

2. At the exit of a nozzle of circular section, the velocity of the steam is 3000 ft/sec and the specific volume of the exit steam is 105 ft^3/lb. Find the exit diameter to pass 600 lb of steam per hour.

3. Dry saturated steam at 115 lb/in^2 abs. is expanded in turbine nozzles to a pressure of 70 lb/in^2 abs. 0·97 dry. Find (i) the velocity at exit. If the area at exit is 2·25 in^2, find (ii) the weight of steam flowing through per hour.

ƒ4. Dry saturated steam at 200 lb/in^2 abs. is expanded in a turbine nozzle to 140 lb/in^2 abs. Assuming expansion to follow the law $PV^n = $ constant, where the value of n is 1·135, calculate:

  (i) the dryness fraction of the steam at exit,
  (ii) the heat drop through the nozzle per lb of steam,
  (iii) the velocity of discharge,
  (iv) the area of nozzle exit in sq. in. per lb of steam discharged per second.

5. Steam leaves the nozzles of an impulse turbine at a velocity of 2800 ft/sec at an angle of 18° to the direction of movement of the blades. The linear velocity of the blades is 1200 ft/sec and the inlet and exit angles of the blades are equal. Neglecting friction of the steam across the blades, calculate (i) the entrance and exit angles of the blades, (ii) the magnitude and direction of the steam at exit.

6. The velocity of the steam from the nozzles of a turbine is 1500 ft/sec, the angle of the nozzles to blade motion is 20° and the blade entrance angle is 33°. Find (i) the linear velocity of the blades so that the steam enters without shock, and (ii) the speed of the rotor in revolutions per minute if the mean diameter of the blade ring is 26 in.

ƒ7. Steam at a velocity of 2500 ft/sec is directed on to the blades of an impulse turbine at an angle of 22° to their motion. The blade speed is 750 ft/sec and the exit angle of the blades is 31°. Friction between the steam and blade surface reduces the relative velocity of the steam by 15%. Calculate (i) the entrance angle of the blades, and (ii) the magnitude and direction of the absolute velocity of the steam at exit from the blades.

*f*8. Steam enters the nozzles of an impulse turbine at a pressure of 300 lb/in^2 abs. with 200 F degrees of superheat, and leaves at 50 lb/in^2 abs. 0·98 dry. The angle of the nozzles is 20° to the direction of motion of the blades and the blade velocity is 1300 ft/sec. Calculate (i) the velocity of the steam leaving the nozzles, and (ii) the inlet angle of the blades.

*f*9. In a single stage impulse turbine, the velocity of the steam from the nozzles is 3000 ft/sec at 18° to the plane of the wheel. The linear velocity of the blades is 1100 ft/sec and the steam consumption is 750 lb per hour. The blades are symmetrical in shape and the blade friction loss is 10%. Find (i) the tangential force on the blades, (ii) the horse power developed, and (iii) the axial thrust.

*f*10. In an impulse turbine the theoretical heat drop through the nozzles is 138·5 Btu/lb of steam and 10% of this is lost in friction in the nozzles. The nozzle angle is 20°, the inlet angle of the blades is 35°, and the absolute velocity of the steam leaving the blades is 680 ft/sec in the direction of the axis of the turbine. Calculate on the basis of one lb of steam supplied per second:

(i) linear velocity of the blades so that there is no shock at steam entry,
(ii) blade angle at exit,
(iii) heat equivalent of the energy lost due to friction of the steam across the blades,
(iv) axial thrust,
(v) horse power supplied,
(vi) efficiency of the blading.

11. At a certain stage of a reaction turbine, the steam leaves the guide blades and enters the moving blades at an absolute velocity of 810 ft/sec at an angle of 23° to the plane of rotation, and the blade speed is 530 ft/sec. Fixed and moving blades have the same inlet and exit angles and the steam flow is 7200 lb per hour. Calculate (i) the inlet angle of the blades, (ii) the force on the blades, (iii) the work done per second at this stage, and (iv) the stage horse power.

*f*12. Air is drawn into a gas turbine working on the constant pressure cycle, at atmospheric pressure and 70°F, and compressed to a pressure of 14 atmospheres. The temperature at the end of the heat supply is 2150°F. Taking expansion and compression to be adiabatic where $\gamma = 1·4$ and $C_V = 0·17$, calculate (i) temperature at end of compression, (ii) temperature at exhaust, (iii) heat supplied per lb at constant pressure, (iv) increase in internal energy per lb from inlet to exhaust, (v) ideal thermal efficiency.

K

CHAPTER 13

COMBUSTION

Combustion of the fuel in the furnaces of boilers and in the cylinders of internal combustion engines is the chemical combination of the combustible elements in the fuel with the oxygen of air. The principal combustibles in all fuels, whether coal or oil, are carbon and hydrogen. Average coal requires a theoretical minimum of about 11 lb of air to completely burn one lb of coal, and oil about 14 lb of air per lb of oil. The air must be intimately mixed with the fuel and the amount of fuel which can be burned depends upon the quantity of air supplied. An excess of the theoretical minimum quantity of air is always necessary, the amount of excess depending upon the design of the furnaces and conditions under which the fuel is burned. If an attempt is made to burn fuel with an insufficient air supply, combustion will not be complete, the first indication of this being black smoke from the funnel; if too much air is supplied an unnecessary amount of heat will be carried away to waste up the funnel; each case represents a loss of efficiency.

The calorific value (heating value) of coal varies according to quality and is in the region of 11000 to 14000 Btu of heat per lb, and the calorific value of oil is about 18000 Btu per lb.

FURNACE DRAUGHT. Natural draught of air to the furnaces is caused by the hot gases rising up the funnel and air from the stokehold rushing into the furnaces to take its place. The intensity of natural draught depends upon the temperature of the funnel gases and the height of the funnel, thus, to obtain a good natural draught, these two evils are necessary; a high temperature of flue gases results in a large amount of heat being carried away to waste and a tall funnel would affect the stability of the ship.

Assisted draught systems are usually installed to enable more air to be supplied to the furnaces, and consequently more fuel to be burned, without relying upon natural conditions.

COMBUSTION 281

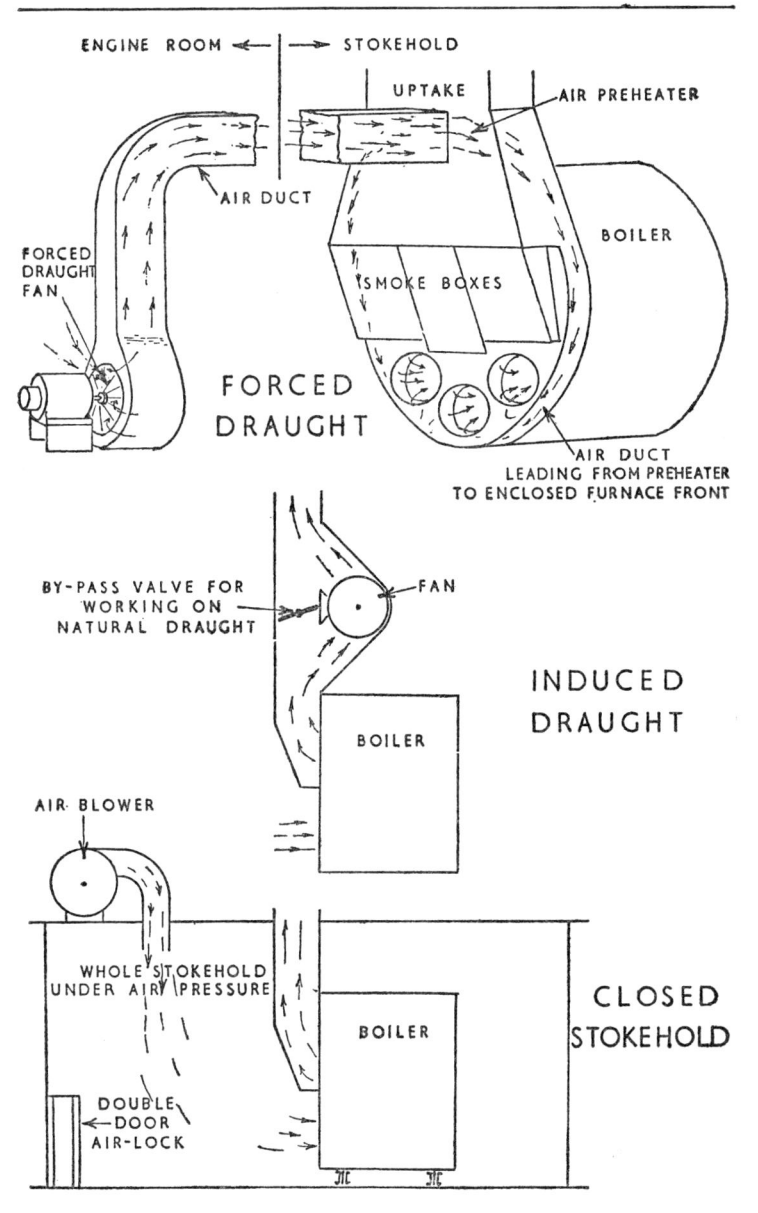

Fig. 110

FORCED DRAUGHT. A common forced draught system is diagrammatically shown in Fig. 110. It consists of a large fan in the engine room, steam engine or electric motor driven, which delivers air along an air duct to an enclosed furnace front. Included in the air ducting range is a tubular heater in the boiler uptake, the hot flue gases pass through the tubes on its way to the funnel and the forced draught air passes around the tubes to become heated on its way to the furnaces. The furnace front is a closed box with air valves to control the air supply.

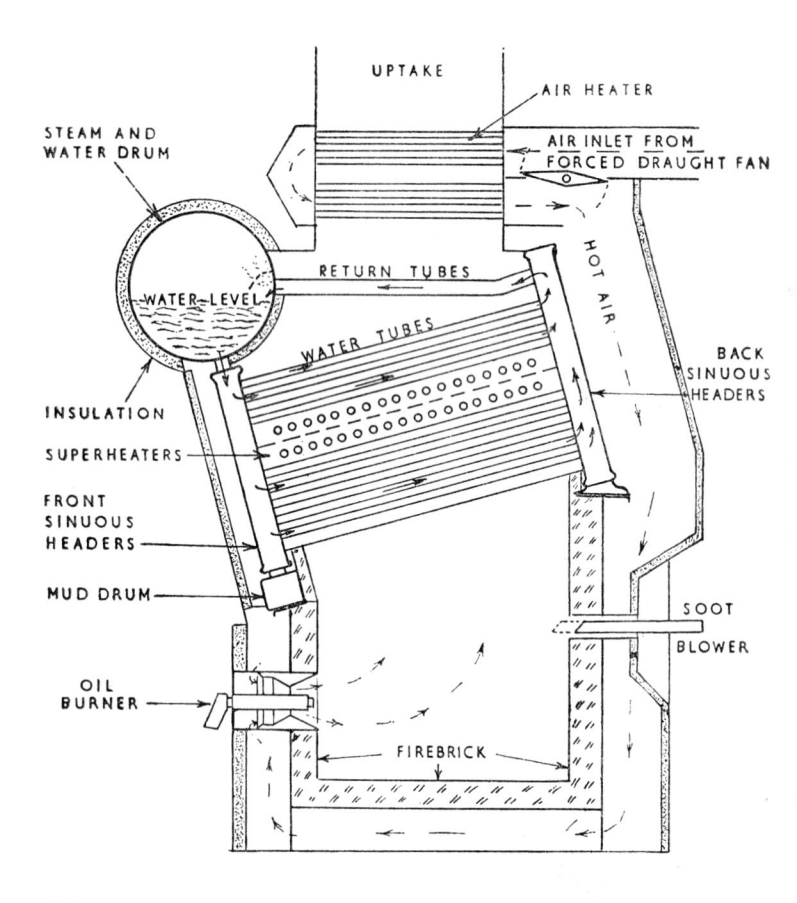

BABCOCK & WILCOX WATER TUBE BOILER

Fig. 111

COMBUSTION

The advantages of a forced draught system over natural draught are: better control of air supply therefore greater control of power; the air is heated prior to admission to the furnaces and this heat being reclaimed from the flue gases means that less heat of the fuel is wasted in heating the furnace air up to combustion temperature and not so much cooling effect on the furnaces; more air can be supplied therefore more fuel can be burned to generate more steam, hence smaller or fewer boilers are required.

Another method of producing more draught is the Induced Draught System. This consists of a large fan installed in the uptake which pulls the gases through the boiler and pushes them up the funnel, thereby inducing more air into the furnaces.

The intensity of the draught is measured by a hydrostatic gauge (Fig. 112) which consists of a glass U-tube containing water, one leg being connected to the fan casing and the other leg open to atmosphere. The difference in water level expresses the draught pressure in inches of water, each inch being approximately equal to one twenty-seventh of a lb per sq. in.

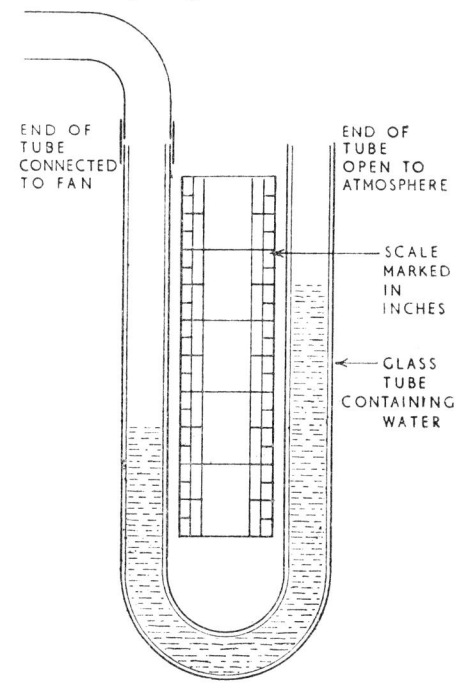

Fig. 112   HYDROSTATIC GAUGE

COAL BURNING

The fire should always be kept clear and bright by shovelling the coal into the furnaces in small quantities and at frequent intervals, cleaned regularly by removing the clinker from the grate and ashes from the ashpits, and the air supply above and below the grate adjusted to obtain best results. Any burnt and sagged firebars must be replaced by new ones at the first convenient opportunity.

From about 16 to 20 lb of coal can be burned per hour per sq. ft of grate with natural draught, and about 50% more than this with forced draught.

The advantages of coal over oil are that a plentiful supply of coal is available in this country, its lower price per ton, and lower cost of installation.

OIL BURNING

A diagrammatic sketch of an oil burning system is shown in Fig. 113. The oil fuel is stored in the double-bottom tanks and pumped from there as required into settling tanks by means of a transfer pump. The settling tanks are situated in the stokehold, two tanks each having a capacity of about 12 hours supply are usually installed so that the oil can stand for about 12 hours to allow any water in the oil to settle to the bottom and be drained off before use. The oil is drawn from the settling tank, through cold filters into the fuel pressure pump which discharges it at a pressure from 80 lb/in^2 upwards (depending upon the design of the system) through a steam operated heater where it is heated to about 200°F (depending upon the class of oil), through hot filters into the supply line to the sprayers in the boiler furnaces.

There is an oil circulating valve on the furnace distribution valve chest of every boiler which allows the oil to be returned to the pump suction so that any cold oil lying in the pipe line is cleared out for hot oil to take its place before attempting to light a fire.

The transfer pump, fuel pressure pump, heater and filters are in duplicate with cross-connections so that either set can be used while the other set is a stand-by.

COMBUSTION 285

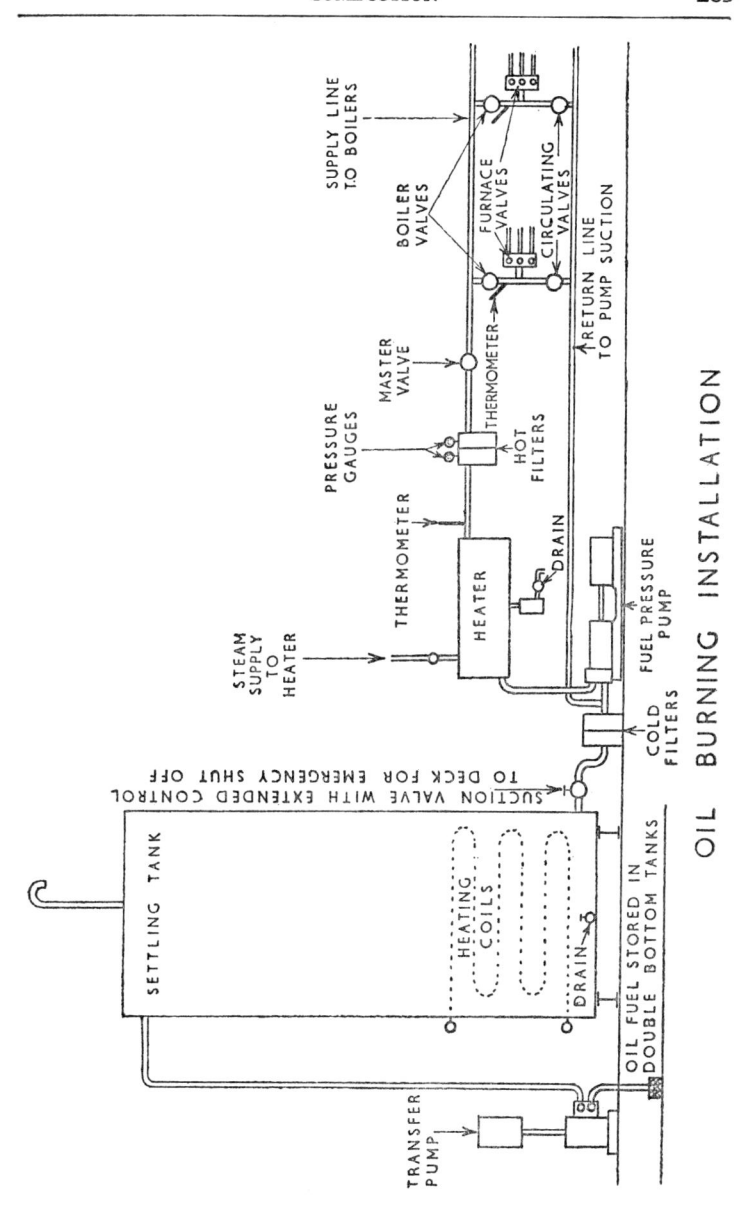

Fig. 113

The burner (Fig. 114), sometimes called a sprayer, is fitted centrally in the furnace front and protrudes into the furnace. The burner is a hollow tube with a diaphragm and nozzle in the end, the diaphragm has small holes drilled through at an angle to the axis so that the oil is broken up into a film or fine particles and acquires a spiral motion to mix readily with the air supplied over the burner as it enters the furnace.

An emergency shut off valve, usually of the quick closing type, is fitted between the settling tank and the cold filters, this valve has

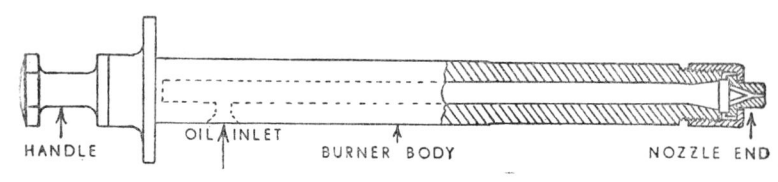

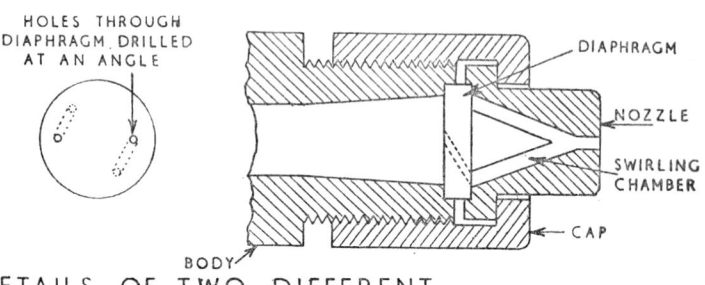

DETAILS OF TWO DIFFERENT TYPES OF NOZZLE ENDS

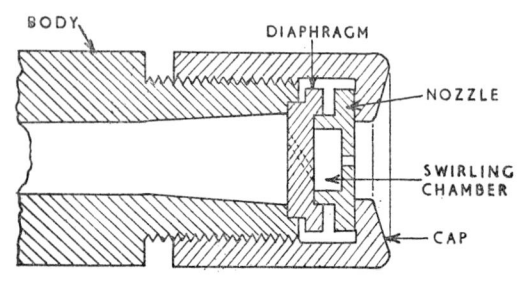

WALLSEND-HOWDEN OIL FUEL BURNER

Fig. 114

either an extended spindle or other remote control to the deck to enable the oil to be shut off in the case of an emergency such as a fire in the stokehold.

ADVANTAGES OF OIL OVER COAL. Less weight of oil is required for a given voyage because of its higher calorific value. Oil is conveniently stowed in the double-bottom tanks and does not require special bunkers, hence less space is taken up. It is much cleaner to handle both when bunkering and working, this is important on passenger ships. Bunkering is quicker and does not require trimming. Oil is under better control, a perfect ratio of air to fuel and good combustion can be maintained with resultant increase of efficiency, absence of smoke, and suitability for varying steam demands. Less stokehold staff required. No problems arise with regard to disposal of ash.

Against the above advantages there is a greater risk of fire, higher cost per ton of oil and higher cost of installation.

CHEMISTRY OF FUEL

An *element* is one single substance which cannot be analysed into two or more substances, and an *atom* is the supposed smallest particle of an element. The main combustible elements in solid and liquid fuels are carbon, hydrogen, and, in some fuels a little sulphur. As a method of short-hand the initial letters of their names are used as symbols to represent the elements, such as H for hydrogen, O for oxygen, etc. The number of atoms present is given by a suffix with the chemical symbol, thus C_2 represents two atoms of carbon, and S_4 represents four atoms of sulphur. If only one atom is present, the suffix 1 is usually omitted.

A *compound* is a chemical mixture of two or more elements, and a *molecule* is the smallest portion of a compound. The composition of a molecule is represented by placing together the chemical symbols of the atoms which constitute the molecule. For example, H_2O represents one molecule of water (or steam) which is composed of two atoms of hydrogen and one atom of oxygen. The number of molecules is given by placing that number in front of the chemical formula. Thus, $3CO_2$ represents three molecules of carbon dioxide, each molecule being composed of one atom of carbon and two atoms of oxygen.

Atoms cannot exist singly in gases, but must be combined with at least one other atom to form a molecule of that gas.

The *atomic weight* is the weight of one atom of an element compared with the weight of an atom of hydrogen. The atomic weight of hydrogen is taken as unity for practical calculations, since it is the lightest elementary gas.

The *molecular weight* of a substance is the weight of one of its molecules relative to the weight of an atom of hydrogen.

A list of the relative weights of the substances involved in calculations on the combustion of fuels is given below.

SUBSTANCE	SYMBOL	ATOMIC WEIGHT	MOLECULAR WEIGHT
Hydrogen	H	1	$H_2 = 1 \times 2 = 2$
Nitrogen	N	14	$N_2 = 14 \times 2 = 28$
Oxygen	O	16	$O_2 = 16 \times 2 = 32$
Carbon	C	12	
Sulphur	S	32	
Water or Steam	H_2O		$(1 \times 2) + 16 = 18$
Carbon monoxide	CO		$12 + 16 = 28$
Carbon dioxide	CO_2		$12 + (16 \times 2) = 44$
Sulphur dioxide	SO_2		$32 + (16 \times 2) = 64$

As previously stated, combustion is the chemical combination of the combustibles in the fuel with oxygen, heat being evolved in the process. The oxygen is obtained from the atmospheric air which is composed of approximately 23% oxygen and 77% nitrogen by weight. The oxygen is the active element in the process of combustion. Nitrogen, being an inert gas, takes no active part, it merely dilutes the products of combustion and, as it absorbs some of the heat, it reduces the temperature of combustion and carries heat away.

From the above proportion of oxygen and nitrogen:
 100 lb of air will produce 23 lb of oxygen

∴ $\frac{100}{23}$ lb ,, ,, ,, ,, 1 lb ,, ,,

Hence, for every lb of oxygen required to cause combustion, $\dfrac{100}{23} = 4\cdot 348$ lb of air must be supplied.

The calorific value of a fuel is the amount of heat evolved during complete combustion of unit weight of the fuel. For gaseous fuels it is stated per unit volume of the gas.

When hydrogen burns it combines with oxygen to form steam (H_2O) and about 62000 Btu of heat are given off for each lb of hydrogen burned.

Carbon, if supplied with sufficient oxygen, will burn completely to carbon dioxide (CO_2) and in doing so will evolve about 14500 Btu per lb of carbon. If there is a deficiency of oxygen, carbon monoxide (CO) will be formed instead of carbon dioxide, and the amount of heat evolved per lb of carbon so burned is comparatively small. Thus there is a great loss of heat when carbon monoxide is produced due to an insufficient air supply to the fuel. Carbon dioxide is often referred to as the gas of complete combustion, and carbon monoxide as the gas of incomplete combustion.

In the burning of sulphur, it chemically combines with oxygen to form sulphur dioxide (SO_2) and about 4000 Btu are given off per lb of sulphur.

Chemical equations represent the proportions in which the elements combine and, by substituting the atomic weights for the chemical symbols in these equations, the required weight of oxygen can be calculated for the burning of each combustible element. In the formation of these chemical equations, the following rules should be observed:

(i) Free elementary gases must contain 2 atoms per molecule. (e.g. O_2, N_2, H_2, etc.)

(ii) The number of atoms of each element must be the same on both sides of the equation.

COMBUSTION OF HYDROGEN:

$$2H_2 + O_2 = 2H_2O$$
$$\text{subs. atomic wts., } 2(1 \times 2) + (16 \times 2) = 2 \times 18$$
$$4 + 32 = 36$$
$$\text{reducing, } 1 + 8 = 9$$

Hence, 1 lb of hydrogen requires 8 lb of oxygen to burn it completely, and this produces 9 lb of steam.

COMBUSTION OF CARBON:

$$C + O_2 = CO_2$$
subs. atomic wts., $\quad 12 + (16 \times 2) = 44$
$\quad\quad\quad\quad\quad\quad\quad 12 + 32 = 44$
reducing, $\quad\quad\quad 1 + 2\frac{2}{3} = 3\frac{2}{3}$

Therefore, 1 lb of carbon requires $2\frac{2}{3}$ lb of oxygen to burn it completely and this produces $3\frac{2}{3}$ lb of carbon dioxide.

COMBUSTION OF SULPHUR:

$$S + O_2 = SO_2$$
subs. atomic wts., $\quad 32 + (16 \times 2) = 64$
$\quad\quad\quad\quad\quad\quad\quad 32 + 32 = 64$
reducing, $\quad\quad\quad 1 + 1 = 2$

Thus, 1 lb of sulphur requires 1 lb of oxygen to burn it completely, and this forms 2 lb of sulphur dioxide.

Some fuels contain a small amount of water and the complete analysis of such fuels includes the splitting up of this water into its elements of hydrogen and oxygen. For instance, suppose we have a fuel composed of 84% carbon, $11\frac{1}{2}$% hydrogen, and $4\frac{1}{2}$% water. The analysis of water is hydrogen and oxygen in the ratio of 1 : 8 by weight. Therefore $\frac{1}{9}$ of the water is hydrogen, and $\frac{8}{9}$ of the water is oxygen.

$\quad\quad\quad \frac{1}{9}$ of $4\frac{1}{2}$% $= \frac{1}{2}$% of hydrogen
$\quad\quad\quad \frac{8}{9}$ of $4\frac{1}{2}$% $= 4$% of oxygen

Therefore the complete analysis of the above fuel would be given as:

$\quad\quad\quad$ Carbon $= 84$%
$\quad\quad\quad$ Hydrogen $= 11\frac{1}{2} + \frac{1}{2} = 12$%
$\quad\quad\quad$ Oxygen $= 4$%

We see then, that when oxygen is given as part of the composition of a fuel, it really exists combined with some of the hydrogen in the form of water, and the full amount of hydrogen quoted is not all available for combustion. The amount of hydrogen which is combined with the oxygen present and which is not available for combustion, is one-eighth of the given amount of oxygen, therefore the hydrogen available for combustion is the remainder:

$$\text{Available hydrogen} = \text{total hydrogen} - \tfrac{1}{8} \text{ of wt. of oxygen}$$
$$= H - \frac{O}{8}$$

From the above we can now write the expressions:

$$\text{Calorific value} = 14500C + 62000 \left(H - \tfrac{O}{8}\right) + 4000S \text{ Btu/lb}$$
$$\text{Oxygen required} = 2\tfrac{2}{3}C + 8\left(H - \tfrac{O}{8}\right) + S \text{ lb oxygen/lb fuel}$$
$$\text{Air required} = \tfrac{100}{23}\left\{2\tfrac{2}{3}C + 8\left(H - \tfrac{O}{8}\right) + S\right\} \text{ lb air/lb fuel}$$

This gives the theoretical weight of air required. In practice about $1\tfrac{3}{4}$ to twice this weight of air would be needed in natural draught boilers, and about $1\tfrac{1}{2}$ times as much for forced draught.

Example. A fuel oil is composed of 86% carbon, 11% hydrogen, 2% oxygen, and 1% impurities. Calculate the calorific value and the theoretical weight of air required to burn 1 lb of the fuel.

In 1 lb of fuel there is:

0·86 lb carbon, 0·11 lb hydrogen, and 0·02 lb oxygen

$$\begin{aligned}
\text{Calorific value} &= 14500C + 62000\left(H - \tfrac{O}{8}\right) \\
&= 14500 \times 0.86 + 62000\left(0.11 - \tfrac{0.02}{8}\right) \\
&= 12470 + 62000 \times 0.1075 \\
&= 12470 + 6665 \\
&= 19135 \text{ Btu/lb Ans. (i)} \\
\text{Air required} &= \tfrac{100}{23}\left\{2\tfrac{2}{3}C + 8\left(H - \tfrac{O}{8}\right)\right\} \\
&= \tfrac{100}{23}\left\{2\tfrac{2}{3} \times 0.86 + 8 \times 0.1075\right\} \\
&= \tfrac{100}{23}\left\{2.293 + 0.86\right\} \\
&= \tfrac{100}{23} \times 3.153 \\
&= 13.71 \text{ lb air/lb fuel. Ans. (ii)}
\end{aligned}$$

HIGHER AND LOWER CALORIFIC VALUES

H_2O formed by the combustion of the hydrogen in the fuel cannot exist as water in the high temperature funnel gases, it must be in the form of steam. The steam passing away in the waste gases carries latent heat with it. Hence, the calorific value as given above is termed the *higher* (or *gross*) *calorific value*. The *lower* (or *net*) *calorific value* is expressed by subtracting from the higher calorific value the heat carried away in the formation of steam. Thus,

L.C.V. = H.C.V. — heat carried away in formation of steam

It is recommended by the Heat Engine Trials Committee that the heat carried away by the steam in the products of combustion should be taken as 1055 Btu/lb, hence,

L.C.V. = H.C.V. — weight of steam formed × 1055

BOMB CALORIMETER

The calorific values of solid and liquid fuels can be found experimentally by means of a bomb calorimeter. There are various different types but all are based on the same principle. The Darroch bomb calorimeter is shown diagrammatically in Fig. 115. It consists of a strong inner vessel known as the 'bomb', which has a rubber jointed cover held firmly down by a cover ring screwed to the top of the bomb. These parts are all made of monel metal. Fitted to the cover are (i) a support to carry the crucible into which a measured sample of fuel is placed, (ii) an oxygen inlet valve through which oxygen is supplied from an oxygen cylinder and shut off when the pressure is about 25 atmospheres, (iii) an electrode (insulated from the cover) to which is attached one end of a fine fuse wire, this fuse dips into the fuel and its other end is connected to the crucible support. The bomb rests on supports inside a calorimeter and a known weight of water is poured into the calorimeter to completely surround and cover the bomb. A stirrer and finely graduated thermometer is immersed in this water jacket. The calorimeter with bomb is usually placed inside an insulated outer vessel (not shown here) to minimise loss of heat. The fuse is heated by a battery (usually 12 volts) which ignites the fuel. The heat given out during combustion of the fuel causes the bomb and its fittings (of known water equivalent) and jacket water, to rise in temperature, uniform heating of the parts being obtained by continual stirring of the water during the experiment. Careful note is taken of the temperature rise and the calorific value of the fuel is calculated thus:

$$\text{Heat given out by the fuel} = \text{Heat absorbed by water} + \text{Heat absorbed by bomb and its fittings}$$

The result is the higher calorific value because the steam formed during combustion condenses inside the bomb and gives up its latent heat to the water jacket and bomb fittings.

COMBUSTION

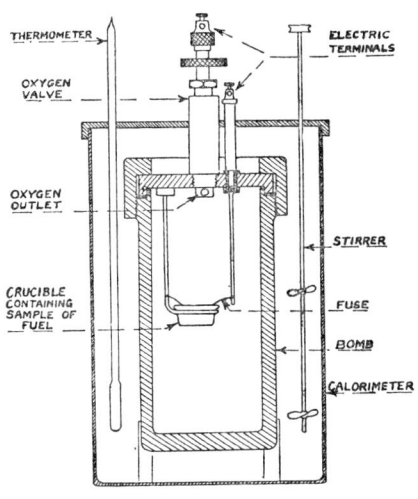

Fig. 115

Example. In an experiment to determine the calorific value of a fuel oil by means of a bomb calorimeter, the weight of the sample of oil was 0·7 gram, and 1·9 litres of water were contained in the calorimeter. The temperature was observed to rise through 3·1C degrees. If the water equivalent of the bomb and its fittings is 420 grams, calculate the calorific value of the oil and express it in (i) gram-calories per gram, (ii) Chu per lb, (iii) Btu per lb.

1·9 litres = 1900 cm^3,   ∴ wt. of water = 1900 grams

Heat given out by fuel = Heat absorbed by water and bomb

$$\text{Wt. of oil} \times \text{C.V.} = \left\{ \begin{array}{l} \text{wt. of} \\ \text{water} \end{array} + \begin{array}{l} \text{water equivalent} \\ \text{of bomb and fittings} \end{array} \right\} \times \text{temp. rise}$$

$$0\cdot7 \times \text{C.V.} = (1900 + 420) \times 3\cdot1$$

$$\text{C.V.} = \frac{2320 \times 3\cdot1}{0\cdot7}$$

= 10280 gram-cal per gram. Ans.
= 10280 Chu per lb

$$\frac{10280 \times 9}{5} = 18504 \text{ Btu per lb}$$

COMPOSITION OF FUNNEL GASES

An estimate of the analysis of the funnel gases can be calculated if the composition of the fuel and the weight of supply air is known, as demonstrated in the following:

Example. A fuel oil is composed of 85% carbon, 12% hydrogen, 2% oxygen, and 1% impurities, and 24 lb of air are supplied per lb of fuel in the boiler furnaces. Estimate the composition of the funnel gases.

Each lb of fuel contains 0·85 lb of carbon, 0·12 lb of hydrogen, 0·02 lb of oxygen, and 0·01 lb of impurities. Assuming the impurities to be deposited in the furnaces, the remainder of the oil passes up the funnel as gases with the air supply. The total weight of funnel gases per lb of fuel burned is,

24 lb air + 0·85 lb C + 0·12 lb H + 0·02 lb O = 24·99 lb
Wt. of nitrogen in 24 lb air = $\frac{77}{100}$ × 24 = 18·48 lb
Wt. of oxygen in 24 lb air = $\frac{23}{100}$ × 24 = 5·52 lb

Weight of oxygen to burn 0·85 lb carbon

$$= 2\tfrac{2}{3} \times 0.85 = 2.267 \text{ lb}$$
Weight of CO_2 formed = 0·85 + 2·267 = 3·117 lb

Hydrogen available for combustion = $H - \dfrac{O}{8}$

$$= 0.12 - \dfrac{0.02}{8} = 0.1175 \text{ lb}$$

Wt. of oxygen to burn hydrogen = 8 × 0·1175 = 0·94 lb
Wt. of H_2O in funnel gases = 9 × 0·12 = 1·08 lb

(note that all the hydrogen will be combined with oxygen in the form of steam in the funnel gases).

Wt. of oxygen required for combustion of carbon and hydrogen

$$= 2.267 + 0.94 = 3.207 \text{ lb}$$
Surplus oxygen = 5·52 − 3·207 = 2·313 lb

COMBUSTION

Hence, the gases formed by the combustion of one lb of oil are:

$$\begin{aligned}
\text{Nitrogen} &= 18\cdot 48 \\
CO_2 &= 3\cdot 117 \\
H_2O &= 1\cdot 08 \\
\text{Free oxygen} &= 2\cdot 313 \\
\text{Total} &= 24\cdot 99 \text{ lb}
\end{aligned}$$

Composition expressed as percentages:

$$\text{Nitrogen} = \frac{18\cdot 48}{24\cdot 99} \times 100 = 73\cdot 94\%$$

$$CO_2 = \frac{3\cdot 117}{24\cdot 99} \times 100 = 12\cdot 48\%$$

$$H_2O = \frac{1\cdot 08}{24\cdot 99} \times 100 = 4\cdot 32\%$$

$$\text{Oxygen} = \frac{2\cdot 313}{24\cdot 99} \times 100 = 9\cdot 26\%$$

CONVERSION FROM VOLUME TO WEIGHT ANALYSIS

Practical tests of funnel gases usually produce the analysis of dry flue gases by volume and this may be converted later into weight analysis.

By Avogadro's law, equal volumes of any gas, at the same temperature and pressure, contain the same number of molecules, therefore, although the *weight* of a molecule of one gas is different to the weight of a molecule of another gas, the *volume* of each molecule is the same, at the same temperature and pressure.

Hence the ratio of the volumes of each gas in the mixture of flue gases can be converted into a ratio of weights by multiplying the volumetric ratio by the molecular weight of the gas, and from this the percentage composition by weight can be calculated.

Example. In a test on a sample of funnel gases from an oil fired boiler, the percentage composition by volume was found to be: $CO_2 = 8.5\%$, Oxygen $= 9.5\%$, CO $= 3\%$, and the remaining 79% was assumed to be nitrogen. Convert these quantities into a weight analysis.

GAS	RATIO OF VOLUMES OF EACH GAS	RATIO OF WEIGHTS OF EACH GAS	RATIO OF WEIGHTS EXPRESSED AS PERCENTAGE
CO_2	8.5	$8.5 \times 44 = 374$	$\frac{374}{2974} \times 100 = 12.58\%$
O_2	9.5	$9.5 \times 32 = 304$	$\frac{304}{2974} \times 100 = 10.22\%$
CO	3.0	$3 \times 28 = 84$	$\frac{84}{2974} \times 100 = 2.82\%$
N_2	79.0	$79 \times 28 = 2212$	$\frac{2212}{2974} \times 100 = 74.38\%$
		Total $= 2974$	

TEST EXAMPLES 13

1. The analysis of a sample of a coal is 81% carbon, 5% hydrogen, 5·6% oxygen, 1% sulphur, and the remainder is ash. Calculate the calorific value of this coal.

2. A certain fuel oil is composed of 85·2% carbon, 12% hydrogen, 1·6% oxygen, and 1·2% impurities. Calculate the calorific value and the theoretical weight of air required per lb of fuel.

3. The constituents of a fuel are 85% carbon, 13% hydrogen, and 2% oxygen. When burning this fuel in a boiler furnace, the air supply is 50% in excess of the theoretical minimum required for complete combustion, the inlet temperature of the air being 87°F and the funnel temperature 535°F. Calculate (i) the calorific value of the fuel, (ii) weight of air supplied per lb of fuel, and (iii) the heat carried away to waste in the funnel gases expressed as a percentage of the heat supplied, taking the specific heat of the flue gases as 0·24.

4. In an experiment to determine the calorific value of a fuel oil by means of a bomb calorimeter, the weight of the sample of fuel was 0·75 gram, weight of water surrounding the bomb 1800 grams, water equivalent of bomb and fittings 470 grams, and the rise in temperature was 3·3 C degrees. Calculate the calorific value of this fuel in Btu/lb.

ƒ5. A fuel oil consists of 84% carbon, 13% hydrogen, 2% oxygen, and the remainder incombustible solid matter. Calculate (i) the calorific value, (ii) the theoretical weight of air required per lb of fuel, and (iii) an estimate of the analysis of the funnel gases if 22 lb of air are supplied per lb of fuel burned.

ƒ6. The analysis of a fuel oil is 85·5% carbon, 11·9% hydrogen, 1·6% oxygen, and 1% impurities. Calculate the percentage CO_2 in the funnel gases when (i) the quantity of air supplied is the minimum for complete combustion, and when the excess air over the minimum is (ii) 25%, (iii) 50%, and (iv) 75%.

CHAPTER 14

REFRIGERATION

The natural flow of heat is from a hot body to a colder body. The function of a refrigeration plant is to act as a heat pump and reverse this process so that rooms can be maintained at low temperatures for the preservation of foodstuffs.

Refrigerating machines can be divided into two classes, (i) those which require a supply of mechanical work which is the vapour compression system, and (ii) those which require a heat supply and work on the absorption system. The former is more efficient and in general use on board ship, the latter is more suited to domestic use, therefore only the vapour compression system will be described here.

The refrigerating agent used in the circuit is a substance which will evaporate at low temperatures. The boiling and condensation points of a liquid depend upon the pressure exerted upon it, for example, if water is under atmospheric pressure it will vaporise at 212°F, if the pressure is 100 lb/in^2 the water will not change into steam until its temperature is 338°F, at 200 lb/in^2 the boiling point is 388°F and so on. The refrigerant used must vaporise at very low temperatures. The boiling point of carbon dioxide at atmospheric pressure is about −109°F (note *minus* 109), by increasing the pressure the temperature at which liquid CO_2 will vaporise (or CO_2 gas will condense) is raised accordingly so that any desired vaporisation and condensation temperature can be attained, within certain limits, by subjecting it to the appropriate pressure.

Some of the agents employed as refrigerants and their more important characteristics are as follows:

Carbon dioxide (CO_2). This is an inert gas, non-poisonous, odourless, and has no corrosive action on the metals. Its natural boiling point is very low which means that it must be run at very high pressures to bring it to the conditions where it will vaporise and condense at the normal temperatures of a refrigerating machine. A further disadvantage is that its critical temperature is about 88°F which falls within the range of sea water temperatures. At its critical temperature and above, it is impossible to liquefy the gas no matter to what pressure it is subjected and, as part of the circuit depends upon condensing the gas in a condenser circulated by sea water, great difficulty is experienced, even with special additional devices, when in high temperature tropical waters.

Ammonia (NH_3). This is a poisonous gas and therefore an ammonia machine should not be open to the engine room but have a compartment of its own so that it can be sealed off in the case of a serious leakage; water will absorb ammonia and therefore a water spray is a good combatant against a leakage. Ammonia will corrode copper and copper alloys and therefore parts in contact with it should be made of such metals as nickel steel and monel metal. Its natural boiling point is about minus 38°F therefore the pressures required throughout the system to obtain the necessary evaporation and liquefaction temperatures are much lower than those required in a CO_2 machine.

Methyl chloride (CH_3Cl). This gas is slightly poisonous, inflammable and, under certain conditions, corrosive. It requires comparatively low pressures throughout the system.

Freon 12 (CCl_2F_2) or Arcton 6, the former is the American name and the latter the British name for the same refrigerant. This also requires only low pressures but has the advantage over methyl chloride in that it is non-inflammable, non-poisonous and non-corrosive.

WORKING CYCLE

Fig. 116 illustrates diagrammatically the essential components of the vapour compression system, which consists of the Compressor driven by a steam engine or electric motor, the Condenser which is circulated by sea water, the Regulator (sometimes referred to as the expansion valve), and the Evaporator which is circulated by brine. It is a completely enclosed circuit, the same quantity of refrigerant passes continually through the system and it only requires to be charged when there are losses due to leakage. Any of the above agents may be used as the refrigerant, the difference in their working being only the pressures throughout the system, therefore in the following description the words 'refrigerant', 'the liquid' and 'the gas', etc. will be used instead of CO_2, ammonia, etc.

The refrigerant is drawn as a gas at low pressure from the evaporator into the compressor where it is compressed to a high pressure and delivered into the coils of the condenser in the state of a superheated gas. As it passes through the condenser coils, the gas is cooled and condensed into a liquid at approximately sea water temperature, the latent heat given up being absorbed by the sea water surrounding the coils and pumped overboard. The liquid, still at a high pressure, passes along to the regulator, this is a valve just partially open to limit the flow through it. The pipe-line from the discharge side of the compressor to the regulating valve is under high pressure but from the regulating valve to the suction side of the compressor is at low pressure due to the regulating valve being just a little way open. As the liquid passes through the regulator from a region of high pressure to a region of low pressure throttling occurs and some of the liquid automatically changes itself into a gas absorbing the required amount of heat to do so from the remainder of the liquid and causing it to fall to a low temperature, this temperature

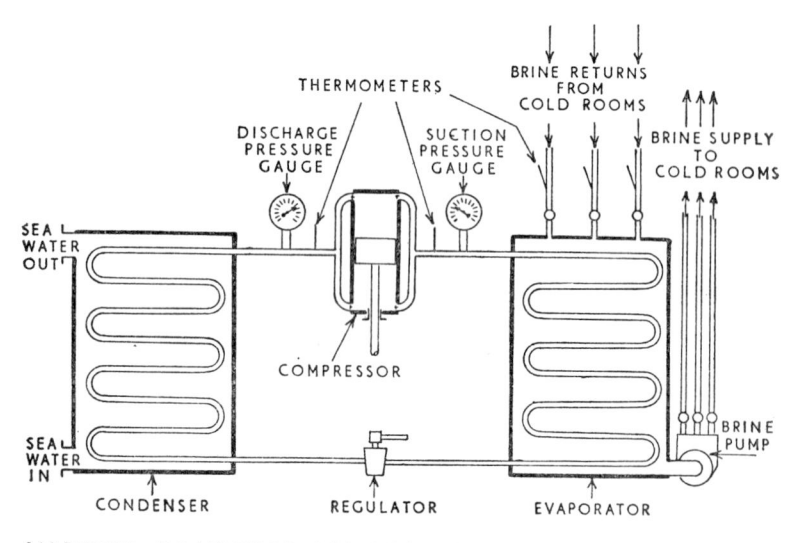

VAPOUR COMPRESSION REFRIGERATING MACHINE

Fig. 116

being regulated by the pressure, so that the refrigerant enters the evaporator at a temperature lower than that of the brine. The liquid (more correctly a mixture of liquid and gas) now passes through the coils of the evaporator where it receives heat to evaporate it into a gas before being taken into the compressor to go through the cycle again. The required heat absorbed by the liquid refrigerant in the coils of the evaporator to cause evaporation is extracted from the brine surrounding the coils resulting in the brine being cooled to a low temperature.

BRINE CIRCULATION

The cold brine from the evaporator is pumped through pipes led around the top of the walls and/or ceiling of the cold rooms where it extracts heat from the rooms before it is returned to the evaporator.

The quantity of brine flowing through the pipes of any one room determines the temperature of that room, therefore the control of the room temperature is by regulating the opening of the valve on the brine return at the evaporator (assuming that the brine supply to the pipes is at a sufficiently low temperature). Each room is provided with a thermometer so that the temperature inside can be taken regularly and recorded, there is also a thermometer at each brine return which gives a good indication to the experienced engineer of the room temperature.

In fruit carrying rooms the air must be continually circulated to prevent stagnant pockets of CO_2 forming around the fruit, therefore the cold air system of cooling is employed. Instead of wrapping the brine pipes around the walls and ceiling inside the rooms, the brine is led into grid-boxes through which the air, drawn from the bottom of the rooms by fans or blowers, is passed over the brine grids and blown back into the rooms via ducting along the ceiling, see Fig. 117.

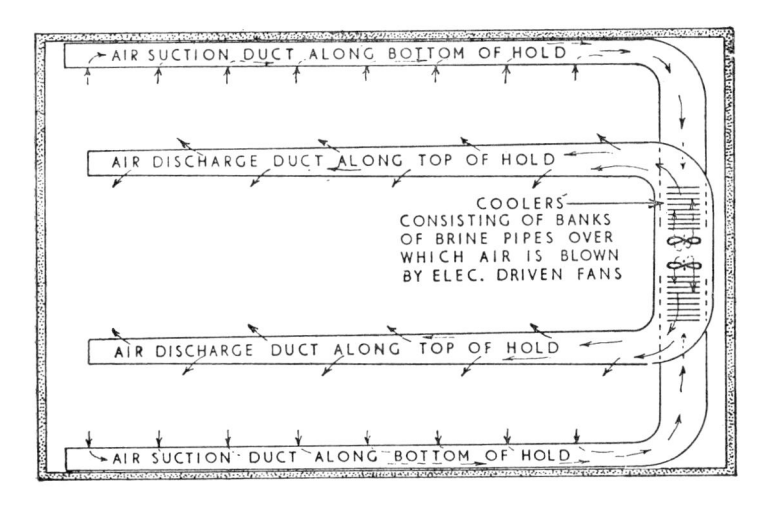

COLD AIR SYSTEM OF COOLING

Fig. 117

The brine is made by dissolving calcium chloride in fresh water, the freezing point depending upon the density, but the liquid should not be too thick as to impede its free flow through the pipes. A satisfactory mixture is obtained by a density of 40 oz. of calcium chloride per gallon of fresh water (measured by a salinometer or hydrometer supplied for this purpose, graduated for a convenient mixing temperature of 60°F). This will produce a brine with a freezing point well below any working temperature.

INSULATION

All cold rooms must be insulated against heat flowing into them from the outside, this is done by building a wood wall around the room leaving a space between it and the steel plating, and filling up this space with some light-weight heat-insulating material such as cork or silicate of cotton. The inside of the wood wall is usually protected against damp by sheet zinc. In some cases a double wall is built to allow an air space between the shell and the insulated wall (see Fig. 118).

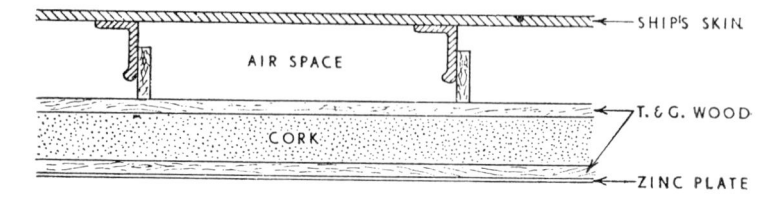

INSULATION OF COLD CHAMBERS

Fig. 118

The hatch covers are also insulated. One method is by constructing them in the form of hollow rectangular boxes or plugs filled with cork. The sides are inclined so that one interlocks another, necessitating the removal of the end plugs (keys) before lifting the others.

TEMPERATURES OF FOOD STORAGE

The temperatures at which the various foodstuffs are stored may vary considerably depending upon varying conditions such as time in storage, whether fruit is taken on board green and is to be almost ripe at time of discharge, and so on. Some average figures are given in Fig. 119.

REFRIGERATION

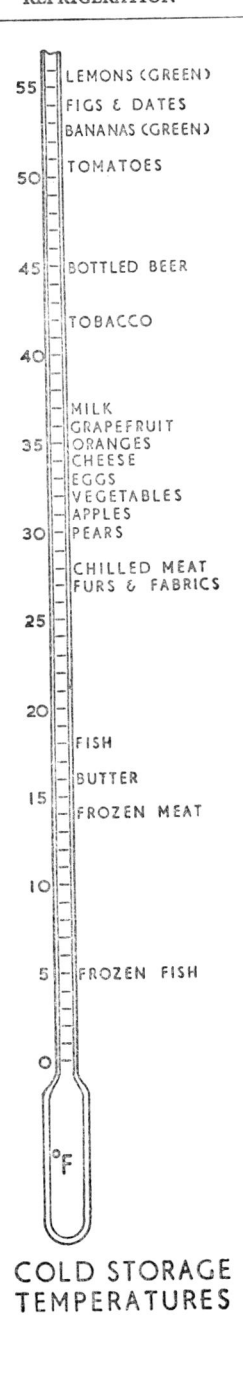

```
55 — LEMONS (GREEN)
   — FIGS & DATES
   — BANANAS (GREEN)
   — TOMATOES
50 —

45 — BOTTLED BEER

   — TOBACCO
40 —

   — MILK
   — GRAPEFRUIT
35 — ORANGES
   — CHEESE
   — EGGS
   — VEGETABLES
   — APPLES
30 — PEARS
   — CHILLED MEAT
   — FURS & FABRICS
25 —

20 —
   — FISH
   — BUTTER
15 — FROZEN MEAT

10 —

 5 — FROZEN FISH

 0
   °F
```

COLD STORAGE TEMPERATURES

Fig. 119

CAPACITY AND PERFORMANCE

The capacity of a refrigerating machine is usually expressed as the tons of ice at 32°F that can be made in 24 hours from water initially at 32°F ('from and at' 32°F). This is assuming that latent heat only (144 Btu/lb) needs to be extracted to freeze the water, already at its freezing point, into ice at the same temperature.

As previously explained, a refrigerating machine is a heat pump, similar to a heat engine working in reverse. Work is put into the machine in order to extract heat and its performance is measured by a coefficient:

$$\text{Coefficient of performance} = \frac{\text{Heat extracted by the machine}}{\text{Heat equivalent of work supplied}}$$

Example. A refrigerating machine is driven by a motor of 3 h.p. output and $2\frac{1}{4}$ tons of ice at 20°F are made per day from water initially at 65°F. Taking the specific heat of ice as 0·5 and latent heat as 144 Btu/lb, calculate the coefficient of performance of the machine and express its capacity in terms of tons of ice per 24 hours from and at 32°F.

Heat extracted per lb
$$= (65 - 32) + 144 + 0·5 (32 - 20) = 183 \text{ Btu/lb}$$

$$\text{Ice made per hour} = \frac{2·25 \times 2240}{24} = 210 \text{ lb/hr}$$

Total heat extracted $= 210 \times 183 = 38430$ Btu/hr

Heat equivalent of work supplied per hour
$$= 3 \text{ h.p.} \times 2545 \text{ Btu/hr}$$

$$\text{Coeff. of performance} = \frac{\text{Heat extracted}}{\text{Heat equivalent of work supplied}}$$

$$= \frac{38430}{3 \times 2545} = 5·032 \text{ Ans. (i)}$$

If water was supplied at 32°F and ice was made at 32°F,
Heat extracted = latent heat only = 144 Btu/lb

$$\therefore \text{Capacity from and at } 32°F = \frac{2 \cdot 25 \times 183}{144}$$

$$= 2 \cdot 86 \text{ tons/day. Ans. (ii)}$$

The ideal heat pump would work on the reversed Carnot cycle as explained in Chapter 6, and the ideal coefficient of performance would be

$$\frac{T_2}{T_1 - T_2}$$

where T_1 and T_2 are the higher and lower absolute temperatures in the cycle.

Calculations involving heat content, dryness fraction, throttling, etc. of refrigerants, are performed in a similar manner to steam problems. The properties of refrigerants, i.e. sensible heat, latent heat, total heat, and so on, at the various pressures within working range are set out in tables just as the properties of steam are set out in steam tables. However, although the total heat in steam is measured above water at 32°F (= 0°C), it is more convenient to measure the total heat in refrigerants above the level of liquid at − 40°F (= − 40°C) as this avoids negative quantities within the normal range of refrigeration.

Example. The dryness fractions of the CO_2 entering and leaving the evaporator of a refrigerating machine are 0·28 and 0·92 respectively. If the latent heat of CO_2 at the evaporator pressure is 125 Btu/lb, calculate (i) the heat absorbed per lb of CO_2 in the evaporator, (ii) the weight of ice at 24°F that would theoretically be made per day from water at 55°F when the flow of CO_2 through the machine is 60 lb per minute. Spec. ht. of ice = 0·5, latent heat = 144 Btu/lb.

Heat absorbed by CO_2 in passing through evaporator
  = Heat in CO_2 leaving — Heat in CO_2 entering

Since there is a change in dryness fraction only, the temperature and sensible heat per lb remains unchanged,

$$\therefore \text{Heat absorbed} = 0.92 \times 125 - 0.28 \times 125$$
$$= 125 (0.92 - 0.28)$$
$$= 80 \text{ Btu/lb. Ans. (i)}$$

Heat to be taken from water to make ice
$$= (55 - 32) + 144 + 0.5 (32 - 24)$$
$$= 171 \text{ Btu/lb}$$

Heat extracted from water = Heat received by CO_2
$$W \times 171 = 60 \times 80$$
$$W = \frac{60 \times 80}{171} \text{ lb/min}$$

$$\frac{60 \times 80 \times 60 \times 24}{171 \times 2240} = 18.05 \text{ tons per day. Ans.}$$

Example. Between the condenser and the regulator of a CO_2 refrigerating machine, the temperature of the liquid CO_2 is 66°F. After the regulator, through the evaporator and up to the compressor suction, the temperature of the CO_2 is 10°F. Find (i) the dryness fraction on entering the evaporator. If 20 lb of refrigerant pass through the circuit every minute and 500 lb of ice at 26°F are made per hour from water at 62°F, find (ii) the dryness fraction at the evaporator exit. Take the specific heat of ice as 0.5, latent heat of ice = 144 Btu/lb, and use the following properties of CO_2:

TEMPERATURE °F	SENSIBLE HEAT Btu/lb	LATENT HEAT Btu/lb
66	60.2	69.5
10	24.0	114.7

Heat before throttling = Heat after throttling
$$60.2 = 24 + q \times 114.7$$
$$q = 0.3156 \text{ Ans. (i)}$$

Heat extracted from water to make ice
$$= 500 [(62 - 32) + 144 + 0.5 (32 - 26)]$$
$$= 500 \times 177 \text{ Btu/hour}$$

Heat absorbed by refrigerant passing through evaporator, (latent heat only is absorbed since temperature remains unchanged)

$$= \text{W lb per hour} \times \text{heat increase per lb}$$
$$= 20 \times 60 \times 114 \cdot 7 \, (q_1 - q_2)$$

Heat absorbed by CO_2 = Heat taken from water
$$20 \times 60 \times 114 \cdot 7 \, (q_1 - q_2) = 500 \times 177$$
$$q_1 - q_2 = 0 \cdot 643$$

This is the increase in dryness fraction, therefore dryness fraction on leaving evaporator
$$= 0 \cdot 3156 + 0 \cdot 643 \quad = 0 \cdot 9586 \text{ Ans. (ii)}$$

TEST EXAMPLES 14

1. The compressor of a refrigerating machine is driven by a motor of 5 h.p. output. If the coefficient of performance is 3·8, calculate the capacity of the machine expressed in tons of ice per day from and at 32°F. Find also the weight of ice at 18°F that this machine can make per day from water at 68°F. Spec. heat of ice = 0·5, latent heat = 144 Btu/lb.

2. In a refrigerating plant, the heat equivalent of the work done on the gas in the compressor is 20 Btu/lb and the total heat in the gas leaving the compressor is 360 Btu/lb. In the condenser, the refrigerant loses 90 Btu/lb to the circulating water. Find (i) the heat extracted from the evaporator per lb of refrigerant, and (ii) the coefficient of performance.

3. An ammonia refrigerating machine is driven by a 7·5 h.p. motor, the coefficient of performance is 4·5, and 3 lb of NH_3 pass through the machine every minute. The temperature of the liquid leaving the condenser is 60°F and at entrance and exit of the evaporator the temperature is — 10°F. Find the dryness fraction of the gas at evaporator entrance and at evaporator exit, taking the specific heat of NH_3 liquid as 1·1, and the value of the latent heat as calculated from:

$$(566 - 0·8t_F) \text{ Btu/lb}$$

4. The total heat in the refrigerant leaving the compressor and entering the condenser is 219 Btu/lb, and on leaving the condenser it is 48·2 Btu/lb. The dryness fraction of the vapour leaving the evaporator is 0·96. The pressure in the evaporator is 19 lb/in^2 and at this pressure the sensible heat is 14·4 and latent heat is 181·8 Btu/lb. Find (i) the dryness fraction on entering the evaporator, (ii) the coefficient of performance, and (iii) the output h.p. of the compressor when the capacity of the machine is one ton of ice per day from and at 32°F, taking latent heat of ice = 144 Btu/lb.

ƒ5. In a CO_2 refrigerating plant in which intermediate liquid cooling is employed, the condenser pressure is 990 lb/in^2 and the gas is liquefied in the condenser without any undercooling. The liquid leaves the condenser and is throttled as it passes through the first regulating valve, and enters the separator at a pressure of 420 lb/in^2. The vapour formed is led back to the compressor and the remaining liquid throttled as it passes through the second regulating valve into the evaporator at a pressure of 275 lb/in². Find (i) the dryness fraction of the CO_2 entering the separator, (ii) the dryness fraction as it enters the evaporator, (iii) the heat extracted in the evaporator expressed in Btu per lb of CO_2 discharged from the compressor if the dryness fraction on leaving the evaporator is 0·92. Find also, for purposes of comparison, the heat extracted in the evaporator for a simple refrigerator circuit with one regulating valve and no separator, assuming the dryness fraction on leaving the evaporator is 0·92 as before. Use the following extract from CO_2 tables:

PRESSURE lb/in²	SENSIBLE HEAT Btu/lb	LATENT HEAT Btu/lb
990	76·0	40·5
420	29·0	109·4
275	15·8	123

*f*6. In an ammonia machine, wet gas leaves the evaporator and enters the compressor at 25 lb/in². It is compressed isentropically and leaves the compressor at 160 lb/in² with a temperature of 100°F. Calculate the dryness fraction of the ammonia on entering the compressor, using the following formula and properties of NH_3:

$$s = C_f \log_\varepsilon \frac{T_R}{420} + \frac{q\, h_{fg}}{T_R} + C_g \log_\varepsilon \frac{T}{T_R}$$

where, s = specific entropy,
 C_f = sp. ht. of liquid ammonia = 1·1,
 C_g = sp. ht. of ammonia vapour = 0·63,
 T_R = temp. of saturation, °F absolute,
 T = temp. of superheated vapour, °F absolute,
 q = dryness fraction,
 h_{fg} = latent heat, Btu/lb.

PRESSURE lb/in²	SAT. TEMP. °F	LATENT HEAT Btu/lb
160	83	496
25	− 8	575

SOLUTIONS TO TEST EXAMPLES 1

1. $C = (140°F - 32) \times \frac{5}{9}$ $\quad$ $C = (41°F - 32) \times \frac{5}{9}$
    $\quad = 108 \times \frac{5}{9}$ $\quad\quad\quad\quad\quad = 9 \times \frac{5}{9}$
    $\quad = 60°C.$ Ans. (i) $\quad\quad\quad = 5°C.$ Ans. (ii)

    $C = (5°F - 32) \times \frac{5}{9}$ $\quad$ $C = (-31°F - 32) \times \frac{5}{9}$
    $\quad = -27 \times \frac{5}{9}$ $\quad\quad\quad\quad = -63 \times \frac{5}{9}$
    $\quad = -15°C.$ Ans. (iii) $\quad\quad = -35°C.$ Ans. (iv)

2. $F = (60°C \times \frac{9}{5}) + 32$ $\quad$ $F = (15°C \times \frac{9}{5}) + 32$
    $\quad = 108 + 32$ $\quad\quad\quad\quad\quad\quad = 27 + 32$
    $\quad = 140°F.$ Ans. (i) $\quad\quad\quad\quad = 59°F.$ Ans. (ii)

    $F = (-10°C \times \frac{9}{5}) + 32$ $\quad$ $F = (-49°C \times \frac{9}{5}) + 32$
    $\quad = -18 + 32$ $\quad\quad\quad\quad\quad\quad = -88.2 + 32$
    $\quad = 14°F.$ Ans. (iii) $\quad\quad\quad\quad = -56.2°F.$ Ans. (iv)

3. $\quad\quad$ Change $C° = 158 \, F° \times \frac{5}{9}$
    $\quad\quad\quad\quad\quad\quad = 87\frac{7}{9} \, C°.$ Ans. (a)

    $\quad\quad$ Difference $F° = 18 \, C° \times \frac{9}{5}$
    $\quad\quad\quad\quad\quad\quad\quad = 32.4 \, F°.$ Ans. (b)

4. $\quad\quad\quad\quad C = (F - 32) \times \frac{5}{9}$
    $\quad\quad\quad\quad F = C,$ substituting for F:
    $\quad\quad\quad\quad C = (C - 32) \times \frac{5}{9}$
    $\quad\quad C \times \frac{9}{5} = C - 32$
    $\quad 1.8C - C = -32$
    $\quad\quad 0.8 \, C = -32$

    $\quad\quad\quad\quad C = \dfrac{-32}{0.8} = -40°$

    $\quad\quad \therefore -40°C = -40°F.$ Ans. (a)

    $\quad\quad\quad\quad F = (C \times \frac{9}{5}) + 32$
    $\quad\quad\quad\quad F = 2 \, C,$ substituting for F:
    $\quad\quad\quad 2C = (C \times \frac{9}{5}) + 32$
    $\quad\quad\quad 2C = 1.8 \, C + 32$

SOLUTIONS TO TEST EXAMPLES 1

$$0.2C = 32$$
$$C = 160$$
$$F = 2C = 320$$
$$\therefore 160°C = 320°F. \text{ Ans. (b)}$$

5. (Btu/lb) × $\frac{5}{9}$ = Chu/lb = gram-calories per gram
  18450 × $\frac{5}{9}$ = 10250 Chu per lb. Ans. (i)
  = 10250 gram-cal per gram. Ans. (ii)

6. Heat to raise temp. = Wt. × spec. ht. × rise in temp.
  = 10 × 0·116 × (1000 — 60)
  = 10 × 0·116 × 940
  = 1090 Btu. Ans. (i)
  1090 × $\frac{5}{9}$ = 605·8 Chu. Ans. (ii)
  1090 × 252 = 274680 gram cal. Ans. (iii)
  274680 ÷ 1000 = 274·68 k cal. Ans. (iv)

7. Total heat = Sensible heat + Latent heat
  = 10 [0·211 (1220 — 60) + 173]
  = 10 [0·211 × 1160 + 173]
  = 10 [244·76 + 173]
  = 10 × 417·76
  = 4177·6 Btu. Ans.

8. Let t = final temperature in °F
  2 pints of water weigh 1·25 × 2 = 2·5 lb

  Working in Btu,
  Heat gained by water = Heat lost by brass
  2·5 × (t — 56) = 5 × 0·094 × (350 — t)

  Dividing both sides by 2·5,

  t — 56 = 2 × 0·094 × (350 — t)
  t — 56 = 65·8 — 0·188 t
  1·188 t = 121·8
  t = 102·5°F. Ans.

9. Let t = initial temperature of copper (= funnel gas temp.)
  Heat lost by copper = Heat gained by water.
  4 × 0·095 × (t — 99) = 5 × (99 — 68)
  0·38t — 37·62 = 155
  0·38t = 192·62
  t = 506·9°F. Ans.

10. Working in gram-calories,
Let V = volume of nickel, in cubic centimetres
then 20V = ,, ,, oil, ,, ,, ,,

Weight of nickel = $8.85 \times V$ grams
Weight of oil = $0.88 \times 20V = 17.6V$ grams
Heat gained by oil = Heat lost by nickel
Wt. × spec. ht. × temp. rise = Wt. × spec. ht. × temp. fall
$17.6V \times 0.48 \times (t - 25) = 8.85V \times 0.109 \times (250 - t)$

V cancels from each side,
$8.448 (t - 25) = 0.9647 (250 - t)$
$8.448 t - 211.2 = 241.2 - 0.9647 t$
$8.448 t + 0.9647 t = 241.2 + 211.2$
$9.4127 t = 452.4$
$t = 48.06°C$. Ans.

11. 1½ gall of water weigh $1.5 \times 10 = 15$ lb
Heat lost by iron = Heat gained by water and vessel
$14 \times$ spec. ht. $\times (212 - 77) = (15 + 1.2) \times (77 - 62)$
$14 \times$ spec. ht. $\times 135 = 16.2 \times 15$

$$\text{spec. heat} = \frac{16.2 \times 15}{14 \times 135}$$

$= 0.1286$ Ans.

12. As all quantities are of the same liquid, the specific heat is the same in each case and, if represented, would cancel from both sides of the heat equation. It is therefore not included in the following equations.

Let the weights be represented by W_A W_B and W_C respectively.

Mixing of A and B:

Heat gained by A = Heat lost by B
$W_A \times (62 - 50) = W_B \times (70 - 62)$
$12 W_A = 8 W_B$
$W_A = \tfrac{2}{3} W_B$

SOLUTIONS TO TEST EXAMPLES 1   313

Mixing of B and C:

Heat gained by B = Heat lost by C
$$W_B \times (82 \cdot 5 - 70) = W_C \times (90 - 82 \cdot 5)$$
$$12 \cdot 5 \, W_B = 7 \cdot 5 \, W_C$$
$$W_C = \tfrac{5}{3} \, W_B$$

Mixing of A and C, let t = resultant temperature:

Heat gained by A = Heat lost by C
$$W_A \times (t - 50) = W_C \times (90 - t)$$

Substituting W_A and W_C in terms of W_B so that they will cancel,

$$\tfrac{2}{3} W \, (t - 50) = \tfrac{5}{3} \, W_B \, (90 - t)$$

Cancelling W_B and multiplying both sides by 3,

$$2 \, (t - 50) = 5 \, (90 - t)$$
$$2t - 100 = 450 - 5t$$
$$7t = 550$$
$$t = 78 \cdot 57°F. \text{ Ans. (i)}$$

Mixing all three:
Heat gained by A + Heat gained by B = Heat lost by C
$$W_A \times (t - 50) + W_B \times (t - 70) = W_C \times (90 - t)$$
$$\tfrac{2}{3} \, W_B \, (t - 50) + W_B \, (t - 70) = \tfrac{5}{3} \, W_B \, (90 - t)$$

Cancelling W_B and multiplying every term by 3,

$$2 \, (t - 50) + 3 \, (t - 70) = 5 \, (90 - t)$$
$$2t - 100 + 3t - 210 = 450 - 5t$$
$$10t = 760$$
$$t = 76°F. \text{ Ans. (ii)}$$

13. Let t = initial temperature of the ice.

Heat gained by ice = Heat lost by water
$$2 \, [0 \cdot 5 \, (32 - t) + 144 + (55 - 32)] = 22 \cdot 5 \times (72 - 55)$$
$$2 \, [16 - 0 \cdot 5t + 144 + 23] = 22 \cdot 5 \times 17$$
$$2 \, [183 - 0 \cdot 5t] = 382 \cdot 5$$
$$366 - t = 382 \cdot 5$$
$$- t = 382 \cdot 5 - 366$$
$$- t = 16 \cdot 5$$
$$t = - 16 \cdot 5°F. \text{ Ans.}$$

14. Heat lost by water in cooling from 62 to 32°F

$$= 15 \times (62 - 32) = 450 \text{ Btu} \quad \ldots \quad \ldots \quad \ldots \quad \text{(i)}$$

Sensible heat required to raise temperature of ice up to melting point of 32°F from its initial temperature of 23°F

$$= 4 \times 0.5 \times (32 - 23)$$
$$= 18 \text{ Btu} \quad \ldots \quad \ldots \quad \ldots \quad \ldots \quad \ldots \quad \ldots \quad \text{(ii)}$$

Latent heat required to melt 4 lb of ice

$$= 4 \times 144 = 576 \text{ Btu} \quad \ldots \quad \ldots \quad \ldots \quad \ldots \quad \text{(iii)}$$

Therefore there is not sufficient heat given out by the water as it cools down to its limit of 32°F to melt all the ice. The whole of the 4 lb of ice will be raised in temperature to 32°F but only part of it will be melted.

$$\text{Heat available to melt ice} = 450 - 18$$
$$= 432 \text{ Btu}$$

Since each lb of ice at 32°F requires 144 Btu to melt it, then weight of ice melted

$$= \frac{432}{144} = 3 \text{ lb}$$

Total weight of water will then be $15 + 3 = 18$ lb
Weight of ice left unmelted $= 4 - 3 = 1$ lb

Hence, final state =
  18 lb of water and 1 lb of ice,
  all at 32°F. Ans.

SOLUTIONS TO TEST EXAMPLES 2

1. Increase in length $= KLt$
    $= 0{\cdot}0000075 \times 25{\cdot}5 \times 12 \times (500 - 64)$
    $= 0{\cdot}0000075 \times 25{\cdot}5 \times 12 \times 436$
    $= 1$ in. Ans.

2. Vol. of sphere $= \frac{\pi}{6} D^3$
    Weight of sphere $= \frac{\pi}{6} \times 6^3 \times 0{\cdot}26$ lb
    Heat supplied $=$ Wt. $\times$ spec. ht. $\times$ temp. rise
    $2000 = \frac{\pi}{6} \times 6^3 \times 0{\cdot}26 \times 0{\cdot}13 \times t$

    $\therefore t = \dfrac{2000 \times 6}{\pi \times 6^3 \times 0{\cdot}26 \times 0{\cdot}13}$

    $= 523{\cdot}1$ F degrees

    Increase in dia. $= KLt$
    $= 0{\cdot}00000618 \times 6 \times 523{\cdot}1$
    $= 0{\cdot}0194$ in. Ans.

3. Internal volume of pipe ($=$ vol. of oil in pipe)
    $= 0{\cdot}7854 \times 1{\cdot}25^2 \times 45 \times 12$
    $= 662{\cdot}7$ cubic inches

    Coefficient of cubical expansion of steel

    $= 3 \times$ coeff. of linear expansion
    $= 3 \times 0{\cdot}0000067$
    $= 0{\cdot}0000201/\text{F}°$

    Increase in vol. of oil $= K_V\text{OIL} \times V \times t$
    Increase in vol. of pipe $= K_V\text{PIPE} \times V \times t$

    Oil overflow $=$ Difference in increase in volume
    $=$ Vol. increase of oil — Vol. increase in pipe
    $= K_V\text{OIL} \times V \times t - K_V\text{PIPE} \times V \times t$
    $= (K_V\text{OIL} - K_V\text{PIPE}) \times V \times t$
    $= (0{\cdot}0005 - 0{\cdot}0000201) \times 662{\cdot}7 \times 50$

$$= 0{\cdot}0004799 \times 662{\cdot}7 \times 50$$
$$= 15{\cdot}9 \text{ cubic inches. Ans.}$$

4. Coefficient of cubical expansion of glass
$$= 3 \times \text{coeff. of linear expansion}$$
$$= 3 \times 0{\cdot}0000085$$
$$= 0{\cdot}0000255/\text{C}°$$

Increase of temperature
$$= 65 - 15$$
$$= 50 \text{ C degrees}$$

Vol. of oil overflow = Apparent (relative) increase in vol. of oil
$$= \text{App. } K_V\text{OIL} \times V \times t$$
$$1{\cdot}57 = \text{App. } K_V\text{OIL} \times 40 \times 50$$

$$\text{App. } K_V\text{OIL} = \frac{1{\cdot}57}{40 \times 50}$$

$$= 0{\cdot}000785/\text{C}°$$

App. K_VOIL = K_VOIL — K_VGLASS
0·000785 = K_VOIL — 0·0000255

∴ Actual K_VOIL = 0·000785 + 0·0000255
$$= 0{\cdot}0008105/\text{C}°. \text{ Ans.}$$

5. Total area of walls, ceiling and floor
$$= 2\,(15 \times 8 + 13{\cdot}5 \times 8 + 15 \times 13{\cdot}5)$$
$$= 2\,(120 + 108 + 202{\cdot}5)$$
$$= 2 \times 430{\cdot}5$$
$$= 861 \text{ square feet.}$$

Area in square centimetres
$$= 861 \times 144 \times 2{\cdot}54^2 \text{ cm}^2$$
$$\text{time} = 60 \text{ seconds}$$

Difference in temp. = $15 - (-5) = 20$ C degrees
Depth of insulation = $6 \times 2{\cdot}54$ cm

$$Q = \frac{kat\theta}{d}$$

SOLUTIONS TO TEST EXAMPLES 2   317

$$= \frac{0.00028 \times 861 \times 144 \times 2.54^2 \times 60 \times 20}{6 \times 2.54}$$

$= 17640$ gram calories

$\dfrac{17640}{252} = 69.98$ Btu. Ans.

6. $Q = \dfrac{kat\theta}{d} \qquad \therefore \theta = \dfrac{Qd}{kat}$

Let θ_1, θ_2 and θ_3 represent the temperature differences across the first wood layer, cork and second wood layer respectively.

Total temperature difference =

$$\theta_1 + \theta_2 + \theta_3 = \frac{Q\,d_1}{k_1 at} + \frac{Q\,d_2}{k_2 at} + \frac{Q\,d_3}{k_3 at}$$

Q, a and t are common because the same quantity of heat passes through each part, across the same area, in the same time.

$$\therefore \text{Total temp. difference} = \frac{Q}{at}\left\{\frac{d_1}{k_1} + \frac{d_2}{k_2} + \frac{d_3}{k_3}\right\}$$

$$52 = \frac{Q}{240 \times 24}\left\{\frac{1}{1.45} + \frac{6}{0.29} + \frac{1}{1.45}\right\}$$

$$= \frac{Q}{240 \times 24}\left\{\frac{0.29 + 8.7 + 0.29}{1.45 \times 0.29}\right\}$$

$$= \frac{Q}{240 \times 24} \times \frac{9.28}{1.45 \times 0.29}$$

$$\therefore Q = \frac{52 \times 240 \times 24 \times 1.45 \times 0.29}{9.28}$$

$= 13570$ Btu per 24 hours. Ans.

7. Heat radiated $= 16 \times 10^{-10} (T_1^4 - T_2^4)$ Btu/ft^2/hr.
where $T_1 = 420 + 460 = 880°F$ abs.
$T_2 = 110 + 460 = 570°F$ abs.

Heat radiated $= 16 \times 10^{-10} (880^4 - 570^4)$
$= 16 \times 10^{-10} (880^2 + 570^2)(880^2 - 570^2)$
$= 16 \times 10^{-10} \times 1099300 \times 449500$
$= 16 \times 10^{-10} \times 10 \cdot 993 \times 10^5 \times 4 \cdot 495 \times 10^5$
$= 16 \times 10 \cdot 993 \times 4 \cdot 495$
$= 790 \cdot 5$ Btu per ft^2 per hr. Ans.

8. Let $T_1 =$ abs. temp. of shell before lagging
$= 450 + 460 = 910°F$ abs.
$T_2 =$ abs. temp. of atmosphere before lagging
$= 120 + 460 = 580°F$ abs.
$T_3 =$ abs. temp. of cleading after lagging
$= 140 + 460 = 600°F$ abs.
$T_4 =$ abs. temp. of atmosphere after lagging
$= 90 + 460 = 550°F$ abs.

Heat radiated from shell before lagging
$= K (T_1^4 - T_2^4)$ Btu per ft^2 per hr.
$= K (T_1^2 + T_2^2)(T_1^2 - T_2^2)$

Heat radiated from cleading after lagging
$= K (T_3^4 - T_4^4)$ Btu per ft^2 per hr.
$= K (T_3^2 + T_4^2)(T_3^2 - T_4^2)$

Heat saved by lagging

$= K [(T_1^2 + T_2^2)(T_1^2 - T_2^2) - (T_3^2 + T_4^2)(T_3^2 - T_4^2)]$

$T_1^2 = 910^2 = 828100$
$T_2^2 = 580^2 = 336400$
$T_1^2 + T_2^2 = 1164500 = 11 \cdot 645 \times 10^5$
$T_1^2 - T_2^2 = 491700 = 4 \cdot 917 \times 10^5$
$T_3^2 = 600^2 = 360000$
$T_4^2 = 550^2 = 302500$
$T_3^2 + T_4^2 = 662500 = 6 \cdot 625 \times 10^5$
$T_3^2 - T_4^2 = 57500 = 0 \cdot 575 \times 10^5$

Heat saved by lagging

$= 16 \times 10^{-10} [11 \cdot 645 \times 4 \cdot 917 \times 10^{10} - 6 \cdot 625 \times 0 \cdot 575 \times 10^{10}]$

SOLUTIONS TO TEST EXAMPLES 2

$$= 16 \times [57{\cdot}27 - 3{\cdot}81]$$
$$= 16 \times 53{\cdot}46 \text{ Btu per ft}^2 \text{ per hr.}$$

Total surface area = area of ends + area of cylinder
$$= \pi D^2 + \pi Dl$$
$$= \pi (4^2 + 4 \times 16)$$
$$= \pi \times 80 \text{ ft}^2$$

Lagged area = $0{\cdot}75 \times \pi \times 80 \text{ ft}^2$
Heat saved = $16 \times 53{\cdot}46 \times 0{\cdot}75 \times \pi \times 80$
$$= 161300 \text{ Btu/hr. Ans. (i)}$$

At 18500 Btu per lb of oil,

$$\text{Weight of oil saved} = \frac{161300}{18500} \text{ lb/hr.}$$

$$= \frac{161300 \times 24}{18500} \text{ lb/day}$$

$$= 209{\cdot}1 \text{ lb/day. Ans. (ii)}$$

9.   95 h.p. = $95 \times 33000 \text{ ft lb/min}$

$$= \frac{95 \times 33000}{778} \text{ Btu}$$

$$= 4029 \text{ Btu/min. Ans. (i)}$$

Heat given to water = Weight × temperature rise

$$\therefore \text{ Weight} = \frac{4029}{18}$$

$$= 223{\cdot}8 \text{ lb/min. Ans. (ii)}$$

10.   Friction force at skin of shaft = μW
$$= 0{\cdot}02 \times 20 \times 2240 \text{ lb}$$
Work lost per rev. = friction force × circumference
$$= 0{\cdot}02 \times 20 \times 2240 \times \pi \times 1{\cdot}25 \text{ ft lb}$$
Work lost per min. = Work per rev. × r.p.m.
$$= 0{\cdot}02 \times 20 \times 2240 \times \pi \times 1{\cdot}25 \times 105 \text{ ft lb/min.}$$

$$\text{h.p. lost} = \frac{0.02 \times 20 \times 2240 \times \pi \times 1.25 \times 105}{33000}$$

$$= 11.19 \text{ h.p. Ans. (i)}$$

Heat generated per min.
$$= \frac{11.19 \times 33000}{778}$$

$$= 474.7 \text{ Btu/min. Ans. (ii)}$$

11.   Heat supplied $= 1750 \times 18700$ Btu/hr.

$$\text{Equivalent h.p. supplied} = \frac{1750 \times 18700 \times 778}{60 \times 33000}$$

$$= \frac{1750 \times 18700}{2545}$$

Shaft h.p. $= 35\%$ of h.p. supplied

$$= \frac{0.35 \times 1750 \times 18700}{2545}$$

$$= 4500 \text{ h.p. Ans.}$$

12.   Friction force at 9 in. radius $= 0.025 \times 24 \times 2240$ lb
Horse power lost due to friction
$$= \frac{0.025 \times 24 \times 2240 \times 2\pi \times 0.75 \times 93}{33000}$$

$$= 17.84 \text{ h.p. Ans. (ai)}$$

Heat generated $=$ Heat equivalent of friction h.p.
$= 17.84 \times 2545$
$= 45420$ Btu/hr. Ans. (aii)

SOLUTIONS TO TEST EXAMPLES 2

Heat absorbed by oil = Wt. × spec. ht. × temp. rise

$$\therefore \text{Weight} = \frac{45420}{0\cdot48 \times 20} \text{ lb}$$

Converting to gallons, 1 gall. weighs $0\cdot88 \times 10 = 8\cdot8$ lb

$\therefore$ Quantity of oil

$$= \frac{45420}{0\cdot48 \times 20 \times 8\cdot8}$$

$$= 537\cdot8 \text{ gal/hr. Ans. (b)}$$

SOLUTIONS TO TEST EXAMPLES 3

1.
$$25\cdot 4 \text{ mm} = 1 \text{ inch}$$

$$710 \text{ mm} = \frac{710}{25\cdot 4} \text{ inches}$$

Atmospheric pressure $= h$ inches $\times\ 0\cdot 491$

$$= \frac{710}{25\cdot 4} \times 0\cdot 491$$

$$= 14\cdot 69 \text{ lb/in}^2. \text{ Ans.}$$

2.
$$P_1 V_1 = P_2 V_2$$
$$215 \times 5 = P_2 \times 12\cdot 5$$

$$P_2 = \frac{215 \times 5}{12\cdot 5}$$

$$= 86 \text{ lb/in}^2 \text{ abs. Ans.}$$

3.
$$P_1 V_1 = P_2 V_2$$
$$14 \times 1 = 110 \times V_2$$

$$V_2 = \frac{14 \times 1}{110}$$

$$= 0\cdot 1273 \text{ ft}^3. \text{ Ans.}$$

4.
$$P_1 V_1 = P_2 V_2$$
$$14\cdot 7 \times V_1 = 600 \times 5$$

$$V_1 = \frac{600 \times 5}{14\cdot 7}$$

$$= 204\cdot 1 \text{ ft}^3. \text{ Ans.}$$

SOLUTIONS TO TEST EXAMPLES 3

5.   Volume of vessel = vol. of 2 hemi-spherical ends + vol. of cyl.

$$\text{Volume of 2 vessels} = 2\left(\tfrac{\pi}{6} \times 1^3 + \tfrac{\pi}{4} \times 1^2 \times 4\right)$$
$$= 2\pi\left(\tfrac{1}{6} + 1\right)$$
$$= 2\pi \times 1\tfrac{1}{6} = 7\tfrac{1}{3}\ \text{ft}^3$$

Since temperature is constant,
$$P_1 V_1 = P_2 V_2$$
$$14.7 \times V_1 = 414.7 \times 7\tfrac{1}{3}$$

$$V_1 = \frac{414.7 \times 22}{14.7 \times 3}$$

$$= 206.8\ \text{ft}^3$$

206·8 cubic feet is the volume of air at 14·7 lb/in² which will compress to 7⅓ cubic feet at 414·7 lb/in². This would be the volume of atmospheric air required if the vessels initially contained no air, however, as they initially contain their own volume of 7⅓ cubic feet of air at 14·7 lb/in² then the volume to be taken from the atmosphere is
$$206.8 - 7.3 = 199.5\ \text{ft}^3.\ \text{Ans.}$$

6.   For constant temp. $P_1 V_1 = P_2 V_2 = P_3 V_3$ etc.

On each suction stroke the volume is increased from the pipe-line volume of 500 ft³ to the combined volume of pipe-line + stroke of pump, which is 500 + 3 = 503 ft³

At beginning of 1st stroke, $P_1 = 14.7$ and $V_1 = 500$
  „ end   „ „ „ $P_2 = $? and $V_2 = 503$

$$P_1 V_1 = P_2 V_2$$

$$P_2 = \frac{14.7 \times 500}{503}\ \text{lb/in}^2$$

and this is also the pressure at the beginning of the 2nd suction stroke when it begins with a volume of 500 ft³.

At the end of the 2nd suction stroke the volume is 503 ft³ and let the pressure then be P_3,

$$P_2 V_2 = P_3 V_3$$

$$P_3 = \frac{14 \cdot 7 \times 500 \times 500}{503 \times 503}$$

$$= 14 \cdot 7 \times \left\{\frac{500}{503}\right\}^2 \text{ lb/in}^2.$$

P_3 is also the pressure at the beginning of the 3rd suction stroke when the volume is 500 ft³. Let the pressure at the end of the 3rd stroke be P_4 when the volume is 503 ft³.

$$P_3 V_3 = P_4 V_4$$

$$P_4 = \frac{14 \cdot 7 \times 500^2 \times 500}{503^2 \times 503}$$

$$= 14 \cdot 7 \times \left\{\frac{500}{503}\right\}^3 \text{ lb/in}^2.$$

Therefore it can be seen that at the end of the nth stroke the pressure will be $14 \cdot 7 \times \left\{\dfrac{500}{503}\right\}^n$

and this is to be 5 lb/in², therefore:

$$14 \cdot 7 \times \left\{\frac{500}{503}\right\}^n = 5$$

$$\left\{\frac{500}{503}\right\}^n = \frac{5}{14 \cdot 7}$$

or

$$\left\{\frac{503}{500}\right\}^n = \frac{14 \cdot 7}{5}$$

SOLUTIONS TO TEST EXAMPLES 3

$$1·006^n = 2·94$$
$$(\log 1·006) \times n = \log 2·94$$
$$0·0026 \times n = 0·4683$$

$$n = \frac{0·4683}{0·0026}$$

$$= 180·1 \text{ suction strokes. Ans.}$$

7. By Charles' law, when pressure is constant,

$$\frac{V_1}{V_2} = \frac{T_1}{T_2}$$

$$V_2 = \frac{V_1 \times T_2}{T_1}$$

$$= \frac{10 \times (600 + 460)}{(70 + 460)}$$

$$= \frac{10 \times 1060}{530}$$

$$= 20 \text{ cubic feet. Ans.}$$

8. By Charles' law, when volume is constant,

$$\frac{P_1}{P_2} = \frac{T_1}{T_2}$$

$$P_2 = \frac{P_1 \times T_2}{T_1}$$

$$= \frac{(81 + 15) \times (195 + 460)}{(64 + 460)}$$

$$= \frac{96 \times 655}{524}$$

$$= 120 \text{ lb/in}^2 \text{ abs.}$$
$$120 - 15 = 105 \text{ lb/in}^2 \text{ gauge. Ans.}$$

9.
$$\frac{P_1 V_1}{T_1} = \frac{P_2 V_2}{T_2}$$

$$\frac{135 \times 0 \cdot 2}{(80 + 460)} = \frac{25 \times 0 \cdot 9}{T_2}$$

$$T_2 = \frac{540 \times 25 \times 0 \cdot 9}{135 \times 0 \cdot 2}$$

$$= 450°F \text{ abs.}$$
$$450 - 460 = -10°F. \text{ Ans.}$$

10. Volume occupied by one lb weight of air

$$= \frac{1}{0 \cdot 0807} = 12 \cdot 39 \text{ cubic feet. Ans. (i)}$$

$$\frac{PV}{T} = R$$

where $P = 14 \cdot 7 \times 144 \text{ lb/ft}^2$
$V = 12 \cdot 39 \text{ ft}^3$
$T = 32 + 460 = 492°F \text{ abs.}$

$$\therefore R = \frac{14 \cdot 7 \times 144 \times 12 \cdot 39}{492}$$

$$= 53 \cdot 3 \text{ ft lb per lb per °F. Ans. (ii)}$$

SOLUTIONS TO TEST EXAMPLES 3

11.
$$PV = wRT$$

$$w = \frac{PV}{RT}$$

$$= \frac{80 \times 144 \times 2}{99\cdot 2 \times (82 + 273)}$$

$$= \frac{80 \times 144 \times 2}{99\cdot 2 \times 355}$$

$$= 0\cdot 6543 \text{ lb. Ans. (i)}$$

$$PV = wRT$$

$$V = \frac{wRT}{P}$$

$$= \frac{2 \times 99\cdot 2 \times (22 + 273)}{200 \times 144}$$

$$= \frac{2 \times 99\cdot 2 \times 295}{200 \times 144}$$

$$= 2\cdot 032 \text{ cubic feet. Ans. (ii)}$$

12. Length of cylindrical part $= 11 - 3 = 8$ feet

Volume of vessel $= \frac{\pi}{6} \times 3^3 + \frac{\pi}{4} \times 3^2 \times 8$
$= \pi (4\cdot 5 + 18)$
$= \pi \times 22\cdot 5 \text{ ft}^3$
$PV = wRT$
$w = \dfrac{PV}{RT}$

$$= \frac{(450 + 15) \times 144 \times \pi \times 22\cdot 5}{53\cdot 3 \times (75 + 460)}$$

$$= 166\cdot 1 \text{ lb. Ans.}$$

13. By Charles' law, when volume is constant,

$$\frac{P_1}{P_2} = \frac{T_1}{T_2} \quad \therefore P_2 = \frac{P_1 \times T_2}{T_1}$$

$P_1 = 400 + 15 = 415$ lb/in² abs.
$T_1 = 60 + 460 = 520°$F abs.
$T_2 = 95 + 460 = 555°$F abs.

$$P_2 = \frac{415 \times 555}{520} = 442\cdot 9 \text{ lb/in}^2 \text{ abs.}$$

$442\cdot 9 - 15 = 427\cdot 9$ lb/in² gauge. Ans. (i)

Heat absorbed = wt. × spec. ht. × temp. rise
= 45 × 0·169 × (95 — 60)
= 45 × 0·169 × 35
= 266·2 Btu. Ans. (ii)

14. Volume of saloon = 42 × 55 × 10 ft³

Volume of air to be supplied every hour
= 2 × 42 × 55 × 10
= 46200 ft³ at 78°F.

Since the volume varies directly as the absolute temperature if the pressure is constant, then the weight per cubic foot varies inversely as the absolute temperature.

Weight of 1 cubic foot of air at 32°F and 14·7 lb/in² = **0·0807 lb**

Weight of 1 cubic foot of air at 78°F and 14·7 lb/in²,

$$= 0\cdot 0807 \times \frac{(32 + 460)}{(78 + 460)} = 0\cdot 07381$$

SOLUTIONS TO TEST EXAMPLES 3   329

Weight of air to be supplied to saloon per hour,
= 46200 × 0·07381
= 3410 lb

Heat extracted = wt. × spec. ht. × temp. fall
= 3410 × 0·2375 × (85 — 78)
= 3410 × 0·2375 × 7
= 5667 Btu per hour. Ans. (i)

Equivalent h.p. = $\dfrac{5667 \times 778}{33000 \times 60}$

= $\dfrac{5667}{2545}$

= 2·227 h.p. Ans. (ii)

15. Work done = Press. (lb/ft^2) × Swept volume (ft^3)
= 500 × 144 × (3·3 — 1·5)
= 500 × 144 × 1·8
= 129600 ft lb. Ans.

16. Work done = $wR(T_2 - T_1)$
= 1·388 × 53·3 × (2752 — 1000)

Note, although T_2 and T_1 are the absolute temperatures, their difference is the same as the difference between the thermometer readings.

Work done = 1·388 × 53·3 × 1752
= 129600 ft lb. Ans.

17. AT CONSTANT VOLUME:
Heat added = $w \times C_V \times (T_2 - T_1)$
= 3 × 0·169 × (468 — 40)
= 3 × 0·169 × 428
= 217 Chu. Ans. (ai)
External work done
= Nil. Ans. (aii)

AT CONSTANT PRESSURE:
$$\text{Heat added} = w \times C_P \times (T_2 - T_1)$$
$$= 3 \times 0.2375 \times 428$$
$$= 304.9 \text{ Chu. Ans. (bi)}$$

$$\text{Head added} = \text{Increase in I.E.} + \text{Ext. work done}$$
$$w \times C_P \times (T_2 - T_1)$$
$$= w \times C_V \times (T_2 - T_1) + \text{Work done}$$
$$304.9 = 217 + \text{Work done}$$
$$\text{Work done} = 304.9 - 217$$
$$= 87.9 \text{ Chu}$$
$$87.9 \times 1400 = 123060 \text{ ft lb. Ans. (bii)}$$

18.   Heat supplied = Increase in I.E. + Ext. work done

Inserting values when the gas is heated at constant volume from T_1 to T_2:
$$w \times C_V \times (T_2 - T_1) = \text{Increase in I.E.} + 0$$

$$\therefore \text{Increase in I.E.} = w \times C_V \times (T_2 - T_1)$$

Inserting values into the energy equation when the gas is heated at constant pressure from T_1 to T_2:

$$w \times C_P \times (T_2 - T_1) = \text{Increase in I.E.} + \frac{P(V_2 - V_1)}{J}$$

$$w \times C_P \times (T_2 - T_1) = w \times C_V \times (T_2 - T_1) + \frac{wR(T_2 - T_1)}{J}$$

$$C_P = C_V + \frac{R}{J}$$

$$\therefore C_P - C_V = \frac{R}{J}$$

SOLUTIONS TO TEST EXAMPLES 4

1. For isothermal expansion,
$$P_1 V_1 = P_2 V_2$$
$$100 \times 0.15 = P_2 \times 0.75$$
$$P_2 = \frac{100 \times 0.15}{0.75}$$
$$= 20 \text{ lb/in}^2 \text{ abs. Ans.}$$

2. For adiabatic compression,
$$P_1 V_1^\gamma = P_2 V_2^\gamma$$
$$14.5 \times 1.2^{1.4} = P_2 \times 0.2^{1.4}$$
$$P_2 = 14.5 \times \left\{\frac{1.2}{0.2}\right\}^{1.4}$$
$$= 14.5 \times 6^{1.4}$$
$$= 178.2 \text{ lb/in}^2 \text{ abs. Ans.}$$

3. $$P_1 V_1^n = P_2 V_2^n$$
$$425 \times V_1^{1.3} = 35 \times 7.5^{1.3}$$
$$V_1^{1.3} = \frac{35 \times 7.5^{1.3}}{425}$$
$$V_1 = 7.5 \times \sqrt[1.3]{\frac{35}{425}}$$
$$\text{or } 7.5 \div \sqrt[1.3]{\frac{425}{35}}$$
$$= 1.099 \text{ cubic feet. Ans.}$$

4. Ratio of compression $= \dfrac{\text{Initial volume}}{\text{Final volume}} = \dfrac{V_1}{V_2} = \dfrac{8.6}{1}$

$$P_1 V_1^{1.36} = P_2 V_2^{1.36}$$
$$14 \times 8.6^{1.36} = P_2 \times 1^{1.36}$$
$$P_2 = 14 \times 8.6^{1.36}$$
$$= 261.2 \text{ lb/in}^2 \text{ abs. Ans. (i)}$$

$$\frac{P_1 V_1}{T_1} = \frac{P_2 V_2}{T_2}$$

$$\frac{14 \times 8 \cdot 6}{(82 + 460)} = \frac{261 \cdot 2 \times 1}{T_2}$$

$$T_2 = \frac{542 \times 261 \cdot 2}{14 \times 8 \cdot 6}$$

$$= 1176°F \text{ abs.}$$

$$1176 - 460 = 716°F. \text{ Ans. (ii)}$$

5. $$P_1 V_1^n = P_2 V_2^n$$
    $$250 \times 2^n = 17 \cdot 5 \times 15^n$$

$$\frac{250}{17 \cdot 5} = \left\{\frac{15}{2}\right\}^n$$

$$\log \frac{250}{17 \cdot 5} = \left\{\log \frac{15}{2}\right\} \times n$$

$$1 \cdot 1549 = 0 \cdot 8751 \times n$$

$$n = \frac{1 \cdot 1549}{0 \cdot 8751}$$

$$= 1 \cdot 32 \text{ Ans.}$$

6. Tabulating the values of P and V with their respective logs:

P	log P	V	log V	
200	2·3010	0·45	$\bar{1}$·6532	= — 0·3468
137	2·1367	0·6	$\bar{1}$·7782	= — 0·2218
95	1·9777	0·8	$\bar{1}$·9031	= — 0·0969
70	1·8451	1·0	0·0000	
50	1·6990	1·3	0·1139	
35	1·5441	1·7	0·2304	

The graph of log P on a base of log V is plotted as in Fig. 120.

Choosing two points on the graph as shown,

SOLUTIONS TO TEST EXAMPLES 4

$$n = \frac{\text{decrease in log P}}{\text{increase in log V}}$$

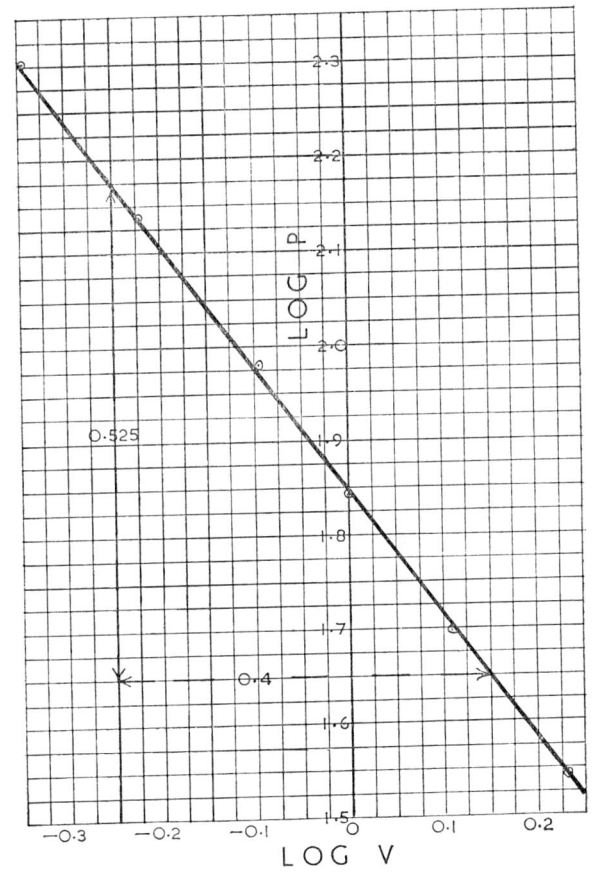

$$= \frac{2\cdot 175 - 1\cdot 65}{0\cdot 15 - (-0\cdot 25)}$$

$$= \frac{0\cdot 525}{0\cdot 4}$$

$$= 1\cdot 31 \text{ Ans.}$$

Fig. 120

7.
$$\frac{T_1}{T_2} = \left\{\frac{P_1}{P_2}\right\}^{\frac{n-1}{n}}$$

$$\frac{n-1}{n} = \frac{1\cdot 35 - 1}{1\cdot 35} = \frac{0\cdot 35}{1\cdot 35} = \frac{0\cdot 07}{0\cdot 27} = \frac{7}{27}$$

$$\frac{T_1}{(80 + 460)} = \left\{\frac{220}{14\cdot 7}\right\}^{\frac{7}{27}}$$

$$T_1 = 540 \times \left\{\frac{220}{14\cdot 7}\right\}^{\frac{7}{27}}$$

$$= 1089°F \text{ abs.}$$
$$1089 - 460 = 629°F. \text{ Ans.}$$

8. For adiabatic expansion, $\dfrac{T_1}{T_2} = \left\{\dfrac{V_2}{V_1}\right\}^{\gamma-1}$

where $\gamma = \dfrac{C_P}{C_V} = \dfrac{0\cdot 24}{0\cdot 17} = 1\cdot 412$

$\gamma - 1 = 1\cdot 412 - 1 = 0\cdot 412$

$$\frac{150 + 460}{35 + 460} = \left\{\frac{V_2}{0\cdot 5}\right\}^{0\cdot 412}$$

$$\left\{\frac{V_2}{0\cdot 5}\right\}^{0\cdot 412} = \frac{610}{495}$$

$$\frac{V_2}{0\cdot 5} = \sqrt[0\cdot 412]{\frac{610}{495}}$$

$$V_2 = 0\cdot 5 \times \sqrt[0\cdot 412]{\frac{610}{495}}$$

$$= 0\cdot 8301 \text{ ft}^3. \text{ Ans.}$$

SOLUTIONS TO TEST EXAMPLES 4

9.
$$\frac{T_1}{T_2} = \left\{\frac{V_2}{V_1}\right\}^{n-1}$$

$$\frac{240 + 460}{30 + 460} = \left\{\frac{3\cdot 5}{1}\right\}^{n-1}$$

$$\frac{700}{490} = 3\cdot 5^{n-1}$$

$$1\cdot 429 = 3\cdot 5^{n-1}$$
$$\log 1\cdot 429 = (\log 3\cdot 5) \times (n-1)$$
$$0\cdot 1549 = 0\cdot 5441\,(n-1)$$
$$0\cdot 1549 = 0\cdot 5441\,n - 0\cdot 5441$$
$$0\cdot 1549 + 0\cdot 5441 = 0\cdot 5441\,n$$
$$0\cdot 6990 = 0\cdot 5441\,n$$

$$n = \frac{0\cdot 6990}{0\cdot 5441}$$

$$= 1\cdot 284 \text{ Ans.}$$

10.
$$\frac{T_1}{T_2} = \left\{\frac{P_1}{P_2}\right\}^{\frac{n-1}{n}}$$

$$\frac{933 + 460}{90 + 460} = \left\{\frac{530}{17}\right\}^{\frac{n-1}{n}}$$

$$\frac{1393}{550} = \left\{\frac{530}{17}\right\}^{\frac{n-1}{n}}$$

$$2\cdot 533 = 31\cdot 18^{\frac{n-1}{n}}$$

$$\log 2\cdot 533 = (\log 31\cdot 18) \times \left(\frac{n-1}{n}\right)$$

$$\log 2 \cdot 533 \times n = (\log 31 \cdot 18) \times (n-1)$$
$$0 \cdot 4036 \times n = 1 \cdot 4939\,(n-1)$$
$$0 \cdot 4036\,n = 1 \cdot 4939\,n - 1 \cdot 4939$$
$$1 \cdot 4939 = 1 \cdot 4939\,n - 0 \cdot 4036\,n$$
$$1 \cdot 4939 = 1 \cdot 0903\,n$$

$$n = \frac{1 \cdot 4939}{1 \cdot 0903}$$

$$= 1 \cdot 37$$

∴ Law of compression is $PV^{1 \cdot 37} = C$. Ans.

11. Ratio of expansion $r = \dfrac{\text{Final volume}}{\text{Initial volume}} = \dfrac{4 \cdot 926}{1 \cdot 5} = 3 \cdot 284$

$\log_\varepsilon 3 \cdot 284 = 1 \cdot 1890$ (from $\log_\varepsilon$ tables)

Work done $= PV \log_\varepsilon r$
$= 60 \times 144 \times 1 \cdot 5 \times 1 \cdot 189$
$= 15410$ ft lb. Ans.

12. FOR ISOTHERMAL EXPANSION,

$$r = \frac{11 \cdot 72}{4} = 2 \cdot 93$$

$\log_\varepsilon 2 \cdot 93 = 1 \cdot 0750$

Work done $= PV \log_\varepsilon r$
$= 120 \times 144 \times 4 \times 1 \cdot 075$
$= 74320$ ft lb. Ans. (i)

FOR ADIABATIC EXPANSION,

$$P_1 V_1^\gamma = P_2 V_2^\gamma$$
$$120 \times 4^{1 \cdot 4} = P_2 \times 11 \cdot 72^{1 \cdot 4}$$

$$P_2 = 120 \times \left\{ \frac{4}{11 \cdot 72} \right\}^{1 \cdot 4}$$

or $120 \div \left\{\dfrac{11\cdot72}{4}\right\}^{1\cdot4}$

$= 120 \div 2\cdot93^{1\cdot4}$
$= 26\cdot64$ lb/in² abs.

Work done $= \dfrac{P_1 V_1 - P_2 V_2}{n - 1}$

$= \dfrac{120 \times 144 \times 4 - 26\cdot64 \times 144 \times 11\cdot72}{1\cdot4 - 1}$

$= \dfrac{144}{0\cdot4}(120 \times 4 - 26\cdot64 \times 11\cdot72)$

$= 360(480 - 312\cdot3)$
$= 360 \times 167\cdot7$
$= 60370$ ft lb. Ans. (ii)

SOLUTIONS TO TEST EXAMPLES 5

1. Mean height of diagram $= 0.605 \div 2.75 = 0.22$ in.

   M.e.p. $= 0.22 \times 300 = 66$ lb/in².

   $$\text{i.h.p.} = \frac{pALN}{33000}$$

   $$= \frac{66 \times 0.7854 \times 6^2 \times 8 \times 175}{33000 \times 12} \times 4$$

   $= 26.38$ i.h.p. Ans.

2. Mechanical efficiency $= \dfrac{\text{b.h.p.}}{\text{i.h.p.}}$

   $$\therefore \text{i.h.p.} = \frac{3000}{0.84} = 3572$$

   $$\text{i.h.p.} = \frac{pALN}{4560}$$

   Let $d =$ dia. of cylinders in millimetres
   then A $= 0.7854 \times (0.1\ d)^2$ cm²
   and L $= 1.25 \times 0.001\ d$ metres

   $$\therefore 3572 = \frac{4.92 \times 0.7854 \times 0.1^2 \times d^2 \times 1.25 \times 0.001d \times 250 \times 6}{4560}$$

   $$d = \sqrt[3]{\frac{3572 \times 4560 \times 100 \times 1000}{4.92 \times 0.7854 \times 1.25 \times 250 \times 6}}$$

   $= 608$

SOLUTIONS TO TEST EXAMPLES 5   339

$$\left.\begin{array}{l}\text{Diameter of cylinders} = 608 \text{ mm.} \\ \text{Stroke} = 1\cdot 25 \times 608 = 760 \text{ mm.}\end{array}\right\} \text{Ans.}$$

3.   Ratio of strokes, Top piston : Bottom piston

$$= 6 : 7\cdot 5$$

$$\left.\begin{array}{l}\text{Stroke of top piston} \\ \quad = \dfrac{6}{6 + 7\cdot 5} \times 2430 = 1080 \text{ mm.} \\ \text{Stroke of bottom piston} \\ \quad = \dfrac{7\cdot 5}{6 + 7\cdot 5} \times 2430 = 1350 \text{ mm.}\end{array}\right\} \text{Ans.(i)}$$

$$\text{i.h.p.} = \dfrac{p\text{ALN}}{4560}$$

$$= \dfrac{6\cdot 5 \times 0\cdot 7854 \times 62\cdot 5^2 \times 2\cdot 43 \times 105 \times 6}{4560}$$

$$= 6695 \text{ i.h.p. Ans. (ii)}$$

$$\begin{aligned}\text{b.h.p.} &= \text{i.h.p.} \times \text{mech. eff.} \\ &= 6695 \times 0\cdot 9 = 6025\cdot 5 \text{ b.h.p. Ans. (iii)}\end{aligned}$$

4. $$\text{M.e.p.} = \dfrac{85 \text{ lb/in}^2}{2\cdot 2046 \times 2\cdot 54^2} \text{ kg/cm}^2$$

$$= \dfrac{85}{14\cdot 22} = 5\cdot 977 \text{ kg/cm}^2$$

(Note that $1 \text{ kg/cm}^2 = 14\cdot 22 \text{ lb/in}^2$)

$$\text{i.h.p.} = \dfrac{p\text{ALN}}{4560}$$

$$= \frac{5 \cdot 977 \times 0 \cdot 7854 \times 75^2 \times 1 \cdot 125 \times 55 \times 8}{4560}$$

$$= 2867 \text{ i.h.p. Ans. (i)}$$

$$\begin{aligned} \text{b.h.p.} &= \text{i.h.p.} \times \text{mechanical efficiency} \\ &= 2867 \times 0 \cdot 86 \\ &= 2466 \text{ b.h.p. Ans. (ii)} \end{aligned}$$

5.   Effective load on brake $= 105 - 16 = 89$ lb
Effective radius $= 24 + 0 \cdot 5 = 24 \cdot 5$ inches

$$\text{b.h.p.} = \frac{\text{Torque (lb ft)} \times 2\pi \times \text{r.p.m.}}{33000}$$

$$= \frac{89 \times 24 \cdot 5 \times 2\pi \times 250}{12 \times 33000}$$

$$= 8 \cdot 65 \text{ b.h.p. Ans. (i)}$$

Heat equivalent $= 8 \cdot 65 \times 1414$ Chu per hour

Heat carried away by water $=$ Weight $\times$ temperature rise
$\therefore 0 \cdot 9 \times 8 \cdot 65 \times 1414 =$ lb per hour $\times 18$

$$\text{gall. of water per hour} = \frac{0 \cdot 9 \times 8 \cdot 65 \times 1414}{18 \times 10}$$

$$= 61 \cdot 14 \text{ gall/hr. Ans. (ii)}$$

6. $$\text{b.h.p.} = \frac{WN}{3000} = \frac{42 \times 2000}{3000} = 28$$

Heat generated at brake $= 1414 \times 28$ Chu per hour
Heat carried away by water $=$ weight $\times$ temperature rise
$= 179 \times 10 \times (37 \cdot 9 - 16 \cdot 4)$
$= 1790 \times 21 \cdot 5$ Chu per hour

SOLUTIONS TO TEST EXAMPLES 5

$$\% \text{ heat carried away} = \frac{1790 \times 21 \cdot 5}{1414 \times 28} \times 100$$

$$= 97 \cdot 21\% \text{ Ans.}$$

7. $\quad \text{b.h.p.} = \dfrac{T \times 2\pi \times \text{r.p.m.}}{33000}$

When speed is 1470 r.p.m., $\text{b.h.p.} = \dfrac{T \times 2\pi \times 1470}{33000}$

$$= 0 \cdot 28 T$$

With all cyls. firing, b.h.p. = $0 \cdot 28 \times 143 = 40 \cdot 04$
  Ans. (i)

With no. 1 cyl. cut out, b.h.p. = $0 \cdot 28 \times 96 \cdot 5 = 27 \cdot 02$

∴ i.h.p. of no. 1 cyl. = $40 \cdot 04 - 27 \cdot 02 = 13 \cdot 02$

With no. 2 cyl. cut out, b.h.p. = $0 \cdot 28 \times 96 = 26 \cdot 88$

∴ i.h.p. of no. 2 cyl. = $40 \cdot 04 - 26 \cdot 88 = 13 \cdot 16$

With no. 3 cyl. cut out, b.h.p. = $0 \cdot 28 \times 95 \cdot 8 = 26 \cdot 83$

∴ i.h.p. of no. 3 cyl. = $40 \cdot 04 - 26 \cdot 83 = 13 \cdot 21$

With no. 4 cyl. cut out, b.h.p. = $0 \cdot 28 \times 96 \cdot 7 = 27 \cdot 08$

∴ i.h.p. of no. 4 cyl. = $40 \cdot 04 - 27 \cdot 08 = 12 \cdot 96$

Total i.h.p. = $13 \cdot 02 + 13 \cdot 16 + 13 \cdot 21 + 12 \cdot 96 = 52 \cdot 35$
  Ans. (ii)

$$\text{Mechanical efficiency} = \frac{\text{b.h.p.}}{\text{i.h.p.}}$$

$$= \frac{40 \cdot 04}{52 \cdot 35} = 0 \cdot 7649 \text{ or } 76 \cdot 49\%$$
  Ans. (iii)

8. Brake thermal efficiency

$$= \frac{2545}{\text{lb fuel/b.h.p. hour} \times \text{C.V.}}$$

$$= \frac{2545}{0.42 \times 18700} = 0.3241 \text{ or } 32.41\% \text{ Ans. (ii)}$$

Indicated thermal efficiency

$$= \frac{\text{brake thermal eff.}}{\text{mechanical eff.}}$$

$$= \frac{0.3241}{0.86} = 0.3769 \text{ or } 37.69\% \text{ Ans. (i)}$$

For each lb of fuel burned:

Total weight of gases = 35 lb air + 1 lb fuel = 36 lb

Heat carried away = wt. × spec. ht. × temp. rise
= 36 × 0.24 × 660 = 5702 Btu/lb fuel

Heat supplied = 18700 Btu/lb fuel

∴ % heat carried away in exhaust gases

$$= \frac{5702}{18700} \times 100$$

$$= 30.49\% \text{ Ans. (iii)}$$

9. Indicated thermal eff. = 100% − (31.7 + 30.8)
= 37.5% Ans. (i)

$$\text{Mechanical eff.} = \frac{\text{b.h.p.}}{\text{i.h.p.}} = \frac{5450}{6650}$$

SOLUTIONS TO TEST EXAMPLES 5

$$= 0.8196 \text{ or } 81.96\% \text{ Ans. (ii)}$$

Overall eff. = Indicated thermal eff. × mech. eff.
$$= 0.375 \times 0.8196$$
$$= 0.3073 \text{ or } 30.73\% \text{ Ans. (iii)}$$

$$\text{Sp. fuel consumption} = \frac{25.7 \times 2240}{24 \times 6650}$$

$$= 0.3607 \text{ lb/i.h.p. hour}$$

$$\text{Indicated thermal eff.} = \frac{2545}{\text{lb fuel/i.h.p. hour} \times \text{C.V.}}$$

$$\therefore \text{Calorific value} = \frac{2545}{0.375 \times 0.3607}$$

$$= 18820 \text{ Btu/lb. Ans. (iv)}$$

10. Indicated thermal eff. $= 100\% - (28.2 + 29.3 + 1.5)$
$$= 41\% \text{ Ans. (i)}$$

Brake thermal eff. = Indicated thermal eff. ×, mech. eff.
$$= 0.41 \times 0.85$$
$$= 0.3485 \text{ or } 34.85\% \text{ Ans. (ii)}$$

$$\text{Brake thermal eff.} = \frac{2545}{\text{lb fuel/b.h.p. hour} \times \text{C.V.}}$$

$$\therefore \text{Specific fuel consumption}$$

$$= \frac{2545}{0.3485 \times 19050}$$

$$= 0.3833 \text{ lb/b.h.p. hour. Ans. (iii)}$$

11. Heat supplied to engine $= 0{\cdot}4 \times 3500 \times 19000$
$= 26{\cdot}6 \times 10^6$ Btu/hour

Heat carried away by lub. oil
$=$ wt. $\times$ spec. ht. $\times$ temp. rise
$= 40 \times 2240 \times 0{\cdot}5 \times (127 - 82)$
$= 2{\cdot}016 \times 10^6$ Btu/hour

Heat carried away as a percentage of heat supplied

$$= \frac{2{\cdot}016 \times 10^6}{26{\cdot}6 \times 10^6} \times 100$$

$= 7{\cdot}578\%$ Ans. (i)

Heat carried away by water
$=$ heat transferred from the oil

Weight $\times$ temp. rise $= 2{\cdot}016 \times 10^6$

$$\therefore \text{Quantity of water} = \frac{2{\cdot}016 \times 10^6}{(89 - 71) \times 2240}$$

$= 50$ tons/hour. Ans. (ii)

12. Specific fuel consumption
$$= \frac{665}{1750} = 0{\cdot}38 \text{ lb/i.h.p. hour}$$

$$\frac{665}{1470} = 0{\cdot}4524 \text{ lb/b.h.p. hour}$$

Indicated thermal eff.
$$= \frac{2545}{\text{lb fuel/i.h.p. hour} \times \text{C.V.}}$$

$$= \frac{2545}{0{\cdot}38 \times 18850}$$

SOLUTIONS TO TEST EXAMPLES 5

$$= 0.3553 \text{ or } 35.53\% \text{ Ans. (i)}$$

$$\text{Brake thermal eff.} = \frac{2545}{\text{lb fuel/b.h.p. hour} \times \text{C.V.}}$$

$$= \frac{2545}{0.4524 \times 18850}$$

$$= 0.2985 \text{ or } 29.85\% \text{ Ans. (ii)}$$

Heat carried away by cooling water = wt. × temp. rise
$$= 1045 \times (140 - 82) \times 60 \text{ Btu/hour}$$

As a percentage of the heat supplied

$$= \frac{1045 \times 58 \times 60}{665 \times 18850} \times 100 = 29.01\%$$

Remaining heat loss, assuming all in exhaust gases
$$= 100 - (35.53 + 29.01) = 35.46\%$$

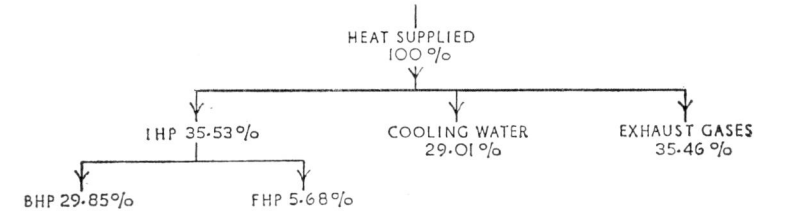

```
            HEAT SUPPLIED
               100 %
      ┌───────────┼───────────┐
      ▼           ▼           ▼
  IHP 35.53%  COOLING WATER  EXHAUST GASES
      │         29.01%          35.46%
  ┌───┴───┐
  ▼       ▼
BHP 29.85%  FHP 5.68%
```

Fig. 121

13. In this case, due to the governor cutting off the supply of gas, the number of power strokes is only the number of explosions which is 123 per minute (if the governor were uncoupled there would be 150 power strokes per minute, i.e., r.p.m. ÷ 2).

$$\text{i.h.p.} = \frac{p\text{ALN}}{33000}$$

$$= \frac{57 \times 0.7854 \times 7^2 \times 1 \times 123}{33000}$$

$$= 8.177 \text{ i.h.p. Ans. (i)}$$

Mechanical efficiency

$$= \frac{\text{b.h.p.}}{\text{i.h.p.}}$$

$$= \frac{5.8}{8.177} = 0.7092 \text{ or } 70.92\% \text{ Ans. (ii)}$$

Indicated thermal effic.

$$= \frac{\text{Heat equivalent of work done in cylinder}}{\text{Heat supplied}}$$

$$= \frac{8.177 \times 2545}{110 \times 472}$$

$$= 0.4009 \text{ or } 40.09\% \text{ Ans. (iii)}$$

Brake thermal effic.

$$= \text{ind. therm. effic.} \times \text{mech. effic.}$$
$$= 0.4009 \times 0.7092$$
$$= 0.2843 \text{ or } 28.43\% \text{ Ans. (iv)}$$

14. $\quad \text{i.h.p.} = \dfrac{p\text{ALN}}{33000}$

For top side of piston:

$$\text{i.h.p.} = \frac{84 \times 0.7854 \times 28^2 \times 54 \times 52.5}{33000 \times 12}$$

SOLUTIONS TO TEST EXAMPLES 5

$$= 370 \cdot 3 \text{ Ans. (i)}$$

For bottom side of piston:

$$\text{i.h.p.} = \frac{71 \times 0 \cdot 7854 \,(28^2 - 10^2) \times 54 \times 52 \cdot 5}{33000 \times 12}$$

$$= 273 \cdot 2 \text{ Ans. (ii)}$$

Total engine i.h.p. $= (370 \cdot 3 + 273 \cdot 2) \times 6 = 3861$ Ans. (iii)
Total b.h.p. $= 3861 \times 0 \cdot 8 = 3089$ Ans. (iv)

SOLUTIONS TO TEST EXAMPLES 6

1.
$$P_1 V_1^{1.34} = P_2 V_2^{1.34}$$
$$14.2 \times 9^{1.34} = P_2 \times 1^{1.34}$$
$$P_2 = 14.2 \times 9^{1.34}$$
$$= 269.8 \text{ lb/in}^2 \text{ abs. Ans. (i)}$$

$$\frac{P_1 V_1}{T_1} = \frac{P_2 V_2}{T_2}$$

$$\frac{14.2 \times 9}{(100 + 460)} = \frac{269.8 \times 1}{T_2}$$

$$T_2 = \frac{560 \times 269.8}{14.2 \times 9}$$

$$= 1182°\text{F abs.}$$
$$1182 - 460 = 722°\text{F. Ans. (ii)}$$

2. Clearance length $= \dfrac{\text{clearance volume}}{\text{area of cylinder}}$

$$= \frac{36 \times 10^3}{0.7854 \times 70^2} \text{ millimetres}$$

$$= 9.354 \text{ mm.}$$

$V_1 =$ clearance + stroke
    $= 9.354 + 75 = 84.354$ mm.

$V_2 =$ clearance $= 9.354$ mm.

$$P_1 V_1^{1.37} = P_2 V_2^{1.37}$$
$$14 \times 84.35^{1.37} = P_2 \times 9.354^{1.37}$$

SOLUTIONS TO TEST EXAMPLES 6   349

$$P_2 = 14 \times \left\{ \frac{84\cdot 35}{9\cdot 354} \right\}^{1\cdot 37}$$

$$= 284\cdot 8 \text{ lb/in}^2 \text{ abs. } \textbf{Ans.}$$

3. Let stroke volume = 100
then clearance volume = 7·5% of 100 = 7·5

$$\therefore V_1 = 100 + 7\cdot 5 = 107\cdot 5$$
$$V_2 = 7\cdot 5$$
$$P_1 V_1^{1\cdot 36} = P_2 V_2^{1\cdot 36}$$
$$1\cdot 1 \times 107\cdot 5^{1\cdot 36} = P_2 \times 7\cdot 5^{1\cdot 36}$$

$$P_2 = 1\cdot 1 \times \left\{ \frac{107\cdot 5}{7\cdot 5} \right\}^{1\cdot 36}$$

$$= 41\cdot 11 \text{ kg/cm}^2 \text{ abs. } \textbf{Ans. (i)}$$

$$\frac{P_1 V_1}{T_1} = \frac{P_2 V_2}{T_2}$$

$$\frac{1\cdot 1 \times 107\cdot 5}{(35 + 273)} = \frac{41\cdot 11 \times 7\cdot 5}{T_2}$$

$$T_2 = \frac{308 \times 41\cdot 11 \times 7\cdot 5}{1\cdot 1 \times 107\cdot 5}$$

$$= 803\cdot 4°\text{C abs.}$$
$$803\cdot 4 - 273 = 530\cdot 4°\text{C. Ans. (ii)}$$

4. BEFORE ALTERATION:

$$V_1 = 87\cdot 5 + 12\cdot 5 = 100$$
$$V_2 = 12\cdot 5$$
$$P_1 V_1^{1\cdot 35} = P_2 V_2^{1\cdot 35}$$
$$14 \times 100^{1\cdot 35} = P_2 \times 12\cdot 5^{1\cdot 35}$$

$$P_2 = 14 \times \left\{ \frac{100}{12 \cdot 5} \right\}^{1 \cdot 35}$$

$$= 14 \times 8^{1 \cdot 35}$$
$$= 231 \cdot 9 \text{ lb/in}^2 \text{ abs. Ans. (i)}$$

AFTER ALTERATION:

$$V_1 = 87 \cdot 5 + 10 = 97 \cdot 5$$
$$V_2 = 10$$
$$P_1 V_1^{1 \cdot 35} = P_2 V_2^{1 \cdot 35}$$
$$14 \times 97 \cdot 5^{1 \cdot 35} = P_2 \times 10^{1 \cdot 35}$$
$$P_2 = 14 \times 9 \cdot 75^{1 \cdot 35}$$
$$= 302 \cdot 9 \text{ lb/in}^2 \text{ abs. Ans. (ii)}$$

5. BEFORE ALTERATION:

$$V_1 = 880 + 80 = 960$$
$$V_2 = 80$$
$$P_1 V_1^{1 \cdot 38} = P_2 V_2^{1 \cdot 38}$$
$$P_1 \times 960^{1 \cdot 38} = 462 \cdot 8 \times 80^{1 \cdot 38}$$

$$P_1 = 462 \cdot 8 \times \left\{ \frac{80}{960} \right\}^{1 \cdot 38}$$

$$= \frac{462 \cdot 8}{12^{1 \cdot 38}}$$

$$= 15 \text{ lb/in}^2 \text{ abs.}$$

AFTER CLEARANCE IS REDUCED BY 5 mm.:

$$V_1 = 880 + 75 = 955$$
$$V_2 = 75$$
$$P_1 V_1^{1 \cdot 38} = P_2 V_2^{1 \cdot 38}$$
$$15 \times 955^{1 \cdot 38} = P_2 \times 75^{1 \cdot 38}$$

$$P_2 = 15 \times \left\{ \frac{955}{75} \right\}^{1 \cdot 38}$$

SOLUTIONS TO TEST EXAMPLES 6 — 351

$$= 502 \cdot 3 \text{ lb/in}^2 \text{ abs.}$$

∴ Increase in final pressure
$$= 502 \cdot 3 - 462 \cdot 8 = 39 \cdot 5 \text{ lb/in}^2. \text{ Ans.}$$

6.
$$\frac{T_2}{T_1} = \left\{ \frac{V_1}{V_2} \right\}^{n-1}$$

$T_1 = 120 + 460 = 580°\text{F abs.}$
$V_1 \div V_2 = 16$
$n - 1 = 1 \cdot 34 - 1 = 0 \cdot 34$

$$\therefore T_2 = 580 \times 16^{0 \cdot 34}$$
$$= 1489°\text{F abs.}$$
$$1489 - 460 = 1029°\text{F. Ans.}$$

7. Volume of air in cylinder at beginning of compression
$$= 0 \cdot 7854 \times 650^2 \times (675 + 65) \text{ mm}^3$$

Converting from cubic millimetres to cubic feet (25·4 mm. = 1 inch),

$$\text{Volume} = \frac{0 \cdot 7854 \times 650^2 \times 740}{25 \cdot 4^3 \times 12^3}$$

$$= 8 \cdot 674 \text{ cubic feet}$$

Taking atmospheric pressure as 14·7 lb/in²,

Pressure of air in cylinder $= 2 + 14 \cdot 7 = 16 \cdot 7 \text{ lb/in}^2$ abs.
Temperature of air in cylinder $= 110 + 460 = 570°\text{F abs.}$

$$\text{Weight varies as } \frac{\text{pressure} \times \text{volume}}{\text{temperature}}$$

∴ Weight of air in cylinder,

$$= 0 \cdot 0807 \times \frac{16 \cdot 7}{14 \cdot 7} \times \frac{8 \cdot 674}{1} \times \frac{492}{570}$$

$$= 0 \cdot 6864 \text{ lb. Ans.}$$

8.
$$\frac{T_2}{T_1} = \left\{\frac{V_1}{V_2}\right\}^{n-1}$$

$V_1 \div V_2 = $ ratio of compression $= r$

$$\frac{(650 + 460)}{(85 + 460)} = r^{0.36}$$

$$\therefore r = \sqrt[0.36]{\frac{1110}{545}}$$

$$= 7.213 \text{ Ans.}$$

9.
$$P_1 V_1^{1.35} = P_2 V_2^{1.35}$$
$$14.5 \times 8.5^{1.35} = P_2 \times 1^{1.35}$$
$$P_2 = 14.5 \times 8.5^{1.35}$$
$$= 260.7 \text{ lb/in}^2 \text{ abs. Ans. (i)}$$

$$\frac{P_1 V_1}{T_1} = \frac{P_2 V_2}{T_2}$$

$$\frac{14.5 \times 8.5}{(110+460)} = \frac{260.7 \times 1}{T_2}$$

$$T_2 = \frac{570 \times 260.7}{14.5 \times 8.5}$$

$$= 1206°F \text{ abs.}$$
$$1206 - 460 = 746°F. \text{ Ans. (ii)}$$

$$\frac{T_3}{T_2} = \frac{P_3}{P_2}$$

$$T_3 = \frac{1206 \times 450}{260.7}$$

$$= 2081°F \text{ abs.}$$
$$2081 - 460 = 1621°F. \text{ Ans. (iii)}$$

10.
$$PV = wRT$$

$$w = \frac{PV}{RT}$$

$$= \frac{14\cdot 4 \times 144 \times 4}{54 \times (120 + 460)}$$

$$= 0\cdot 265 \text{ lb. Ans. (i)}$$

$$P_1 V_1^{1\cdot 37} = P_2 V_2^{1\cdot 37}$$
$$P_2 = 14\cdot 4 \times 10^{1\cdot 37}$$
$$= 337\cdot 6 \text{ lb/in}^2 \text{ abs. Ans. (iia)}$$

$$\frac{P_1 V_1}{T_1} = \frac{P_2 V_2}{T_2}$$

$$T_2 = \frac{580 \times 337\cdot 6}{14\cdot 4 \times 10}$$

$$= 1359°\text{F abs.}$$

$$1359 - 460 = 899°\text{F. Ans. (iib)}$$

Heat received = wt. × spec. ht. × temp. rise
90 = 0·265 × 0·17 × temp. rise

$$\text{Temp. rise} = \frac{90}{0\cdot 265 \times 0\cdot 17} = 1998 \text{ F degrees}$$

$$\therefore T_3 = 899 + 1998 = 2897°\text{F. Ans. (iiib)}$$
$$\frac{P_3}{P_2} = \frac{T_3}{T_2}$$

$$P_3 = \frac{337\cdot 6 \times (2897 + 460)}{(899 + 460)}$$

$$= 833\cdot 7 \text{ lb/in}^2 \text{ abs. Ans. (iiia)}$$

11.

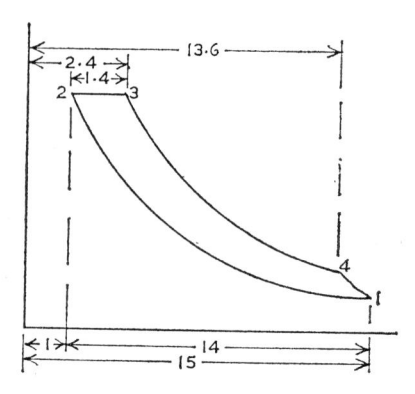

Fig. 122

Let $V_1 = 15$, and $V_2 = 1$
then, stroke volume $= 15 - 1 = 14$
Fuel burning period $= \frac{1}{10} \times 14 = 1.4$

$$\therefore V_3 = 1.4 + 1 = 2.4$$
$$V_4 = 1 + \tfrac{9}{10} \times 14 = 13.6$$
$$T_1 = 105 + 460 = 565°F \text{ abs.}$$

COMPRESSION PERIOD:

$$\frac{T_2}{T_1} = \left\{ \frac{V_1}{V_2} \right\}^{n-1}$$

$$T_2 = 565 \times 15^{0.34} = 1419°F \text{ abs.}$$

$\therefore$ Temperature at end of compression
$\quad = 1419 - 460 = 959°F.$ Ans. (i)

BURNING PERIOD:

$$\frac{T_3}{T_2} = \frac{V_3}{V_2}$$

$$T_3 = 1419 \times 2.4 = 3406°F \text{ abs.}$$

$\therefore$ Temperature at end of combustion
$\quad = 3406 - 460 = 2946°F.$ Ans. (ii)

SOLUTIONS TO TEST EXAMPLES 6

EXPANSION PERIOD:

$$\frac{T_4}{T_3} = \left\{\frac{V_3}{V_4}\right\}^{n-1}$$

$$T_4 = 3406 \times \left\{\frac{2\cdot 4}{13\cdot 6}\right\}^{0\cdot 34}$$

$$= 1889°F \text{ abs.}$$

∴ Temperature at end of expansion
$$= 1889 - 460 = 1429°F. \text{ Ans. (iii)}$$

12. CONSTANT VOLUME CYCLE:

$$\text{Air Standard Efficiency} = 1 - \frac{1}{r^{\gamma-1}}$$

$$= 1 - \frac{1}{7^{0\cdot 4}}$$

$$= 1 - 0\cdot 4592$$
$$= 0\cdot 5408 \text{ or } 54\cdot 08\% \text{ Ans. (i)}$$

DIESEL CYCLE:

Referring to Fig. 37, $V_1 = 14$, and $V_2 = 1$
∴ stroke $= 14 - 1 = 13$
Fuel burning period $= 6\%$ of $13 = 0\cdot 78$
$V_3 = 1 + 0\cdot 78 = 1\cdot 78$

$$\text{Fuel cut-off ratio} = \rho = \frac{V_3}{V_2} = 1\cdot 78$$

$$\text{Air Standard Efficiency} = 1 - \frac{1}{\gamma} \times \frac{1}{r^{\gamma-1}} \left\{\frac{\rho^\gamma - 1}{\rho - 1}\right\}$$

$$= 1 - \frac{1}{1\cdot 4} \times \frac{1}{14^{0\cdot 4}} \left\{ \frac{1\cdot 78^{1\cdot 4} - 1}{1\cdot 78 - 1} \right\}$$

$$= 1 - \frac{1}{1\cdot 4} \times \frac{1}{14^{0\cdot 4}} \times \frac{1\cdot 242}{0\cdot 78}$$

$$= 1 - 0\cdot 3959$$
$$= 0\cdot 6041 \text{ or } 60\cdot 41\% \text{ Ans. (ii)}$$

13. $P_1 V_1^{1\cdot 35} = P_2 V_2^{1\cdot 35}$ $15 \times 14^{1\cdot 35} = P_2 \times 1^{1\cdot 35}$

$$P_2 = 15 \times 14^{1\cdot 35} = 528\cdot 8 \text{ lb/in}^2 \text{ abs. Ans. (i)}$$

$$\frac{P_1 V_1}{T_1} = \frac{P_2 V_2}{T_2} \qquad \frac{15 \times 14}{(125 + 460)} = \frac{528\cdot 8 \times 1}{T_2}$$

$$T_2 = \frac{585 \times 528\cdot 8}{15 \times 14} = 1473°F \text{ abs.}$$

∴ Temperature at end of compression
$$= 1473 - 460 = 1013°F. \text{ Ans. (ii)}$$

$$\text{Stroke} = 1\cdot 2 \times 20 = 24 \text{ inches}$$

$$\text{Stroke volume} = \frac{0\cdot 7854 \times 20^2 \times 24}{1728} = 4\cdot 364 \text{ ft}^3$$

Stroke volume $= V_1 - V_2 = 4\cdot 364$

Since $\dfrac{V_1}{V_2} = 14$ $\qquad V_2 = \dfrac{V_1}{14} \qquad$ substituting:

$$V_1 - \frac{V_1}{14} = 4\cdot 364, \qquad \frac{14 V_1 - V_1}{14} = 4\cdot 364$$

$$\frac{13}{14} V_1 = 4\cdot 364 \qquad V_1 = \frac{4\cdot 364 \times 14}{13} = 4\cdot 7 \text{ft}^3$$

SOLUTIONS TO TEST EXAMPLES 6   357

From $PV = wRT$, $w = \dfrac{PV}{RT}$

$$= \dfrac{15 \times 144 \times 4.7}{53.3 \times 585} = 0.3256 \text{ lb. Ans. (iii)}$$

Work done during compression

$$= \dfrac{P_1 V_1 - P_2 V_2}{n-1} = \dfrac{wR(T_1 - T_2)}{n-1}$$

$$= \dfrac{0.3256 \times 53.3 (585 - 1473)}{(1.35 - 1)}$$

$$= -44020 \text{ ft lb}$$

$$\dfrac{-44020}{778} = -56.58 \text{ Btu}$$

i.e., Work done ON the air = 44020 ft lb ⎫
                            or 56.58 Btu ⎬ Ans. (iv)

Change of internal energy = $W \times C_V \times (T_2 - T_1)$

= $0.3256 \times 0.169 \times 888$ = 48.87 Btu. Ans. (v)

Heat exchanged = Change in I.E. + work done
              = 48.87 + (−56.58)
              = −7.71 Btu

i.e., Heat exchanged *from* air *to* cylinder walls during compression = 7.71 Btu. Ans. (vi)

14. Referring to Fig. 38, data given:

$P_1 = 14.5$, $V_1 = V_5 = 10.7$, $T_1 = 90 + 460$
                                        $= 550$
$V_2 = V_3 = 1$, $P_3 = P_4 = 600$, $T_4 = 2900 + 460$
                                          $= 3360$

$$\gamma = \frac{C_P}{C_V} = \frac{0.238}{0.17} = 1.4$$

COMPRESSION PERIOD:

$$P_1 V_1^\gamma = P_2 V_2^\gamma \qquad P_2 = \frac{14.5 \times 10.7^{1.4}}{1^{1.4}}$$

$$= 400.5 \text{ lb/in}^2 \text{ abs.}$$

$$\frac{P_1 V_1}{T_1} = \frac{P_2 V_2}{T_2} \qquad T_2 = \frac{550 \times 400.5 \times 1}{14.5 \times 10.7}$$

$$= 1420°\text{F abs.}$$

$$1420 - 460 = 960°\text{F.}$$

PART COMBUSTION AT CONSTANT VOLUME:

Absolute temperature varies as absolute pressure,

$$\therefore T_3 = T_2 \times \frac{P_3}{P_2} = \frac{1420 \times 600}{400.5} = 2127°\text{F abs.}$$

$$2127 - 460 = 1667°\text{F}$$

PART COMBUSTION AT CONSTANT PRESSURE:

Volume varies as absolute temperature,

$$\therefore V_4 = V_3 \times \frac{T_4}{T_3} = \frac{1 \times 3360}{2127} = 1.58$$

EXPANSION PERIOD:

$$P_4 V_4^{1.4} = P_5 V_5^{1.4} \qquad P_5 = \frac{600 \times 1.58^{1.4}}{10.7^{1.4}} = 41.2 \text{ lb/in}^2 \text{ abs.}$$

SOLUTIONS TO TEST EXAMPLES 6

Since $V_5 = V_1$, then $\dfrac{T_5}{T_1} = \dfrac{P_5}{P_1}$

$$T_5 = \frac{550 \times 41 \cdot 2}{14 \cdot 5} = 1563°\text{F abs.}$$

$1563 - 460 = 1103°\text{F}$

Hence, pressures and temperatures required:

$$\left.\begin{array}{l} P_2 = 400 \cdot 5 \text{ lb/in}^2 \text{ abs.,} \quad t_2 = 960°\text{F} \\ \phantom{P_2 = 400 \cdot 5 \text{ lb/in}^2 \text{ abs.,}} \quad t_3 = 1667°\text{F} \\ P_5 = 41 \cdot 2 \text{ lb/in}^2 \text{ abs.,} \quad t_5 = 1103°\text{F} \end{array}\right\} \text{Ans. (i)}$$

Ideal thermal efficiency

$$= 1 - \frac{\text{Heat rejected}}{\text{Heat supplied}}$$

$$= 1 - \frac{C_V (T_5 - T_1)}{C_V (T_3 - T_2) + C_P (T_4 - T_3)}$$

$$= 1 - \frac{(T_5 - T_1)}{(T_3 - T_2) + \gamma (T_4 - T_3)}$$

$$= 1 - \frac{1563 - 550}{(2127 - 1420) + 1 \cdot 4 (3360 - 2127)}$$

$$= 1 - 0 \cdot 4164$$
$$= 0 \cdot 5836 = 58 \cdot 36\% \text{ Ans. (ii)}$$

SOLUTIONS TO TEST EXAMPLES 7

1.
$$\text{Clearance length} = \frac{\text{clearance volume}}{\text{area of cylinder}}$$

$$= \frac{900}{0.7854 \times 25^2} = 1.833 \text{ cm}$$

$$V_1 = 35 + 1.833 = 36.833 \text{ cm}$$
$$P_1 V_1^{1.25} = P_2 V_2^{1.25}$$
$$14.3 \times 36.833^{1.25} = 60 \times V_2^{1.25}$$

$$V_2 = 36.833 \times \sqrt[1.25]{\frac{14.3}{60}}$$

$$= 11.69 \text{ cm}$$

$$\text{Compression period} = V_1 - V_2$$
$$= 36.83 - 11.69$$
$$= 25.14 \text{ cm. Ans.}$$

2. Let c = clearance length, in centimetres
V_1 = stroke + clearance = $16 + c$
$V_2 = (16 + c) - 12 = 4 + c$

$$P_1 V_1^{1.3} = P_2 V_2^{1.3}$$
$$14 \times (16 + c)^{1.3} = 70 \times (4 + c)^{1.3}$$

$$\left\{\frac{16 + c}{4 + c}\right\}^{1.3} = \frac{70}{14}$$

$$\frac{16 + c}{4 + c} = \sqrt[1.3]{5}$$

SOLUTIONS TO TEST EXAMPLES 7   361

$$16 + c = 3\cdot 449\ (4 + c)$$
$$16 + c = 13\cdot 796 + 3\cdot 449 c$$
$$2\cdot 204 = 2\cdot 449 c$$
$$c = 0\cdot 9 \text{ cm}$$

Clearance volume = area × length
$$= 0\cdot 7854 \times 8^2 \times 0\cdot 9$$
$$= 45\cdot 24 \text{ cc. Ans.}$$

3. $PV = wRT$ $\therefore w = \dfrac{PV}{RT}$

Hence, weight varies directly as the absolute pressure and volume, and inversely as the absolute temperature.

$$\therefore \text{Weight in vessel} = 0\cdot 0807 \times \frac{40}{1} \times \frac{20}{1} \times \frac{(0 + 273)}{(26 + 273)}$$

$$= 58\cdot 95 \text{ lb. Ans.}$$

4. $PV = wRT$ $\therefore V = \dfrac{wRT}{P}$

Hence, volume varies directly as the weight and absolute temperature, and inversely as the absolute pressure.

$\therefore$ Volume of 2 lb weight at atmospheric pressure and 75°F

$$= 1 \times \frac{2}{0\cdot 0807} \times \frac{(75 + 460)}{(32 + 460)}$$

$$= 26\cdot 95 \text{ cubic feet. Ans. (i)}$$

$$P_1 V_1^{1\cdot 2} = P_2 V_2^{1\cdot 2}$$
$$1 \times 26\cdot 95^{1\cdot 2} = 10 \times V_2^{1\cdot 2}$$

$$V_2 = \frac{26\cdot 95}{{}^{1\cdot 2}\sqrt{10}}$$

$$= 3\cdot 956 \text{ cubic feet. Ans. (ii)}$$

$$\frac{P_1 V_1}{T_1} = \frac{P_2 V_2}{T_2}$$

$$\frac{1 \times 26.95}{(75 + 460)} = \frac{10 \times 3.956}{T_2}$$

$$T_2 = \frac{535 \times 10 \times 3.956}{26.95}$$

$$= 785.4°F \text{ abs.}$$

∴ Temperature of the compressed air
$$= 785.4 - 460 = 325.4°F. \text{ Ans. (iii)}$$

5. $\quad 1.5 \text{ ft}^3 \times 1728 = 2592 \text{ in}^3$

ISOTHERMAL COMPRESSION:

$$P_1 V_1 = P_2 V_2$$
$$14.5 \times 2592 = 116 \times V_2$$

$$V_2 = \frac{14.5 \times 2592}{116} = 324 \text{ in}^3$$

Volume delivered per stroke = V_2 — clearance
$= 324 - 75 = 249$ cubic inches. Ans. (i)

ADIABATIC COMPRESSION:

$$P_1 V_1^{1.4} = P_2 V_2^{1.4}$$
$$14.5 \times 2592^{1.4} = 116 \times V_2^{1.4}$$

$$V_2 = 2592 \times \sqrt[1.4]{\frac{14.5}{116}}$$

$$= 586.8 \text{ in}^3$$

Volume delivered per stroke
$= 586.8 - 75 = 511.8$ cubic inches. Ans. (ii)

SOLUTIONS TO TEST EXAMPLES 7   363

6. Stroke volume $= 0.7854 \times 14^2 \times 20 = 3079$ cc

WHEN CLEARANCE IS 10% OF STROKE VOLUME:

Clearance volume $= 10\%$ of $3079 = 307.9$ cc

$V_1 =$ stroke volume $+$ clearance volume
$= 3079 + 307.9 = 3386.9$ cc

$$P_1 V_1^{1.28} = P_2 V_2^{1.28}$$
$$14.2 \times 3387^{1.28} = 100 \times V_2^{1.28}$$

$$V_2 = 3387 \times \sqrt[1.28]{\frac{14.2}{100}}$$

$$= 737 \text{ cc}$$

Volume delivered per stroke
$= 737 - 307.9 = 429.1$ cc. Ans. (i)

WHEN CLEARANCE IS 5% OF STROKE VOLUME:

Clearance volume $= 5\%$ of $3079 = 153.95$ cc
$V_1 = 3079 + 153.95 = 3232.95$
$$P_1 V_1^{1.28} = P_2 V_2^{1.28}$$
$$14.2 \times 3233^{1.28} = 100 \times V_2^{1.28}$$
$$V_2 = 3233 \times \sqrt[1.28]{\frac{14.2}{100}}$$

$$= 703.6 \text{ cc}$$

Volume delivered per stroke
$= 703.6 - 153.9 = 549.7$ cc. Ans. (ii)

7. Clearance length $= 7\%$ of $15 = 1.05$ inches
$V_1 = 15 + 1.05 = 16.05$ inches
$V_2 = 16.05 - 10.5 = 5.55$ inches

$$P_1 V_1^n = P_2 V_2^n$$
$$14.5 \times 16.05^n = 57 \times 5.55^n$$

$$\left\{\frac{16·05}{5·55}\right\}^n = \frac{57}{14·5}$$

$$2·892^n = 3·931$$
$$n \times (\log 2·892) = \log 3·931$$
$$n \times 0·4612 = 0·5945$$

$$n = \frac{0·5945}{0·4612}$$

$$= 1·289 \text{ Ans.}$$

8. Neglecting clearance,

$$V_1 = \frac{0·7854 \times 10^2 \times 15}{1728} = 0·6819 \text{ ft}^3$$

When $n = 1·2$:

$$P_1 V_1^{1·2} = P_2 V_2^{1·2}$$
$$14·7 \times 0·6819^{1·2} = 80 \times V_2^{1·2}$$

$$V_2 = 0·6819 \times \sqrt[1·2]{\frac{14·7}{80}}$$

$$= 0·1662 \text{ ft}^3$$

$$\text{Work done} = \frac{n}{n-1}(P_2 V_2 - P_1 V_1)$$

$$= \frac{1·2}{0·2}(80 \times 144 \times 0·1662 - 14·7 \times 144 \times 0·6819)$$

$$= 6 \times 144 (13·3 - 10·02)$$
$$= 6 \times 144 \times 3·28 = 2834 \text{ ft lb/stroke}$$

$$\text{Air h.p.} = \frac{2834 \times 200}{33000} = 17·18 \text{ h.p. Ans. (i)}$$

SOLUTIONS TO TEST EXAMPLES 7

When $n = 1.3$:
$$14.7 \times 0.6819^{1.3} = 80 \times V_2^{1.3}$$

$$V_2 = 0.6819 \times \sqrt[1.3]{\frac{14.7}{80}}$$

$$= 0.1852 \text{ ft}^3$$

$$\text{Work done} = \frac{1.3}{0.3} \times 144 \,(80 \times 0.1852 - 14.7 \times 0.6819)$$

$$= \frac{13}{3} \times 144 \,(14.82 - 10.02)$$

$$= 2995 \text{ ft lb/stroke}$$

$$\% \text{ Extra power} = \frac{2995 - 2834}{2834} \times 100$$

$$= 5.68\% \text{ Ans. (ii)}$$

9. Volume of air taken into compressor from atmosphere

$$= \frac{0.7854 \,(7.5^2 - 2^2) \times 6}{1728} \times 0.9 \times 140$$

$$= 17.96 \text{ cubic feet per minute}$$

Since the final temperature is the same as the initial temperature, the volume of air delivered is inversely proportional to the absolute pressure:

$$P_1 V_1 = P_2 V_2$$
$$14.7 \times 17.96 = 414.7 \times V_2$$
$$V_2 = 0.6365 \text{ ft}^3/\text{min. Ans.}$$

10. Volume of free air (at 15 lb/in² abs. from the atmosphere) to make 800 ft³ at 440 lb/in² abs., at the same temperature

$$= 800 \times \frac{440}{15} \text{ cubic feet}$$

Volume of free air to make 800 ft³ at 290 lb/in²

$$= 800 \times \frac{290}{15} \text{ cubic feet}$$

∴ Volume of free air to supply the difference

$$= \frac{800}{15}(440 - 290)$$

$$= \frac{800 \times 150}{15} = 8000 \text{ cubic feet} \quad \ldots \quad \text{(i)}$$

Volume of free air dealt with by compressor per minute

$$= \frac{0.7854\,(14^2 - 3^2) \times 12}{1728} \times 0.92 \times 170$$

$$= 159.5 \text{ cubic feet per minute} \quad \ldots \quad \text{(ii)}$$

∴ Time $= \dfrac{8000}{159.5} = 50.17$ minutes. Ans.

11. Volume of three air bottles
$$= 3\left\{\frac{\pi}{6} \times 1.5^3 + \frac{\pi}{4} \times 1.5^2 \times 6\right\}$$

$$= 3 \times \frac{\pi}{2} \times 1.5^2 \left\{\frac{1.5}{3} + \frac{6}{2}\right\}$$

$$= 37.12 \text{ cubic feet}$$

To produce 37·12 ft³ of air at 615 lb/in² abs., volume of free air required (assuming same temperature)

$$= 37 \cdot 12 \times \frac{615}{15} = 1522 \text{ ft}^3$$

Volume of free air to be dealt with by compressor

$$= 1522 - 37 \cdot 12 = 1484 \cdot 88 \text{ ft}^3 \quad \ldots \text{ (i)}$$

Volume of free air taken into compressor per minute

$$= \frac{0 \cdot 7854 \, (10 \cdot 75^2 - 3 \cdot 25^2) \times 10}{1728} \times 0 \cdot 9 \times 110$$

$$= 47 \cdot 25 \text{ ft}^3/\text{minute} \quad \ldots \quad \ldots \quad \ldots \text{ (ii)}$$

$$\therefore \text{Time} = \frac{1484 \cdot 88}{47 \cdot 25} = 31 \cdot 43 \text{ minutes. Ans.}$$

12. $$\frac{T_2}{T_1} = \left\{ \frac{V_1}{V_2} \right\}^{n-1}$$

$$T_2 = (75 + 460) \left\{ \frac{0 \cdot 8}{0 \cdot 2} \right\}^{0 \cdot 2}$$

$$= 535 \times 4^{0 \cdot 2} = 706°\text{F abs.}$$

∴ Temperature at end of compression
$$= 706 - 460 = 246°\text{F. Ans. (i)}$$

Volume delivered per minute to reservoirs

$$= 0 \cdot 2 \times \frac{535}{706} \times 200 = 30 \cdot 32 \text{ ft}^3. \text{ Ans. (ii)}$$

From $PV = wRT$, $\qquad w = \dfrac{PV}{RT}$

Weight of air taken in per minute

$$= \dfrac{14\cdot 7 \times 144 \times 0\cdot 8 \times 200}{53\cdot 3 \times 535}$$

$$= 11\cdot 88 \text{ lb/min. Ans. (iii)}$$

Heat gained by water = Heat lost by air
$$W \times 0\cdot 98 \times 20 = 11\cdot 88 \times 0\cdot 2375 \times (246 - 75)$$
$$W = 24\cdot 6 \text{ lb/min. Ans. (iv)}$$

13.

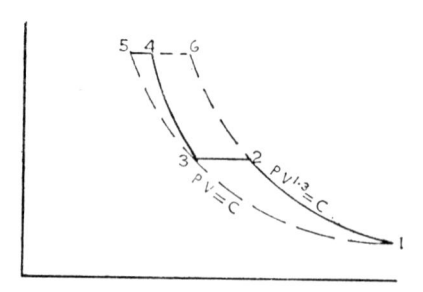

Fig. 123

$$PV = wRT \qquad V_1 = \dfrac{wRT}{P}$$

$$= \dfrac{0\cdot 1 \times 53\cdot 3 \times 520}{14 \times 144} = 1\cdot 375 \text{ ft}^3$$

FIRST STAGE COMPRESSION:

$$P_1 V_1^{1\cdot 3} = P_2 V_2^{1\cdot 3}$$
$$14 \times 1\cdot 375^{1\cdot 3} = 42 \times V_2^{1\cdot 3}$$

$$V_2 = \dfrac{1\cdot 375}{\sqrt[1\cdot 3]{3}} = 0\cdot 5905 \text{ ft}^3$$

SOLUTIONS TO TEST EXAMPLES 7

Effect of cooling at constant pressure to the initial temperature is to reduce the volume to V_3, that is, to the same volume had it been compressed isothermally.

$$P_1 V_1 = P_3 V_3$$
$$14 \times 1\cdot 375 = 42 \times V_3$$
$$V_3 = \frac{14 \times 1\cdot 375}{42} = 0\cdot 4583 \text{ ft}^3$$

% decrease in volume at end of first stage

$$= \frac{0\cdot 5905 - 0\cdot 4583}{0\cdot 5905} \times 100$$

$$= 22\cdot 39 \% \text{ Ans. (ia)}$$

SECOND STAGE COMPRESSION:

$$P_3 V_3{}^{1\cdot 3} = P_4 V_4{}^{1\cdot 3}$$
$$42 \times 0\cdot 4583^{1\cdot 3} = 105 \times V_4{}^{1\cdot 3}$$

$$V_4 = \frac{0\cdot 4583}{\sqrt[1\cdot 3]{2\cdot 5}} = 0\cdot 2266 \text{ ft}^3$$

$$P_3 V_3 = P_5 V_5$$

$$V_5 = \frac{42 \times 0\cdot 4583}{105} = 0\cdot 1833 \text{ ft}^3$$

% decrease in volume at end of second stage

$$= \frac{0\cdot 2266 - 0\cdot 1833}{0\cdot 2266} \times 100$$

$$= 19\cdot 11 \% \text{ Ans. (ib)}$$

SINGLE STAGE COMPRESSION:

$$P_1 V_1^{1.3} = P_6 V_6^{1.3}$$

$$V_6 = \frac{1 \cdot 375}{\sqrt[1.3]{7 \cdot 5}} = 0 \cdot 2918 \text{ ft}^3$$

% decrease in volume from V_6 to V_5

$$= \frac{0 \cdot 2918 - 0 \cdot 1833}{0 \cdot 2918} \times 100$$

$$= 37 \cdot 18\% \text{ Ans. (ii)}$$

SOLUTIONS TO TEST EXAMPLES 8

1. From steam tables,
   $P = 135 \text{ lb/in}^2$, $\quad h_f = 321 \cdot 9$, $\quad h_{fg} = 871 \cdot 5$ Btu/lb

   Total heat per lb steam, measured above water at 32°F
   $= h_f + qh_{fg} = 321 \cdot 9 + 0 \cdot 96 \times 871 \cdot 5$
   $= 1158 \cdot 5$ Btu/lb

   Heat in feed water, measured above 32°F
   $= 180 - 32 = 148$ Btu/lb
   $\therefore$ Heat supplied $= 1158 \cdot 54 - 148$
   $= 1010 \cdot 5$ Btu/lb. Ans.

2. From steam tables,
   $P = 115 \text{ lb/in}^2$, $h_{fg} = 881 \cdot 5$ Btu/lb

   Heat per lb of wet steam $+ 35 \cdot 3 =$ Heat per lb of dry steam
   $h_f + qh_{fg} + 35 \cdot 3 = h_f + h_{fg}$
   $q \times 881 \cdot 5 = 881 \cdot 5 - 35 \cdot 3$
   $q \times 881 \cdot 5 = 846 \cdot 2$
   $q = 0 \cdot 96$ Ans.

3. From steam tables,
   $P = 245 \text{ lb/in}^2$, $\quad h_f = 374 \cdot 2$, $\quad h_{fg} = 827 \cdot 7$ Btu/lb

   Heat per lb steam, above water at 32°F
   $= h_f + qh_{fg} = 374 \cdot 2 + 0 \cdot 97 \times 827 \cdot 7$
   $= 1177 \cdot 1$ Btu/lb

   Heat per lb feed water, above 32°F
   $= 180 - 32 = 148$ Btu/lb

   Heat required to produce steam from water at 180°F
   $= 1177 \cdot 1 - 148 = 1029 \cdot 1$ Btu/lb. Ans. (i)

   Feed water at 230°F, being 50 F degrees higher, requires 50 Btu less heat per lb.

   Percentage less heat

$$= \frac{50}{1029 \cdot 1} \times 100 = 4 \cdot 86\% \text{ Ans. (ii)}$$

4. From superheat steam tables,
    $P = 400$ lb/in², supht. 200°, $\quad h = 1331 \cdot 6$ Btu/lb

    From saturation steam tables,
    $P = 0 \cdot 9$ lb/in², $\quad h_f = 66 \cdot 2$, $\quad h_{fg} = 1038 \cdot 1$ Btu/lb

    Heat in exhaust steam $= h_f + q h_{fg}$
    $= 66 \cdot 2 + 0 \cdot 88 \times 1038 \cdot 1 = 979 \cdot 7$ Btu/lb

    Heat drop through turbine $= 1331 \cdot 6 - 979 \cdot 7$
    $= 351 \cdot 9$ Btu/lb. Ans. (i)

    Total heat used by turbine $= 2000 \times 351 \cdot 9$ Btu per hour

    $$\text{Horse power equivalent} = \frac{2000 \times 351 \cdot 9 \times 778}{60 \times 33000}$$

    $$\text{or } \frac{2000 \times 351 \cdot 9}{2545}$$

    $$= 276 \cdot 5 \text{ h.p. Ans. (ii)}$$

5. From superheat steam tables,
    $P = 300$ lb/in² supht. 180°, $\quad h = 1312 \cdot 6$ Btu/lb

    From saturation steam tables,
    $P = 300$ lb/in², $h_f = 394$, $h_{fg} = 809 \cdot 8$ Btu/lb,
    $\qquad\qquad\qquad\qquad\qquad\qquad v_g = 1 \cdot 543$ ft³/lb

    Heat in superheated steam
    $\qquad\qquad = 1312 \cdot 6$ Btu/lb

    Heat in wet steam $= h_f + q h_{fg}$
    $= 394 + 0 \cdot 98 \times 809 \cdot 8 = 1187 \cdot 6$ Btu/lb

    Heat supplied in superheaters
    $\qquad\qquad = 1312 \cdot 6 - 1187 \cdot 6 = 125$ Btu/lb. Ans. (i)

Specific volume of wet steam $= qv_g$
$$= 0.98 \times 1.543 = 1.512 \text{ ft}^3/\text{lb. Ans. (iia)}$$

Specific volume of superheated steam

$$v = \frac{1.248\,(1312.6 - 835.2)}{300} + 0.0123$$

$$= 1.998 \text{ ft}^3/\text{lb. Ans. (iib)}$$

6. From saturation steam tables,
$P = 250 \text{ lb/in}^2$, $\quad h_f = 376.1$, $\quad h_{fg} = 826 \text{ Btu/lb}$

From superheat steam tables,
$P = 250 \text{ lb/in}^2$, supht. 200°, $\quad h = 1318.5 \text{ Btu/lb}$

Heat to form wet steam, 0·95 dry
$$= (h_f + qh_{fg}) - (\text{feed temp.} - 32)$$
$$= 376.1 + 0.95 \times 826 - (205 - 32)$$
$$= 987.8 \text{ Btu/lb}$$

Extra heat to dry the steam
$$= 0.05 \times 826 = 41.3 \text{ Btu/lb}$$

$$\% \text{ extra} = \frac{41.3}{987.8} \times 100 = 4.181\% \text{ Ans. (i)}$$

Heat to form superheated steam
$$= 1318.5 - (205 - 32) = 1145.5 \text{ Btu/lb}$$

Extra heat to superheat the steam from its initial wet condition
$$= 1145.5 - 987.8 = 157.7 \text{ Btu/lb}$$

$$\% \text{ extra} = \frac{157.7}{987.8} \times 100 = 15.96\% \text{ Ans. (ii)}$$

7. From steam tables,
   $P = 120 \text{ lb/in}^2$, $v_g = 3 \cdot 729 \text{ ft}^3/\text{lb}$
   $P = 60 \text{ lb/in}^2$, $v_g = 7 \cdot 175 \text{ ft}^3/\text{lb}$

   Volume of one lb of wet steam at 120 lb/in²
   $= 0 \cdot 94 \times 3 \cdot 729 = 3 \cdot 505 \text{ ft}^3$

   $$P_1 V_1^{1 \cdot 12} = P_2 V_2^{1 \cdot 12}$$
   $$120 \times 3 \cdot 505^{1 \cdot 12} = 60 \times V_2^{1 \cdot 12}$$
   $$V_2 = 3 \cdot 505 \times {}^{1 \cdot 12}\sqrt{2} = 6 \cdot 508 \text{ ft}^3$$

   Since the volume of one lb of dry steam at 60 lb/in² is 7·175 ft³, then, dryness fraction $= \dfrac{6 \cdot 508}{7 \cdot 175} = 0 \cdot 907$ Ans.

8. From saturation steam tables,
   $P = 200 \text{ lb/in}^2$, $h_f = 355 \cdot 5$, $h_{fg} = 844$, $h_g = 1199 \cdot 5$ Btu/lb

   From superheat steam tables,
   $P = 200 \text{ lb/in}^2$, supht. 80°, $h = 1247$ Btu/lb

   Total heat before mixing = Total heat after
   $$1247 + (355 \cdot 5 + q \times 844) = 2 \times 1199 \cdot 5$$
   $$q \times 844 = 796 \cdot 5$$
   $$q = 0 \cdot 9439 \text{ Ans.}$$

9. From steam tables,
   When $P = 115 \text{ lb/in}^2$ abs., $v_g = 3 \cdot 88 \text{ ft}^3/\text{lb}$
   When $P = 80 \text{ lb/in}^2$ abs., $v_g = 5 \cdot 472 \text{ ft}^3/\text{lb}$

   Volume of 1 lb of steam at 115 lb/in², 0·95 dry
   $= 0 \cdot 95 \times 3 \cdot 88 = 3 \cdot 686 \text{ ft}^3/\text{lb}$

   Weight of 8 ft³ of wet steam
   $= \dfrac{8}{3 \cdot 686} = 2 \cdot 171 \text{ lb}$

   Volume of 2·171 lb of dry steam at 80 lb/in²
   $= 2 \cdot 171 \times 5 \cdot 472 = 11 \cdot 88 \text{ ft}^3$

WHEN EXPANSION FOLLOWS THE LAW PV = C:

$$115 \times 8 = 80 \times V_2 \quad \therefore V_2 = 11 \cdot 5 \text{ ft}^3$$

$$\therefore q = \frac{11 \cdot 5}{11 \cdot 88} = 0 \cdot 9681 \text{ Ans. (i)}$$

WHEN EXPANSION FOLLOWS THE LAW $PV^{1 \cdot 14} = C$:

$$115 \times 8^{1 \cdot 14} = 80 \times V_2^{1 \cdot 14}$$

$$V_2 = 8 \times \sqrt[1 \cdot 14]{\frac{115}{80}} = 11 \text{ ft}^3$$

$$\therefore q = \frac{11}{11 \cdot 88} = 0 \cdot 926. \text{ Ans. (ii)}$$

10. From superheat steam tables (Chu version),
P = 300 lb/in², supht. 160 C°, $h = 761 \cdot 5$ Chu/lb

From saturation steam tables (Chu version),
P = 180 lb/in², $h_f = 192 \cdot 3$, $h_{fg} = 473 \cdot 3$ Chu/lb,
$v_g = 2 \cdot 534$ ft³/lb

Volume of superheated steam
$$= \frac{2 \cdot 246 \, (761 \cdot 5 - 464)}{300} + 0 \cdot 0123$$

$$= 2 \cdot 239 \text{ ft}^3/\text{lb}$$

Since the steam is cooled at *constant volume*, the volume of one lb of the lower pressure steam is the same at 2·239 cubic feet. This is *less* than the volume of one lb of dry saturated steam at 180 lb/in², therefore it must now be wet, and its dryness fraction is:

$$q = \frac{2 \cdot 239}{2 \cdot 534} = 0 \cdot 8835. \text{ Ans. (i)}$$

Total heat in lower pressure steam = $h_f + qh_{fg}$
= 192·3 + 0·8835 × 473·3   = 610·5 Chu/lb

∴ Heat lost by steam = 761·5 — 610·5
= 151 Chu/lb. Ans. (ii)

11. From steam tables,
P = 320 lb/in², T_R = 883°F abs., v_g = 1·448 ft³/lb
P = 1·5 lb/in², v_g = 228 ft³/lb

Temperature of superheated steam
= saturation temp. + degrees of superheat
= 883 + 250 = 1133°F abs.

Specific volume of superheated steam

$$= 1\cdot 448 \times \frac{1133}{883} = 1\cdot 859 \text{ ft}^3/\text{lb}$$

Weight of steam passing through engine, i.e., **weight of 5 cubic feet at 1·859 cubic feet per lb**

$$= \frac{5}{1\cdot 859} = 2\cdot 69 \text{ lb}$$

Specific volume of exhaust steam = $q\, v_g$
= 0·9 × 228 = 205·2 ft³/lb

∴ Total volume of exhaust steam
= 2·69 × 205·2 = 552·1 cubic feet. Ans.

12. From steam tables,
P = 74 lb/in², h_f = 276·5, h_{fg} = 906 Btu/lb

One gallon water weighs 10 lb, ∴ 15 gall = 150 lb

Heat lost by steam = Heat gained by water

3[0·95 × 906 + 276·5 — (t — 32)] = 150 (t — 55)

Where t = final temperature of the water.

Dividing both sides by 3 and simplifying:
1169·2 — t = 50t — 2750
3919·2 = 51t
t = 76·85°F. Ans.

Alternatively, as explained in text:
Heat in steam and water before mixing = Heat after mixing
$$3(276{\cdot}5 + 0{\cdot}95 \times 906) + 150(55 - 32) = 153(t - 32)$$
from which $t = 76{\cdot}85°F$ as before'

13. From steam tables,
$P = 2$ lb/in^2, $h_f = 94$, $h_{fg} = 1022{\cdot}2$ Btu/lb

Dryness fraction $= 1 - 0{\cdot}12 = 0{\cdot}88$
Heat lost by steam = Heat gained by circ. water
$$94 + 0{\cdot}88 \times 1022{\cdot}2 - (105 - 32) = W(84 - 53)$$
$$920{\cdot}5 = W \times 31$$
$$W = 29{\cdot}69 \text{ lb Ans. (i)}$$

Each lb of circulating water is heated 31 F degrees therefore heat carried away by 500 tons
$$= 500 \times 2240 \times 31 \text{ Btu per hour}$$
Equivalent h.p.
$$= \frac{500 \times 2240 \times 31 \times 778}{60 \times 33000}$$
or
$$\frac{500 \times 2240 \times 31}{2545}$$

$$= 13640 \text{ h.p. Ans. (ii)}$$

14. From steam tables,
$P = 15$ lb/in^2, $h_f = 181{\cdot}2$, $h_{fg} = 970$ Btu/lb

Weight of steam blown in $= 21{\cdot}5 - 20 = 1{\cdot}5$ lb

Taking the initial and final temperatures of the vessel the same as the water contained in it,

Heat lost by steam = Heat gained by water and vessel

$$1{\cdot}5[q \times 970 + 181{\cdot}2 - (133 - 32)] = 20(133 - 60) + 0{\cdot}5(133 \times 60)$$

$$1{\cdot}5[970q + 80{\cdot}2] = 20{\cdot}5 \times 73$$

$$1455q + 120{\cdot}3 = 1496{\cdot}5$$
$$1455q = 1376{\cdot}2$$
$$q = 0{\cdot}9458 \text{ Ans.}$$

15. Taking the nearest to given values in the steam tables,

h @ 140 lb/in² supht. 140° = 1271·2
h @ 120 lb/in² supht. 140° = 1267·2

Difference for 20 lb/in² = 4·0
Difference for 5 lb/in² = $\frac{5}{20} \times 4$ = 1·0 Btu
∴ h @ 125 lb/in² supht. 140° =
= 1267·2 + 1 = 1268·2 ... ... (i)

h @ 140 lb/in² supht. 120° = 1260·6
h @ 120 lb/in² supht. 120° = 1256·8

Difference for 20 lb/in² = 3·8
Difference for 5 lb/in² = $\frac{5}{20} \times 3\cdot8$ = 0·95 Btu
∴ h @ 125 lb/in² supht. 120°
= 1256·8 + 0·95 = 1257·75 ... ... (ii)

From (i) and (ii),
h @ 125 lb/in² supht. 140° = 1268·2
h @ 125 lb/in² supht. 120° = 1257·75

Difference for 20° = 10·45
Difference for 10° = 5·225

∴ h @ 125 lb/in² supht. 130° = 1257·75 + 5·225
= 1262·975 say 1263 Btu/lb. Ans. (i)

Heat lost by steam = Heat gained by water
2 [1263 − (t − 32)] = 50 (t − 57)
1295 − t = 25t − 1425
2720 = 26t
t = 104·6°F. Ans. (ii)

16. From steam tables,
P = 14·696 lb/in², h_g = 1150·7 Btu/lb

Total equivalent weight of water
= 10 lb of water + 2 lb W.E. of vessel = 12 lb

Heat lost by steam
= Heat gained by water, vessel and ice
1150·7 − (t − 32) = 12 (t − 60) + 2·5[0·5 (32 − 20) + 144 + (t − 32)]

SOLUTIONS TO TEST EXAMPLES 8 379

$$1182 \cdot 7 - t = 12t - 720 + 295 + 2 \cdot 5t$$
$$1607 \cdot 7 = 15 \cdot 5t$$
$$t = 103 \cdot 7°F. \text{ Ans.}$$

17. From steam tables,
    $P = 35$ lb/in^2, $h_g = 1167 \cdot 6$ Btu/lb

    Working in percentages (i.e. units of 100),

    Let 100 lb steam be supplied to engine from boilers,
    „ x lb   „   „ tapped off
    then $(100 - x)$ lb passes through condenser to hotwell.

    $$\left. \begin{array}{c} \text{Total heat before mixing} \\ \text{in } x \text{ lb steam \& } (100 - x) \text{ lb water} \end{array} \right\} = \left\{ \begin{array}{c} \text{Total heat after mixing} \\ \text{in 100 lb of feed water} \end{array} \right.$$

    $$x \times 1167 \cdot 6 + (100 - x)(105 - 32) = 100 (210 - 32)$$
    $$1167 \cdot 6x + 7300 - 73x = 17800$$
    $$1094 \cdot 6x = 10500$$
    $$x = 9 \cdot 594$$

    ∴ 9·594% of steam is tapped off. Ans.

18. From steam tables,
    $P = 44$ lb/in^2, $h_f = 242$, $h_{fg} = 930 \cdot 3$ Btu/lb

    Treating heater and hotwell as one combined unit,

    Let 1 lb of steam be supplied by boiler,
    0·1 lb „ „ is tapped off for the heater,
    0·9 lb „ „ passes to engine then as water to hotwell.

    Total heat before mixing = Total heat after mixing
    $$0 \cdot 1 (242 + q \times 930 \cdot 3) + 0 \cdot 9 (109 - 32) = 1 \times (214 - 32)$$
    $$24 \cdot 2 + 93 \cdot 03q + 69 \cdot 3 = 182$$
    $$93 \cdot 03q = 88 \cdot 5$$
    $$q = 0 \cdot 9513 \text{ Ans.}$$

19. From steam tables,
    $P = 215$ lb/in^2, $h_f = 362 \cdot 1$, $h_{fg} = 838 \cdot 3$ Btu/lb
    $P = 115$ lb/in^2, $h_f = 309 \cdot 2$, $h_{fg} = 881 \cdot 5$ Btu/lb

    Heat/lb after throttling = Heat/lb before throttling
    $$309 \cdot 2 + q \times 881 \cdot 5 = 362 \cdot 1 + 0 \cdot 97 \times 838 \cdot 3$$
    $$q \times 881 \cdot 5 = 866 \cdot 1$$
    $$q = 0 \cdot 9826 \text{ Ans.}$$

20. From steam tables,
P = 180 lb/in², h_f = 346·1, h_{fg} = 851·9 Btu/lb
P = 17 lb/in², t_F = 219·5°F, h_g = 1153·5 Btu/lb

Degree of superheat = 241·5 — 219·5 = 22 F degrees
Heat/lb before throttling = Heat/lb after throttling
346·1 + q × 851·9 = 1153·5 + 0·5 × 22
q × 851·9 = 818·4
q = 0·9607 Ans.

21. From steam tables,
P = 210 lb/in², h_f = 360, h_{fg} = 840·1 Btu/lb
P = 16 lb/in², t_F = 216·3°F, h_g = 1152·4 Btu/lb

Steam in calorimeter is superheated by
230·7 — 216·3 = 14·4 F degrees

Dryness fraction by separator,

$$q_1 = \frac{W}{W + w} = \frac{20}{21·1}$$

Dryness fraction by throttling calorimeter,

Heat/lb before throttling = Heat/lb after throttling
360 + q_2 × 840·1 = 1152·4 + 0·5 × 14·4
q_2 × 840·1 = 799·6
q_2 = 0·9515

True dryness fraction of sample,
$$q = q_1 \times q_2$$
$$= \frac{20 \times 0·9515}{21·2}$$
$$= 0·902. \text{ Ans.}$$

22. From steam tables, for a steam temperature of 100°F, absolute temperature is 559·7°F, the pressure is 0·95 lb/in² abs. and the specific volume is 350·5 ft³/lb.

Volume of one lb of wet steam at this pressure and temperature and dryness fraction 0·86

SOLUTIONS TO TEST EXAMPLES 8

$$= 0.86 \times 350.5 = 301.43 \text{ ft}^3/\text{lb}$$

Let V = volume of condenser, then weight of steam in V cubic feet of space $= \dfrac{V}{301.43} = 0.003317 \text{ V lb} \quad \ldots \quad (i)$

Partial press. due to air = Total press. — Steam press.
$$= 1.15 - 0.95 = 0.2 \text{ lb/in}^2$$

$$PV = wRT, \therefore w = \dfrac{PV}{RT}$$

$$= \dfrac{0.2 \times 144 \times V}{53.3 \times 559.7}$$

$$= 0.0009656 \text{ V lb}\ldots \quad \ldots \quad (ii)$$

Weight ratio of steam to air
$$= 0.003317 \text{ V} : 0.0009656 \text{ V}$$

Dividing throughout by 0.0009656 V:
Ratio steam to air = 3.436 : 1 Ans.

23. WHEN PRESSURE IS 25 LB/IN² GAUGE = 39.7 LB/IN² ABSOLUTE:

From Centigrade steam tables, when the temperature of saturated steam is 128°C, the absolute temperature is 401.2°C, absolute pressure is 37 lb/in² and the specific volume is 11.29 ft³/lb.

$$\therefore \text{ Weight of steam in 300 ft}^3 = \dfrac{300}{11.29}$$

$$= 26.57 \text{ lb. Ans. (ia)}$$

Partial press. due to air = 39.7 — 37 = 2.7 lb/in² abs.

$$PV = wRT, \therefore w = \dfrac{PV}{RT}$$

$$= \frac{2 \cdot 7 \times 144 \times 300}{96 \times 401 \cdot 2}$$

$$= 3 \cdot 029 \text{ lb. Ans. (ib)}$$

WHEN PRESSURE IS 105 LB/IN^2 GAUGE = 119·7 LB/IN^2 ABSOLUTE:

From Centigrade steam tables, when the temperature of saturated steam is 170°C, absolute temperature is 443·2°C, absolute pressure is 115 lb/in^2 and specific volume is 3·88 ft^3/lb

$$\therefore \text{ Weight of steam } = \frac{300}{3 \cdot 88}$$

$$= 77 \cdot 32 \text{ lb. Ans. (iia)}$$

Partial press. due to air = 119·7 — 115 = 4·7 lb/in^2 abs.

$$\text{Weight of air } = \frac{PV}{RT}$$

$$= \frac{4 \cdot 7 \times 144 \times 300}{96 \times 443 \cdot 2}$$

$$= 4 \cdot 772 \text{ lb. Ans. (iib)}$$

SOLUTIONS TO TEST EXAMPLES 9

1. From steam tables, Btu version,
   $P = 250$ lb/in², $T_R = 860 \cdot 7°F$ abs., $h_{fg} = 826$ Btu/lb

   $$s = \log_\varepsilon \frac{860 \cdot 7}{492} + \frac{0 \cdot 95 \times 826}{860 \cdot 7}$$

   $= 0 \cdot 5593 + 0 \cdot 9116$
   $= 1 \cdot 4709$ Ans.

2. From steam tables, Chu version,
   $t_C = 192°C$, $T_K = 465 \cdot 2°C$ abs., $h_{fg} = 471 \cdot 1$ Chu/lb

   $$s = \log_\varepsilon \frac{465 \cdot 2}{273} + \frac{0 \cdot 9 \times 471 \cdot 1}{465 \cdot 2}$$

   $= 0 \cdot 533 + 0 \cdot 9111$
   $= 1 \cdot 4441$ Ans.

3. From steam tables, Btu version,
   $P = 180$ lb/in², $T_R = 832 \cdot 8°F$ abs., $h_{fg} = 851 \cdot 9$ Btu/lb

   Abs. temp. of superheated steam $= 832 \cdot 8 + 200 = 1032 \cdot 8$

   $$s = \log_\varepsilon \frac{832 \cdot 8}{492} + \frac{851 \cdot 9}{832 \cdot 8} + 0 \cdot 58 \log_\varepsilon \frac{1032 \cdot 8}{832 \cdot 8}$$

   $= 0 \cdot 5264 + 1 \cdot 023 + 0 \cdot 58 \times 0 \cdot 2152$
   $= 0 \cdot 5264 + 1 \cdot 023 + 0 \cdot 1248$
   $= 1 \cdot 6742$ Ans.

4. From steam tables,
   $P = 80$ lb/in², $T_R = 771 \cdot 7°F$ abs., $h_{fg} = 901 \cdot 9$ Btu/lb
   $P = 3$ lb/in², $T_R = 601 \cdot 2°F$ abs., $h_{fg} = 1013 \cdot 2$ Btu/lb

   Entropy of exhaust steam = Entropy of supply steam

   $$\log_\varepsilon \frac{601 \cdot 2}{492} + \frac{q \times 1013 \cdot 2}{601 \cdot 2} = \log_\varepsilon \frac{771 \cdot 7}{492} + \frac{901 \cdot 9}{771 \cdot 7}$$

$$\frac{q \times 1013 \cdot 2}{601 \cdot 2} = \log_\varepsilon \frac{771 \cdot 7}{601 \cdot 2} + \frac{901 \cdot 9}{771 \cdot 7}$$

$$1 \cdot 686\, q = 0 \cdot 2497 + 1 \cdot 169$$
$$1 \cdot 686\, q = 1 \cdot 4187$$
$$q = 0 \cdot 8414 \text{ Ans.}$$

5. From steam tables,
$P = 250$ lb/in^2, $T_R = 860 \cdot 7°$F abs., $h_{fg} = 826$ Btu/lb
$P = 25$ lb/in^2, $T_R = 699 \cdot 7°$F abs., $h_{fg} = 952 \cdot 5$ Btu/lb

$$T = 860 \cdot 7 + 100 = 960 \cdot 7°\text{F abs.}$$

s at end of expansion $= s$ at beginning of expansion

$$\log_\varepsilon \frac{699 \cdot 7}{492} + \frac{q \times 952 \cdot 5}{699 \cdot 7}$$
$$= \log_\varepsilon \frac{860 \cdot 7}{492} + \frac{826}{860 \cdot 7} + 0 \cdot 613 \log_\varepsilon \frac{960 \cdot 7}{860 \cdot 7}$$

$$\frac{q \times 952 \cdot 5}{699 \cdot 7} = \log_\varepsilon \frac{860 \cdot 7}{699 \cdot 7} + \frac{826}{860 \cdot 7} + 0 \cdot 613 \log_\varepsilon \frac{960 \cdot 7}{860 \cdot 7}$$

$$1 \cdot 361\, q = 0 \cdot 2071 + 0 \cdot 9596 + 0 \cdot 0674$$
$$1 \cdot 361\, q = 1 \cdot 2341$$
$$q = 0 \cdot 9067 \text{ Ans.}$$

6. From steam tables,
$P = 195$ lb/in^2, $T_R = 839 \cdot 4°$F abs., $h_{fg} = 846$, $h_g = 1199 \cdot 2$ Btu/lb
$P = 100$ lb/in^2, $T_R = 787 \cdot 5°$F abs., $h_{fg} = 889 \cdot 7$, $h_g = 1188 \cdot 2$ Btu/lb

As throttling is a constant total heat process:

Heat/lb after throttling = Heat/lb before throttling
$1188 \cdot 2 + 0 \cdot 55 \times °$supht. $= 1199 \cdot 2$

$$\therefore \text{ degree of superheat} = \frac{1199 \cdot 2 - 1188 \cdot 2}{0 \cdot 55}$$

$$= 20 \text{ F degrees. Ans. (i)}$$

Temp. at 100 lb/in² = 787·5 + 20 = 807·5°F abs.

$$s @ 100 = \log_\varepsilon \frac{787 \cdot 5}{492} + \frac{889 \cdot 7}{787 \cdot 5} + 0 \cdot 55 \log_\varepsilon \frac{807 \cdot 5}{787 \cdot 5}$$

$$= 0 \cdot 4704 + 1 \cdot 129 + 0 \cdot 0138$$
$$= 1 \cdot 6132$$

$$s @ 195 = \log_\varepsilon \frac{839 \cdot 4}{492} + \frac{846}{839 \cdot 4}$$

$$= 0 \cdot 5342 + 1 \cdot 008$$
$$= 1 \cdot 5422$$

Gain in s = s @ 100 lb/in² — s @ 195 lb/in²
= 1·6132 — 1·5422
= 0·071 Ans. (ii)

SOLUTIONS TO TEST EXAMPLES 10

1. When crank is on T.D.C. the valve is open to lead at the top and open to exhaust at the bottom. The displacement of the valve from mid-position, i.e. the distance it has moved down from its mid-position is,

   At the top : steam lap + lead
   At the bottom : exhaust lap + opening to exhaust

   These two are equal, therefore,

   Top steam lap + Top lead = Bot. exh. lap + Bot. opening to exh.
   Top steam lap + $\frac{1}{8}$ = $\frac{3}{16}$ + $1\frac{7}{16}$
   Top steam lap = $1\frac{5}{8}$ − $\frac{1}{8}$ = $1\frac{1}{2}$ in. Ans. (i)

   When crank is on B.D.C. the valve is open to lead at the bottom and open to exhaust at the top. The displacement of the valve from mid-position is the same as before, i.e. $1\frac{5}{8}$ in.

   ∴ Top exh. lap + Top opening to exh. = $1\frac{5}{8}$
   Top exh. lap = $1\frac{5}{8}$ − $1\frac{1}{2}$ = $\frac{1}{8}$ inch. Ans. (ii)

2. $$\sin \alpha = \frac{\text{steam lap + lead}}{\text{half travel}}$$

   $$= \frac{1\cdot 75 + 0\cdot 25}{3\cdot 25} = 0\cdot 6154$$

   ∴ Angle of advance = 37° 59′ Ans. (i)
   Steam lap + M.P.O. to steam = half travel
   M.P.O. to steam = 3·25 − 1·75
   = 1·5 in. Ans. (ii)

3. Half travel = steam lap + M.P.O. to steam
   = 1·5 + 1 = 2·5 in.

   ∴ Valve travel = 2 × 2·5 = 5 in. Ans. (i)

SOLUTIONS TO TEST EXAMPLES 10   387

$$\sin \alpha = \frac{\text{steam lap} + \text{lead}}{\text{half travel}}$$

$$= \frac{1 \cdot 5 + 0 \cdot 125}{2 \cdot 5} = 0 \cdot 65$$

∴ Angle of advance = 40° 33′ Ans. (ii)

4. Steam lap + M.P.O. to steam = ½ Travel
∴ Steam lap = 3 − 1·3 = 1·7 in. Ans. (i)

$$\sin \alpha = \frac{\text{steam lap} + \text{lead}}{\text{half travel}}$$

∴ Lead = half travel × sin α − steam lap
       = 3 × sin 38° 19′ − 1·7
       = 0·16 in. Ans. (ii)

5.
$$\sin \alpha = \frac{\text{steam lap} + \text{lead}}{\text{half travel}}$$

Lead = half travel × sin α − steam lap
     = 4 × sin 33° 22′ − 2
     = 0·2 in. Ans. (a)

Steam lap + M.P.O. = ½ Travel

∴ M.P.O. = 4 − 2 = 2 in. Ans. (b)

Referring to Fig. 79,

$$\sin \theta = \frac{\text{steam lap}}{\text{half travel}} = \frac{2}{4} = 0 \cdot 5$$

∴ θ = 30°
  φ = 180 − α − θ
    = 180 − 33° 22′ − 30° = 116° 38′

Hence, at cut off, crank angle from T.D.C.
= 116° 38′ Ans. (ci)

Angle of pre-admission
$$= \alpha - \theta$$
$$= 33° 22' - 30° = 3° 22' \text{ Ans. (cii)}$$

6. Steam lap + M.P.O. = ½ Travel
∴ Steam lap = ½ T − M.P.O. ... ... (i)

$$\sin \alpha = \frac{\text{steam lap} + \text{lead}}{\frac{1}{2}\text{ travel}}$$

∴ Steam lap = $\sin \alpha \times \frac{1}{2}$ T − lead ... (ii)

Equating (i) and (ii),

$$\frac{1}{2} \text{T} - \text{M.P.O.} = \sin \alpha \times \frac{1}{2} \text{T} - \text{lead}$$
$$\frac{1}{2} \text{T} - 1·25 = 0·6108 \times \frac{1}{2} \text{T} - 0·18$$
$$\frac{1}{2} \text{T} (1 - 0·6108) = 1·25 - 0·18$$
$$\frac{1}{2} \text{T} \times 0·3892 = 1·07$$
$$\frac{1}{2} \text{T} = 2·75$$

∴ Travel of valve = $2 \times 2·75 = 5·5$ in. Ans.

7. Referring to Fig. 80,

$$\sin \alpha = \frac{\text{steam lap} + \text{lead}}{\text{half travel}}$$

$$= \frac{1·75 + 0·125}{3·5} = 0·5356$$

∴ $\alpha = 32° 23'$
$\theta = \phi - (90 - \alpha) = 77 - (90 - 32° 23') = 19° 23'$

Displacement of valve from mid-position
$= r \cos \theta = 3·5 \times \cos 19° 23' = 3·3$ in.

Port opening when crank is 77° past T.D.C.
$= 3·3 -$ steam lap $= 3·3 - 1·75 = 1·55$ in. Ans. (i)

For M.P.O. to steam, valve is at bottom of its travel, angle ϕ will then be $90 - \alpha$
$= 90 - 32° 23' = 57° 37'$ Ans. (ii)

SOLUTIONS TO TEST EXAMPLES 10   389

8.
$$\sin \alpha = \frac{\text{steam lap} + \text{lead}}{\text{half travel}}$$

$$= \frac{1 \cdot 7 + 0 \cdot 1}{3} = 0 \cdot 6$$

$$\therefore \alpha = 36° \, 52'$$

Referring to Fig. 80,

When crank is 90° past dead centre, $\theta = \alpha = 36° \, 52'$

Displacement of valve from mid-position
$$= r \cos \theta = 3 \times \cos 36° \, 52' = 2 \cdot 4 \text{ in.}$$

When crank is 90° past TOP centre:

Top port opening to steam
$$= \text{valve displacement} - \text{top steam lap}$$
$$= 2 \cdot 4 - 1 \cdot 7 = 0 \cdot 7 \text{ in. Ans. (i)}$$

Bottom port opening to exhaust
$$= \text{valve displ.} - \text{bot. exh. lap}$$
$$= 2 \cdot 4 - 0 \cdot 2 = 2 \cdot 2 \text{ in. Ans. (ii)}$$

When crank is 90° past BOTTOM centre:
Bottom port opening to steam
$$= \text{valve displ.} - \text{bot. steam lap}$$
$$= 2 \cdot 4 - 1 \cdot 6 = 0 \cdot 8 \text{ in. Ans. (iii)}$$

Top port opening to exhaust
$$= \text{valve displ.} - \text{top exh. lap}$$
$$= 2 \cdot 4 - 0 \cdot 15 = 2 \cdot 25 \text{ in. Ans. (iv)}$$

9.
$$\text{Half travel} = \text{Steam lap} + \text{M.P.O. to steam}$$
$$= 1 \cdot 75 + 1 = 2 \cdot 75 \text{ in.}$$

$$\therefore \text{Travel of valve} = 2 \times 2 \cdot 75 = 5 \cdot 5 \text{ in. Ans. (i)}$$

$$\sin \alpha = \frac{\text{steam lap} + \text{lead}}{\text{half travel}}$$

$$= \frac{1 \cdot 75 + 0 \cdot 17}{2 \cdot 75} = 0 \cdot 6982$$

∴ Angle of advance = 44° 17′ Ans. (ii)

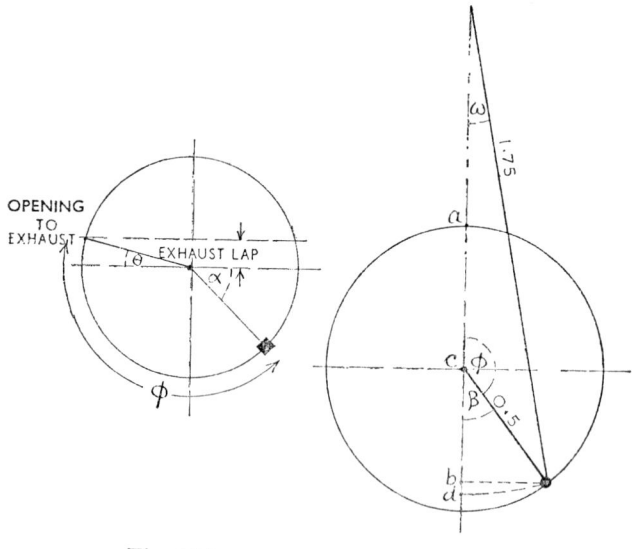

Fig. 124

Fig. 125

Referring to Fig. 124,

$$\text{Sin } \theta = \frac{\text{Exhaust lap}}{\text{half travel}} = \frac{0 \cdot 375}{2 \cdot 75} = 0 \cdot 1363$$

∴ θ = 7° 50′

Crank angle from T.D.C. when valve opens to exhaust
= φ = 180 + θ − α = 180 + 7° 50′ − 44° 17′
= 143° 33′
Referring to Fig. 125,
β = 180 − φ = 180 − 143° 33′ = 36° 27′

Neglecting angularity of connecting rod, distance travelled by piston from T.D.C. = ab

SOLUTIONS TO TEST EXAMPLES 10

Let stroke = 1, then radius of crank circle = 0·5,
$$ab = 0·5 + 0·5 \times \cos 36° 27'$$
$$= 0·5 + 0·4022$$
$$= 0·9022 \text{ of the stroke. Ans. (iiia)}$$

Con. rod length = 3·5 × crank
$$= 3·5 \times 0·5 = 1·75$$

$$\frac{0·5}{\sin \omega} = \frac{1·75}{\sin \phi}$$

$$\sin \omega = \frac{0·5 \times \sin 143° 33'}{1·75} = 0·1697$$

$$\therefore \omega = 9° 46'$$

Effect of angularity of connecting rod = bd
$$bd = 1·75 - 1·75 \times \cos 9° 46'$$
$$= 1·75 - 1·724 = 0·026$$
$$ad = ab + bd = 0·9022 + 0·026$$
$$= 0·9282 \text{ of the stroke. Ans. (iiib)}$$

SOLUTIONS TO TEST EXAMPLES 11

1. Mean height of diagram $= 2 \cdot 24 \div 3 \cdot 2 = 0 \cdot 7$ in.
Mean effective pressure $= 0 \cdot 7 \times 100 = 70 \, \text{lb/in}^2$. **Ans. (i)**

$$\text{i.h.p. of H.P.} = \frac{p\text{ALN}}{33000}$$

$$= \frac{70 \times 0 \cdot 7854 \times 24^2 \times 3 \cdot 5 \times 98 \times 2}{33000}$$

$$= 658 \cdot 3 \text{ i.h.p. Ans. (ii)}$$

Total i.h.p.$= 4 \times 658 \cdot 3$
$= 2633 \cdot 2$ i.h.p. **Ans. (iii)**

2. Sum of mid-ordinates $= 9 \cdot 3$ in.
Mean height $= 9 \cdot 3 \div 10 = 0 \cdot 93$ in.
M.e.p. $= 0 \cdot 93 \times 40 = 37 \cdot 2 \, \text{lb/in}^2$. **Ans. (i)**

$$\text{i.h.p.} = \frac{p\text{ALN}}{33000}$$

$$= \frac{37 \cdot 2 \times 0 \cdot 7854 \times 35^2 \times 3 \times 110 \times 2}{33000}$$

$$= 715 \cdot 8 \text{ i.h.p. Ans. (ii)}$$

3. Let stroke $= 1$

$$\text{Ratio of expansion} = \frac{\text{Volume at end of expansion}}{\text{Volume at beginning of expansion}}$$

$$= \frac{1}{0 \cdot 34} = 2 \cdot 941$$

Admission pressure $= 95 + 15 = 110$ lb/in^2 abs.
Back pressure $= 10 + 15 = 25$ lb/in^2 abs.

Admission area of diagram $= 110 \times 0.34$
Expansion ,, ,, ,, $= PV \log_\varepsilon r$
$= 110 \times 0.34 \times \log_\varepsilon 2.941$

Gross area $=$ Admission area $+$ Expansion area
$= (110 \times 0.34) + (110 \times 0.34 \times 1.0787)$
$= 110 \times 0.34 (1 + 1.0787)$
$= 110 \times 0.34 \times 2.0787$
$= 77.75$

Mean height $=$ area $\div$ length

Since the length is unity, then,
Mean gross pressure $= 77.75$ lb/in^2 abs. Ans. (i)

Mean effective pressure $=$ mean gross press. — back press.
$= 77.75 - 25$
$= 52.75$ lb/in^2. Ans. (ii)

4. Let stroke $= 1$, then clearance $= 0.05$

$$\text{Ratio of expansion} = \frac{1 + 0.05}{0.45 + 0.05} = \frac{1.05}{0.5} = 2.1$$

Admission area $= 120 \times 0.45 = 54$
Expansion area $= PV \log_\varepsilon r$
$= 120 \times 0.5 \times \log_\varepsilon 2.1$
$= 44.51$

Gross area $= 54 + 44.51 = 98.51$
Theor. mean gross press.
$= 98.51 \div 1 = 98.51$ lb/in^2 abs.

Theor. mean eff. press.
$= 98.51 - 35 = 63.51$ lb/in^2. Ans. (i)

Actual mean eff. press.
$= 0.72 \times 63.51 = 45.73$ lb/in^2. Ans. (ii)

5. Let stroke = 1, then clearance = 0·08

Volume at beginning of expansion = 0·4 + 0·08 = 0·48
" " end " " = 1 + 0·08 = 1·08

Ratio of expansion = 1·08 ÷ 0·48 = 2·25

Vol. at beginning of compression = 1·08 — 0·9 = 0·18
" " end " " = 0·08

Ratio of compression = 0·18 ÷ 0·08 = 2·25

Compression period:
$P_1 V_1 = P_2 V_2$
$90 \times 0·18 = P_2 \times 0·08$, ∴ $P_2 = 202·5$

Press. at end of compression
= 202·5 lb/in^2 abs. Ans. (i)

Admission area = 225 × 0·4 = 90
Expansion area = PV log$_\varepsilon$ r
= 225 × 0·48 × log$_\varepsilon$ 2·25
= 87·58
Gross area = 90 + 87·58 = 177·58

Exhaust area = 90 × 0·9 = 81

Compression area = PV log$_\varepsilon$ r
= 90 × 0·18 × log$_\varepsilon$ 2·25
= 13·14

Total back pressure area = 81 + 13·14 = 94·14

Effective area = gross area — back press. area
= 177·58 — 94·14
= 83·44
Mean height = area ÷ length
= 83·44 ÷ 1 = 83·44
∴ Mean effective press. = 83·44 lb/in^2. Ans. (ii)

6. $\quad$ i.h.p. $= \dfrac{p\text{ALN}}{33000}$

i.h.p. of H.P. $= \dfrac{75 \times 0·7854 \times 24^2 \times 3·5 \times 108 \times 2}{33000} = 777·3$

$$\text{i.h.p. of I.P.} = \frac{30 \times 0.7854 \times 38^2 \times 3.5 \times 108 \times 2}{33000} = 779.4$$

$$\text{i.h.p. of L.P.} = \frac{12 \times 0.7854 \times 60^2 \times 3.5 \times 108 \times 2}{33000} = 777.3$$

$$\text{M.e.p. of H.P. referred to L.P.} = \frac{75 \times 24^2}{60^2} = 12 \text{ lb/in}^2$$

$$\text{M.e.p. of I.P. referred to L.P.} = \frac{30 \times 38^2}{60^2} = 12.03 \text{ lb/in}^2$$

M.e.p. of L.P. referred to L.P. $= 12 \text{ lb/in}^2$

M.e.p. all referred to L.P.
$= 12 + 12.03 + 12 = 36.03 \text{ lb/in}^2$. Ans. (ii)

Total i.h.p.
$= 777.3 + 779.4 + 777.3 = 2334$ i.h.p. Ans. (iii)

As a check on the above:

Total i.h.p.
$$= \frac{36.03 \times 0.7854 \times 60^2 \times 3.5 \times 108 \times 2}{33000} = 2334 \text{ i.h.p.}$$

7. $\text{M.e.p. of H.P. referred to L.P.} = \dfrac{78.8 \times 1}{6.5} = 12.13 \text{ lb/in}^2$

$\text{M.e.p. of I.P. referred to L.P.} = \dfrac{31.7 \times 2.5}{6.5} = 12.19 \text{ lb/in}^2$

M.e.p. of L.P. referred to L.P. $= \qquad 12.5 \text{ lb/in}^2$

Total referred mean pressure $= 36.82 \text{ lb/in}^2$
Ans. (i)

Area of L.P. piston $= 0.7854 \times 27^2 \times 6.5 \text{ in}^2$

$$\text{Total i.h.p.} = \frac{36 \cdot 82 \times 0 \cdot 7854 \times 27^2 \times 6 \cdot 5 \times 45 \times 85 \times 2}{33000 \times 12}$$

$$= 2646 \text{ i.h.p. Ans. (ii)}$$

8. $\quad$ M.e.p. in L.P. $= 45 \div 4 = 11 \cdot 25$ lb/in^2

$$\text{M.e.p. in H.P.} = \frac{11 \cdot 25 \times 75^2}{27^2} = 86 \cdot 82 \text{ lb/in}^2$$

$$\text{M.e.p. in 1st I.P.} = \frac{11 \cdot 25 \times 75^2}{38^2} = 43 \cdot 82 \text{ lb/in}^2$$

$$\text{M.e.p. in 2nd I.P.} = \frac{11 \cdot 25 \times 75^2}{53^2} = 22 \cdot 53 \text{ lb/in}^2$$

$$\text{Total i.h.p.} = \frac{45 \times 0 \cdot 7854 \times 75^2 \times 800}{33000}$$

$$= 4819 \text{ Ans.}$$

9. $\quad$ Ratio of powers, $\quad$ H.P. : I.P. : L.P.
$\qquad\qquad\qquad\quad = 110 : 105 : 100$
$\qquad\qquad\qquad\quad = 1 \cdot 1 : 1 \cdot 05 : 1$

Sum of ratios $= 1 \cdot 1 + 1 \cdot 05 + 1 = 3 \cdot 15$

L.P. develops $\dfrac{1}{3 \cdot 15}$ of the total power

$$\text{M.e.p. in L.P.} = \frac{1}{3 \cdot 15} \times 34 \cdot 5 = 10 \cdot 95 \text{ lb/in}^2 \left.\begin{array}{l}\\\\\\\\\\\\\\\end{array}\right\}$$

$$\text{M.e.p. in I.P.} = \frac{10 \cdot 95 \times 76^2}{46^2} \times \frac{105}{100} = 31 \cdot 39 \text{ lb/in}^2 \quad \text{Ans.}$$

$$\text{M.e.p. in H.P.} = \frac{10 \cdot 95 \times 76^2}{28^2} \times \frac{110}{100} = 88 \cdot 78 \text{ lb/in}^2$$

SOLUTIONS TO TEST EXAMPLES 11   397

10. Referred mean press.

$$= \left\{ \frac{P}{R}(1 + \log_\varepsilon R) - P_b \right\} \times f_o$$

$$= \left\{ \frac{220}{13\cdot 5}(1 + 2\cdot 6027) - 3 \right\} \times 0\cdot 7$$

$$= 38\cdot 99 \text{ lb/in}^2$$

$$\text{i.h.p.} = \frac{p\text{ALN}}{33000}$$

$$3500 = \frac{38\cdot 99 \times 0\cdot 7854 \times D^2 \times 720}{33000}$$

$$D = \sqrt{\frac{3500 \times 33000}{38\cdot 99 \times 0\cdot 7854 \times 720}}$$

$$= 72\cdot 38 \text{ in.}$$

$$R = \frac{\text{volume of L.P. cyl.}}{\text{vol. of H.P. up to cut off}}$$

$$13\cdot 5 = \frac{72\cdot 38^2}{d^2 \times 0\cdot 5}$$

$$d = \sqrt{\frac{72\cdot 38^2}{13\cdot 5 \times 0\cdot 5}} = 27\cdot 86 \text{ in.}$$

I.P. dia. $= \sqrt{d \times D}$
$= \sqrt{27\cdot 86 \times 72\cdot 38} = 44\cdot 9$ in.

Practical cylinder diameters would be:
28, 45 and 72 in. Ans.

11. Total effective steam load on piston
$$= 105 \times 0.7854 \times 21^2 = 36370 \text{ lb}$$

$$\text{Accelerating force} = \frac{\text{weight} \times \text{acceleration}}{g}$$

∴ Force to accelerate reciprocating parts

$$= \frac{1200 \times 116}{32 \cdot 2} = 4323 \text{ lb}$$

Total downward thrust on crosshead
= Steam load + Wt. of parts — accel. force
= 36370 + 1200 — 4323
= 33247 lb. Ans. (i)

Let ϕ = angle of con. rod to centre-line of engine (Refer to Vol. II Chapter 13 for revision)

$$\frac{\text{con. rod length}}{\sin 45} = \frac{\text{crank length}}{\sin \phi}$$

$$\sin \phi = \frac{0.7071 \times 1.5}{6} = 0.1768$$

∴ $\phi = 10° 11'$

$$\frac{\text{Piston force}}{\text{Thrust in con. rod}} = \cos \phi$$

$$\text{Thrust in con. rod} = \frac{33247}{\cos 10° 11'} = 33780 \text{ lb}$$

Perpendicular distance of line of action of thrust in connecting rod to shaft centre
= crank length × sin (10° 11' + 45°)
= 1·5 × sin 55° 11' = 1·231 ft

SOLUTIONS TO TEST EXAMPLES 11   399

Turning moment = Thrust in con. rod × perp. distance
= 33780 × 1·231
= 41600 lb ft. Ans. (ii)

Turning moment = crank effort × crank length

∴ Crank effort = $\dfrac{41600}{1·5}$ = 27730 lb. Ans. (iii)

12. From steam tables,
At 200 lb/in², superheat 100°, $s = 1·6111$
At 2 lb/in², $s_g = s_f + qs_{fg}$
= 0·1749 + q (1·9200 — 0·1749)
= 0·1749 + 1·7451 q

Expansion is isentropic, therefore:
0·1749 + 1·7451 q = 1·6111
1·7451 q = 1·4362
q = 0·823. Ans. (i)

Rankine Efficiency = $\dfrac{h_{g1} - h_{g2}}{h_{g1} - h_{f2}}$

= $\dfrac{1258·2 - (94 + 0·823 \times 1022·2)}{1258·2 - 94}$

= $\dfrac{1258·2 - 935·2}{1258·2 - 94}$

= $\dfrac{323}{1164·2}$

= 0·2774 or 27·74% Ans. (ii)

13. Shaft dia. $= \sqrt[3]{\dfrac{22\cdot 5 \times 225 \times 68^2}{1110\left(2 + \frac{68^2}{25\cdot 5^2}\right)}}$

$= \sqrt[3]{\dfrac{22\cdot 5 \times 225 \times 68^2}{1110 \times 9\cdot 112}}$

$= 13\cdot 23$ in. Ans.

14. From steam tables, saturation temperature of steam at 250 lb/in² abs. is 401°F, therefore the steam has (581 — 401) = 180 F degrees of superheat.

From steam tables,
P = 250 lb/in² supht. 180°, $h = 1307\cdot 7$ Btu/lb

Heat given to steam in boiler
= Heat in steam — Heat in feed water
= 1307·7 — (190 — 32)
= 1149·7 Btu/lb

Thermal efficiency

$= \dfrac{\text{Heat given to steam}}{\text{Heat supplied to boiler}}$

$= \dfrac{2500 \times 1149\cdot 7}{193 \times 18200}$

$= 0\cdot 8183$ or $81\cdot 83\%$ Ans. (i)

Equivalent evaporation from and at 212°F

$= \dfrac{2500 \times 1149\cdot 7}{193 \times 970\cdot 6}$

$= 15\cdot 34$ lb steam/lb of fuel. Ans. (ii)

SOLUTIONS TO TEST EXAMPLES 11   401

15. From steam tables, Chu version,
h_{fg} at 100°C = 539·22 Chu/lb

If Centigrade steam tables are not to hand, this value can be obtained from 970·6 Btu/lb × $\tfrac{5}{9}$ = 539·22 Chu/lb
$$539·22 \text{ Chu/lb}$$
$$= 539·22 \text{ calories per gram}$$
$$= 539·22 \text{ kilocalories per kilogram}$$
C.V. of fuel = 10 kilocalories per gram
= 10 × 1000 kilocalories per kilogram

Therefore, when one kilogram of fuel is burned:

Heat given to steam = 15 × 539·22 kilocalories
Heat supplied by fuel = 1 × 10 × 1000 kilocalories

Hence,

$$\text{Thermal efficiency} = \frac{\text{Heat given to steam}}{\text{Heat supplied}}$$

$$= \frac{15 \times 539·22}{10 \times 1000}$$

$$= 0·8088 \text{ or } 80·88\% \text{ Ans.}$$

16. W = a + bP: 41300 = a + b × 3000 ... (i)
29500 = a + b × 2000 ... (ii)
─────────────────────
·11800 =   b × 1000

∴ b = 11·8

From (ii),   29500 = a + 11·8 × 2000
a = 29500 — 23600
= 5900

∴ Willans' law is, W = 5900 + 11·8 P. Ans. (a)

When developing 2800 i.h.p., steam consumption =

$$W = 5900 + 11\cdot8 \times 2800$$
$$= 5900 + 33040$$
$$= 38940 \text{ lb per hour. Ans. (i)}$$

$$\frac{38940}{2800} = 13\cdot91 \text{ lb/i.h.p. hour. Ans. (ii)}$$

SOLUTIONS TO TEST EXAMPLES 12

1. $$v = \sqrt{2\,gJ \times \text{heat drop}}$$
   $$= \sqrt{2 \times 32 \cdot 2 \times 778 \times 57}$$
   $$= 1690 \text{ ft/sec. Ans.}$$

2. Let d = dia. of exit in inches

   Discharge (ft^3/sec) = area (ft^2) × velocity (ft/sec)

   $$\frac{600 \times 105}{3600} = \frac{0 \cdot 7854\,d^2}{144} \times 3000$$

   $$d = \sqrt{\frac{600 \times 105 \times 144}{3600 \times 0 \cdot 7854 \times 3000}}$$

   $$= 1 \cdot 034 \text{ in. Ans.}$$

3. From steam tables,
   P = 115 lb/in^2, h_g = 1190·7 Btu/lb
   P = 70 lb/in^2, h_f = 272·7, h_{fg} = 908·7, v_g = 6·20

   Heat drop = 1190·7 − (272·7 + 0·97 × 908·7)
   = 36·6 Btu/lb

   $$v = \sqrt{2\,gJ \times \text{heat drop}}$$
   $$= \sqrt{2 \times 32 \cdot 2 \times 778 \times 36 \cdot 6}$$
   $$= 1354 \text{ ft/sec. Ans. (i)}$$

   Spec. vol. of exit steam = $q\,v_g$
   = 0·97 × 6·206 ft^3/lb

   Discharge (ft^3/sec) = area (ft^2) × velocity (ft/sec)

   $$= \frac{2 \cdot 25}{144} \times 1354$$

   ∴ Weight discharged $= \dfrac{2 \cdot 25 \times 1354 \times 3600}{144 \times 0 \cdot 97 \times 6 \cdot 206}$

   = 12660 lb per hour. Ans.

4. From steam tables,
P = 200 lb/in², h_g = 1199·5 Btu/lb, v_g = 2·29 ft³/lb
P = 140 lb/in², h_f = 324·9, h_{fg} = 869·1, v_g = 3·222

$$P_1 V_1^{1.135} = P_2 V_2^{1.135}$$

For one lb:
$$200 \times 2.29^{1.135} = 140 \times V_2^{1.135}$$

$$V_2 = 2.29 \times \sqrt[1.135]{\frac{200}{140}}$$

$$= 3.135 \text{ ft}^3/\text{lb}$$

Since the volume of dry steam at 140 lb/in² is 3·222 ft³/lb,

then, $q = \dfrac{3.135}{3.222} = 0.9729$ Ans. (i)

Heat drop = 1199·5 — (324·9 + 0·9729 × 869·1)

= 28·9 Btu/lb. Ans. (ii)

$$v = \sqrt{2\,gJ \times \text{heat drop}}$$
$$= \sqrt{2 \times 32.2 \times 778 \times 28.9}$$
$$= 1203 \text{ ft/sec. Ans. (iii)}$$

Discharge (ft³/sec) = area (ft²) × velocity (ft/sec)

$$1 \times 3.135 = \frac{\text{area (in}^2)}{144} \times 1203$$

$$\therefore \text{Area} = \frac{3.135 \times 144}{1203}$$

$$= 0.3753 \text{ in}^2. \text{ Ans. (iv)}$$

5. Referring to Fig. 104,

$$V_A = 2800 \times \sin 18° = 865.4 \text{ ft/sec}$$
$$V_W = 2800 \times \cos 18° = 2663 \text{ ft/sec}$$
$$x = V_W - u = 2663 - 1200 = 1463 \text{ ft/sec}$$

SOLUTIONS TO TEST EXAMPLES 12   405

$$\tan \theta_1 = \frac{V_A}{x} = \frac{865 \cdot 4}{1463} = 0 \cdot 5914$$

∴ Entrance and exit angles = 30° 36'. Ans. (i)

$$v_W = x - u = 1463 - 1200 = 263$$

$$\tan \phi = \frac{v_W}{v_A} = \frac{263}{865 \cdot 4} = 0 \cdot 3040$$

∴ $\phi = 16° 55'$, and $\beta = 90 - 16° 55' = 73° 5'$

$$v = \frac{v_A}{\cos \phi} = \frac{865 \cdot 4}{\cos 16° 55'} = 904 \cdot 4$$

∴ Velocity of exit steam = 904·4 ft/sec.
at 16° 55' to axis, or 73° 5' to blade movement } Ans. (ii)

6.   Referring to Fig. 104,

Angle between u and V_R = 180 — 33 = 147°

Angle at apex of velocity diagram
= 33 — 20 = 13°

By sine rule: $\dfrac{V}{\sin 147} = \dfrac{u}{\sin 13}$

$$\therefore u = \frac{1500 \times 0 \cdot 2250}{0 \cdot 5446} = 619 \cdot 7 \text{ ft/sec. Ans. (i)}$$

$$\text{Rotational speed} = \frac{619 \cdot 7 \times 60 \times 12}{\pi \times 26}$$

= 5462 r.p.m. Ans. (ii)

7. $V_A = 2500 \times \sin 22° = 936·5$ ft/sec
$V_W = 2500 \times \cos 22° = 2318$ ft/sec
$x = 2318 - 750 = 1568$ ft/sec

$$\text{Tan } \theta_1 = \frac{936·5}{1568} = 0·5973$$

∴ Entrance angle of blades = 30° 51′ Ans. (i)

$$V_R = \frac{V_A}{\sin \theta_1} = \frac{936·5}{\sin 30° 51'} = 1827 \text{ ft/sec}$$

$v_R = 0·85 \times 1827 = 1552$ ft/sec
$v_W = v_R \cos 31° - u$
$\quad = 1552 \times 0·8572 - 750 = 580$

$v_A = v_R \sin 31° = 1552 \times 0·5150 = 799·5$

$$\text{Tan } \beta = \frac{v_A}{v_W} = \frac{799·5}{580} = 1·378$$

∴ $\beta = 54° 2'$

$$v = \frac{v_W}{\cos \beta} = \frac{580}{\cos 54° 2'} = 987·2$$

∴ velocity at exit from blades = 987·2 ft/sec,
at 54° 2′ to blade motion,
or 35° 58′ to axis of turbine } Ans. (ii)

8. From steam tables,
P = 300 lb/in² suphtd. 200°, $h = 1323·4$ Btu/lb
P = 50 lb/in², $h_f = 250·2$, $h_{fg} = 924·6$

Heat drop thro' nozzles =
$1323·4 - (250·2 + 0·98 \times 924·6) = 167·1$ Btu/lb

Velocity at nozzle exit = $\sqrt{2 \times 32·2 \times 778 \times 167·1}$
$= 2894$ ft/sec. Ans. (i)

SOLUTIONS TO TEST EXAMPLES 12   407

$$V_A = 2894 \times \sin 20° = 989 \cdot 7 \text{ ft/sec}$$
$$V_W = 2894 \times \cos 20° = 2719 \text{ ft/sec}$$
$$x = 2719 - 1300 = 1419 \text{ ft/sec}$$

$$\text{Tan } \theta_1 = \frac{989 \cdot 7}{1419} = 0 \cdot 6974$$

∴ Entrance angle of blades = 34° 54′ **Ans. (ii)**

9.
$$V_W = 3000 \times \cos 18° = 2853$$
$$x = 2853 - 1100 = 1753$$

Since $\theta_2 = \theta_1$ and friction factor = 0·9, then,

$$v_W + u = 0 \cdot 9 \times 1753$$
$$v_W = 0 \cdot 9 \times 1753 - 1100 = 478$$
$$\therefore V_C = V_W + v_W = 2853 + 478 = 3331 \text{ ft/sec}$$

$$\text{Tangential force on blades} = \frac{WV_C}{g}$$

$$= \frac{750 \times 3331}{3600 \times 32 \cdot 2} = 21 \cdot 55 \text{ lb. Ans. (i)}$$

$$\text{Horse power} = \frac{\text{force on blades} \times \text{blade velocity}}{550}$$

$$= \frac{21 \cdot 55 \times 1100}{550} = 43 \cdot 1 \text{ h.p. Ans. (ii)}$$

Change of axial velocity of steam = $V_A - v_A$

$$V_A = 3000 \times \sin 18° = 927 \text{ ft/sec}$$
Since $\theta_2 = \theta_1$, then $v_A = 0 \cdot 9 V_A$

$$V_A - v_A = V_A - 0 \cdot 9 V_A = 0 \cdot 1 V_A = 92 \cdot 7 \text{ ft/sec}$$

$$\text{Axial thrust} = \frac{W}{g}(V_A - v_A)$$

$$= \frac{750 \times 92 \cdot 7}{3600 \times 32 \cdot 2}$$

$$= 0 \cdot 5998 \text{ lb. Ans. (iii)}$$

10. Velocity at nozzle exit $= \sqrt{2 \times 32 \cdot 2 \times 778 \times 0 \cdot 9 \times 138 \cdot 5}$
 $= 2500$ ft/sec

$V_A = 2500 \times \sin 20° = 855$,,
$V_W = 2500 \times \cos 20° = 2349$,,

$$x = \frac{V_A}{\tan \theta_1} = \frac{855}{\tan 35} = 1221$$

Blade speed $= u = V_W - x$
 $= 2349 - 1221 = 1128$ ft/sec. Ans. (i)

Since $\beta = 90°$, $\tan \theta_2 = \dfrac{v}{u} = \dfrac{680}{1128} = 0 \cdot 6029$

∴ Exit angle of blades $= 31° \; 5'$ Ans. (ii)

$$V_R = \frac{V_A}{\sin \theta_1} = \frac{855}{\sin 35} = 1490$$

$$v_R = \frac{v}{\sin \theta_2} = \frac{680}{\sin 31° \; 5'} = 1317$$

Loss of kinetic energy of the steam across the blades

$$= \frac{W}{2g}(V_R^2 - v_R^2)$$

Heat equivalent $= \dfrac{1 \times (1490^2 - 1317^2)}{2 \times 32 \cdot 2 \times 778}$

$= 9 \cdot 692$ Btu/lb. Ans. (iii)

Axial thrust $= \dfrac{W}{g}(V_A - v_A)$

$= \dfrac{1 \times (855 - 680)}{32 \cdot 2}$

$= 5 \cdot 434$ lb/lb of steam. Ans. (iv)

Horse power $= \dfrac{W V_C u}{g \times 550}$

Since the steam leaves the turbine axially, i.e. at 90° to the blade movement, there is no velocity of whirl at exit,

hence, $V_C = V_W$

$\therefore$ h.p. $= \dfrac{1 \times 2349 \times 1128}{32 \cdot 2 \times 550}$

$= 149 \cdot 6$ h.p./lb steam. Ans. (v)

Blade efficiency $= \dfrac{2 u V_C}{V^2} = \dfrac{2 \times 1128 \times 2349}{2500^2}$

$= 0 \cdot 848$ or $84 \cdot 8\%$ Ans. (vi)

11. $V_W = 810 \times \cos 23° = 745 \cdot 6$ ft/sec
       $x = 745 \cdot 6 - 530 = 215 \cdot 6$,,
       $V_A = 810 \times \sin 23° = 316 \cdot 5$,,

$\text{Tan } \theta_1 = \dfrac{V_A}{x} = \dfrac{316 \cdot 5}{215 \cdot 6} = 1 \cdot 469$

∴ Blade inlet angle = 55° 45′ Ans. (i)

Since the combined vector diagram of inlet and outlet velocities is symmetrical (Fig. 107), the total velocity of whirl is:

$V_C = 2 \times 745 \cdot 6 - 530$
or $745 \cdot 6 + 215 \cdot 6 = 961 \cdot 2$ ft/sec

7200 lb steam per hour = 2 lb/sec

$$\text{Force on blades} = \frac{W V_C}{g}$$

$$= \frac{2 \times 961 \cdot 2}{32 \cdot 2} = 59 \cdot 69 \text{ lb. Ans. (ii)}$$

Work done per sec = force × blade speed
= $59 \cdot 69 \times 530 = 31630$ ft lb/sec. Ans. (iii)

$$\text{Horse power} = \frac{31630}{550} = 57 \cdot 51 \text{ h.p. Ans. (iv)}$$

12. $C_P = C_V \times \gamma = 0 \cdot 17 \times 1 \cdot 4 = 0 \cdot 238$
$T_1 = 70 + 460 = 530°$F abs.
$T_3 = 2150 + 460 = 2610°$F abs.

$$\frac{T_2}{T_1} = \left\{ \frac{P_2}{P_1} \right\}^{\frac{\gamma-1}{\gamma}} \quad \text{where} \quad \frac{P_2}{P_1} = \frac{14}{1} = 14$$

$$\text{and} \quad \frac{\gamma - 1}{\gamma} = \frac{1 \cdot 4 - 1}{1 \cdot 4} = \frac{0 \cdot 4}{1 \cdot 4} = \frac{2}{7}$$

∴ $T_2 = 530 \times 14^{2/7} = 1127°$F abs.

Temperature at end of compression
= 1127 − 460 = 667°F. Ans. (i)

$$\frac{T_1}{T_2} = \frac{T_4}{T_3}$$

(because ratios of expansion and compression are equal)

SOLUTIONS TO TEST EXAMPLES 12

$$\therefore T_4 = \frac{530 \times 2610}{1127} = 1227°F. \text{ abs}$$

Temperature at end of expansion
$$= 1227 - 460 = 767°F. \text{ Ans. (ii)}$$

Heat supplied at constant pressure
$$= W \times C_P \times (T_3 - T_2)$$
$$= 1 \times 0.238 \times (2610 - 1127)$$
$$= 353 \text{ Btu/lb. Ans. (iii)}$$

Increase in internal energy from T_1 to T_4
$$= W \times C_V \times (T_4 - T_1)$$
$$= 1 \times 0.17 \times (1227 - 530)$$
$$= 118.5 \text{ Btu/lb. Ans. (iv)}$$

Ideal thermal efficiency

$$= 1 - \frac{T_4 - T_1}{T_3 - T_2}, \quad \text{or } 1 - \frac{T_1}{T_2}, \quad \text{or } 1 - \frac{T_4}{T_3}$$

as explained in text.

$$1 - \frac{T_4}{T_3} = 1 - \frac{1227}{2610}$$

$$= 0.53 \text{ or } 53\% \text{ Ans. (v)}$$

SOLUTIONS TO TEST EXAMPLES 13

1. C.V. $= 14500\,C + 62000\,(H - \frac{O}{8}) + 4000\,S$
$= 14500 \times 0.81 + 62000\,(0.05 - \frac{0.056}{8}) + 4000 \times 0.01$
$= 11745 + 62000 \times 0.043 + 40$
$= 11745 + 2666 + 40$
$= 14451$ Btu/lb. Ans.

2. C.V. $= 14500\,C + 62000\,(H - \frac{O}{8})$
$= 14500 \times 0.852 + 62000\,(0.12 - \frac{0.016}{8})$
$= 12354 + 62000 \times 0.118$
$= 12354 + 7316$
$= 19670$ Btu/lb. Ans. (i)

Wt. of air $= \frac{100}{23}\{2\frac{2}{3}C + 8\,(H - \frac{O}{8})\}$
$= \frac{100}{23}\{2\frac{2}{3} \times 0.852 + 8 \times 0.118\}$
$= \frac{100}{23}\{2.272 + 0.944\}$
$= \frac{100}{23} \times 3.216$
$= 13.98$ lb air/lb fuel. Ans. (ii)

3. Available hydrogen $= H - \frac{O}{8} = 0.13 - \frac{0.02}{8} = 0.1275$ lb
C.V. $= 14500\,C + 62000\,(H - \frac{O}{8})$
$= 14500 \times 0.85 + 62000 \times 0.1275$
$= 20230$ Btu/lb. Ans. (i)

Theoretical air reqd. $= \frac{100}{23}\{2\frac{2}{3}C + 8\,(H - \frac{O}{8})\}$
$= \frac{100}{23}\{2\frac{2}{3} \times 0.85 + 8 \times 0.1275\}$
$= 14.29$ lb

Actual air $= 1.5 \times 14.29 = 21.44$ lb air/lb fuel. Ans. (ii)

Products of combustion per lb of fuel burned
$= 21.44$ lb air $+ 1$ lb fuel $= 22.44$ lb

Heat carried away in flue gases
$=$ wt. $\times$ spec. ht. $\times$ temp. rise
$= 22.44 \times 0.24 \times (535 - 87)$
$= 2413$ Btu/lb fuel

% heat lost $= \dfrac{2413}{20230} \times 100 = 11.92\%$ Ans. (iii)

SOLUTIONS TO TEST EXAMPLES 13

4. Heat given out by fuel
   = Heat absorbed by water and bomb
   Wt. of oil × C.V.
   = (Wt. of water + W.E. of bomb) × temp. rise

   $0.75 \times$ C.V. $= (1800 + 470) \times 3.3$

   $$\text{C.V.} = \frac{2270 \times 3.3}{0.75}$$

   = 9986 gram-cal/gram
   $9986 \times \tfrac{9}{5} = 17970$ Btu/lb. Ans.

5. Available hydrogen
   $= H - \tfrac{O}{8}$
   $= 0.13 - \tfrac{0.02}{8} = 0.1275$ lb
   C.V. $= 14500 \times 0.84 + 62000 \times 0.1275$
   $= 12180 + 7905 = 20085$ Btu/lb. Ans. (i)

   Theo. wt. of air
   $= \tfrac{100}{23}(2\tfrac{2}{3} \times 0.84 + 8 \times 0.1275)$
   $= \tfrac{100}{23} \times 3.26 = 14.17$ lb air/lb fuel. Ans. (ii)

   Weight of gases passing up funnel per lb of fuel burned
   = 22 lb air + 0.84 lb C + 0.13 lb H + 0.02 lb O
   = 22.99 lb

   Wt. of nitrogen in 22 lb air $= 0.77 \times 22 = 16.94$ lb
   Wt. of oxygen in 22 lb air $= 0.23 \times 22 = 5.06$ lb

   Weight of oxygen to burn 0.84 lb carbon
   $= 2\tfrac{2}{3} \times 0.84 = 2.24$ lb

   Wt. of CO_2 formed $= 0.84 + 2.24 = 3.08$ lb

   Weight of oxygen to burn available hydrogen
   $= 8 \times 0.1275 = 1.02$ lb

   Wt. of H_2O formed $= 9 \times 0.13 = 1.17$ lb

   Weight of oxygen used for combustion
   $= 2.24 + 1.02 = 3.26$ lb

   Surplus oxygen $= 5.06 - 3.26 = 1.8$ lb

∴ Composition of funnel gases:

$$\text{Nitrogen} = \frac{16 \cdot 94}{22 \cdot 99} \times 100 = 73 \cdot 68\%$$

$$CO_2 = \frac{3 \cdot 08}{22 \cdot 99} \times 100 = 13 \cdot 4\%$$

$$H_2O = \frac{1 \cdot 17}{22 \cdot 99} \times 100 = 5 \cdot 09\%$$

$$\text{Oxygen} = \frac{1 \cdot 8}{22 \cdot 99} \times 100 = 7 \cdot 83\%$$

Ans.

6. 0·855 lb carbon requires $2^2 \times 0 \cdot 855 = 2 \cdot 28$ lb oxygen
    CO_2 formed $= 0 \cdot 855 + 2 \cdot 28 = 3 \cdot 135$ lb

Hydrogen available for combustion

$$= 0 \cdot 119 - \frac{0 \cdot 016}{8} = 0 \cdot 117 \text{ lb}$$

0·117 lb hydrogen requires $8 \times 0 \cdot 117 = 0 \cdot 936$ lb oxygen

Total oxygen required per lb of fuel
$= 2 \cdot 28 + 0 \cdot 936 = 3 \cdot 216$ lb

Minimum weight of air required per lb of fuel
$= \frac{100}{23} \times 3 \cdot 216 = 13 \cdot 98$ lb

When the air supply is minimum:

Weight of products of combustion per lb of fuel burned
$= 13 \cdot 98$ lb air $+ 1$ lb fuel $- 0 \cdot 01$ lb impurities
$= 13 \cdot 98 + 0 \cdot 99 = 14 \cdot 97$ lb

$$\% CO_2 = \frac{3 \cdot 135}{14 \cdot 97} \times 100 = 20 \cdot 94\% \text{ Ans. (i)}$$

SOLUTIONS TO TEST EXAMPLES 13

When the air supply is 25% excess:

Weight of products of combustion per lb of fuel burned
$$= 1{\cdot}25 \times 13{\cdot}98 + 0{\cdot}99 = 18{\cdot}47 \text{ lb}$$

$$\% \ CO_2 = \frac{3{\cdot}135}{18{\cdot}47} \times 100 = 16{\cdot}97\% \text{ Ans. (ii)}$$

When 50% excess air is supplied:

Weight of products of combustion per lb of fuel burned
$$= 1{\cdot}5 \times 13{\cdot}98 + 0{\cdot}99 = 21{\cdot}96 \text{ lb}$$

$$\% \ CO_2 = \frac{3{\cdot}135}{21{\cdot}96} \times 100 = 14{\cdot}28\% \text{ Ans. (iii)}$$

When 75% excess air is supplied:

Weight of products of combustion per lb of fuel burned
$$= 1{\cdot}75 \times 13{\cdot}98 + 0{\cdot}99 = 25{\cdot}45 \text{ lb}$$

$$\% \ CO_2 = \frac{3{\cdot}135}{25{\cdot}45} \times 100 = 12{\cdot}31\% \text{ Ans. (iv)}$$

SOLUTIONS TO TEST EXAMPLES 14

1.
$$\text{One h.p. hour} = 2545 \text{ Btu}$$
$$\text{Heat equivalent of 5 h.p.} = 5 \times 2545 \text{ Btu/hour}$$

Heat extracted = Co-eff. of performance × heat equiv. of work done
$$= 3{\cdot}8 \times 5 \times 2545 \text{ Btu/hour}$$

Producing ice at 32°F from water at 32°F, heat to be extracted from each lb = latent heat only = 144 Btu

∴ Capacity of machine

$$= \frac{3{\cdot}8 \times 5 \times 2545 \times 24}{144 \times 2240}$$
$$= 3{\cdot}598 \text{ tons per day. Ans. (i)}$$

To make ice at 18°F from water at 68°F, heat extracted,
$$(68 - 32) + 144 + 0{\cdot}5(32 - 18) = 187 \text{ Btu/lb}$$

∴ Weight of ice that can be made

$$= 3{\cdot}598 \times \frac{144}{187} = 2{\cdot}771 \text{ tons per day. Ans. (ii)}$$

2. Total heat in refrigerant:

Leaving compressor and entering condenser = 360 Btu/lb.
   ,, condenser ,, ,, evaporator = 360 — 90 = 270 Btu/lb.
   ,, evaporator ,, ,, compressor = 360 — 20 = 340 Btu/lb.

∴ Heat absorbed by refrigerant in evaporator
$$= 340 - 270 = 70 \text{ Btu/lb. Ans. (i)}$$

$$\text{Coeff. of performance} = \frac{\text{Heat extracted}}{\text{Heat equiv. of work done}}$$

$$= \frac{70}{20} = 3{\cdot}5 \text{ Ans. (ii)}$$

SOLUTIONS TO TEST EXAMPLES 14   417

3. Heat extracted = Coeff. of perf. × heat equiv. of work done
    = 4·5 × 7·5 × 2545 Btu per hour

This is equal to the heat absorbed by the refrigerant in the evaporator. Expressing this in Btu per lb of refrigerant:

$$= \frac{4·5 \times 7·5 \times 2545}{3 \times 60} = 477·2 \text{ Btu/lb}$$

Latent heat of the vapour at $-10°F$
    $= 566 - 0·8(-10) = 574$ Btu/lb
∴ Increase in dryness fraction through evaporator

$$= \frac{477·2}{574} = 0·8314$$

Total heat before throttling
    = Total heat after throttling
Sensible heat at 60°F = (sensible + latent heat) at
    $-10°F$

$$h_{f1} = h_{f2} + q\, h_{fg2}$$
$$h_{f1} - h_{f2} = q \times 574$$
$$1·1[60 - (-10)] = q \times 574$$
$$1·1 \times 70 = q \times 574$$
$$q = 0·1342$$

∴ Dryness fraction, entering evaporator = 0·1342 ⎫
                                                  ⎬ Ans.
Leaving evaporator = 0·1342 + 0·8314 = 0·9656 ⎭

4. Total heat in refrigerant leaving condenser and entering evaporator = $14·4 + q \times 181·8 = 48·2$
    $q \times 181·8 = 33·8$
    $q = 0·1859$ Ans. (i)

Total heat in refrigerant leaving evaporator and entering compressor = $14·4 + 0·96 \times 181·8$
    = 188·9 Btu/lb

Heat absorbed in evaporator = 188·9 − 48·2 = 140·7 Btu/lb
Work done in compressor = 219 − 188·9 = 30·1 Btu/lb

$$\text{Coeff. of performance} = \frac{\text{Heat extracted}}{\text{Work done}}$$

$$= \frac{140 \cdot 7}{30 \cdot 1} = 4 \cdot 674 \text{ Ans. (ii)}$$

Heat extracted from water to make ice = 144 Btu/lb

$$= \frac{1 \times 2240 \times 144}{24} \text{ Btu/hour}$$

Work done in compressor to extract this heat

$$= \frac{\text{heat extracted}}{\text{coeff. of performance}}$$

$$= \frac{1 \times 2240 \times 144}{24 \times 4 \cdot 674} \text{ Btu/hour}$$

one h.p. = 2545 Btu per hour, therefore,

$$\text{Output horse power} = \frac{1 \times 2240 \times 144}{24 \times 4 \cdot 674 \times 2545}$$

$$= 1 \cdot 127 \text{ h.p. Ans. (iii)}$$

5. Through first regulating valve:
Heat before throttling = Heat after throttling
sensible heat (liquid only) @ 990 lb/in²
$$= (\text{sensible} + \text{latent heat}) \text{ @ } 420 \text{ lb/in}^2$$
$$76 = 29 + q \times 109 \cdot 4$$
$$q = 0 \cdot 4296. \text{ Ans. (i)}$$

Through second regulating valve:
Heat before throttling = Heat after throttling
sensible heat (liquid only) @ 420 lb/in²
$$= (\text{sensible} + \text{latent heat}) \text{ @ } 275 \text{ lb/in}^2$$
$$29 = 15 \cdot 8 + q \times 123$$
$$q = 0 \cdot 1074. \text{ Ans. (ii)}$$

SOLUTIONS TO TEST EXAMPLES 14

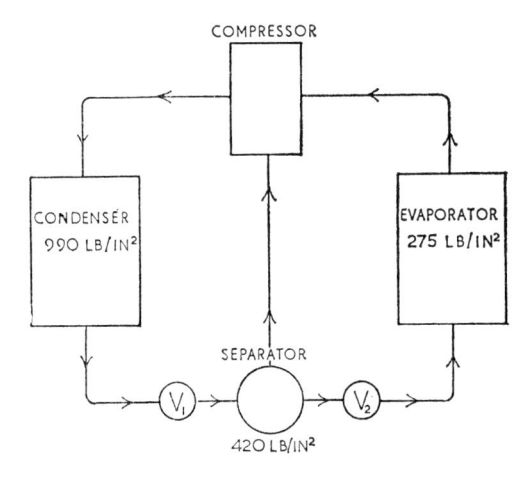

```
                    COMPRESSOR
                   ┌──────────┐
         ┌────←────│          │────←────┐
         │         └──────────┘         │
         │              ↑               │
   ┌─────┴────┐                    ┌────┴─────┐
   │ CONDENSER│                    │EVAPORATOR│
   │ 990 LB/IN²│                   │275 LB/IN²│
   │          │                    │          │
   └─────┬────┘                    └────↑─────┘
         │          SEPARATOR           │
         └──→─(V₁)──→─◯──→─(V₂)──→──────┘
                   420 LB/IN²
```

Fig. 126

Heat extracted per lb of CO_2 passing through **evaporator**
$= (0{\cdot}92 - 0{\cdot}1074) \times 123 = $ **99·95 Btu**

For one lb of gas discharged from compressor, one lb of liquid leaves condenser and 0·4296 lb is flashed off as gas in separator and returned to compressor, therefore weight of CO_2 remaining in separator to be passed on to evaporator
$= 1 - 0{\cdot}4296 = 0{\cdot}5704$ lb

Hence, heat extracted in evaporator per lb of gas leaving compressor

$= 0{\cdot}5704 \times 99{\cdot}95 = $ **57·02 Btu. Ans. (iii)**

If no separator, through the one regulating valve only:

Heat before throttling = Heat after throttling
$$76 = 15{\cdot}8 + q \times 123$$
$$q = 0{\cdot}4895$$

Heat extracted in evaporator
$= (0{\cdot}92 - 0{\cdot}4895) \times 123 = $ **52·95 Btu/lb. Ans. (iv)**

6. Entropy before compression

$$= 1 \cdot 1 \log_\varepsilon \frac{452}{420} + \frac{q \times 575}{452}$$

Entropy after compression

$$= 1 \cdot 1 \log_\varepsilon \frac{543}{420} + \frac{496}{543} + 0 \cdot 63 \log_\varepsilon \frac{560}{543}$$

Entropy before and after are equal for isentropic compression. Note that in equating the two, $\log_\varepsilon 420$ cancels from each side.

$1 \cdot 1 \times 6 \cdot 1137 + 1 \cdot 273q$
$\qquad = 1 \cdot 1 \times 6 \cdot 2971 + 0 \cdot 9135 + 0 \cdot 63 \times 0 \cdot 0309$
$6 \cdot 7251 + 1 \cdot 273q$
$\qquad = 6 \cdot 9268 + 0 \cdot 9135 + 0 \cdot 0195$
$\quad 1 \cdot 273q = 1 \cdot 1347$
$\qquad q = 0 \cdot 8917$ Ans.

EXAMPLE OF
DATA SUPPLIED AT EXAMINATIONS

METALS	WEIGHT IN POUNDS		SPECIFIC GRAVITY	SPECIFIC HEAT
	PER CUBIC INCH	PER CUBIC FOOT		
Cast Iron	0·26	450	7·21	0·13
Wrought Iron	0·281	486	7·78	0·113
Steel	0·283	490	7·86	0·116
Copper	0·317	548	8·77	0·095
Brass	0·303	524	8·4	0·094
Lead	0·412	712	11·4	0·029
Mercury	0·491	849	13·6	0·033

Volume of sphere = $\frac{\pi}{6} \times D^3$
Volume of cone = Area of base × ⅓ height
One nautical mile = 6080 feet
One metre = 39·37 inches
One litre = 0·22 gallon
One kilogram = 2·2 lb
Latent heat of water = 144 Btu per lb
Specific heat of ice = 0·5

JOULES EQUIVALENT:
778 foot pounds = 1 Btu
One cubic foot of fresh water weighs 62·5 lb
One cubic foot of sea water weighs 64 lb
One cubic foot contains 6·25 gallons.

NOTE. In the selection of examination questions to follow, certain data required for the solution of a problem may not be given with the question if it is included in the above memoranda. The student is expected to be capable of extracting the relevant matter as required, such as the specific weight of the material, or its specific gravity, specific heat, etc.

SELECTION OF EXAMINATION QUESTIONS
SECOND CLASS

1. A boiler on test was found to have an equivalent evaporation from and at 100°C of 14·65 kilograms per kilogram of fuel. If the calorific value of the fuel was 10·17 kilo-calories per gram, find the thermal efficiency of the boiler.

2. A motor was tested by coupling it to a Froude dynamometer and the brake load was 25·5 lb at 2300 r.p.m. A steady flow of water passed through the brake and was raised in temperature from 15·2°C to 48·7°C. If the horse-power absorbed by the brake is $\dfrac{WN}{4500}$ where W is the brake load in lb and N is the speed in r.p.m. and assuming that 98% of the heat generated at the brake was carried away by the cooling water, find the quantity of water passing through the dynamometer per minute.

3. One lb of superheated steam of pressure 215 lb/in² abs. and temperature 600°F is expanded adiabatically until the pressure is 15 lb/in² abs., and the work done during expansion is 177200 ft lb. Taking the mean specific heat of the superheated steam as 0·56, calculate the final dryness fraction of the steam.

4. The dummy clearance of a turbine is 0·072 in. when at 60°F. The rotor is 10 feet long, is made of steel and the casing of cast iron, the coefficients of linear expansion being 0·0000067 and 0·0000061 per F° respectively. Find the clearance when at the running temperature of 360°F.

5. An ordinary D slide valve has a travel of $6\tfrac{1}{2}$ inches. The exhaust lap at the top is $+\tfrac{1}{8}$ inch, and the lead at the bottom is $\tfrac{3}{16}$ inch. The exhaust openings at the instants the crank passes top and bottom centres are $1\tfrac{7}{8}$ and 2 inches respectively. Find the steam and exhaust laps at the bottom of the valve.

6. Derive a formula to convert a temperature reading from the Fahrenheit to the Centigrade scale.
   Determine the temperature when the Fahrenheit reading is exactly twice the Centigrade reading, and the temperature when the two readings are the same.
   Also convert −13°F to Centigrade.

SECOND CLASS EXAMINATION QUESTIONS

7. One lb of wet saturated steam at 100 lb/in² abs. and 0·9 dry is expanded according to the law $PV^{1·3} = C$ until the pressure is 50 lb/in² abs. Calculate the final dryness fraction of the steam.

8. A four-cylinder two-stroke single-acting internal combustion engine develops 400 i.h.p. when the mean effective pressure in each cylinder is 85 lb/in² and the speed is 270 r.p.m. If the stroke is 25% greater than the cylinder diameters, calculate the stroke and diameter of cylinders.

9. 1000 cubic inches of wet steam at 200 lb/in² abs. and 0·95 dry passes through a reducing valve and is throttled to a pressure of 50 lb/in² abs. Calculate the dryness fraction and the volume (in cubic feet) after throttling.

10. A slide valve has $8\tfrac{1}{2}$ inches travel. The exhaust laps are $\tfrac{1}{4}$ inch at top and $\tfrac{1}{8}$ inch at bottom. The steam laps are $2\tfrac{3}{4}$ inches at top and $2\tfrac{5}{8}$ inches at bottom. The top lead is $\tfrac{1}{8}$ inch and the bottom lead $\tfrac{1}{4}$ inch. What is the exhaust port opening when the crank is passing (a) the top centre, (b) the bottom centre?

11. From a boiler working at 185 lb/in² abs., a sample was drawn through the salinometer cock and the weight of the sample in the salinometer pot was 1·5 lb. Find the weight of water taken from the boiler. Explain why the densities of the water in the pot and the water in the boiler are different.

12. The mean effective pressure in the cylinders of a 4-cylinder two-stroke single-acting internal combustion engine is 88·6 lb/in² when running at 170 r.p.m. The diameter of the cylinders is 24·5 in. and the stroke is 51·7 in. Find the i.h.p. If the mechanical efficiency is 75%, find the b.h.p.

    Sketch a typical indicator diagram for a two-stroke Diesel engine, marking thereon the important pressures.

13. A boiler working at 220 lb/in² abs. produces 18000 lb of steam per hour from feed water at 220°F, the dryness fraction of the steam being 0·97. The calorific value of the fuel used was 14400 Btu/lb and the boiler efficiency 87%. Calculate the weight of fuel used per day, in tons.

14. 15 lb of crushed ice at 20°F are mixed with 25 lb of water at 38°F. Find the final temperature and physical state of the mixture.

15. An oil engine uses 0·45 lb of fuel per b.h.p. per hour of calorific value 19000 Btu/lb. The mechanical efficiency is 85%. Find the thermal efficiency (a) based upon the i.h.p., (b) based upon the b.h.p.   (continued next page)

If 28 lb of air are supplied per lb of fuel, air inlet temp. 60°F and exhaust temp. 760°F, find the percentage heat loss in the exhaust gases in relation to the heat supplied to the engine; take the specific heat of the gases as 0·25 and neglect the latent heat carried away, due to the formation of steam during combustion.

16. A glass tube of small uniform bore is closed at one end and open at the other; it contains air imprisoned by a column of mercury 4 centimetres long. When the tube is held vertically with the closed end at the bottom, the length of the air column is 21·78 cm, and when held vertically with sealed end at the top, the length of the air column is 24·2 cm. Calculate the atmospheric pressure in centimetres of mercury, inches of mercury, and lb/in².

17. The cylinder diameters of a triple expansion engine are 24·5, 41·5 and 70 inches respectively, and the stroke is 48 in. Find (a) the mean pressure referred to the L.P. and (b) the mean effective pressure in each cylinder, when running at 76 r.p.m. and developing a total i.h.p. of 2470. Assume each cylinder to develop the same power.

18. The analysis by volume of a sample of funnel gas is: carbon dioxide = 10%, carbon monoxide = 1·7%, oxygen = 8·1% and nitrogen = 80·2%. Convert this into a percentage analysis by weight, given that the molecular weights are, CO_2 = 44, CO = 28, O_2 = 32, and N_2 = 28.

19. Dry saturated steam at 200 lb/in² abs. is supplied to a turbine, the pressure of the exhaust is 1·25 lb/in² abs. and the hotwell temperature is 90°F. If the quantity of heat given up by the steam in the condenser is 4·7 times the heat drop through the turbine, find the dryness fraction of the exhaust steam.

20. An engine develops 925 i.h.p. when running at 99 r.p.m. and the mean effective pressure is 89 lb/in². If the mean effective pressure is reduced to 79 lb/in², find (i) the engine speed in r.p.m., (ii) the i.h.p. developed, taking the m.e.p. to vary as (speed)$^{2·1}$.

21. If the total heat contained in one kilogram of wet steam at a pressure of 180 lb/in² abs. is 1311·9 Chu, calculate its dryness fraction from the data given below. Calculate also the additional heat required, in Chu and in Btu, to dry this steam at the same temperature and pressure.
P = 180 lb/in² abs., h_f = 192·3 Chu/lb, h_{fg} = 473·3 Chu/lb

SECOND CLASS EXAMINATION QUESTIONS 425

22. In an eight-cylinder single-acting four-stroke I.C. engine, the diameter of the cylinders is 620 mm and the stroke is 1300 mm. The mean effective pressures average out at 93·2 lb/in² per cylinder when running at 119·6 r.p.m., and the mechanical efficiency is 88%. Calculate the i.h.p. and b.h.p. Sketch a typical indicator diagram for a 4-stroke diesel engine.

23. A turbine installation is supplied with superheated steam and incorporates a feed-water heater operated by bleeding from one of the turbines. From the following particulars of the total heat in the working fluid measured above 32°F at the cardinal stages of the cycle, calculate the ideal thermal efficiency of the installation (a) with, (b) without, the bleeding system in operation.

   At engine stop valve ... ... 1357·6 Btu/lb
   At bleeding point ... ... 1118·66 ,,
   At exhaust 904   ,,
   Weight of steam bled ... ... 0·1388 lb
   Heat in feed water without bleeding 47·42 Btu/lb
   Heat in feed water with bleeding... 196·1 ,,

24. Three Scotch boilers produce 27470 lb of steam per hour, at a pressure of 240 lb/in² abs. from feed water at 200°F, and the temperature of the steam at the superheater outlet is 746°F. The gross calorific value of the fuel is 19200 Btu/lb and the boiler efficiency based upon the gross calorific value is 76%. Find the oil consumption in 24 hours. Take the spec. ht. of superheated steam as 0·55.

25. (i) State the laws connecting pressure, volume and temperature of a perfect gas.
    (ii) Calculate the value of the gas constant R for one lb weight of air from the information that 10 lb weight of air at a temperature of 32°F and pressure 14·7 lb/in² occupy a volume of 124 cubic feet. Express the value per F degree and per C degree, stating clearly the units of the constant.

26. Superheated steam at 450 lb/in² abs. and at 750°F is supplied to a turbine, and exhausted at 0·85 lb/in² abs. and 0·9 dry. Calculate the equivalent horse power supplied when the consumption of steam is 22000 lb per hour.
    Take the mean specific heat of superheated steam at this pressure and temperature range as 0·655.

27. State the advantages of compounding in a steam engine. In a compound steam engine having two cylinders, cut off takes place at 0·5 stroke in the H.P. cylinder and the steam is expanded 7 times. Calculate (a) the ratio of the cylinder diameters, (b) the

terminal pressure in the L.P. cylinder if the initial steam pressure to the H.P. cylinder is 150 lb/in^2 abs.

28. The specific fuel consumption of a diesel engine is 0·43 lb per b.h.p. per hour. The heat losses to the cooling water and exhaust gases are 29% and 31% respectively, of the total heat supplied, and the mechanical efficiency is 78%. Calculate the brake thermal efficiency and the calorific value of the fuel.

29. The consumption of steam per i.h.p. hour of a triple expansion engine was 14·5 lb, the exhaust being at 2 lb/in^2 abs. and 0·8 dry. After superheaters were fitted to the boiler the steam consumption was reduced to 13 lb per i.h.p. hour, the exhaust being at 2 lb/in^2 abs. and 0·87 dry. Find, for before and after conversion to superheating:

(a) the heat given up in the condenser per lb of steam condensed;

(b) the heat lost in the condenser per i.h.p. per minute;

(c) the equivalent horse power lost per i.h.p.

Assume that the condensate from the condenser is at the same temperature as the exhaust steam.

30. Define Specific Heat. A piece of steel was plunged into 50 times its own volume of oil of specific gravity 0·95, the initial temperature of the steel was 540°F and the resultant temperature of the mixture 42°F. If the initial temperature of the oil was 18·8°F, find its specific heat.

31. If the rate of heat transmission through the walls of condenser tubes is taken as 2 × 10^6 gram calories per hour per square metre of surface area per Centigrade degree difference in temperature between outside and inside, find the equivalent horse power transmitted per square foot for a temperature difference of 5 Centigrade degrees.

32. The air in a ship's saloon is maintained at 65°F, and is changed twice every hour from the outside atmosphere which is at 45°F. The saloon is 60 ft by 80 ft by 9 ft high. Find the kilowatt loading on the generator to warm this air. Note: One cubic foot of air at 14·7 lb/in^2 abs. and 32°F weighs 0·0807 lb. The specific heat of air may be taken as 0·24.

33. A liquid of specific gravity 0·8, specific heat 0·6 and temperature 80°F is mixed with another liquid of specific gravity 0·82, specific heat 0·45 and temperature 110°F in the ratio of one of the first to three of the second by volume. Find the resulting temperature.

34. 12 lb of wet steam at 250 lb/in² abs. is found to require a further 960 Btu to completely dry the steam and then a further 1360 Btu to superheat it to 600°F. From this data determine the initial dryness fraction of the steam and also the mean specific heat of the superheated steam.

35. The clearance space in a compressor cylinder is equivalent to a linear clearance of 45 mm. When the air is compressed isothermally, the air pressure in the cylinder is 44 lb/in² gauge when the piston has travelled 225 mm from the beginning of its stroke, and 134 lb/in² gauge at 400 mm of the stroke. Take atmospheric pressure as 14·7 lb/in² and calculate the length of the stroke.

36. The diameter of a single stage air compressor is 13 cm, its stroke is 18 cm and the clearance volume is 70 cc. The pressure in the cylinder at the end of the suction stroke is 14 lb/in² abs. and the delivery pressure is constant at 65 lb/in² abs. Find for what fraction of the stroke the maximum load is exerted on the compressor, if the law of compression is $PV^{1\cdot 2}$ = constant.

37. The indicator cards given are from one cylinder of a triple expansion engine. The diameters of the cylinders are 30 inches, 44 inches and 80 inches, and the stroke is 51 inches. The revolutions are 70 per minute. If each engine indicates the same power, and the coal consumption is 1·6 lb per i.h.p. hour, find the amount of coal burnt per day.

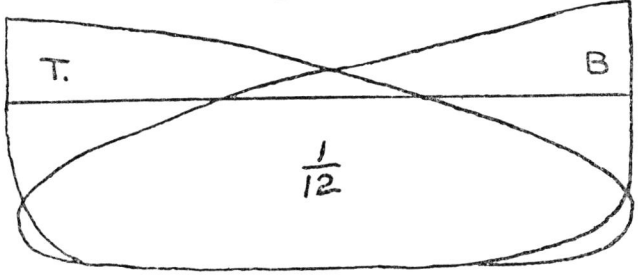

38. State the expression connecting eccentricity of an eccentric sheave and travel of the slide valve. The eccentricity of an eccentric is $3\tfrac{1}{4}$ in. and the angle of advance is 30°, if the slide valve has an outside steam lap of $1\tfrac{9}{16}$ in., find (a) lead of the valve, (b) maximum port opening to steam, (c) crank angle at cut off, (d) crank angle at admission.

39. An air vessel contains 60 lb weight of air at a temperature of 63°F, and at a pressure of 500 lb/in² gauge. Owing to a fire in the vicinity the pressure rises to 650 lb/in² gauge. Taking the

specific heat of air at constant volume to be 0·17, find how many units of heat have been given to the air.

40. The lead of an ordinary slide valve is 0·15 inch and the maximum port opening to steam is 1·612 inches. The angle of advance of the eccentric is 36° 8′. Calculate the travel of the valve.

41. An indicator card taken off a double-acting steam engine was divided into ten sections and the mid-ordinates measured as follows: 0·58, 0·8, 0·9, 0·92, 0·89, 0·73, 0·53, 0·42, 0·37 and 0·11 inches respectively. The scale of the spring in the indicator was one inch to 60 lb/in². The diameter of the engine piston was 6 in., length of stroke 6·5 in. and speed 120 r.p.m. Find (a) the mean effective pressure, (b) work done per revolution, (c) indicated horse power.

42. A two-stage tandem air-compressor has cylinders $2\frac{1}{2}$ inches and $9\frac{1}{2}$ inches diameter, and a stroke of 8 inches. It is used to fill a blast bottle, with hemispherical ends, 12 inches diameter and 6 feet long overall. Air is taken into the compressor at 15 lb/in² absolute, pressure in bottle is to be 900 lb/in² by gauge. Assuming revolutions to be constant at 120 per minute, and volumetric efficiency to be 0·9, find the time it would take to fill the bottle.

43. A ship has 50000 cu ft available for bunker space. Fuel can be (i) oil of specific gravity 0·9 and calorific value 19000 Btu/lb; (ii) coal which stows at 48 cu ft to the ton and having a calorific value of 14000 Btu/lb. Find for each fuel, the available energy in horse-power-hours and state which has the advantage as regards the amount of heat stored.

44. A bar one inch diameter and 5 feet long is heated through 150 C degrees and then clamped at its ends. On cooling down to its original temperature, the bar is found to have shortened by 0·07 inch from its heated length. If the coefficient of linear expansion of the material is 0·000011 per degree C, and $E = 30 \times 10^6$ lb/in², find the pulling force on the clamps.

45. Steam at a pressure of 185 lb/in² gauge is admitted to a steam engine cylinder and cut off at half stroke. The clearance volume is equal to 5% of the stroke volume. Assuming hyperbolic expansion of the steam after cut-off, and exhaust to commence at the end of the stroke, sketch the diagram representing the work done in the cylinder and calculate the mean gross pressure of the steam on the piston during one stroke.

SECOND CLASS EXAMINATION QUESTIONS 429

46. The high pressure cylinder of a triple expansion engine is 18 inches diameter. The stroke is 3 feet and the connecting rod is 7 feet long. The boiler pressure is 180 lb/in^2 gauge and the back pressure on the H.P. piston 70 lb/in^2 gauge. Calculate the thrust in the connecting rod in lb, and the twisting moment in the shaft in lb ft due to the H.P. engine when the crank is 30 degrees from top centre.

47. The pressure in a boiler is released from 230 to 14·7 lb/in^2 abs. by lifting the safety valves with the easing gear. Estimate the weight of water left in the boiler expressed as a percentage of the original weight.

48. 115 cubic feet of wet steam at 40 lb/in^2 abs. are blown into 160 lb of water at a temperature of 60°F, and the resulting temperature of the mixture is 135°F. Calculate the dryness fraction of the steam.

49. 56 kilograms of ice at — 10°C were placed in a tank containing 250 kilograms of fresh water at 30°C, and 10 kilograms of superheated steam at 600 lb/in^2 abs. and 650°F were blown into the mixture. Find the final temperature of the water, taking the specific heat of the superheated steam to be 0·6.

50. The cylinder diameters of a triple expansion engine are 24 in., 40 in., and 64 in. The mean pressure referred to the L.P. is 33 lb/in^2. The I.P. engine generates 3% more power than the L.P. engine and 5% more than the H.P. engine. Find the mean effective pressure in each cylinder.

51. A boiler contains 40 tons of water. After 8 tons of fresh water are pumped in and then 8 tons of the boiler water blown out, the density is found to be 14·5 ounces per gallon. If it were practicable to blow the 8 tons out of the boiler first and then pump in 8 tons of fresh water, what would be the final density of the water in the boiler?

52. The mechanical efficiency of a diesel engine is 74%. The heat loss to the cooling water is 27·5%, and 28% is lost in the exhaust gases. The calorific value of the fuel used is 19,000 Btu/lb. Find the lb of fuel used per b.h.p. per hour.

53. An air compressor takes in air at 14·7 lb/in^2 abs. and discharges it at 26 lb/in^2 abs. Neglecting clearance, calculate the percentage of the inward stroke travelled by the piston when the discharge valves open, assuming the compression to be (i) isothermal, (ii) adiabatic, taking the value of γ as 1·4.

54. The travel of a slide valve is 8·125 in. and the angle of advance of the eccentric is 32°. When the crank has moved through 74° from the dead centre, the steam port opening is 1·875 in. Calculate the steam lap and the lead of the valve.

55. A fuel tank 8 feet 6 inches diameter and 6 feet high when filled with oil would supply fuel for 24 hours. The specific gravity of the oil is 0·92 and the fuel consumption of the engine 0·4 lb per i.h.p. per hour. Find the i.h.p. of the engine. If the calorific value of the fuel is 18500 Btu/lb find also the indicated thermal efficiency of the engine.

56. A shaft 26 inches diameter runs at 95 r.p.m. The coefficient of friction between the shaft and the bearings is 0·021 and the load on the bearings is 92 tons. Find the horse power lost in friction at this speed and the heat units generated at the bearings per minute.

57. During a trial run on a diesel engine the following data were taken:

$$\begin{aligned}
\text{i.h.p. developed} &= 3500 \\
\text{Flow of cooling water} &= 47\cdot5 \text{ tons per hour} \\
\text{Inlet temp. of cooling water} &= 60°F \\
\text{Outlet temp. of cooling water} &= 127°F \\
\text{Fuel consumption} &= 1120 \text{ lb per hour} \\
\text{C.V. of fuel} &= 19300 \text{ Btu/lb}
\end{aligned}$$

Calculate the indicated thermal efficiency of the engine and the percentage of the total heat of the fuel carried away by the cooling water.

58. A fuel oil is composed of 84·6% carbon, 11·4% hydrogen, 0·4% sulphur, 2·4% oxygen and 1·2% impurities. Calculate the calorific value of the oil and the theoretical weight of air required to burn one lb, taking the values:

	C.V. Btu/lb	Atomic weight
Hydrogen	62000	1
Carbon	14500	12
Sulphur	4000	32
Oxygen	—	16

Composition of air by weight: 23% oxygen, 77% nitrogen.

59. A compound turbine power plant is supplied with steam at 250 lb/in^2 abs. superheated 200 F degrees. It is expanded in the H.P. turbine from which it leaves dry and saturated at 28 lb/in^2 abs. At this point some steam is bled off to the feed heaters, the remaining steam passing on to the L.P. turbine where it is

expanded to 0·8 lb/in² abs. having a dryness fraction of 0·89. Calculate the amount of steam bled off expressed as a percentage of the steam supplied, so that the same quantity of work is done in each turbine.

60. 15 kilograms of ice at −10°C are converted into dry saturated steam at 100 lb/in². By reference to the following extract from steam tables and taking the latent heat of ice as 80 Chu/lb and spec. ht. 0·5, find the total heat supplied in Chu, and Btu, and calculate the time taken in minutes if the heat supply is equivalent to 25 horse power.

P lb/in²	h_f Chu/lb	h_{fg} Chu/lb	h_g Chu/lb
100	165·8	494·3	660·1

61. The temperature of the flame in the furnace of a boiler is 2250°F and the temperature of the furnace is 380°F. The quantity of heat conducted through the metal of the furnace to the water in the boiler is $7\frac{1}{2}\%$ of the heat radiated from the flame to the furnace metal. If the quantity of heat radiated, Q, in Btu per square foot of surface per hour, from a body at T°F absolute, is given by $Q = kT^4$, where $k = 16 \times 10^{-10}$ find the equivalent evaporation from and at 212°F per square foot of plate per hour.

62. 25 lb of ice at 32°F are dropped into 100 lb of water at 55°F contained in a vessel whose water equivalent is 10 lb. How many lb of steam at atmospheric pressure having 200 F° of superheat must be blown into the water to raise its temperature 60°F above its original temperature? Specific heat of superheated steam = 0·48.

63. In an opposed piston engine, the stroke of each piston varies inversely as the weights of the moving parts which are in the ratio of 6 to 7·5. Find the strokes of the top and bottom pistons if the combined stroke is 108 inches. The cylinder diameter is 27 inches, mean effective pressure 97 lb/in², speed 100 r.p.m., find the brake horse power taking the mechanical efficiency to be 92%.

64. Air is drawn into a cylinder at an absolute pressure corresponding to 381 millimetres of mercury, and compressed to 8·5 atmospheres. Find the clearance volume in the cylinder as a percentage of the distance travelled by the piston, when this pressure is reached, assuming isothermal compression, and taking the barometer reading to be 30 inches of mercury.

65. Owing to a leaking condenser, the hotwell water density is 0·11 of the sea density. Blowing out is resorted to, to keep the boiler water at a certain density. The density of the water blown out of the boiler is 0·8 of the sea density, the feed water temperature is 130°F and the pressure of the steam is 185 lb/in^2 abs. Find the percentage increase in consumption due to blowing down.

66. The total horse power generated by a compound engine is 969, and the mean effective pressure in the H.P. cylinder is 67 lb/in^2. If the diameter of the L.P. piston is 2·2 times the diameter of the H.P. piston, and 1·15 times as much power is developed in the L.P. cylinder as in the H.P. cylinder, find the mean effective pressure in the L.P.

67. Show that the co-efficient of cubical expansion is, within very small limits, equal to three times the coefficient of linear expansion. A rectangular block of steel 33 ft in length, 8 in. in breadth and 6 in. in depth, has its temperature raised through 500 F°. Calculate the percentage increase in volume, taking the coefficient of linear expansion of steel as 6·7 × 10^{-6} per F°.

68. During an experimental run of a single cylinder steam engine, the following data were observed. Diameter of cylinder 14 inches, stroke 20 inches, mean effective pressure 45 lb/in^2, average speed 165 r.p.m., load on brake 600 lb, effective radius of brake 5 feet. Weight of steam condensed per hour 1950 lb. Find (a) i.h.p., (b) b.h.p., (c) mechanical efficiency of engine, (d) weight of steam used per i.h.p. per hour.

69. A single-acting three-stage tandem air compressor with pistons 3, 13½ and 15 inches diameter, and stroke 15 inches, runs at 140 r.p.m. and the volumetric efficiency is 88%. Find the volume of air discharged in 30 minutes at a gauge pressure of 1000 lb/in^2, assuming the suction to be at atmospheric pressure (15 lb/in^2).

70. In a CO_2 refrigerating machine, the refrigerant leaves the condenser as liquid at a pressure of 850 lb/in^2 without any undercooling, passes through the regulating valve where it is throttled to 250 lb/in^2 and enters the evaporator at this pressure. The dryness fraction of the vapour leaving the evaporator is 0·98. If the rate of flow of the CO_2 circulating through the machine is 16 lb per minute, calculate, by reference to the extract from CO_2 tables given below, (i) the dryness fraction of the CO_2 entering the evaporator, (ii) the heat

extracted in the evaporator per minute, (iii) the volume of vapour entering the compressor per minute.

POSITION IN CYCLE	PRESSURE lb/in²	SENSIBLE HEAT Btu/lb	LATENT HEAT Btu/lb	SPEC. VOLUME ft³/lb
Condenser	850	206	—	—
Evaporator	250	156·2	126	0·36

71. A compressor cylinder has a stroke of 28 centimetres, and the volume when delivery begins is 550 cubic centimetres. The cylinder has a clearance volume equal to 5% of the stroke volume. The pressure at the beginning of compression is 13·5 lb/in² absolute, and at the end of compression 68 lb/in² abs. If the law of the compression curve is $PV^{1·35}$ = constant, find the cylinder diameter.

72. A closed vessel contains 500 cc of gas under pressure. The pressure is measured by a U tube containing mercury, one end of the tube being attached to the vessel and the other end open to the atmosphere. The difference in level of the mercury is 8 inches. More mercury is now poured into the open end until the difference in level is 14 inches. Find the percentage change in volume of the gas, if the temperature does not change. The mercury barometer stands at 30 inches.

73. A cylindrical air reservoir 22 ft long contains 1200 lb weight of air at a pressure of 165 lb/in² gauge and temperature 75°F. Calculate the volume and diameter of the reservoir, given that one cubic foot of air at atmospheric pressure and at 32°F weighs 0·0807 lb, and taking atmospheric pressure as 14·7 lb/in². Find also, the volume of atmospheric air at 75°F required to fill the reservoir up to the pressure of 165 lb/in² gauge.

74. An eight-cylinder double-acting four-stroke heavy oil engine develops 4182 b.h.p. when running at 110 r.p.m. The cylinders are $26\frac{3}{4}$ in. dia. and the stroke is 55 in. The mechanical efficiency is 0·82, and the ratio of the power developed in the top and bottom of the cylinder is 10 : 7. If the piston rod is 9 in. diameter, calculate the mean indicated pressure on the top and on the underside of the piston.

75. In a heavy oil engine having a mechanical efficiency of 0·78, the heat carried away by the exhaust gases and cooling water are 28% and 30% respectively of that contained in the fuel. If 60% of the heat in the exhaust gases is recovered in a waste heat boiler, calculate:

(a) What proportion of the available heat is represented by the power transmitted to the shaft.

(b) How many heat units are lost per lb of fuel consumed, if the calorific value is 19,500 Btu/lb.

76. A steam engine drives a 100-kilowatt generator, and is supplied with steam from an oil fired boiler. If the overall efficiency of the plant is 7%, calculate the fuel consumption per kilowatt hour, taking the calorific value of the fuel as 19500 Btu/lb.

77. 10 cubic feet of steam at 200 lb/in^2 abs. and 0·95 dry passes through a reducing valve and throttled to 50 lb/in^2 abs. Assuming no heat losses, determine the dryness fraction and volume of the steam at its reduced pressure.

78. In a four-stroke oil engine the exhaust valve lift rod is in contact with its cam for 124° of a revolution of the cam shaft. If the speed of the crank shaft is 385 r.p.m., find the time the exhaust valve is open during one cycle, and for what percentage of the cycle the valve is closed.

79. The stroke of a gas engine is 18 in. and the clearance volume is equivalent to a linear clearance of 3·5 in. A plate is attached to the piston so that the clearance is reduced to 3 in. If the gas and air mixture is taken in at 14 lb/in^2 abs. and the index of the law of compression is 1·33, find the final compression pressure before and after the alteration. Show by a sketch the effect of the alteration on the final compression pressure.

80. A solid cast iron sphere 5 in. diameter is cooled in a refrigerating chamber to — 14°F and then immersed into 2 lb of water at 34°F. Find the resulting temperature of sphere and water, and the weight of ice formed.

81. During a trial run on a single cylinder four-stroke gas engine of stroke 15 in. and cylinder diameter 7·5 in., the following data were obtained: Mean effective pressure 84 lb/in^2, speed 320 r.p.m., explosions 125 per minute, brake horse power 12·2, gas consumption 240 cu ft per hour, calorific value of gas 470 Btu per cu ft. Calculate (a) indicated thermal efficiency, (b) brake thermal efficiency, (c) mechanical efficiency.

Find also what the m.e.p. would be to develop the same power at the same speed with the governor removed.

SECOND CLASS EXAMINATION QUESTIONS 435

82. The swept volume of the cylinder of an engine is 4·63 cubic feet per stroke, and the mean piston speed is 1100 feet per minute. The stroke is 22 inches, the mechanical efficiency is 0·78, and 300 b.h.p. is developed. Find the number of Btu expended in work done per hour per square inch of piston area.

83. In a four stroke diesel engine the maximum lift of the exhaust valve is 2 inches. It opens at 32° before the bottom dead centre, and closes at 18° after the top dead centre. The valve is full open from 18° before the bottom dead centre to 6° before the top dead centre. If the engine runs at 120 revolutions per minute, find (a) the time the valve is open, (b) the time it is full open, (c) the mean velocity of opening, (d) the mean velocity of closing.

84. An auxiliary engine, running at full load, consumes 640 kilograms of fuel per day. At half load the consumption per horse power is 18% more than at full load. Assuming the increase in consumption per horse power varies directly as the reduction of load, calculate the consumption at three-quarters of full load.

85. The diameter of the cast iron cylinder of an engine is 12 in. at 60°F and the diameters of the steel piston at 80°F are 11·92 in. at the crown and 11·98 in. at mid-depth. Calculate the radial clearances at the crown and mid-depth of the piston under running conditions when the crown temperature is 560°F, mid-depth temperature 160°F, and cylinder temperature 460°F. Take the coefficients of linear expansion as 0·0000061 per F° for cast iron, and 0·0000067 per F° for steel.

86. One pound weight of oil gas, of calorific value 18500 Btu/lb is mixed with 23 lb of air at 80°F and completely burnt under constant volume conditions. If the initial pressure was 20 lb/in^2 absolute, what is the pressure when combustion is complete?
The specific heat at constant volume is 0·169.

87. A steam pipe, 60 feet long and 4 inches external diameter, connects a boiler and an engine. At the boiler end of the pipe the steam pressure is 180 lb/in^2 abs. and the steam is dry and saturated. At the engine end of the pipe the steam pressure is 165 lb/in^2 abs. and the steam is 0·95 dry.
Find (a) the heat lost by radiation, etc. per minute, when 54 lb weight of steam pass through the pipe per minute, (b) the loss of heat per square foot of pipe surface per minute.

88. The mean area of the indicator diagrams from a steam reciprocating engine is 5·175 sq. inches, the spring used being 1 inch = 72 lb. The length of the diagram represents the stroke of the engine to a scale of 1 inch = 1 foot. The cylinder is 26 inches diameter and the engine runs at 75 r.p.m. Find the indicated horse power.

89. The scavenge ports of a two-stroke diesel engine are just covered when the piston is 800 mm from the top of the stroke. The contents of the cylinder at this instant are at 3 lb/in^2 by gauge and 105°F. The piston diameter is 700 mm and the clearance is equivalent to 70 mm. Find the weight of air taken in per cycle, if the scavenge efficiency is 0·95. Note: 1 cu ft of air at 32°F and 14·7 lb/in^2 weighs 0·0807 lb.

90. The error in marking off an eccentric keyway was $\frac{1}{2}$ inch in advance of the correct position. The intended angle of advance was 30°. The shaft was 12 inches diameter; valve travel 6 inches; maximum port opening to steam $1\frac{3}{4}$ inches. Find the lead the valve will have, and how much must be added to the lap in order to make the lead the same amount as was originally intended.

91. A steel shaft is 10 inches diameter and has a brass sleeve shrunk on it. They are at 60°F. Find the temperature to which both must be raised so that the diameter of the sleeve is 0·02 inch more than that of the shaft. Neglect strains due to shrinkage. The coefficient of linear expansion for brass is 0·0000136 per °F, and for steel 0·0000067 per °F.

92. A boiler working at 215 lb/in^2 abs. generates 14000 lb of steam per hour. The steam leaves the boiler stop valve dry and saturated and then passes through superheater tubes. The temperature of the flue gases entering the nests of superheaters is 1525°F and leaves at 1281°F. The fuel consumption is 1500 lb per hour and 24 lb of air are supplied per lb of fuel burned. Find (i) the temperature of the superheated steam, (ii) the weight of injection water at 70°F required to desuperheat each lb of steam. Take the values:

    Spec. ht. of superheated steam = 0·55
    Spec. ht. of flue gases = 0·25

93. A glass vessel is filled with mercury and it contains 30 cubic inches at a temperature of 20°C. Calculate the volume of mercury, in cubic centimetres, which will overflow when heated to 50°C. Take the coefficient of cubical expansion of mercury as 0·00018 per C°, and coefficient of linear expansion of glass as 0·0000085 per C°.

SECOND CLASS EXAMINATION QUESTIONS 437

94. A slide valve has 1·43 inches steam lap and 0·21 inch lead. The angle of advance is 29° 40'. The steam port is 26·5 inches broad and 2·75 inches deep. What is the greatest area of opening to steam?

95. A pump, the cylinder of which is 3 inches diameter, is connected to a tank containing 3 cu ft of air at 50 lb/in^2 abs. If the pressure is reduced 0·5 lb/in^2 by one stroke of the pump, find the stroke assuming the index of the expansion law is 1·3.

96. The diameters of a cylinder and its piston valve liner are 30 in. and 12½ in. respectively, and the mean piston speed is 700 feet per minute. If one third of the area through the cylinder ports of the liner is obstructed by diagonal guide bars, find the depth of these ports to limit the speed of the exhaust steam to 8500 feet per minute.

97. A turbine is supplied with steam at a pressure of 500 lb/in^2 with 360 F degrees of superheat, and exhausts at 1 lb/in^2 abs. If the ideal developed horse power is 5500 for a steam consumption of 10 lb per second, find the dryness fraction of the exhaust steam.

98. 2 lb of ice are converted into superheated steam at a pressure of 68 lb/in^2 abs. and a temperature of 400°F. The specific heat of superheat is 0·48. Describe the changes that occur to the ice during its conversion into steam, and give the quantity of heat supplied to effect the change at each stage. The initial temperature of the ice was 22°F.

99. One cubic foot of nitrogen at 14·7 lb/in^2 abs. and 0°C weighs 0·078 lb. Calculate (a) the weight of 0·4 cu ft of nitrogen at 85 lb/in^2 abs. and 61°C, and (b) the volume of this weight of nitrogen at 20 lb/in^2 abs. and 15°C.

100. Steam leaves the nozzles of a single-stage impulse turbine at a velocity of 3600 ft per second, the angle of the jet being 20° to the direction of movement of the blades. If the blade speed is 1000 ft per sec, find (a) the velocity of whirl, (b) the inlet angle of the blades, and (c) the magnitude and direction of the absolute velocity of the steam at exhaust. Assume exit and inlet angles of the blades to be equal, and neglect frictional losses.

101. 300 grams of cast iron, specific heat 0·13 are held in the funnel gases until the temperature is the same as that of the gases. The cast iron is then placed in 0·16 litre of water and the final Centigrade temperature of the water becomes twice the initial Centigrade temperature. If no water is evaporated, find the maximum possible temperature of the gases.

102. Twelve cubic feet of steam at 250 lb/in² abs., superheated 200 F degrees, is expanded in an engine to a pressure of 2 lb/in² abs. and its final condition is 0·9 dry. Calculate the work done in ft lb and the final volume of the steam. Assume that the volume of superheated steam varies directly as its absolute temperature, and that the work done is equal to the heat drop.

103. Steam at a pressure of 57 lb/in² gauge is supplied to an engine cylinder and cut off at 0·6 of the stroke. The back pressure is 9 lb/in² gauge. Find the mean effective pressure during the stroke, neglecting clearance, assuming expansion according to the law $PV = C$, and atmospheric pressure as 15 lb/in².

104. A solid brass sphere absorbs 1098 Btu when heated, and its diameter increases by 0·021 inch. If the coefficient of linear expansion is 0·0000105 per F°, find the original diameter of the sphere.

105. A 4-cylinder opposed piston two-stroke internal combustion engine has cylinders 600 mm diameter, the long stroke is 1356 mm and the short stroke is 954 mm. The mean effective pressure in the cylinders is 7·24 kilograms per sq. centimetre when running at 85 r.p.m. Taking the mechanical efficiency as 0·86, calculate the brake horse power.

106. Define Latent Heat. Water is heated in a sealed vessel to 368·4°F. An escape valve in the top is now opened, the temperature falls to 212°F, and it is found that one sixth of the volume of the water evaporates. From this experiment, determine the latent heat of steam at atmospheric pressure.

107. The ratio of compression in a diesel engine is 15 to 1. If the temperature of the air at the beginning of compression is 140°F, calculate the temperature at the end of compression, assuming the law of compression to be $PV^{1 \cdot 3} = C$.

108. A vertical cylindrical boiler, 6 ft diameter, contains 2 tons of water at 60°F. Steam at 100 lb/in² abs. and 0·9 dry is blown in until the temperature of the water in the boiler is 120°F. Calculate the resultant rise of water level, taking the specific weights of water as:
        62·45 lb/ft³ at 60°F
        61·4 lb/ft³ at 120°F

109. Steam enters a desuperheater at 400 lb/in² abs. with 320 F degrees of superheat, and leaves at the same pressure as dry saturated steam. The temperature of the water injected into the desuperheater is 100°F. Calculate (i) the weight of injection

water used per lb of steam desuperheated, and (ii) the percentage change in volume, from that occupied by one lb of superheated steam to the volume occupied by the dry saturated steam resulting from the mixture of one lb of superheated steam and its injected water. Use the following formula for the specific volume of superheated steam:

$$v = 1{\cdot}248\,(h - 835{\cdot}2)/P + 0{\cdot}0123 \text{ ft}^3/\text{lb}$$
where h = total heat Btu/lb of superheated steam
P = absolute pressure, lb/in^2

110. The i.h.p. of an engine can be found by multiplying the mean effective pressure in kilograms per sq cm by the revolutions per minute and by a constant. Calculate the constant for an 8-cylinder 4-stroke single acting diesel engine, which has cylinders 740 mm dia. and a stroke of 1150 mm, and use the constant to find the i.h.p. of this engine when the mean effective pressure is 5·82 kilograms per sq cm, and the speed 115·1 r.p.m.

111. At the entrance of a nozzle, the velocity of the steam is 1500 ft per sec, and specific volume 4·429 cu ft per lb. At exit the velocity is 5000 ft per sec, and specific volume 118·6 cu ft per lb. If 2500 lb of steam pass through per hour, find the entrance and exit diameters.

112. In a 2-stroke 6-cylinder diesel engine, the fuel and starting air valves are cam operated. The fuel is admitted from 5° before T.D.C. to 48° after T.D.C., when running ahead, and from 5° before T.D.C. to 38° after T.D.C. when running astern. Starting air is admitted 8° after T.D.C. for both ahead and astern running. Allowing 10° starting air overlap, find the angles between the centre lines of the cams.

113. The mean effective pressure all referred to the L.P. cylinder of a triple expansion steam engine is 35 lb/in^2 when developing 3000 brake horse power, the mechanical efficiency being 78%. If the L.P. cylinder diameter is 88 inches, find the mean piston speed in feet per minute.

114. In a boiler furnace, 18 lb of air are supplied per lb of fuel, the calorific value of the fuel being 14500 Btu/lb. The products of combustion enter the air preheater at 710°F, and leave at 530°F, the air enters the preheater at the stokehold temperature of 80°F. Taking the specific heat of the funnel gases as 0·25 and specific

heat of air as 0·24, find the temperature of the air as it leaves the preheater, and the saving of heat expressed as a percentage of the calorific value of the fuel.

115. A mercury barometer tube has a bore cross-sectional area of $\frac{1}{3}$ sq inch. When a perfect vacuum exists in the space above the mercury, the barometer stands at 30 inches, and when $\frac{1}{3}$ cu inch of atmospheric air is admitted to the vacuum space the mercury level falls 5 inches. Calculate the initial volume of the vacuum space above the mercury.

116. The temperature of three liquids, A, B and C, of different weights and specific heats, are 40°, 100° and 150°F respectively. When A and B are mixed together the resulting temperature is 75°F and when B and C are mixed together the resulting temperature is 130°F. Find the resultant temperature when A and C are mixed together.

117. The clearance volume of a steam engine cylinder is equal to 6·7% of the piston swept volume. Steam is admitted at a pressure of 210 lb/in² gauge, and cut off at 0·34 of the stroke. Find the terminal gauge pressure and the ratio of expansion, assuming expansion according to the law $P \times V$ = constant, and that exhaust does not commence before the end of the stroke.

118. In an ordinary slide valve with outside steam admission, the top steam lap is 1·95 in., bottom steam lap 1·75 in., top lead 0·15 in., top exhaust lap — 0·125 in., bottom exhaust lap + 0·125 in., and the travel of the valve is $7\frac{1}{8}$ in. Find the opening to exhaust when the crank is (a) 45° past T.D.C., (b) 45° past B.D.C.

119. The work done on the gas in the compressor of a CO_2 refrigerating machine increases the heat content in the gas by 15 Btu/lb, and the heat in the CO_2 on entering and leaving the condenser is 276 and 206 Btu/lb respectively. The rate of flow of the refrigerant through the circuit is 90 lb per minute. If the dryness fractions on entering and leaving the evaporator are 0·37 and 0·76 respectively, calculate (i) the latent heat of the CO_2 at the evaporator pressure, (ii) the capacity of the machine in tons of ice per day from and at 32°F.

120. Dry saturated steam at 100 lb/in² abs. is supplied to an engine and expanded isentropically to 1·5 lb/in² abs. By reference to the following extract from the steam tables, and using the given

formula to calculate the entropys of steam, find the dryness fraction at the end of expansion.

P	T_K	h_{fg}
lb/in²	°C	Chu/lb
1·5	319·64	571·16
100	437·6	494·3

$$s = \log_e \frac{T_K}{273} + \frac{q\, h_{fg}}{T_K}$$

where T_K = absolute temp. of sat. steam in °C
h_{fg} = latent heat in Chu/lb
q = dryness fraction

SOLUTIONS TO
SECOND CLASS EXAMINATION QUESTIONS

1. Latent heat at atmos. pressure
   $$= 970 \cdot 6 \times \tfrac{5}{9} \text{ calories per gram}$$
   $$= 970 \cdot 6 \times \tfrac{5}{9} \text{ kilo-calories per kilogram}$$

   For one kilogram of fuel burned:
   Heat given to steam $= 14 \cdot 65 \times 970 \cdot 6 \times \tfrac{5}{9}$ kilo-calories
   Heat put into boiler $= 10 \cdot 17 \times 1000$ kilo-calories

   $$\text{Thermal effy. of boiler} = \frac{\text{Heat given to steam}}{\text{Heat put into boiler}}$$

   $$= \frac{14 \cdot 65 \times 970 \cdot 6 \times 5}{10 \cdot 17 \times 1000 \times 9} = 0 \cdot 777$$

   or 77·7% Ans.

2. $$\text{b.h.p.} = \frac{WN}{4500} = \frac{25 \cdot 5 \times 2300}{4500} = 13 \cdot 03$$

   $$\text{Heat generated at brake} = \frac{13 \cdot 03 \times 33000}{1400} \text{ Chu per min}$$

   Heat carried away by water $= 98\%$ of heat at brake

   $$W \times (48 \cdot 7 - 15 \cdot 2) = \frac{98 \times 13 \cdot 03 \times 33000}{100 \times 1400}$$

   $$\therefore W = \frac{98 \times 13 \cdot 03 \times 33000}{100 \times 1400 \times 33 \cdot 5}$$

   $$= 8 \cdot 985 \text{ lb per min. Ans.}$$

SOLUTIONS TO SECOND CLASS EXAMINATION QUESTIONS 443

3. From steam tables:
$P = 215$ lb/in², $t_F = 387 \cdot 9°F$, $h_g = 1200 \cdot 4$ Btu/lb
$P = 15$ lb/in², $h_f = 181 \cdot 2$ Btu/lb, $h_{fg} = 970$ Btu/lb
Degree of superheat $= 600 - 387 \cdot 9 = 212 \cdot 1$ F°

Heat equivalent of work done $= \dfrac{177200}{778} = 227 \cdot 8$ Btu/lb

Heat drop in steam $=$ Heat equivalent of work done

$(1200 \cdot 4 + 0 \cdot 56 \times 212 \cdot 1) - (181 \cdot 2 + q \times 970) = 227 \cdot 8$
$970\, q = 910 \cdot 2$
$q = 0 \cdot 9383$
Ans.

4. Relative coefficient of expansion
$= 0 \cdot 0000067 - 0 \cdot 0000061$
$= 0 \cdot 0000006$ per F°

Relative linear expansion
$= 0 \cdot 0000006 \times 10 \times 12 \times (360 - 60)$
$= 0 \cdot 0216$ in.

Running clearance $= 0 \cdot 072 - 0 \cdot 0216 = 0 \cdot 0504$ in. Ans.

5. For all problems of this nature the student should make sketches to illustrate the data given, i.e., for this question we need three sketches: (i) Valve in mid-position showing $\frac{1}{8}$ inch exhaust lap at top; (ii) Valve open to lead on the bottom (piston is now on bottom centre) showing $\frac{3}{16}$ inch lead on the bottom and 2 inches open to exhaust at the top; (iii) Valve open to lead at the top (piston is now on top centre) showing $1\frac{7}{8}$ inches open to exhaust at the bottom.

From (i) to (ii) valve has moved up from mid-position a distance of $\frac{1}{8} + 2 = 2\frac{1}{8}$ in., the bottom edge of the valve is now $\frac{3}{16}$ in. open to steam, so it must have originally overlapped the port by $2\frac{1}{8} - \frac{3}{16} = 1\frac{15}{16}$ in.

Steam lap at bottom $= 1\frac{15}{16}$ in. Ans. (i)

As from (i) to (ii) the valve had moved up $2\frac{1}{8}$ in. from mid-position, so it must have moved down from mid-position from (i) to (iii) the same amount of $2\frac{1}{8}$ in.
∴ Bottom exhaust lap $= 2\frac{1}{8} - 1\frac{7}{8} = \frac{1}{4}$ in. Ans. (ii)

6. Fahrenheit reading is (F — 32) degrees above the freezing point of water. As 9F° is equal to 5C° then this is (F — 32) × $\frac{5}{9}$ degrees above the freezing point on the Centigrade scale, and because the freezing point on the Centigrade scale is 0°, then Centigrade reading = (F — 32) × $\frac{5}{9}$

$$
\begin{aligned}
\text{(a) When F} &= 2C, \quad C = (2C - 32) \times \tfrac{5}{9} \\
\tfrac{9}{5} C &= 2C - 32 \\
2C - 1 \cdot 8 C &= 32 \\
0 \cdot 2 C &= 32 \\
C &= 160° = 320°F. \text{ Ans. (a)}
\end{aligned}
$$

$$
\begin{aligned}
\text{(b) When F} &= C, \quad C = (C - 32) \times \tfrac{5}{9} \\
\tfrac{5}{9} C &= C - 32 \\
1 \cdot 8 C - C &= -32 \\
0 \cdot 8 C &= -32 \\
C &= -40° = -40°F. \text{ Ans. (b)}
\end{aligned}
$$

(c) $C = (-13 - 32) \times \tfrac{5}{9} = -45 \times \tfrac{5}{9} = -25°C$. **Ans. (c)**

7. From steam tables,
P = 100 lb/in², v_g = 4·434 ft³/lb
P = 50 lb/in², v_g = 8·516 ft³/lb

Volume of one lb of steam at 100 lb/in², 0·9 dry,
$$= 0 \cdot 9 \times 4 \cdot 434 = 3 \cdot 991 \text{ ft}^3$$
$$P_1 V_1^{1 \cdot 3} = P_2 V_2^{1 \cdot 3}$$
$$100 \times 3 \cdot 991^{1 \cdot 3} = 50 \times V_2^{1 \cdot 3}$$
$$V_2 = 3 \cdot 991 \times \sqrt[1 \cdot 3]{\tfrac{100}{50}}$$
$$= 6 \cdot 801 \text{ ft}^3$$

Spec. vol. of dry sat. steam at 50 lb/in² is 8·516 ft³/lb, therefore dryness fraction of the steam after expansion

$$= \frac{6 \cdot 801}{8 \cdot 516} = 0 \cdot 7987 \text{ Ans.}$$

8. Let D = dia. of cylinder, in inches,
then 1·25 D in. = stroke

$$\text{i.h.p.} = \frac{p A L N}{33000}$$

SOLUTIONS TO SECOND CLASS EXAMINATION QUESTIONS 445

$$400 = \frac{85 \times 0.7854 \times D^2 \times 1.25\,D \times 270}{33000 \times 12} \times 4$$

$$D = \sqrt[3]{\frac{400 \times 12 \times 33000}{85 \times 0.7854 \times 1.25 \times 270 \times 4}}$$

= 12·07 in. dia. Ans.

Stroke = 1·25 × 12·07 = 15·09 in. Ans.

9. From steam tables,
P = 200 lb/in², h_f = 355·5, h_{fg} = 844, v_g = 2·29
P = 50 lb/in², h_f = 250·2, h_{fg} = 924·6, v_g = 8·516

Heat/lb before throttling = Heat/lb after throttling
355·5 + 0·95 × 844 = 250·2 + q × 924·6
907·1 = 924·6 q
q = 0·9809 Ans. (i)

Vol. per lb of the wet steam at 200 lb/in²
= 0·95 × 2·29 ft³

$$\therefore \text{Weight of steam} = \frac{1000}{1728 \times 0.95 \times 2.29} \text{ lb}$$

Vol. per lb of the wet steam at 50 lb/in²
= 0·9809 × 8·516 ft³

∴ Final vol. of throttled steam
= weight (lb) × vol. per lb

$$= \frac{1000 \times 0.9809 \times 8.516}{1728 \times 0.95 \times 2.29}$$

= 2·222 cubic feet. Ans. (ii)

10. When crank is on top centre, distance valve has moved **down** from its mid-position is:

Top steam lap + Top lead = Bottom exhaust lap + Exhaust opening at bottom
$2\frac{3}{4} + \frac{1}{8} = \frac{1}{8}$ + Exhaust opening
∴ Exhaust opening = $2\frac{3}{4}$ in. Ans. (a)

When crank is on bottom centre, distance valve has moved up from its mid-position is:

$$\begin{matrix}\text{Bottom}\\\text{steam lap}\end{matrix} + \begin{matrix}\text{Bottom}\\\text{lead}\end{matrix} = \begin{matrix}\text{Top}\\\text{exhaust lap}\end{matrix} + \begin{matrix}\text{Exhaust opening}\\\text{at top}\end{matrix}$$

$$2\tfrac{5}{8} + \tfrac{1}{4} = \tfrac{1}{4} + \text{Exhaust opening}$$

$$\therefore \text{Exhaust opening} = 2\tfrac{5}{8} \text{ in. Ans. (b)}$$

11. From steam tables,
$P = 185$ lb/in^2, $h_f = 348\cdot5$ Btu/lb
$P = 14\cdot7$ lb/in^2, $h_f = 180\cdot1$, $h_g = 1150\cdot7$ Btu/lb

Let W lb = wt. of water taken from boiler
then, (W—1·5) = wt. flashed off as atmospheric steam

$$\begin{matrix}\text{Heat in W lb water}\\\text{taken from boiler}\end{matrix} = \begin{matrix}\text{Heat in 1·5 lb}\\\text{water in pot}\end{matrix} + \begin{matrix}\text{Heat in steam}\\\text{flashed off}\end{matrix}$$

$$W \times 348\cdot5 = 1\cdot5 \times 180\cdot1 + (W - 1\cdot5) \times 1150\cdot7$$
$$348\cdot5\ W = 270\cdot15 + 1150\cdot7\ W - 1726\cdot05$$
$$1455\cdot9 = 802\cdot2\ W$$
$$W = 1\cdot814 \text{ lb. Ans.}$$

When the pressure is relieved, 1·814 — 1·5 = 0·314 lb of water is evaporated into steam and passes away leaving its solids behind. The solids that were in 1·814 lb of boiler water are now concentrated in 1·5 lb of water in the salinometer pot, therefore the density is higher.

12. $\text{i.h.p.} = \dfrac{p\text{ALN}}{33000}$

$$= \dfrac{88\cdot6 \times 0\cdot7854 \times 24\cdot5^2 \times 51\cdot7 \times 170}{33000 \times 12} \times 4$$

$$= 3709.\ \text{Ans. (i)}$$

$\text{b.h.p.} = 0\cdot75 \times 3709$
$= 2782.\ \text{Ans. (ii)}$

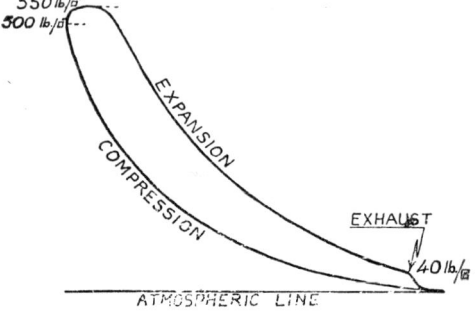

SOLUTIONS TO SECOND CLASS EXAMINATION QUESTIONS 447

13. Total heat in 1 lb of steam $= h_f + q\, h_{fg}$
$= 364 \cdot 2 + 0 \cdot 97 \times 836 \cdot 5$
$= 1175 \cdot 6$ Btu measured above 32°F.

The feed water is at 220°F, and assuming its sensible heat above $32°F = 220 - 32 = 188$ Btu/lb

∴ heat to form 1 lb of steam
$= 1175 \cdot 6 - 188 = 987 \cdot 6$ Btu

Steam formed per 1 lb of fuel $= \dfrac{14400 \times 0 \cdot 87}{987 \cdot 6}$

$= 12 \cdot 68$ lb

Fuel used per hour $= \dfrac{18000}{12 \cdot 68}$ lb

Fuel used per day $= \dfrac{18000 \times 24}{12 \cdot 68 \times 2240} = 15 \cdot 21$ tons. **Ans.**

14. It is obvious from the figures that there is not sufficient heat in the water to completely melt the ice, the final condition will be a mixture of ice and water all at 32°F.

Let x lb of ice be melted:

$\left.\begin{array}{l}\text{Heat lost by the water} \\ \text{to reduce its temperature} \\ \text{from 38° to 32°F.}\end{array}\right\} = \left\{\begin{array}{l}\text{Heat absorbed by the 15 lb of ice to raise its} \\ \text{temperature from 20° to 32°F, plus the latent heat} \\ \text{to convert } x \text{ lb of ice into water at 32°F.}\end{array}\right.$

$25 \times (38 - 32) = 15 \times 0 \cdot 5 \times (32 - 20) + x \times 144$
$25 \times 6 = 90 + 144x$

$x = \dfrac{150 - 90}{144} = 0 \cdot 4167$ lb

∴ Final temperature of mixture $= 32°F$. **Ans. (i)**

Final state $= 25 \cdot 4167$ lb water and $14 \cdot 5833$ lb ice. **Ans. (ii)**

15. Heat equivalent of one horse power hour

$= \dfrac{33000 \times 60}{778} = 2545$ Btu

Thermal efficiency on b.h.p.

$$= \frac{2545}{\text{lb fuel per b.h.p. hour} \times \text{C.V.}}$$

$$= \frac{2545}{0.45 \times 19000} = 0.2976, \text{ or } 29.76\%. \quad \text{Ans. (b)}$$

Thermal efficiency on i.h.p. $= 0.2976 \times \frac{100}{85}$
$= 0.3502$, or 35.02%. Ans. (a)

For each lb of fuel burnt:

Total weight of gases $= 28$ lb air $+ 1$ lb fuel $= 29$ lb

Heat carried away $=$ Weight $\times$ spec. heat $\times$ rise in temp.

$= 29 \times 0.25 \times (760 - 60)$
$= 5075$ Btu

Heat supplied to cylinders by 1 lb fuel $= 19000$ Btu

∴ Percentage heat carried away by exhaust gases

$$= \frac{5075}{19000} \times 100 = 26.71\%. \quad \text{Ans. (c)}$$

16.

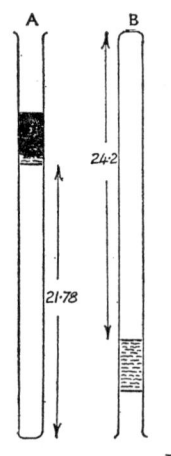

Let $a =$ cross sectional area of bore of tube in sq cm

Let $H =$ atmospheric pressure in cm of mercury

Pressure of trapped air in A
$= (H + 4)$ cm of mercury

Volume of trapped air in A
$= 21.78 \times a$ cu cm

Pressure of trapped air in B
$= (H - 4)$ cm of mercury

Volume of trapped air in B
$= 24.2 \times a$ cu cm

Since temperature is constant:
$P_1 \times V_1 = P_2 \times V_2$
$(H+4) \times 21.78a = (H-4) \times 24.2a$
$21.78H + 87.12 = 24.2H - 96.8$
$2.42H = 183.92$
$H = 76$ centimetres. Ans. (i)
$\frac{76}{2.54} = 29.92$ in. Ans. (ii)
$29.92 \times 0.491 = 14.69$ lb/in^2. Ans. (iii)

SOLUTIONS TO SECOND CLASS EXAMINATION QUESTIONS 449

17. $$\text{i.h.p.} = \frac{p\text{ALN}}{33000}$$

Total i.h.p. is 2470 when p is the mean effective pressure referred to the L.P. and A is the area of the L.P. cylinder.

$$2470 = \frac{p \times 0.7854 \times 70^2 \times 4 \times 76 \times 2}{33000}$$

∴ Mean pressure referred to L.P.

$$= \frac{2470 \times 33000}{0.7854 \times 70^2 \times 4 \times 152}$$

$$= 34.83 \text{ lb/in}^2. \text{ Ans.}$$

$$\left. \begin{array}{l} \text{m.e.p. in L.P.} = \dfrac{34.83}{3} = 11.61 \text{ lb/in}^2 \\[1em] \text{m.e.p. in I.P.} = \dfrac{11.61 \times 70^2}{41.5^2} = 33.05 \text{ lb/in}^2 \\[1em] \text{m.e.p. in H.P.} = \dfrac{11.61 \times 70^2}{24.5^2} = 94.78 \text{ lb/in}^2 \end{array} \right\} \text{Ans.}$$

18.

Gases in mixture	% by volume	Molecular weight	Ratio of weights of gases in mixture	% by weight
CO_2	10	44	440	14·71
CO	1·7	28	47·6	1·591
O_2	8·1	32	259·2	8·661
N_2	80·2	28	2246	75·04
			Total 2992·8	

Gas % by weight: $CO_2 = 14.71$; $CO = 1.59$; $O_2 = 8.66$; $N_2 = 75.04$. Ans.

19. From steam tables,
P = 200 lb/in², $h_g = 1199.5$ Btu/lb
P = 1·25 lb/in², $h_f = 77.3$, $h_{fg} = 1031.7$ Btu/lb

Heat in steam exhausted from turbine
$= 77 \cdot 3 + 1031 \cdot 7\, q$ Btu/lb

Heat drop through turbine
$= 1199 \cdot 5 - (77 \cdot 3 + 1031 \cdot 7\, q)$
$= 1122 \cdot 2 - 1031 \cdot 7\, q$ Btu/lb

Heat given up in condenser
$= 4 \cdot 7\,(1122 \cdot 2 - 1031 \cdot 7\, q)$ Btu/lb ... ... (i)

Also, heat given up in condenser
= latent heat (condensing) + sensible heat (undercooling)
$= 1031 \cdot 7\, q + [77 \cdot 3 - (90 - 32)]$
$= 1031 \cdot 7\, q + 19 \cdot 3$ Btu/lb ... ... (ii)

From (i) and (ii),
$4 \cdot 7\,(1122 \cdot 2 - 1031 \cdot 7\, q) = 1031 \cdot 7\, q + 19 \cdot 3$
$q = 0 \cdot 8936$. Ans.

20. M.e.p. $\propto$ (r.p.m.)$^{2 \cdot 1}$

$$\therefore \left\{\frac{\text{r.p.m.}_2}{\text{r.p.m.}_1}\right\}^{2 \cdot 1} = \frac{\text{m.e.p.}_2}{\text{m.e.p.}_1}$$

$$\left\{\frac{\text{r.p.m.}_2}{99}\right\}^{2 \cdot 1} = \frac{79}{89}$$

$$\text{r.p.m.}_2 = 99 \times \sqrt[2 \cdot 1]{\frac{79}{89}}$$

$= 93 \cdot 53$ r.p.m. Ans. (i)

From, i.h.p. $= \dfrac{p\text{ALN}}{33000}$

i.h.p. $\propto$ m.e.p. $\times$ r.p.m.

$$\therefore \frac{\text{i.h.p.}_2}{\text{i.h.p.}_1} = \frac{p_2 \times \text{r.p.m.}_2}{p_1 \times \text{r.p.m.}_1}$$

$$\text{i.h.p.}_2 = 925 \times \frac{79 \times 93 \cdot 53}{89 \times 99}$$

$= 775 \cdot 6$ i.h.p. Ans. (ii)

SOLUTIONS TO SECOND CLASS EXAMINATION QUESTIONS 451

21. Taking 2·2 lb = 1 kilogram

Total heat = 1311·9 ÷ 2·2 = 596·3 Chu/lb

$$h_f + q \times h_{fg} = \text{Total heat}$$
$$192 \cdot 3 + q \times 473 \cdot 3 = 596 \cdot 3$$
$$q \times 473 \cdot 3 = 404$$
$$q = 0 \cdot 8535 \text{ Ans. (i)}$$

Wetness = 1 — 0·8535 = 0·1465

Extra heat to dry the steam
   = 0·1465 × 473·3 = 69·34 Chu/lb

69·34 × 2·2 = 152·5 Chu/kilogram. Ans. (ii)
152·5 × 9/5 = 274·5 Btu/kilogram. Ans. (iii)

22. Cyl. dia. = $\dfrac{620}{25 \cdot 4}$ in.

Stroke = $\dfrac{1300}{25 \cdot 4 \times 12}$ ft

i.h.p. = $\dfrac{pALN}{33000}$

$$= \frac{93 \cdot 2 \times 0 \cdot 7854 \times 620^2 \times 1300 \times 119 \cdot 6}{33000 \times 25 \cdot 4^2 \times 25 \cdot 4 \times 12 \times 2} \times 8$$

= 2698 i.h.p. Ans. (i)
b.h.p. = 0·88 × 2698
     = 2373 b.h.p. Ans. (ii)
(See Chapter 5 for typical indicator diagram)

23. Ideal thermal efficiency = $\dfrac{\text{Heat drop through engine}}{\text{Heat supplied}}$

(a) with bleeding system in operation:

Steam through engine up to bleeding point = 1 lb
Heat drop up to this point = 1357·6 — 1118·66 = 238·94 Btu

Steam through engine from bleeding point to exhaust
$= 1 - 0.1388 = 0.8612$ lb

Heat drop to this point
$= 0.8612 (1118.66 - 904) = 184.8$ Btu
Total heat drop
$= 238.94 + 184.8 = 423.74$ Btu
Heat supplied
$= 1357.6 - 196.1 = 1161.5$ Btu

$$\therefore \text{Ideal thermal efficiency} = \frac{423.74}{1161.5} = 0.3647 \text{ Ans. (a)}$$

(b) Without bleeding system:

$$\text{Ideal thermal efficiency} = \frac{1357.6 - 904}{1357.6 - 47.42}$$

$$= \frac{453.6}{1310.18} = 0.3462 \text{ Ans. (b)}$$

24.   From steam tables,
$P = 240$ lb/in^2, $t_F = 397.4°$F, $h_g = 1201.7$ Btu/lb
Degree of superheat $= 746 - 397.4 = 348.6$ F°

Heat to form one lb of superheated steam from feed at 200°F
$= (1201.7 + 0.55 \times 348.6) - (200 - 32)$
$= 1225.4$ Btu/lb

Heat required per day $= 1225.4 \times 27470 \times 24$ Btu
Available heat from 1 lb of fuel $= 0.76 \times 19200$ Btu

$$\therefore \text{Fuel per day} = \frac{1225.4 \times 27470 \times 24}{0.76 \times 19200 \times 2240}$$

$$= 24.72 \text{ tons/day. Ans.}$$

25.   See Chapter 3 for laws of perfect gases.

From, $PV = wRT$

$$R = \frac{14.7 \times 144 \times 124}{10 \times (32 + 460)}$$

though
SOLUTIONS TO SECOND CLASS EXAMINATION QUESTIONS 453

$= 53.34$ ft lb per lb per F° Ans. (i)
$53.34 \times \frac{9}{5} = 96$ ft lb per lb per C° Ans. (ii)

26. From steam tables,
$P = 450$ lb/in², $\quad t_F = 456.3°F, \quad h_g = 1205.6$ Btu/lb
$P = 0.85$ lb/in², $\quad h_f = 64.3, \quad h_{fg} = 1039.2$ Btu/lb

Heat in steam supplied to turbine
$= 1205.6 + 0.655 (750 - 456.3)$
$= 1398$ Btu/lb

Heat in steam exhausted from turbine
$= 64.3 + 0.9 \times 1039.2$
$= 999.6$ Btu/lb

Heat given to turbine $= 1398 - 999.6$
$= 398.4$ Btu/lb

Total power supplied $= 22000 \times 398.4 \times 778$ ft lb per hour

Equivalent horse power $= \dfrac{22000 \times 398.4 \times 778}{33000 \times 60}$

$= 3443$ h.p. Ans.

27. See Chapter 10 for advantages of compounding.

Ratio of expansion

$= \dfrac{\text{Final volume}}{\text{Initial volume}} = \dfrac{\text{Volume of L.P.}}{\text{Volume of H.P. up to cut off}}$

$= \dfrac{(\text{Dia. of L.P.})^2}{(\text{Dia. of H.P.})^2 \times \text{cut off in H.P.}}$

$\therefore 7 = \dfrac{D^2}{d^2 \times 0.5}$

$\therefore \dfrac{D^2}{d^2} = 3.5, \quad \therefore \dfrac{D}{d} = \sqrt{3.5} = 1.871$

Ratio of cylinder diameters $= 1 : 1.871$. Ans. (a)

$P_1 V_1 = P_2 V_2 \quad \therefore 150 \times 1 = P_2 \times 7$

Final pressure $= \dfrac{150}{7} = 21.43$ lb/in² abs.
Ans. (b)

28. Indicated thermal effic. %
$\quad$ = % heat turned into work in cylinders
$\quad$ = 100% supplied — % losses
$\quad$ = 100 — 29 — 31 = 40%

Brake thermal effic.
$\quad$ = Indicated thermal effic. × mech. effic.
$\quad$ = 0·4 × 0·78 = 0·312
$\quad$ = 31·2% Ans. (i)

Also,

Brake thermal effic.
$$= \frac{\text{Heat equivalent of one b.h.p. hour}}{\text{Wt. fuel/b.h.p. hour} \times \text{C.V.}}$$

$$\text{C.V.} = \frac{2545}{0·43 \times 0·312}$$

$\quad$ = 18970 Btu/lb. Ans. (ii)

29. From steam tables,
$P = 2$ lb/in², $\qquad h_{fg} = 1022·2$ Btu/lb

Before conversion:

Heat given up in condenser per lb of steam
$\quad$ = 0·8 × 1022·2 = 817·76 Btu. Ans. (a)

Heat given up in condenser per h.p. minute
$$= \frac{14·5}{60} \times 817·76 = 197·625 \text{ Btu. Ans. (b)}$$

$$\text{Equivalent h.p. lost} = \frac{197·625 \times 778}{33000} = 4·66 \text{ Ans. (c)}$$

After conversion:
Ans. to (a) = 0·87 × 1022·2 = 889·314 Btu
$\quad$,, $\quad$ (b) = $\frac{13}{60}$ × 889·314 = 192·68 Btu

$$\text{,, } \quad (c) = \frac{192·68 \times 778}{33000} = 4·542 \text{ h.p.}$$

SOLUTIONS TO SECOND CLASS EXAMINATION QUESTIONS 455

30. The specific heat of a substance is the heat required to raise unit weight of the substance through unit temperature.

From data given at examinations specific gravity of steel = 7·86, specific heat of steel = 0·116
Let V cu ft = volume of steel, then 50V = volume of oil

Heat lost by steel = Heat gained by oil

wt. × sp. heat × change of temp. = wt. × sp. ht. × change of temp.
V × 62·5 × 7·86 × 0·116 × (540–42) = 50 × V × 62·5 × 0·95 × sp ht. × (42–18·8)

$$7·86 \times 0·116 \times 498 = 50 \times 0·95 \times \text{sp. ht.} \times 23·2$$

$$\therefore \text{ Specific heat of oil} = \frac{7·86 \times 0·116 \times 498}{50 \times 0·95 \times 23·2}$$

$$= 0·412 \text{ Ans.}$$

31. 2×10^6 gram calories per hour $= \dfrac{2 \times 10^6}{252}$ Btu per hr

$$1 \text{ square metre} = \frac{39·37^2}{144} = 10·76 \text{ sq ft}$$

$\therefore$ Heat transmission per sq ft per Centigrade degree

$$= \frac{2 \times 10^6}{252 \times 10·76} \text{ Btu per hour}$$

$$\text{Equivalent h.p. for 5 C}° = \frac{2 \times 10^6 \times 5}{252 \times 10·76} \times \frac{778}{60 \times 33000}$$

$$= 1·449. \text{ Ans.}$$

32. Weight of one cu ft of air at 65°F

$$= 0·0807 \times \frac{32 + 460}{65 + 460} = 0·07563 \text{ lb.}$$

Weight of air supplied per hour
= 60 × 80 × 9 × 0·07563 × 2 = 6535 lb

Heat supplied = Weight × Spec. Heat × Rise in temp.
= 6535 × 0·24 × 20 = 31360 Btu per hour

3412 Btu per hour = 1 kilowatt hour

∴ Electrical loading = $\dfrac{31360}{3412}$ = 9·19 kilowatts. Ans.

Alternatively: Horse power equivalent of heat supplied

= $\dfrac{31360 \times 778}{60 \times 33000}$

746 watts = 1 horse power = 0·746 kilowatt

∴ Electrical loading = $\dfrac{31360 \times 778 \times 0·746}{60 \times 33000}$

= 9·19 kilowatts

33. Let there be 1 cc of the first liquid. Its weight is 0·8 gram. Let there be 3 cc of the second liquid. It weighs 3 × 0·82 = 2·46 grams.

The ratio of the weights is 0·8 to 2·46 whatever may be the unit of weight taken. We may assume 8 lb of the first liquid and 24·6 lb of the second, and the ratio is unaltered.

Heat gained by first = Heat lost by second
8 × 0·6 × (T − 80) = 24·6 × 0·45 × (110 − T)
15·87 T = 1601·7
T = 100·9°F. Ans.

34. From steam tables,
P = 250 lb/in², t_F = 401°F, h_{fg} = 826 Btu/lb

One lb of wet steam requires
960 ÷ 12 = 80 Btu to dry it,

SOLUTIONS TO SECOND CLASS EXAMINATION QUESTIONS 457

∴ latent heat of 1 lb of this wet steam
$$= 826 - 80 = 746 \text{ Btu}$$

$$\therefore q = \frac{746}{826} = 0.9031 \text{ Ans. (a)}$$

One lb of the dried steam requires
$1360 \div 12 = 113.3$ Btu to superheat it,
Wt. × mean sp. ht. × rise in temp. = heat supplied
$$1 \times x \times (600 - 401) = 113.3$$
$$199x = 113.3$$
$$x = 0.5694 \text{ Ans. (b)}$$

35. Let S = stroke in millimetres
$$P_1 V_1 = P_2 V_2$$
$$(44+14.7)(S+45-225) = (134 + 14.7)(S + 45 - 400)$$
$$58.7 (S - 180) = 148.7 (S - 355)$$
$$58.7 S - 10560 = 148.7 S - 52780$$
$$90 S = 42220$$
$$S = 469.1 \text{ mm. Ans.}$$

36. 70 cc clearance is equivalent to $\dfrac{70}{13^2 \times 0.7854}$

$$= 0.5274 \text{ cm of stroke}$$

$v_1 = 18 + 0.5274 = 18.5274, p_1 = 14, p_2 = 65$
$$p_1 v_1^{1.2} = p_2 v_2^{1.2}$$
$$14 \times 18.5274^{1.2} = 65 \times v_2^{1.2}$$
Log $14 + 1.2 \log 18.5274 = \log 65 + 1.2 \log v_2$

∴ $1.2 \log v_2 = \log 14 + 1.2 \log 18.5274 - \log 65$
$\log v_2 = 0.7122$ and $v_2 = 5.155$ cm

Delivery occurs for $5.155 - 0.5274 = 4.6276$ cm of stroke

,, ,, ,, $\dfrac{4.6276}{18} = 0.2571$ stroke. Ans.

37.

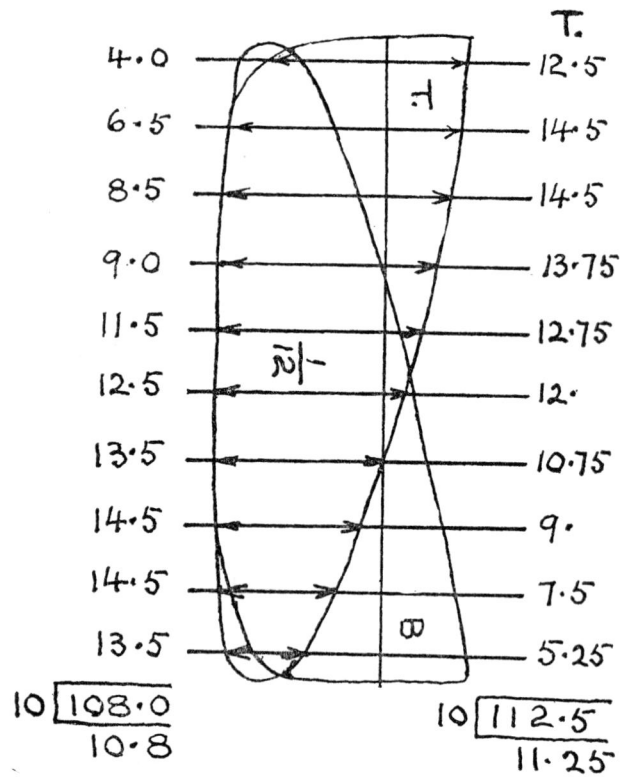

```
       T.
 4·0 ──────╮╭──── 12·5
 6·5 ◄──────────── 14·5
 8·5 ◄──────────── 14·5
       ──────►
 9·0 ◄──────────── 13·75
       ──────►
11·5 ◄──────────── 12·75
       ──────►
12·5 ◄────1/12──── 12·
       ──────►
13·5 ◄──────────── 10·75
       ──────►
14·5 ◄──────────── 9·
       ──────►
14·5 ◄──────────── 7·5
       ──────►     B
13·5 ──────╯╰──── 5·25
10│108·0          10│112·5
   10·8              11·25
```

M.E.P. = 11 LBS. SQ. IN.

$$\text{i.h.p.} = \frac{p\text{ALN}}{33000}$$

$$= \frac{11 \times 80^2 \times 0\cdot7854 \times 51 \times 2 \times 70}{33000 \times 12}$$

$$= 997 \text{ i.h.p.}$$

i.h.p. of whole engine $= 997 \times 3 = 2991$

$$\text{Consumption per day} = \frac{2991 \times 1\cdot6 \times 24}{2240}$$

$$= 51\cdot27 \text{ tons. Ans.}$$

38. The eccentricity of an eccentric is the distance from the shaft centre to the geometrical centre of the sheave, it is the throw and is equal to half the travel of the valve.

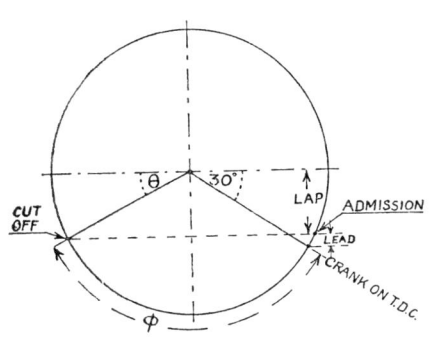

$$\frac{\text{Lap} + \text{lead}}{\frac{1}{2}\text{ travel}} = \text{sine of angle of advance}$$

$$1\tfrac{9}{16} + \text{lead} = 3\tfrac{1}{4} \times \sin 30°$$

$$\therefore \text{lead} = 3\tfrac{1}{4} \times 0.5 - 1\tfrac{9}{16}$$

$$= \tfrac{1}{16} \text{ in. Ans. (a)}$$

$$\text{Steam lap} + \text{M.P.O.} = \tfrac{1}{2} \text{ travel}$$

$$\therefore \text{M.P.O.} = 3\tfrac{1}{4} - 1\tfrac{9}{16}$$

$$= 1\tfrac{11}{16} \text{ in. Ans. (b)}$$

$$\frac{\text{Steam lap}}{\frac{1}{2}\text{ travel}} = \sin \theta$$

$$\sin \theta = \frac{1\tfrac{9}{16}}{3\tfrac{1}{4}} = \frac{25 \times 4}{16 \times 13} = \frac{25}{52} = 0.4808$$

$$\therefore \theta = 28° 44'$$
$$\phi = 180° - 30° - 28° 44' = 121° 16'$$

∴ Crank is 121° 16′ from dead centre when cut off takes place. Ans. (c)

$$30° - 28° 44' = 1° 16'$$

∴ Crank is 1° 16′ before dead centre when admission takes place. Ans. (d)

39. Volume remains constant, then absolute pressure varies as the absolute temperature.

$$\frac{P_1}{T_1} = \frac{P_2}{T_2} \quad \therefore T_2 = \frac{(460 + 63) \times (650 + 15)}{(500 + 15)}$$

$$= 675.3°\text{F abs.}$$
$$t_2 = 675.3 - 460 = 215.3°\text{F}$$

Heat given to air = Wt. × spec. heat × rise in temp.
$$= 60 \times 0.17 \times (215.3 - 63)$$
$$= 1553 \text{ Btu. Ans.}$$

40. Sine of angle of advance $= \dfrac{\text{Lap} + \text{lead}}{\frac{1}{2}\text{ travel}}$

∴ Lap $= \frac{1}{2}$ travel $\times \sin 36° 8' - 0·15$
Also, $\frac{1}{2}$ travel $=$ lap $+$ max. port opening.

Substitute value of lap:
∴ $\frac{1}{2}$ travel $= \frac{1}{2}$ travel $\times \sin 36° 8' - 0·15 + 1·612$
$\frac{1}{2}$ travel $(1 - \sin 36° 8') = 1·462$

$$\tfrac{1}{2}\text{ travel} = \dfrac{1·462}{0·4103} \qquad \text{travel} = \dfrac{1·462 \times 2}{0·4103}$$

$$= 7·126 \text{ in. Ans.}$$

41. Sum of mid-ordinates $= 6·25$ in.
Mean height $= 6·25 \div 10 = 0·625$ in.
∴ Mean effective pressure $= 0·625 \times 60 = 37·5$ lb/in^2.
Ans. (a)

Mean force on piston $=$ m.e.p. $\times$ area of piston
$= 37·5 \times 0·7854 \times 6^2$ lb

Work done per rev. $=$ mean force $\times$ distance
(i.e. two strokes)

$= 37·5 \times 0·7854 \times 6^2 \times \tfrac{6·5}{12} \times 2$
$= 1148$ ft lb. Ans. (b)

i.h.p. $=$ work done per minute $\div$ 33000

$$= \dfrac{1148 \times 120}{33000} = 4·177 \text{ Ans. (c)}$$

42. Volume of bottle $= \left\{ \tfrac{11}{14} \times 1^2 \times 5 \right\} + \left\{ \dfrac{\pi}{6} \times 1^3 \right\}$

$= 4·4518$ cu ft

If temperature of air is the same after being compressed and cooled, as the atmosphere, then $pv =$ constant, and 4·4518 cu ft of air at 915 lb/in^2 abs. is equal to

$$\dfrac{4·4518 \times 915}{15} \text{ cu ft at 15 lb/in}^2 \text{ abs.}$$

SOLUTIONS TO SECOND CLASS EXAMINATION QUESTIONS 461

= 271·6 cu ft

Volume of free air to be pumped = 271·6 — 4·4518 = 267·1428 cu ft, because we assume bottle to contain 4·4518 cu ft of atmospheric air before pumping begins.

Volume of free air taken into L.P. compressor per minute
$$= \tfrac{11}{14}(9\tfrac{1}{2}^2 - 2\tfrac{1}{2}^2) \times \tfrac{1}{144} \times \tfrac{8}{12} \times 0.9 \times 120$$

$$= \frac{11 \times 12 \times 7 \times 8 \times 0.9 \times 120}{14 \times 144 \times 12} = 33 \text{ cu ft}$$

Then time to pump up bottle $= \dfrac{267 \cdot 1428}{33}$

= 8·095 minutes. Ans.

43. One horse-power hour $= \dfrac{33000 \times 60}{778} = 2545$ Btu

Weight of oil = 50000 × 62·5 × 0·9 lb

Available energy $= \dfrac{50000 \times 62 \cdot 5 \times 0 \cdot 9 \times 19000}{2545}$ h.p. hours

= 21000000
or 21 × 10⁶ horse-power hours

Weight of coal $= \dfrac{50000}{48}$ tons

Available energy $= \dfrac{50000 \times 2240 \times 14000}{48 \times 2545}$

= 12830000
or 12·83 × 10⁶ horse-power hours

Potential energy of oil = 21,000,000 h.p. hr. ⎫
   ,,   ,, ,, coal = 12,830,000 ,, ,, ⎬ Ans.
                                            ⎭

Oil has the advantage over coal

44.   Free expansion $= KLT$
$$= 0.000011 \times 60 \times 150$$
$$= 0.099 \text{ in.}$$

When cooled, total stretch of bar
$$= 0.099 - 0.07 = 0.029 \text{ in.}$$

Strain = extension ÷ length
Stress = strain × E

$$= \frac{0.029 \times 30 \times 10^6}{60}$$

$$= 14500 \text{ lb/in}^2$$

Load = stress × area

∴ Pull in bar $= 14500 \times 0.7854 \times 1^2$
$$= 11390 \text{ lb. Ans.}$$

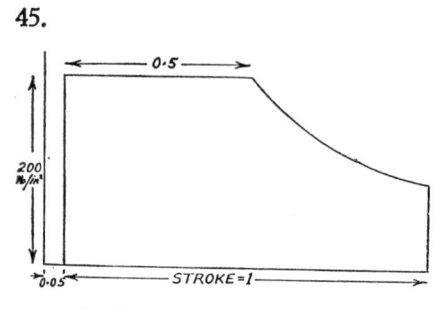

45.

Let stroke = 1
Ratio of expansion
$$= \frac{1.05}{0.55} = 1.909$$

$\log_e$ from hyperbolic logarithm tables
$= 0.6465$
(a close approximation to this could be obtained by multiplying the common log by 2·3)

Area representing work done during admission of steam
$$= 200 \times 0.5 = 100$$

Area representing work done during expansion of steam
$$= PV \log_e r = 200 \times 0.55 \times 0.6465 = 71.1$$
Gross area $= 100 + 71.1 = 171.1$
Mean height = Area ÷ length
$$= 171.1 \div 1$$
$$= 171.1$$

∴ Mean gross pressure = 171·1 lb/in² abs. Ans.

SOLUTIONS TO SECOND CLASS EXAMINATION QUESTIONS 463

46.

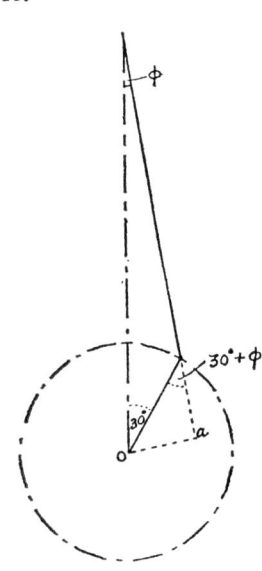

Load on piston $= \frac{11}{14} \times 18^2 \times (180-70)$ lb

$$\frac{1\cdot5}{\text{Sin }\phi} = \frac{7}{\text{Sin }30°}$$

$$\therefore \text{Sin }\phi = \frac{1\cdot5 \times 0\cdot5}{7} = 0\cdot1071$$

$$\therefore \phi = 6° 9', \text{ and } 30° + \phi = 36° 9'$$

Thrust in connecting rod

$$= \frac{\text{load on piston}}{\text{Cos } 6° 9'}$$

$$= \frac{\frac{11}{14} \times 18^2 \times 110}{0\cdot9942} = 28160 \text{ lb}$$

Twisting moment = Load on connecting rod $\times \overline{ao}$
= 28160 $\times$ 1·5 sin 36° 9'
= 28160 $\times$ 1·5 $\times$ 0·5899
= 24900 ft lb. Ans.

47. From steam tables,
P = 230 lb/in², $h_f =$ 368·3 Btu/lb
P = 14·7 lb/in², $h_f =$ 180·1, $h_{fg} =$ 970·6 Btu/lb

Let original weight of water in boiler = 100
Let weight of water evaporated into steam = x

Sensible heat rejected by 100 lb of water in falling in pressure from 230 to 14·7 } = { Latent heat absorbed by x lb of water in evaporating at atmospheric pressure

$$100 (368\cdot3 - 180\cdot1) = x \times 970\cdot6$$
$$18820 = x \times 970\cdot6$$
$$x = 19\cdot38$$

Weight of water evaporated = 19·38%
$\therefore$ Weight of water remaining = 100 — 19·38 = 80·62%
Ans.

48. From steam tables,
$P = 40$ lb/in^2, $\quad h_f = 236 \cdot 1, \quad h_{fg} = 934 \cdot 4, \quad v_g = 10 \cdot 5$

Total heat before mixing = Total heat after

Let w = weight of steam blown in,

$w(236 \cdot 1 + 934 \cdot 4\, q) + 160(60 - 32) = (160 + w)(135 - 32)$
$w(236 \cdot 1 + 934 \cdot 4\, q) + 4480 = 16480 + 103\, w$
$w(236 \cdot 1 + 934 \cdot 4\, q) - 103\, w = 12000 \quad \ldots \quad \ldots \quad (i)$

Also, volume of one lb of wet steam at 40 lb/in^2, of dryness fraction $q = q \times 10 \cdot 5$ ft^3/lb

$$\therefore \text{ wt. of 115 ft}^3 \text{ of wet steam} = \frac{115}{10 \cdot 5\, q} \text{ lb}$$

Substituting this value for w into (i),

$$\frac{115}{10 \cdot 5\, q}(236 \cdot 1 + 934 \cdot 4\, q) - \frac{103 \times 115}{10 \cdot 5\, q} = 12000$$

Multiplying throughout by $\dfrac{10 \cdot 5\, q}{115}$,

$$236 \cdot 1 + 934 \cdot 4\, q - 103 = \frac{12000 \times 10 \cdot 5\, q}{115}$$

$236 \cdot 1 - 103 = 1096\, q - 934 \cdot 4\, q$
$133 \cdot 1 = 161 \cdot 6\, q$
$q = 0 \cdot 8237$ Ans.

49. $30°C = 86°F$, and $-10°C = 14°F$

From steam tables,
$P = 600$ lb/in^2, $\quad t_F = 486 \cdot 2, \quad h_g = 1204 \cdot 2$

Degrees of superheat $= 650 - 486 \cdot 2 = 163 \cdot 8$ F°

Superheat $= 0 \cdot 6 \times 163 \cdot 8 = 98 \cdot 28$ Btu/lb

Total heat of superheated steam $= 1204 \cdot 2 + 98 \cdot 28$
$= 1302 \cdot 48$ Btu/lb

SOLUTIONS TO SECOND CLASS EXAMINATION QUESTIONS 465

Heat lost by steam = Heat gained by water + Heat gained by ice

Let T = final temperature

$10 \times 2 \cdot 2[1302 \cdot 48-(T-32)]=250 \times 2 \cdot 2(T-86)+56 \times 2 \cdot 2[0 \cdot 5(32-14)+144+(T-32)]$

$10(1334 \cdot 48 - T) = 250(T - 86) + 56(121 + T)$
$28068 \cdot 8 = 316 T$
$T = 88 \cdot 84°F.$ Ans.

50. I.P. generates 1·03 times L.P. power, I.P. generates 1·05 times H.P. power

$\therefore$ H.P. generates $\dfrac{1 \cdot 03}{1 \cdot 05}$ = 0·981 times L.P. power

Ratio of powers, as L.P. : I.P. : H.P. : : 1 : 1·03 : 0·981
$1 + 1 \cdot 03 + 0 \cdot 981 = 3 \cdot 011$

$\therefore$ L.P. generates $\dfrac{1}{3 \cdot 011}$ of the total power

I.P. generates $\dfrac{1 \cdot 03}{3 \cdot 011}$ of the total power

H.P. generates $\dfrac{0 \cdot 981}{3 \cdot 011}$ of the total power

Mean effective pressure in L.P. = $\dfrac{1}{3 \cdot 011} \times 33$

= 10·96 lb/in². Ans.

Mean effective pressure in I.P. = $10 \cdot 96 \times \dfrac{64^2}{40^2} \times 1 \cdot 03$

= 28·9 lb/in². Ans.

Mean effective pressure in H.P. = $10 \cdot 96 \times \dfrac{64^2}{24^2} \times 0 \cdot 981$

= 76·47 lb/in². Ans.

51. Let x be the boiler density at first, then:
$$(40 \times x) + (8 \times 0) - (8 \times 14 \cdot 5) = 40 \times 14 \cdot 5$$
$$\therefore 40x = 48 \times 14 \cdot 5$$
$$x = 17 \cdot 4 \text{ oz per gall}$$

Now let y be the final density of the boiler water in the second case:
$$(40 \times 17 \cdot 4) - (8 \times 17 \cdot 4) + (8 \times 0) = 40 \times y$$
$$32 \times 17 \cdot 4 = 40y$$
$$13 \cdot 92 = y$$

$\therefore$ Final density would be 13·92 oz per gall. Ans.

52. Power developed in cylinder $= 100 - 27 \cdot 5 - 28 = 44 \cdot 5 \%$ of the heat supplied

$\therefore$ thermal efficiency $= 44 \cdot 5 \%$ or $0 \cdot 445$

Overall efficiency $=$ mechanical efficiency $\times$ thermal efficiency
$$= 0 \cdot 74 \times 0 \cdot 445$$
$$= 0 \cdot 3293$$

$$\text{Efficiency} = \frac{\text{heat got out}}{\text{heat supplied}}$$

$$\therefore 0 \cdot 3293 = \frac{\frac{33000 \times 60}{778}}{W \times 19000} = \frac{2545}{W \times 19000}$$

$\therefore$ W, the weight of fuel used per b.h.p. per hour

$$= \frac{2545}{0 \cdot 3293 \times 19000}$$

$$= 0 \cdot 4067 \text{ lb. Ans.}$$

53. Working in percentages, let stroke $= 100$
$$P_1 V_1 = P_2 V_2$$
$$14 \cdot 7 \times 100 = 26 \times V_2$$
$$V_2 = 56 \cdot 53$$

SOLUTIONS TO SECOND CLASS EXAMINATION QUESTIONS 467

∴ Distance travelled by piston when discharge valves open
= 100 — 56·53 = 43·47% Ans. (i)

$$P_1 V_1^{1·4} = P_2 V_2^{1·4}$$
$$14·7 \times 100^{1·4} = 26 \times V_2^{1·4}$$

$$V_2 = 100 \times \sqrt[1·4]{\frac{14·7}{26}}$$

$$= 66·55$$

∴ Distance travelled by piston when discharge valves open
= 100 — 66·55 = 33·45% Ans. (ii)

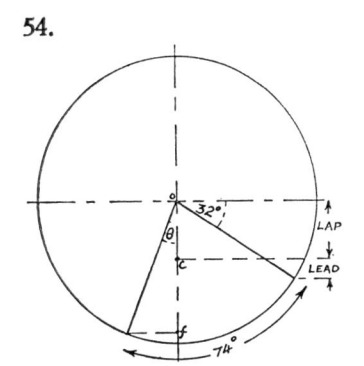

54.

$\theta = 32 + 74 — 90 = 16°$

$of = 4\frac{1}{16} \times \cos 16°$

$= 3·904$ in.

$oc = $ steam lap
$= of — cf$
$= 3·904 — 1·875$

$= 2·029$ in. Ans. (i)

$$\frac{\text{Lap} + \text{Lead}}{\frac{1}{2} \text{ travel}} = \text{Sine of angle of advance}$$

∴ $2·029 + \text{lead} = 4\frac{1}{16} \times \sin 32°$
$\text{lead} = 2·153 — 2·029$
$= 0·124$ in. Ans. (ii)

55. Weight of fuel used in 24 hours
$= \frac{11}{14} \times (8·5)^2 \times 6 \times 62·5 \times 0·92 = \text{i.h.p.} \times 0·4 \times 24$ lb

∴ i.h.p. $= \dfrac{11 \times 8·5^2 \times 6 \times 62·5 \times 0·92}{14 \times 0·4 \times 24} = 2040$ Ans. (i)

Thermal efficiency $= \dfrac{\frac{33000 \times 60}{778}}{0·4 \times 18500} = \dfrac{2545}{0·4 \times 18500}$

$= 0·344$ or $34·4\%$ Ans. (ii)

56. Force applied at circumference to turn shaft
$$= 0{\cdot}021 \times 92 \times 2240 \text{ lb}$$

Work done in turning shaft one revolution
$$= 0{\cdot}021 \times 92 \times 2240 \times \pi \times \tfrac{26}{12} \text{ ft lb}$$

Work done per minute
$$= 0{\cdot}021 \times 92 \times 2240 \times \pi \times \tfrac{26}{12} \times 95 \text{ ft lb}$$

$$\therefore \text{Horse power} = \frac{0{\cdot}021 \times 92 \times 2240 \times \pi \times 26 \times 95}{33000 \times 12}$$

$$= 84{\cdot}82 \text{ h.p. Ans. (i)}$$

Heat units generated per minute

$$= \frac{84{\cdot}82 \times 33000}{778} = 3597 \text{ Btu per min. Ans. (ii)}$$

57. Heat equivalent of i.h.p. of engine
$$= 3500 \times 2545 \text{ Btu per hour}$$
Heat in fuel consumed $= 1120 \times 19300$ Btu per hour
$$\therefore \text{Thermal efficiency} = \frac{3500 \times 2545}{1120 \times 19300} = 0{\cdot}4121, \text{ or } 41{\cdot}21\% \quad \text{Ans.}$$

Heat carried away in the cooling water
$$= (127 - 60) \times 47{\cdot}5 \times 2240 \text{ Btu per hour}$$
$$= 67 \times 47{\cdot}5 \times 2240 \text{ Btu}$$

$\therefore$ Per cent of total heat carried away

$$= \frac{67 \times 47{\cdot}5 \times 2240 \times 100}{1120 \times 19300} = 32{\cdot}97\% \quad \text{Ans.}$$

58. Available hydrogen $= H - \tfrac{O}{8} = 0{\cdot}114 - \tfrac{0{\cdot}024}{8} = 0{\cdot}111$ lb
C.V. $= 14500\, C + 62000\, (H - \tfrac{O}{8}) + 4000\, S$
$$= 14500 \times 0{\cdot}846 + 62000 \times 0{\cdot}111 + 4000 \times 0{\cdot}004$$
$$= 12267 + 6882 + 16$$
$$= 19165 \text{ Btu/lb. Ans. (i)}$$

SOLUTIONS TO SECOND CLASS EXAMINATION QUESTIONS   469

$$\begin{aligned}
\text{Air reqd.} &= \tfrac{100}{23}\{2\tfrac{2}{3}C + 8(H - \tfrac{O}{8}) + S\} \\
&= \tfrac{100}{23}\{2\tfrac{2}{3} \times 0.846 + 8 \times 0.111 + 0.004\} \\
&= \tfrac{100}{23}\{2.256 + 0.888 + 0.004\} \\
&= \tfrac{100}{23} \times 3.148 \\
&= 13.69 \text{ lb air/lb fuel. Ans. (ii)}
\end{aligned}$$

59. From superheat steam tables,
$P = 250$ lb/in^2 superheat 200 F°, $h = 1318.5$ Btu/lb

From saturation steam tables,
$P = 28$ lb/in^2, $h_g = 1163.3$ Btu/lb
$P = 0.8$ lb/in^2, $h_f = 62.4$, $h_{fg} = 1040.3$ Btu/lb

Heat drop through H.P. turbine
$= 1318.5 - 1163.3 = 155.2$ Btu/lb

Heat drop through L.P. is to be the same,
Let x lb pass through,
$$\begin{aligned}
x[1163.3 - (62.4 + 0.89 \times 1040.3)] &= 155.2 \\
x \times 175 &= 155.2 \\
x &= 0.8869 \text{ lb}
\end{aligned}$$

∴ Steam bled off $= 1 - 0.8869 = 0.1131$ lb
$= 11.31\%$ Ans.

60. Heat supplied $= 15 \times 2.2 (0.5 \times 10 + 80 + 660.1)$
$= 24588$ Chu. Ans. (i)
$24588 \times \tfrac{9}{5} = 44258$ Btu. Ans. (ii)

$$25 \text{ h.p.} = \frac{25 \times 2545}{60} \text{ Btu per minute}$$

$$\therefore \text{Time} = \frac{44258 \times 60}{25 \times 2545} = 41.75 \text{ minutes. Ans. (iii)}$$

61. $2250 + 460 = 2710°$F abs.
$380 + 460 = 840°$F abs.

Heat radiated from one body at $T_1°$ to another body at $T_2°$
$= k(T_1^4 - T_2^4)$
$= 16 \times 10^{-10} \times (2710^2 + 840^2)(2710^2 - 840^2)$
$= 16 \times 10^{-10} \times 8050700 \times 6639300$
$= 85530$ Btu per sq ft of surface per hour

Heat conducted = 0·075 × 85530
 = 6415 Btu

lb of water evaporated from and at 212°F

$$= \frac{6415}{970 \cdot 6} = 6 \cdot 609 \text{ lb per sq ft surface per hour. Ans.}$$

62. Final temperature = 55 + 60 = 115°F

Since the water equivalent of the vessel is 10 lb, then there is, in effect,
100 + 10 = 110 lb of water

Let x lb of steam be required

Heat lost by steam
 = Heat gained by (Water + Vessel + Ice)
$x\{(200 \times 0 \cdot 48) + 970 \cdot 6 + (212 - 115)\}$
 = 110 (115 — 55) + 25 (144 + 115 — 32)
1163·6x = 6600 + 5675

$$x = \frac{12275}{1163 \cdot 6} = 10 \cdot 55 \text{ lb. Ans.}$$

63. 6 + 7·5 = 13·5

$$\frac{6}{13 \cdot 5} \text{ of } 108 = 48 \text{ in. Also } \frac{7 \cdot 5}{13 \cdot 5} \text{ of } 108 = 60 \text{ in.}$$

The weight of the moving parts in connection with the top piston will be greater than for the bottom piston.

∴ the top piston has 48 in. travel
,, bottom ,, 60 ,, ,,

The engine is a two stroke

$$\text{b.h.p.} = \frac{97 \times 27^2 \times 0 \cdot 7854 \times 108 \times 100 \times 0 \cdot 92}{33000 \times 12}$$

= 1394. Ans.

SOLUTIONS TO SECOND CLASS EXAMINATION QUESTIONS 471

64. Compression pressure = 8·5 × 30 × 25·4 mm of mercury

Let the stroke volume be 100, this also represents the stroke

Let the clearance volume be x

$$P_1 V_1 = P_2 V_2$$
$$381 (100 + x) = 8·5 \times 30 \times 25·4 x$$
$$38100 + 381x = 6477x$$

$$x = \frac{38100}{6096} = 6·25$$

Clearance = 6·25% of stroke. Ans.

65. Fraction of the feed blown out

$$= \frac{\text{Feed density}}{\text{Boiler density}} = \frac{0·11}{0·8} = \frac{11}{80}$$

This means that if 80 lb are fed in, 11 lb will be blown out and 80 — 11 = 69 lb will be left to form steam. Or, in order to form 1 lb of steam when blowing down, $\frac{80}{69}$ lb of feed must be taken, $\frac{11}{69}$ lb will be blown out, and the remainder will form steam.

From tables,
P = 185 lb/in², h_f = 348·5, h_{fg} = 849·9

The heat to form 1 lb of steam before blowing was resorted to
$$= 348·5 - (130 - 32) + 849·9$$
$$= 250·5 + 849·9 = 1100·4 \text{ Btu}$$

When blowing down, the heat to form 1 lb of steam will be
$$\tfrac{80}{69} \times 250·5 + 849·9 \text{ Btu}$$

The extra heat required will be the sensible heat to $\frac{11}{69}$ lb of water which is blown out.
$$\tfrac{11}{69} \times 250·5 = 39·93 \text{ Btu}$$

The daily consumption will increase in direct proportion to the extra heat required to form 1 lb of steam

∴ % increase in consumption

$$= \frac{39·93}{1100·4} \times 100 = 3·628\% \text{ Ans.}$$

Q

66.   Since the piston speed of each engine is the same, then Horse power $\propto$ Mean pressure $\times$ Area of piston.

$$\therefore \frac{\text{Horse power}}{\text{Mean press.} \times \text{Area}} = \text{constant}$$

Let the horse power of H.P. be 1, then that of L.P. is 1·15
Let the area of H.P. be 1, then that of L.P. is $(2·2)^2 = 4·84$

$$\frac{1}{67 \times 1} = \frac{1·15}{\text{M.E.P.} \times 4·84}$$

$$\therefore \text{M.E.P. of L.P.} = \frac{67 \times 1·15}{4·84} = 15·92 \text{ lb/in}^2. \text{ Ans.}$$

67.   For proof that the co-efficient of cubical expansion is three times co-efficient of linear expansion, see Chapter 2.

Let V = original volume of block
Increase in volume = 3 KVT

$$\% \text{ increase} = \frac{3 \text{ KVT}}{V} \times 100 = 3 \text{ KT} \times 100$$

$$= 3 \times 6·7 \times 10^{-6} \times 500 \times 100 = 1·005\% \text{ Ans.}$$

68.   b.h.p. $= \dfrac{600 \times 2\pi \times 5 \times 165}{33000} = 94·25$ Ans. (b)

i.h.p. $= \dfrac{45 \times 14^2 \times 0·7854 \times 20 \times 330}{33000 \times 12} = 115·5$ Ans. (a)

Mechanical effic. $= \dfrac{\text{b.h.p.}}{\text{i.h.p.}} = \dfrac{94·25}{115·5} = 0·816$

or 81·6% Ans. (c)

Steam per i.h.p. hour
$$= \frac{1950}{115·5} = 16·88 \text{ lb. Ans. (d)}$$

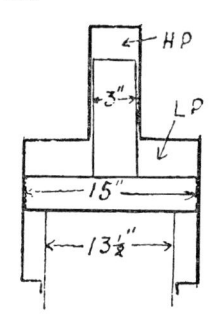

SOLUTIONS TO SECOND CLASS EXAMINATION QUESTIONS   473

69. Assuming that the air is delivered at the same temperature as it entered the L.P. stage, and that atmospheric pressure is 15 lb/in² abs., then:

Volume of atmospheric air taken in, in 30 minutes

$$= \frac{(15^2 - 3^2)\,0.7854 \times 15 \times 0.88}{1728} \times 140 \times 30 \text{ cu ft}$$

Volume $\alpha \dfrac{1}{\text{Pressure}}$, when temperature is constant

∴ Volume at 1015 lb/in² abs.

$$= \frac{(15^2 - 3^2) \times 0.7854 \times 15 \times 0.88}{1728} \times 140 \times 30 \times \frac{15}{1015}$$

= 80·43 cu ft. Ans.

70. On passing through regulating valve:
Total heat before throttling = Total heat after
$$206 = 156.2 + 126q$$
$$126q = 49.8$$
$$q = 0.3952$$

∴ CO_2 enters evaporator with a dryness fraction of 0·3952.
Ans. (i)

On passing through evaporator:
Heat extracted = Heat absorbed by CO_2
  = Heat on leaving — Heat on entering
  = 0·98 × 126 — 0·3952 × 126
  = 0·5848 × 126
  = 73·69 Btu/lb

Heat extracted by 16 lb of CO_2
  = 16 × 73·69
  = 1179 Btu/min. Ans. (ii)

Volume per lb of vapour leaving evaporator
$= 0.98 \times 0.36$
Volume of 16 lb $= 16 \times 0.98 \times 0.36 = 5.644$ cu ft/min.

Ans. (iii)

71. Let v_1 be the initial volume
$$13.5 \times v_1{}^{1\cdot 35} = 68 \times (550)^{1\cdot 35}$$

By logs.
$$\log 13.5 + 1\cdot 35 \log v_1 = \log 68 + 1\cdot 35 \log 550$$

$$\therefore \log v_1 = \frac{\log 68 + 1\cdot 35 \log 550 - \log 13\cdot 5}{1\cdot 35}$$

Putting in the values, and working out
$\log v_1 = 3\cdot 2605$
$\therefore v_1 = 1822$ cc

Now this includes the clearance volume, i.e. it is $1\cdot 05 \times$ stroke volume

$$\therefore \text{Stroke vol.} = \frac{1822}{1\cdot 05} = 1736 \text{ cc}$$
$d^2 \times 0\cdot 7854 \times 28 = 1736$

$$\therefore \text{Diameter} = \sqrt{\frac{1736}{0\cdot 7854 \times 28}} = 8\cdot 8 \text{ cm. Ans.}$$

72. $p_1 = 30 + 8 = 38$ in. of mercury
$p_2 = 30 + 14 = 44$,, ,, ,,
$p \times v =$ constant when the temperature does not change

$\therefore 38 \times 500 = 44 \times v_2$

$$v_2 = \frac{38 \times 500}{44} = 431\cdot 8 \text{ cc}$$

Change of volume $= 500 - 431\cdot 8 = 68\cdot 2$ cc

$$\text{Change per cent} = \frac{68\cdot 2}{500} \times 100 = 13\cdot 64 \% \text{ Ans.}$$

SOLUTIONS TO SECOND CLASS EXAMINATION QUESTIONS 475

Alternatively:
$$v_2 = \frac{38}{44} \text{ of } 500 \text{ cc}$$

$$\text{Change of volume} = 500 - \frac{38}{44} \times 500 = \frac{6}{44} \times 500 \text{ cc}$$

$$\therefore \text{Change per cent} = \frac{6}{44} \times 500 \times \frac{100}{500} = 13.64\% \text{ Ans.}$$

73. Abs. press. of air in reservoir $= 165 + 14.7 = 179.7$ lb/in^2
Abs. temp. of air in reservoir $= 75 + 460 = 535°$F

$$PV = wRT, \qquad R = \frac{PV}{wT}$$

$$\frac{P_1 V_1}{w_1 T_1} = \frac{P_2 V_2}{w_2 T_2}$$

$$\frac{14.7 \times 144 \times 1}{0.0807 \times 492} = \frac{179.7 \times 144 \times V_2}{1200 \times 535}$$

$$V_2 = \frac{1200 \times 535 \times 14.7}{179.7 \times 0.0807 \times 492}$$

$$= 1322 \text{ ft}^3. \text{ Ans. (i)}$$

$$\text{Volume} = 0.7854 \times d^2 \times \text{length}$$

$$d = \sqrt{\frac{1322}{0.7854 \times 22}}$$

$$= 8.75 \text{ ft} = 8 \text{ ft } 9 \text{ in. Ans. (ii)}$$

Volume of atmospheric air to make 1322 ft^3 of compressed air at the same temperature

$$= 1322 \times \frac{179.7}{14.7} = 16160 \text{ ft}^3. \text{ Ans. (iii)}$$

74.
$$\text{i.h.p.} = \frac{4182}{0.82} = 5100$$

Power developed on top side of pistons

$$= \frac{10}{(10 + 7)} \text{ of } 5100 = 3000 \text{ i.h.p.}$$

Power developed on under side of pistons
$$= 5100 - 3000 = 2100 \text{ i.h.p.}$$

For top side:
$$\frac{3000}{8} = \frac{p \times (26.75)^2 \times 0.7854 \times 55 \times \tfrac{110}{2}}{33000 \times 12}$$

from which mean indicated pressure
 $= 87.35$ lb/in². Ans.

For under side:
$$\frac{2100}{8} = \frac{p \times [(26.75)^2 - 9^2] \times 0.7854 \times 55 \times \tfrac{110}{2}}{33000 \times 12}$$

from which mean indicated pressure
 $= 68.93$ lb/in². Ans.

Alternatively for second part:
H.P. α mean pressure $\times$ effective area

$$\therefore \frac{\text{H.P.}}{p \times a} = \text{constant}$$

$$\frac{3000}{87.35 \times (26.75)^2} = \frac{2100}{p \times [(26.75)^2 - 9^2]}$$

from which $p = 68.93$ lb/in² as before.

75. Loss due to exhaust gases $= 28\%$
 ,,   ,,   cooling water $= 30\%$
$$\text{total} = 58\%$$

SOLUTIONS TO SECOND CLASS EXAMINATION QUESTIONS 477

If there are no other losses then i.h.p. represents $100 - 58 = 42\%$ of the available heat.

b.h.p. represents 0·78 of $42\% = 32·76\%$. Ans. (a)

$42 - 32·76 = 9·24\%$, which may be regarded as friction loss. If 60% of the heat in the exhaust gases is recovered then 40% is lost.

$$40\% \text{ of } 28\% = 11·2\%$$

Loss due to exhaust gases = 11·2%
,, ,, cooling water = 30%
,, ,, friction = 9·24%
total = 50·44%

$50·44\%$ of $19500 = 9836$ Btu. Ans. (b)

76. Heat equivalent of 1 kilowatt hour = 3412 Btu

Let x lb of oil be required,

$$\text{then } 0·07 = \frac{3412}{x \times 19500}$$

$$x = \frac{3412}{0·07 \times 19500} = 2·5 \text{ lb. Ans.}$$

77. From steam tables:
$P = 200 \quad h_f = 355·5 \quad h_{fg} = 844 \quad v_g = 2·29$
$P = 50 \quad h_f = 250·2 \quad h_{fg} = 924·6 \quad v_g = 8·516$

Let $q = $ dryness fraction of reduced pressure steam.

As no work is done as the steam passes through the reducing valve, then no heat is lost, therefore:

Total heat before throttling = Total heat after throttling
$355·5 + 0·95 \times 844 = 250·2 + q \times 924·6$
$1157·3 = 250·2 + 924·6\, q$
$907·1 = 924·6\, q$
$q = 0·9809$ Ans. (i)

Volume of one lb of wet steam at 200 lb/in² and 0·95 dry
$= 2·29 \times 0·95$ cu ft/lb

$$\text{Weight of 10 cu ft} = \frac{10}{2 \cdot 29 \times 0 \cdot 95} \text{ lb}$$

As the weight remains constant, this is also the weight of the reduced pressure steam.

Volume of one lb. of wet steam at 50 lb/in² and 0·9809 dry
$$= 8 \cdot 516 \times 0 \cdot 9809 \text{ cu ft /lb}$$

∴ Total volume of reduced pressure steam
$$= \frac{10 \times 8 \cdot 516 \times 0 \cdot 9809}{2 \cdot 29 \times 0 \cdot 95} = 38 \cdot 4 \text{ cubic feet. Ans. (ii)}$$

78. The speed of the cam shaft of a 4-stroke engine is one-half that of the crank shaft.

Speed of cam shaft $= 385 \div 2 = 192 \cdot 5$ r.p.m.

$$\text{Time for one rev.} = \frac{60}{192 \cdot 5} \text{ second}$$

$$\therefore \text{Time for } 124° = \frac{124}{360} \times \frac{60}{192 \cdot 5} = 0 \cdot 1074 \text{ sec.} \quad \text{Ans. (a)}$$

One revolution of the cam shaft = one cycle
Valve is closed for 360° — 124° = 236°
And as a percentage of one cycle $= \frac{236}{360} \times 100 = 65 \cdot 55\%$
Ans. (b)

79. The law of compression is $PV^{1 \cdot 33} = $ constant
Before alteration, $14 \times (18 + 3 \cdot 5)^{1 \cdot 33} = P_2 \times (3 \cdot 5)^{1 \cdot 33}$.
By logs, Log $14 + 1 \cdot 33 \log 21 \cdot 5 = \text{Log } P_2 + 1 \cdot 33 \log 3 \cdot 5$

$$\therefore \text{Log } P_2 = \log 14 + 1 \cdot 33 (\log 21 \cdot 5 - \log 3 \cdot 5)$$
$$= 1 \cdot 1461 + 1 \cdot 33 \times 0 \cdot 7883 = 2 \cdot 1945$$
$$P_2 = 156 \cdot 5 \text{ lb/in}^2. \text{ Ans.}$$

After alteration, $14 \times (18 + 3)^{1 \cdot 33} = P_2 \times (3)^{1 \cdot 33}$

By logs, Log $14 + 1 \cdot 33 \log 21 = \log P_2 + 1 \cdot 33 \log 3$

SOLUTIONS TO SECOND CLASS EXAMINATION QUESTIONS 479

$$\therefore \text{Log } P_2 = \log 14 + 1\cdot33 \ (\log 21 - \log 3)$$
$$= 1\cdot1461 + 1\cdot33 \times 0\cdot8451 = 2\cdot2700$$
$$P_2 = 186\cdot2 \text{ lb/in}^2. \text{ Ans.}$$

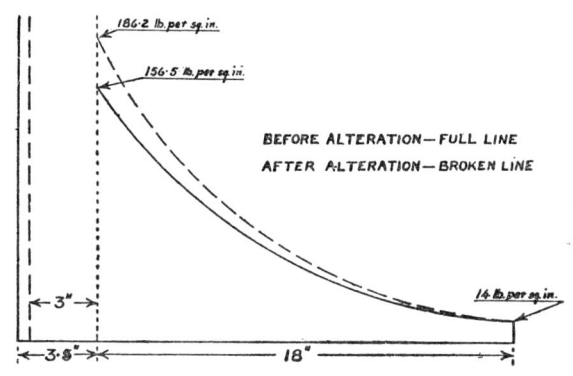

BEFORE ALTERATION — FULL LINE
AFTER ALTERATION — BROKEN LINE

80. Volume of sphere $= \frac{\pi}{6} D^3$
Weight of sphere $= \frac{\pi}{6} \times 5^3 \times 0\cdot26 = 17\cdot02$ lb

Heat to be given to sphere to raise its temperature to 32°F
$$= \text{Wt.} \times \text{spec. ht.} \times \text{temp. rise}$$
$$= 17\cdot02 \times 0\cdot13 \times [32 - (-14)]$$
$$= 17\cdot02 \times 0\cdot13 \times 46$$
$$= 101\cdot7 \text{ Btu} \quad \ldots \quad \ldots \quad \ldots \quad \ldots \quad (i)$$

Heat required to be taken from 2 lb of water to freeze it at 32°F
$$= 2 \ [(34 - 32) + 144]$$
$$= 292 \text{ Btu} \quad \ldots \quad \ldots \quad \ldots \quad \ldots \quad (ii)$$

We see from (i) and (ii) that the whole 2 lb of water will be cooled to 32°F but only a portion of it will be frozen. Let w lb = weight of ice formed at 32°F, leaving $(2 - w)$ lb of water at 32°F,

Heat taken from water = Heat absorbed by sphere

Sensible heat to cool   Latent heat   Total heat absorbed
2 lb water from 34°   +   to freeze w lb   =   by sphere from —14°F
to 32°F   of water at   to 32°F
                              32°F

$$2\ (34 - 32) + w \times 144 = 101\cdot7$$
$$w \times 144 = 97\cdot7$$
$$w = 0\cdot6784 \text{ lb}$$

$\therefore$ Final temp. of mixture = 32°F $\left.\begin{array}{c} \\ \\ \end{array}\right\}$ Ans.
Weight of ice formed = 0·6784 lb

81. $$\text{i.h.p.} = \frac{84 \times 7\cdot5^2 \times 0\cdot7854 \times 15 \times 125}{33000 \times 12} = 17\cdot58$$

Heat turned into work in cylinder $= 17\cdot58 \times 2545$ Btu per hour

Heat put into cylinder $= 240 \times 470$ Btu per hour

Thermal effic. based on i.h.p.

$$= \frac{\text{Heat equivalent of work in cylinder}}{\text{Heat put into cylinder}}$$

$$= \frac{17\cdot58 \times 2545}{240 \times 470} = 0\cdot3967 \text{ or } 39\cdot67\% \quad \text{Ans. (a)}$$

Thermal effic. based on b.h.p.

$$= \frac{\text{Heat equivalent of power at brake}}{\text{Heat put into cylinder}}$$

$$= \frac{12\cdot2 \times 2545}{240 \times 470} = 0\cdot2753 \text{ or } 27\cdot53\% \quad \text{Ans. (b)}$$

$$\text{Mechanical effic.} = \frac{\text{b.h.p.}}{\text{i.h.p.}} = \frac{12\cdot2}{17\cdot58} = 0\cdot694 \text{ or } 69\cdot4\% \quad \text{Ans. (c)}$$

If the governor were removed, there would be one power stroke every two revolutions, that is $\frac{320}{2} = 160$

Power α power strokes $\times$ m.e.p., and if the power is to be constant, then power strokes $\times$ m.e.p. must be constant.

$\therefore 125 \times 84 = 160 \times$ m.e.p.

$$\text{m.e.p.} = \frac{125 \times 84}{160} = 65\cdot625 \text{ lb/in}^2. \quad \text{Ans. (d)}$$

82. $$\text{Indicated horse power} = \frac{300}{0\cdot78}$$

Heat equivalent of the indicated horse power

$$= \frac{300}{0\cdot78} \times 2545 \text{ Btu per hour}$$

SOLUTIONS TO SECOND CLASS EXAMINATION QUESTIONS **481**

$$\text{Piston area} = \frac{\text{Swept volume}}{\text{Stroke}} = \frac{4\cdot 63 \times 1728}{22} \text{ sq in.}$$

∴ heat equivalent of the work done per sq in. of piston area per hour

$$= \frac{\text{Heat equivalent of i.h.p.}}{\text{Piston Area}}$$

$$= \frac{300 \times 2545}{0\cdot 78} \times \frac{22}{4\cdot 63 \times 1728}$$

$$= 2692 \text{ Btu. Ans.}$$

83. The exhaust valve is open for $32° + 180° + 18° = 230°$ of the crankshaft movement.

$$\text{Time for one revolution of crank} = \frac{60}{120} = \tfrac{1}{2} \text{ second}$$

∴ Valve is open for $\dfrac{230}{360}$ of $\tfrac{1}{2} = 0\cdot 3194$ second. Ans. (a)

The valve is full open for $18° + 180° — 6°$
$= 192°$ of the crank movement

∴ Valve is full open for $\dfrac{192}{360}$ of $\tfrac{1}{2} = 0\cdot 2667$ second. Ans. (b)

The valve opens during $32° — 18° = 14°$ of the crank movement

∴ Time of opening $\dfrac{14}{360}$ of $\tfrac{1}{2}$ second, and during this time the valve lifts 2 in., or $\tfrac{1}{6}$ ft

Time of opening × Average velocity of opening = Valve lift
∴ Average velocity of opening

$$= \tfrac{1}{6} \div \left\{ \frac{14}{360} \times \frac{1}{2} \right\} = 8\tfrac{4}{7} \text{ ft per sec. Ans. (c)}$$

The valve closes during 6° + 18° = 24° of the crank movement

∴ Average velocity of closing

$$= \tfrac{1}{6} \div \left\{ \frac{24}{360} \times \frac{1}{2} \right\} = 5 \text{ ft per sec. Ans. (d)}$$

84. Increase in consumption per h.p. α Reduction in load

Increase in consumption per h.p. from
Full to Half load = 18%
Full to ¾ load = 9%

Consumption per day at full load = 640 kilograms

∴ Consumption per day at ¾ load = 640 × ¾ × $\frac{109}{100}$

= 523·2 kilograms. Ans.

85. Increase in diameter from workshop temperature to running temperature

= Coeff. of linear exp. × original dia. × temp. rise

For the cylinder, from 60°F to 460°F:

Increase in dia. = 0·0000061 × 12 × 400
= 0·02928 in.
∴ Running dia. = 12 + 0·02928 = 12·02928 in.

For the piston crown, from 80°F to 560°F:

Increase in dia. = 0·0000067 × 11·92 × 480
= 0·03832 in.
∴ Running dia. = 11·92 + 0·03832 = 11·95832 in.

For the piston at mid-depth, from 80°F to 160°F:

Increase in dia. = 0·0000067 × 11·98 × 80
= 0·006421 in.
∴ Running dia. = 11·98 + 0·006421 = 11·986421 in.

RUNNING CLEARANCES
At crown of piston = 12·02928 − 11·95832
= 0·07096 in. on dia., or 0·03548 in. on radius. Ans. (i)

At mid-depth of piston = 12·02928 − 11·98642
= 0·04286 in. on dia., or 0·02143 in. on radius. Ans. (ii)

SOLUTIONS TO SECOND CLASS EXAMINATION QUESTIONS   483

86. 18500 Btu are given to $1 + 23 = 24$ lb of gases, under constant volume conditions.

Heat given = Weight × Change of temp. × Specific heat

∴ Change of temperature

$$= \frac{18500}{24 \times 0{\cdot}169} = 4561 \text{ F degrees}$$

Initial temperature $= 80 + 460 = 540°\text{F abs.}$
Final temperature $= 540 + 4561 = 5101°\text{F abs.}$

Now $\dfrac{\text{Absolute pressure}}{\text{Absolute temp.}}$ is constant when the volume is constant

∴ $\dfrac{20}{540} = \dfrac{\text{Final pressure}}{5101}$

Final pressure $= \dfrac{20 \times 5101}{540} = 188{\cdot}9 \text{ lb/in}^2$ abs. **Ans.**

87. From steam tables,
$P = 180 \text{ lb/in}^2$, $h_g = 1198 \text{ Btu/lb}$
$P = 165 \text{ lb/in}^2$, $h_f = 338{\cdot}6$, $h_{fg} = 858$ **Btu/lb**

Heat lost during passage along pipe
$= 54 [1198 - (338{\cdot}6 + 0{\cdot}95 \times 858)]$
$= 2392{\cdot}2$ Btu. Ans. (a)

Surface area of the pipe
$= \dfrac{\pi \times 4}{12} \times 60 = 62{\cdot}832 \text{ ft}^2$

∴ Heat lost per minute per sq ft of pipe surface

$$= \frac{2392{\cdot}2}{62{\cdot}832} = 38{\cdot}07 \text{ Btu. Ans. (b)}$$

88. Let the stroke of the engine = L ft
then length of diagram = L in.

$$\text{Mean height of diagram} = \frac{5{\cdot}175}{L} \text{ in.}$$

$$\text{Mean effective pressure} = \frac{5{\cdot}175 \times 72}{L} \text{ lb/in}^2$$

$$\text{i.h.p.} = \frac{pALN}{33000}$$

$$= \frac{5{\cdot}175 \times 72 \times 0{\cdot}7854 \times 26^2 \times L \times 75 \times 2}{L \times 33000}$$

$$= 899 \text{ i.h.p. Ans.}$$

89. Volume enclosed when scavenge ports are closed
$$= (700)^2 \times 0{\cdot}7854 \times (800 + 70) \text{ mm}^3$$

$$= \frac{(700)^2 \times 0{\cdot}7854 \times 870}{(25{\cdot}4)^3 \times 1728} = 11{\cdot}83 \text{ ft}^3$$

∴ Volume of air taken in $= 11{\cdot}83 \times 0{\cdot}95 = 11{\cdot}24$ ft^3

Now 0·0807 lb of air at 14·7 lb per sq in. and 492°F abs. occupies 1 ft^3

∴ ,, ,, ,, ,, 17·7 ,, ,, ,, 492°F abs. ,, $\dfrac{14{\cdot}7}{17{\cdot}7}$,,

and ,, ,, ,, 17·7 ,, ,, ,, 565°F abs. ,, $\dfrac{14{\cdot}7 \times 565}{17{\cdot}7 \times 492}$

$$= 0{\cdot}9535 \text{ ft}^3$$

∴ 0·0807 lb of air under the cylinder conditions occupies 0·9535 ft^3

$$\therefore \text{Weight of air} = 0{\cdot}0807 \times \frac{11{\cdot}24}{0{\cdot}9535} = 0{\cdot}951 \text{ lb. Ans.}$$

90. Radius of shaft = 6 in.

$$\frac{\frac{1}{2}}{6} = \frac{1}{12} \text{radian} = \frac{360}{12 \times 2\pi} = 4{\cdot}775 \text{ degrees} = 4°\ 47'$$

∴ Actual angle of advance of the eccentric
$$= 30° + 4°\ 47' = 34°\ 47'$$

SOLUTIONS TO SECOND CLASS EXAMINATION QUESTIONS 485

Steam lap = Half travel — max. p.o. to steam
= $3 - 1\frac{3}{4} = 1\frac{1}{4}$ in.

Lap + lead = Half travel × sine of angle of advance
= $3 \times \sin 34° 47' = 1\cdot711$ in.

∴ lead the valve has = $1\cdot711 - 1\cdot25 = 0\cdot461$ in. Ans.

The intended angle of advance was 30°
∴ Lap + intended lead = $3 \times \sin 30° = 1\cdot5$ in.
Intended lead = $1\cdot5 - 1\cdot25 = 0\cdot25$ in.

∴ Amount to add to lap = $0\cdot461 - 0\cdot25 = 0\cdot211$ in. Ans.

91. Let $t°$ be the increase in temperature in F degrees.
Increase in diameter of sleeve = $10 \times t \times 0\cdot0000136$
Increase in diameter of shaft = $10 \times t \times 0\cdot0000067$

∴ $(10 \times t \times 0\cdot0000136) - (10 \times t \times 0\cdot0000067) = 0\cdot02$
$10 \times t \times 0\cdot0000069 = 0\cdot02$

$$t = \frac{0\cdot02}{0\cdot000069} = 290 \text{ F}°$$

Final temp. = $60 + 290 = 350°$F. Ans.

92. From steam tables,
P = 215 lb/in², $t_F = 387\cdot9°$F, $h_g = 1200\cdot4$ Btu/lb

Let t = temperature of superheated steam

Heat gained by steam in superheaters
= $14000 \times 0\cdot55 \times (t - 387\cdot9)$ Btu per hour

Weight of flue gases per lb of fuel
= 1 lb fuel + 24 lb air = 25 lb

Heat lost by flue gases in superheaters
= $25 \times 1500 \times 0\cdot25 \times (1525 - 1281)$
= $25 \times 1500 \times 0\cdot25 \times 244$ Btu per hour

Heat gained by steam = Heat lost by flue gases
$14000 \times 0\cdot55 (t - 387\cdot9) = 25 \times 1500 \times 0\cdot25 \times 244$

$$t = \frac{25 \times 1500 \times 0\cdot25 \times 244}{14000 \times 0\cdot55} + 387\cdot9$$

= $297 + 387\cdot9 = 684\cdot9°$F. Ans. (i)

Let w = wt. of injection water per lb of steam

Total heat entering desuperheater = Total heat leaving
$$(1200\cdot 4 + 0\cdot 55 \times 297) + w(70-32) = (1+w) \times 1200\cdot 4$$
$$1200\cdot 4 + 163\cdot 4 + 38w = 1200\cdot 4 + 1200\cdot 4w$$
$$163\cdot 4 = 1162\cdot 4\, w$$
$$w = 0\cdot 1406 \text{ lb. Ans. (ii)}$$

93. Coeff. of cubical expansion = 3 × coeff. of linear exp.
    = 3 × 0·0000085 for glass = 0·0000255/C°

    Increase in volume = coeff. of cubical exp. × orig. vol.
    × temp. rise

    ∴ Apparent (or relative) cubical expansion of mercury
    = $(0\cdot 00018 - 0\cdot 0000255) \times 30 \times 2\cdot 54^3 \times 30$
    = 2·278 cm³. Ans.

94. $$\dfrac{\text{Lap + lead}}{\text{Half travel}} = \sin 29° \, 40'$$

    ∴ half travel = $(1\cdot 43 + 0\cdot 21) \div \sin 29° \, 40' = 3\cdot 313$ in.
    Max. port opening = $3\cdot 313 - 1\cdot 43 = 1\cdot 883$ in.
    ∴ Greatest area of opening = $1\cdot 883 \times 26\cdot 5 = 49\cdot 9$ in². Ans.

95. $$P_1 V_1^n = P_2 V_2^n$$
    $$50 \times 3^{1\cdot 3} = 49\cdot 5 \times V_2^{1\cdot 3}$$

    $$V_2 = 3 \times \sqrt[1\cdot 3]{\dfrac{50}{49\cdot 5}}$$

    = 3·024 cu ft

    Increase in Volume = $3\cdot 024 - 3 = 0\cdot 024$ cu ft
    = $0\cdot 024 \times 1728$ cu in.

    ∴ Stroke of pump = $\dfrac{0\cdot 024 \times 1728}{0\cdot 7854 \times 3^2}$ = 5·866 in. Ans.

96. Let d inches be the depth of the ports in the liner
    The free passage through the ports is:
    $\tfrac{2}{3}$ × circumference of the liner × depth of the ports
    The ratio of maximum piston speed to mean piston speed is
    $\pi : 2$

SOLUTIONS TO SECOND CLASS EXAMINATION QUESTIONS 487

∴ maximum piston speed $= \frac{\pi}{2} \times 700$ ft per minute

Area of cylinder × max. piston speed = Area through ports × steam speed

$$\frac{\pi}{4} \times 30^2 \times \frac{\pi}{2} \times 700 = \frac{2}{3} \times \pi \times 12\frac{1}{2} \times d \times 8500$$

$$d = \frac{\pi \times 30^2 \times \pi \times 700 \times 3}{4 \times 2 \times 2 \times \pi \times 12\frac{1}{2} \times 8500}$$

$$= 3\cdot 49 \text{ in. Ans.}$$

97. From superheat steam tables,
P = 500 lb/in², supht. 360F°, $\quad h = 1427$ Btu/lb

From saturation steam tables,
P = 1 lb/in², $\quad h_f = 69\cdot 7$, $\quad h_{fg} = 1036\cdot 1$ Btu/lb

Heat drop in steam = Heat equivalent of work done
$[1427 - (69\cdot 7 + 1036\cdot 1\, q)] \times 10 \times 3600 = 5500 \times 2545$

$$1357\cdot 3 - 1036\cdot 1\, q = \frac{5500 \times 2545}{10 \times 3600}$$

$$-1036\cdot 1\, q = 388\cdot 9 - 1357\cdot 3$$
$$1036\cdot 1\, q = 968\cdot 4$$
$$q = 0\cdot 9348 \text{ Ans.}$$

98. The first heat given to the ice raises its temperature from 22°F to 32°F. This is Sensible Heat, and the specific heat of ice is 0·5

Therefore, Sensible Heat given to the ice
$= 2\,(32 - 22) \times 0\cdot 5 = 10$ Btu

When the ice has reached the temperature of 32°F, Latent Heat must be given to change its physical state from solid into liquid. Whilst any ice remains unmelted the temperature of the resulting water remains at 32°F. The Latent Heat of water is 144 Btu/lb

∴ $2 \times 144 = 288$ Btu must be supplied to melt the ice.

The volume occupied by the water is slightly less than the volume of the ice.

When all the ice has melted, further application of heat causes the temperature of the water to rise. At first the volume of the water decreases a little until 39°F is reached, afterwards the volume increases.

From steam tables,
$P = 68$ lb/in^2, $\quad h_f = 270\cdot7$, $\quad h_{fg} = 910\cdot1$

The heat supplied to raise the temperature to boiling point is Sensible Heat, and the quantity is $2 \times 270\cdot7 = 541\cdot4$ Btu.

The water has now reached the temperature at which it is changed into steam. Latent heat is required to effect the change, and the volume of the steam formed is much greater than that of the water. At the pressure of 68 lb/in^2 abs., it is about 400 times as great.

The Latent Heat of formation is $2 \times 910\cdot1 = 1820\cdot2$ Btu

When all the water has been evaporated, the temperature of the steam is now raised above 301°F. The steam becomes superheated, and the heat to superheat is Sensible Heat.

The amount required is $2 (400 - 301) \times 0\cdot48 = 95$ Btu. Steam, in practice, is always superheated at constant pressure.

The volume increases, and it is approximately $\dfrac{400 + 460}{301 + 460}$

$= 1\cdot13$ times the volume it occupied as saturated steam.

The total heat supplied
$= 10 + 288 + 541\cdot4 + 1820\cdot2 + 95$
$= 2754\cdot6$ Btu

99. From $PV = wRT$, $\quad R = \dfrac{PV}{wT}$

$\therefore R = \dfrac{14\cdot7 \times 144 \times 1}{0\cdot078 \times 273}$

$= 99\cdot4$ ft lb per lb per C°

$$w = \frac{PV}{RT} = \frac{85 \times 144 \times 0.4}{99.4 \times 334} = 0.1475 \text{ lb. Ans. (a)}$$

$$V = \frac{wRT}{P} = \frac{0.1475 \times 99.4 \times 288}{20 \times 144} = 1.466 \text{ ft}^3 \text{ Ans. (b)}$$

100.

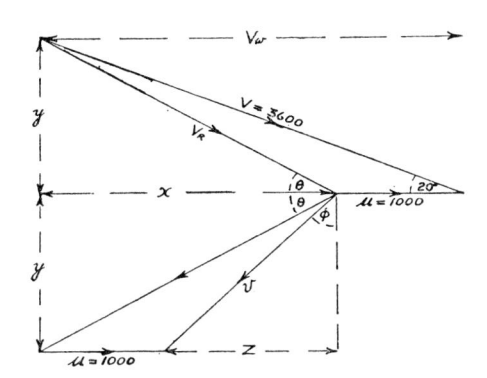

$Vw = 3600 \times \cos 20° = 3383$ ft per sec
$y = 3600 \times \sin 20° = 1231$
$x = 3383 - 1000 = 2383$

$$\text{Tan } \theta = \frac{1231}{2383} = 0.5166$$

∴ Inlet angle of blades = 27° 19′ Ans. (b)

$Z = 2383 - 1000 = 1383$

$$\text{Tan } \phi = \frac{1383}{1231} = 1.124 \quad \phi = 48° 20′$$

$$v = \frac{Z}{\sin \phi} = \frac{1383}{0.747} = 1852$$

∴ Velocity of exit = 1852 ft per sec at 48° 20′ to axis.
Ans. (c)

Velocity of whirl at entrance = Vw = 3383 ft/sec
Velocity of whirl at exit = Z = 1383 ft/sec
Total velocity of whirl = 3383 + 1383
= 4766 ft/sec. Ans. (a)

101. Let T°C be the temperature of the gases, and the temperature of the iron before being placed in the water.

Since no water is evaporated its highest temperature will be 100°C and its initial temperature $= \frac{100}{2} = 50°$C.

Heat lost by iron = Heat gained by water
$$300 \,(T - 100) \times 0{\cdot}13 = 0{\cdot}16 \times 1000 \,(100 - 50)$$

$$T - 100 = \frac{0{\cdot}16 \times 1000 \times 50}{300 \times 0{\cdot}13} = 205$$

$$\therefore T = 205 + 100 = 305°C. \text{ Ans.}$$

102. From superheat steam tables,
$P = 250$ lb/in² supht. 200 F°, $\qquad h = 1318{\cdot}5$ Btu/lb

From saturation steam tables,
$P = 250$ lb/in², $\quad t_F = 401°$F, $\quad v_g = 1{\cdot}844$ ft³/lb
$P = 2$ lb/in², $\quad h_f = 94, \quad h_{fg} = 1022{\cdot}2, \quad v_g = 173{\cdot}7$

Heat drop $= 1318{\cdot}5 - (94 + 0{\cdot}9 \times 1022{\cdot}2)$
$\qquad\qquad = 304{\cdot}52$ Btu/lb

Spec. vol. of superheated steam $= 1{\cdot}844 \times \dfrac{601 + 460}{401 + 460}$
$\qquad\qquad = 2{\cdot}272$ ft³/lb

Weight of supply steam $= \dfrac{12}{2{\cdot}272} = 5{\cdot}282$ lb

Equivalent work done $= 5{\cdot}282 \times 304{\cdot}52 \times 778$
$\qquad\qquad = 1251000$ ft lb. Ans. (i)

Spec. vol. of expanded steam $= 0{\cdot}9 \times 173{\cdot}7$ ft³/lb
$\therefore$ Final vol. of the steam $= 5{\cdot}282 \times 0{\cdot}9 \times 173{\cdot}7$
$\qquad\qquad = 825{\cdot}6$ ft³. Ans. (ii)

103. Admission press. $= 57 + 15 = 72$ lb/in² abs.
Back press. $= 9 + 15 = 24$ lb/in² abs.
Let stroke $= 1$,

Ratio of expansion $= \dfrac{\text{Final vol.}}{\text{Initial vol.}} = \dfrac{1}{0{\cdot}6} = 1{\cdot}667$

SOLUTIONS TO SECOND CLASS EXAMINATION QUESTIONS 491

$$\log_\varepsilon 1\cdot 667 = 0\cdot 511$$
(from hyperbolic log tables)

Admission area of diagram $= 72 \times 0\cdot 6 = 43\cdot 2$
Expansion area of diagram $= PV \log_\varepsilon r$
$= 72 \times 0\cdot 6 \times 0\cdot 511$
$= 22\cdot 08$

Gross area $= 43\cdot 2 + 22\cdot 08 = 65\cdot 28$
Mean height $=$ area $\div$ length
$\therefore$ Mean gross press. $= 65\cdot 28 \div 1 = 65\cdot 28$ lb/in^2
Mean effective press. $=$ Mean gross press. $-$ back press.
$= 65\cdot 28 - 24$
$= 41\cdot 28$ lb/in^2. Ans.

104. Change in dia. $= K \times$ original dia. $\times$ change in temp.
$0\cdot 021 = 0\cdot 0000105 \times D \times$ temp. change

$$\therefore \text{temp. change} = \frac{0\cdot 021}{0\cdot 0000105 \times D}$$

Heat supplied $=$ Weight $\times$ sp. ht. $\times$ change in temp.

$$1098 = \tfrac{\pi}{6} D^3 \times 0\cdot 303 \times 0\cdot 094 \times \frac{0\cdot 021}{0\cdot 0000105 \times D}$$

$$\therefore D = \sqrt{\frac{1098 \times 6 \times 0\cdot 0000105}{\pi \times 0\cdot 303 \times 0\cdot 094 \times 0\cdot 021}}$$

$= 6\cdot 068$ in. Ans.

105. Combined stroke $= 1356 + 954$
$= 2310$ mm $= 2\cdot 31$ metres

Cylinder dia. $= 600$ mm $= 60$ cm

$$\text{i.h.p.} = \frac{p\text{ALN}}{4560} \times 4$$

$$= \frac{7\cdot 24 \times 0\cdot 7854 \times 60^2 \times 2\cdot 31 \times 85}{4560} \times 4$$

$= 3526$
b.h.p. $= 0\cdot 86 \times 3526 = 3032$. Ans.

106. Latent heat is the heat given to or taken from a substance which causes a change of physical state while its temperature remains unchanged.

In order that the water in the vessel shall attain a temperature of 368·4°F it must be subjected to a pressure of 170 lb/in² abs. It is assumed that this pressure is due to air contained in the vessel and that no steam is formed until the escape valve is opened; this reduces the pressure to atmospheric pressure and the temperature falls to 212°F causing some water to evaporate owing to release of sensible heat.

From steam tables,
P = 170 lb/in², t_F = 368·4°F, h_f = 341·2 Btu/lb
P = 14·7 lb/in², t_F = 212°F, h_f = 180·1 Btu/lb

Let h_{fg} = latent heat at atmospheric press. in Btu/lb

Total heat before release = Total heat after
$1 \times 341·2 = (1 - \frac{1}{6}) \times 180·1 + \frac{1}{6} \times (180·1 + h_{fg})$
$341·2 = 150·1 + 30 + \frac{1}{6} h_{fg}$
$161·1 = \frac{1}{6} h_{fg}$
$h_{fg} = 966·6$ Btu/lb. Ans.

107. $P_1 V_1^{1·3} = P_2 V_2^{1·3}$
$P_1 \times 15^{1·3} = P_2 \times 1^{1·3}$
$P_1 \times 33·8 = P_2$

$$\frac{P_1 V_1}{T_1} = \frac{P_2 V_2}{T_2} \therefore \frac{P_1 \times 15}{(140 + 460)} = \frac{P_1 \times 33·8 \times 1}{T_2}$$

$T_2 = \dfrac{600 \times 33·8}{15} = 1352°F$ abs.

∴ Final temperature = 1352 − 460 = 892°F. Ans.

108. From steam tables,
P = 100 lb/in², h_f = 298·5, h_{fg} = 889·7 Btu/lb

Original weight of water in boiler = 4480 lb

Let w = lb weight of steam blown in

Heat lost by steam = Heat gained by water
$w [(298·5 + 0·9 \times 889·7) − (120 − 32)] = 4480 (120 − 60)$
$w \times 1011 = 4480 \times 60$
$w = 265·9$ lb

Final weight of water in boiler
= 4480 + 265·9 = 4746 lb

SOLUTIONS TO SECOND CLASS EXAMINATION QUESTIONS 493

Original volume of water in boiler
$$= \frac{4480}{62 \cdot 45} = 71 \cdot 74 \text{ ft}^3$$

Final volume of water in boiler
$$= \frac{4746}{61 \cdot 4} = 77 \cdot 29 \text{ ft}^3$$

Increase in volume $= 77 \cdot 29 - 71 \cdot 74 = 5 \cdot 55$ ft^3

$$\text{Increase in height} = \frac{\text{Increase in volume}}{\text{Area of water level}}$$

$$= \frac{5 \cdot 55}{0 \cdot 7854 \times 6^2} \times 12 \text{ in.}$$

$$= 2 \cdot 356 \text{ in. Ans.}$$

109. From superheat steam tables,
P = 400 lb/in^2 supht. 320F°, $\qquad h = 1397 \cdot 5$ Btu/lb

From saturation steam tables,
P = 400 lb/in^2, $\quad h_g = 1205 \cdot 5$ Btu/lb, $\quad v_g = 1 \cdot 161$ ft^3/lb

Let $w =$ weight of injection water per lb of superheated steam
Heat before mixing = Heat after mixing
$1397 \cdot 5 + w(100 - 32) = (1 + w) \times 1205 \cdot 5$
$\qquad 1397 \cdot 5 + 68w = 1205 \cdot 5 + 1205 \cdot 5w$
$\qquad\qquad\quad 1137 \cdot 5w = 192$
$\qquad\qquad\qquad\quad w = 0 \cdot 1688$ lb. Ans. (i)

Volume of one lb of superheated steam

$$= \frac{1 \cdot 248 (1397 \cdot 5 - 835 \cdot 2)}{400} + 0 \cdot 0123 = 1 \cdot 766 \text{ ft}^3$$

Volume of $(1 + 0 \cdot 1688)$ lb of dry saturated steam
$= 1 \cdot 1688 \times 1 \cdot 161 = 1 \cdot 357$ ft^3

% change of volume
$$= \frac{1 \cdot 766 - 1 \cdot 357}{1 \cdot 766} \times 100 = 23 \cdot 16 \%. \text{ Ans. (ii)}$$

110.
$$\text{i.h.p.} = \frac{p\text{ALN}}{4560}$$

$$= \frac{p \times 0.7854 \times 74^2 \times 1.15 \times \text{r.p.m.}}{4560 \times 2} \times 8$$

$$= 4.338 \times p \times \text{r.p.m.}$$

∴ For this engine, constant is 4·338. Ans. (i)

$$\text{i.h.p.} = 4.338 \times 5.82 \times 115.1$$
$$= 2906. \text{ Ans. (ii)}$$

111.
$$\text{Area in sq. ft} = \frac{\text{Quantity in cu ft per sec}}{\text{velocity in ft per sec}}$$

$$\text{At entrance: Area} = \frac{4.429 \times 2500}{3600 \times 1500} \text{ sq ft}$$

$$\therefore \text{ dia. at entrance} = \sqrt{\frac{4.429 \times 2500 \times 4}{3600 \times 1500 \times \pi}} \times 12 \text{ in.}$$

$$= 0.6131 \text{ in. Ans.}$$

$$\text{Dia. at exit} = \sqrt{\frac{118.6 \times 2500 \times 4}{3600 \times 5000 \times \pi}} \times 12 \text{ in.}$$

$$= 1.738 \text{ in. Ans.}$$

112. Since it is a 2-stroke engine, the speed of the cam shaft is the same as that of the crank shaft, and the position of opening and closing of the valves relative to the crank shaft must be the same as for the cam shaft.

$$\text{Centre-line of ahead fuel cam} = \frac{5 + 48}{2} \text{ from } 5° \text{ before T.D.C.}$$

$$= 26\tfrac{1}{2}° \text{ from } 5° \text{ before T.D.C.}$$

$$= 21\tfrac{1}{2}° \text{ from T.D.C.}$$

SOLUTIONS TO SECOND CLASS EXAMINATION QUESTIONS 495

Centre-line of astern fuel cam $= \dfrac{5 + 38}{2}$ from 5° before T.D.C.

$= 21\tfrac{1}{2}°$ from 5° before T.D.C.

$= 16\tfrac{1}{2}°$ from T.D.C.

∴ Angle between centre-lines of ahead and astern fuel cams
$= 21\tfrac{1}{2} + 16\tfrac{1}{2} = 38$ degrees. Ans. (a)

Angle between cranks of a 2-stroke 6-cylinder engine

$$= \dfrac{360°}{\text{No. of cylinders}} = \dfrac{360}{6} = 60°$$

∴ Starting air is open for $60° + 10° = 70°$

Angle between centre-lines of ahead and astern starting air cams $= (\tfrac{70}{2} + 8) \times 2 = 43 \times 2 = 86$ degrees. Ans. (b)

113. Mech. Effic. $= \dfrac{\text{b.h.p.}}{\text{i.h.p.}}$ ∴ i.h.p. $= \dfrac{3000}{0\cdot78}$

i.h.p. $= \dfrac{p\text{ALN}}{33000}$

but, L × N = mean piston speed in ft per min.

∴ Mean piston speed $= \dfrac{3000 \times 33000 \times 4}{0\cdot78 \times 35 \times \pi \times 88^2}$

$= 596$ ft per min. Ans.

114. Working on one lb of fuel burned:
Weight of air passing through preheater = 18 lb
Weight of funnel gases passing through preheater
$= 18 + 1 = 19$ lb

Heat gained by air = Heat lost by gases

$18 \times 0\cdot24 \times$ temp. rise $= 19 \times 0\cdot25 \times (710 - 530)$

$$\therefore \text{Temp. rise of air} = \frac{19 \times 0{\cdot}25 \times 180}{18 \times 0{\cdot}24} = 198°F$$

$\therefore$ Temp. of air leaving preheater $= 198 + 80 = 278°F$.
Ans. (a)

$$\text{Heat given to air} = 18 \times 0{\cdot}24 \times 198$$

$$\therefore \% \text{ saving of heat} = \frac{18 \times 0{\cdot}24 \times 198}{14500} \times 100$$

$$= 5{\cdot}897\%. \text{ Ans. (b)}$$

115. Let $x =$ length of vacuum space above mercury level

$P_1 \times V_1 = P_2 \times V_2$ where:

$P_1 =$ pressure of air before admission to tube $= 30$ in. of mercury
$V_1 =$ volume „   „   „   „   „   „ $= \frac{1}{3}$ cu in.
$P_2 =$ pressure „   „   „   after „   „   „ $= 5$ in. of mercury
$V_2 =$ volume „   „   „   „   „   „ $= \frac{1}{3}(5+x)$ cu in.

$$\therefore 30 \times \tfrac{1}{3} = 5 \times \tfrac{1}{3}(5+x)$$
$$\tfrac{30}{5} = 5 + x$$
$$x = 1 \text{ in.}$$

$\therefore$ Initial volume of vacuum space $= 1 \times \tfrac{1}{3} = \tfrac{1}{3}$ cu in. Ans.

116. Weight $\times$ specific heat $=$ water equivalent
Let water equivalent of liquid A $=$ A
„ „ „ „ B $=$ B
„ „ „ „ C $=$ C

Mixing of A and B:
Heat gained by A $=$ Heat lost by B
A $\times (75 - 40) =$ B $\times (100 - 75)$
$\tfrac{35}{25} \times$ A $=$ B

Mixing of B and C:
Heat gained by B $=$ Heat lost by C
B $\times (130 - 100) =$ C $\times (150 - 130)$
B $\times 30 =$ C $\times 20$
$\tfrac{35}{25} \times$ A $\times 30 =$ C $\times 20$

$$\frac{35 \times 30}{25 \times 20} A = C$$

$$2 \cdot 1 \, A = C$$

Mixing of A and C:
Heat gained by A = Heat lost by C
$$A \times (T - 40) = C \times (150 - T)$$
$$A \times (T - 40) = 2 \cdot 1 \times A \times (150 - T)$$
$$T - 40 = 315 - 2 \cdot 1 \, T$$
$$3 \cdot 1 \, T = 355$$
$$T = 114 \cdot 5°F. \text{ Ans.}$$

117. Assuming atmospheric pressure = 15 lb/in^2
Admission pressure = 210 + 15 = 225 lb/in^2 abs.
Let stroke volume = 1
Volume at beginning of expansion = 0·34 + 0·067 = 0·407
Volume at end of expansion = 1 + 0·067 = 1·067

$$\therefore \text{Ratio of expansion} = \frac{1 \cdot 067}{0 \cdot 407} = 2 \cdot 621 \quad \text{Ans. (b)}$$

$$\text{Terminal pressure} = \frac{225}{2 \cdot 621} = 85 \cdot 84 \text{ lb/in}^2 \text{ abs.}$$

$$= 70 \cdot 84 \text{ lb/in}^2 \text{ gauge. Ans. (a)}$$

118. Steam lap + lead = 1·95 + 0·15 = 2·1 in.

$$\text{Sine angle of advance} = \frac{\text{steam lap + lead}}{\frac{1}{2} \text{ travel}}$$

$$= \frac{2 \cdot 1}{3\frac{9}{16}} = \frac{2 \cdot 1 \times 16}{57} = 0 \cdot 5893$$

$\therefore$ angle of advance = 36° 6′

When crank is 45° past its dead centre, angular displacement of eccentric from mid-position point = 36° 6′ + 45° = 81° 6′.

Displacement of valve from mid-position = $3\frac{9}{16} \times \sin 81° 6′$
= 3·521 in.

Exhaust opening = Valve displacement − exhaust lap.
When crank is 45° past T.D.C., opening to exhaust at bottom of
valve = 3·521 − 0·125
= 3·396 in. Ans. (a)

When crank is 45° past B.D.C., opening to exhaust at top of
valve = 3·521 − (−0·125)
= 3·646 in. Ans. (b)

119. Work done in compressor increases the heat content in the gas by 15 Btu/lb, therefore heat in refrigerant leaving the evaporator (and entering compressor) is 15 Btu/lb less than leaving compressor and entering condenser.

$$= 276 - 15 = 261 \text{ Btu/lb}$$

There is no gain or loss of heat in the CO_2 on being throttled through the expansion valve (regulator), therefore the heat in the CO_2 at entrance to the evaporator is the same as on leaving the condenser, = 206 Btu/lb.

As the CO_2 passes through the evaporator there is a change in dryness fraction only, therefore there is no change in temperature and consequently no change in sensible heat. The increase of (261 − 206) = 55 Btu/lb is latent heat causing evaporation of some of the CO_2.

Let h_{fg} = latent heat in Btu to evaporate one lb,

Heat absorbed in evaporator = latent heat received
$$55 = (q_2 - q_1) \times h_{fg}$$
$$55 = (0.76 - 0.37) \times h_{fg}$$
$$55 = 0.39 \, h_{fg}$$
$$h_{fg} = 141 \text{ Btu/lb. Ans. (i)}$$

Heat taken from water to make ice
= 55 × 90 × 60 × 24 Btu/day

Weight of ice formed from and at 32°F

$$= \frac{55 \times 90 \times 60 \times 24}{144 \times 2240}$$

= 22·1 tons per day. Ans. (ii)

SOLUTIONS TO SECOND CLASS EXAMINATION QUESTIONS

120.   Entropy after expansion = Entropy before expansion

$$\log_\varepsilon \frac{319\cdot 64}{273} + \frac{q \times 571\cdot 16}{319\cdot 64} = \log_\varepsilon \frac{437\cdot 6}{273} + \frac{494\cdot 3}{437\cdot 6}$$

$$\frac{q \times 571\cdot 16}{319\cdot 64} = \log_\varepsilon \frac{437\cdot 6}{273} - \log_\varepsilon \frac{319\cdot 64}{273} + \frac{494\cdot 3}{437\cdot 6}$$

$$\frac{q \times 571\cdot 16}{319\cdot 64} = \log_\varepsilon \frac{437\cdot 6}{319\cdot 64} + \frac{494\cdot 3}{437\cdot 6}$$

$1\cdot 787\, q = 0\cdot 3142 + 1\cdot 129$
$1\cdot 787\, q = 1\cdot 4432$
$ q = 0\cdot 8076$ Ans.

SELECTION OF EXAMINATION QUESTIONS
FIRST CLASS

1. During a Morse test on a 4-cylinder I.C. engine, the speed was kept constant at 1200 r.p.m. With all cylinders working, it developed 26·2 b.h.p. and the specific fuel consumption was 0·55 lb of fuel per b.h.p. hour. With each cylinder cut out in turn the average torque developed by the three remaining cylinders was 80 lb ft. Calculate the i.h.p. of the engine and the indicated thermal efficiency, taking the calorific value of the fuel as 18500 Btu/lb.

2. 18 kilograms of ice at — 5°C are mixed with 19·5 kilograms of water at 89°C. Determine the resulting temperature in degrees F. What weight of water (in pounds) must be taken from the mixture, heated to boiling point and then poured back to cause a rise in temperature of 10 F degrees?

3. Steam is supplied to the H.P. cylinder of a double-acting steam engine at 190 lb/in^2 gauge and cut off at 0·5 stroke, the back pressure being 68 lb/in^2 gauge. The diameter of the cylinder is 26 inches, stroke 4 ft 6 in. and the clearance is equal to one twentieth of the stroke volume. Find (a) the hypothetical mean effective pressure, (b) the actual mean effective pressure assuming a diagram factor of 0·7, and (c) the indicated horse power when running at 75 r.p.m.

4. A mixture of gas and air was compressed in a gas engine cylinder, the stroke being 18 inches and the clearance equal to 3·4 inches. The mixture was taken in at atmospheric pressure, and at the end of compression the pressure was 161 lb/in^2 gauge. Find the law connecting pressure and volume, also the pressure when the piston had completed 12 inches of its stroke. Assume atmospheric pressure = 15 lb/in^2.

5. Owing to a leaky condenser the feed water density is 3·5 oz per gallon. The boilers are kept at 18 oz per gallon by blowing down. If the steam pressure is 195 lb/in^2 abs., the feed temperature is 160°F, and the coal consumption was 20 tons per day before the leakage started, what is the consumption now?

FIRST CLASS EXAMINATION QUESTIONS

6. A Diesel engine cylinder is 650 mm diameter and the stroke is 1200 mm, the clearance being equal to 10% of the stroke volume. The law of compression is $PV^{1\cdot 4}$ = constant and the final compression pressure is 528 lb/in² abs. If the depth of the clearance is decreased by 1 mm, find the alteration in the final compression pressure.

7. Steam is supplied to the H.P. cylinder of a steam engine at 225 lb/in² abs., cut off takes place at 0·55 of the stroke and it is to be assumed that the steam expands to the end of the stroke. The clearance volume is equal to 3·7% of the stroke volume and the back pressure is 84 lb/in² abs. Compression begins at 0·95 of the exhaust stroke. If the steam expands and is compressed according to the law $PV = k$ (area under curve $= PV \log_e r$), sketch and dimension the theoretical indicator diagram and find (a) the pressure at end of compression, (b) the mean effective pressure by the method of algebraic addition of areas.

8. The pressure of the steam supplied to a quadruple expansion engine is 225 lb/in² abs., the back pressure on the L.P. is 2 lb/in² abs., and the ratio of expansion throughout the engine is 14. The cylinder volume ratios are, $1 : 9^{\frac{1}{3}} : 9^{\frac{2}{3}} : 9$, and the cut-off points in the cylinders are, 0·62 in the 1st I.P., 0·63 in the 2nd I.P. and 0·75 in the L.P. Find the cut-off fraction in the H.P. and the initial steam pressure to each cylinder.

9. Explain the meaning of the term 'Equivalent Evaporation' and state why it is evaluated on boiler trials. Two boilers, A and B, undergo tests. A generates 9·5 lb of dry saturated steam at 250 lb/in² abs. per lb of fuel consumed, from feed water at 210°F. B generates 8·9 lb of dry saturated steam at 215 lb/in² abs. per lb of fuel, from feed water at 160°F. The thermal efficiency of B is 72% and the calorific value of the fuel used in this boiler is 10% higher than that of the fuel used in boiler A. Find (i) the calorific value of the fuel used in A, (ii) the thermal efficiency of A, and (iii) the equivalent evaporation of each boiler.

10. In a reaction turbine the steam leaves the guide blades at a velocity of 250 ft/sec, the exit angle being 20°. The inlet angle of the moving blades is 38°. Find the mean blade speed. If the blades are 6 in. long and the turbine rotates at 500 r.p.m., find the rotor drum diameter.

11. The ratio of the cylinder volumes of a triple expansion engine are 1 : 2·35 : 6·75. The diameter of the H.P. is 33·5 in., stroke 4 ft 6 in. and the mean effective pressures, when the engine runs at 75 r.p.m., are 78·8, 39 and 13·7 lb/in² respectively. Find (a)

the mean pressure all referred to the L.P., (b) the total i.h.p. developed, (c) the percentage powers developed in the cylinders compared to the H.P.

12. A triple expansion engine has cylinders 26 in., 42 in. and 71 in. diameter. The mean effective pressures are 73, 27 and 10·5 lb/in^2 respectively. The revolutions are 72 per minute. Find the new speed if the vacuum drops back 4 inches.

Note.—Mean effective pressure varies directly as (r.p.m.)^2.

13. Superheated steam at a pressure of 220 lb/in^2 abs. and at 700°F is expanded to 20 lb/in^2 abs. Calculate the dryness fraction of the steam after expansion, assuming that the expansion is isentropic and using the following formula for the specific entropy of steam.

$$s = \log_\varepsilon \frac{T_R}{492} + \frac{q h_{fg}}{T_R} + C_P \log_\varepsilon \frac{T}{T_R}$$

where, s = entropy per lb of steam
T_R = saturation temp. of the steam, in °F abs.
h_{fg} = latent heat, Btu/lb
T = temperature of superheated steam, in °F abs.
C_P = mean specific heat of superheated steam, in this case to be taken as 0·6.

14. Gas is compressed in an I.C. engine according to the law $PV^{1\cdot35}$ = constant. The initial and final temperatures of the gas are 100°F and 520°F. Find the compression ratio.

15. Steam at 600 lb/in^2 abs. and 0·98 dry, passes through a superheater, and leaves at 580 lb/in^2 abs., at the temperature of 800°F. Find (a) the heat supplied per lb in the superheater, (b) the final volume per pound.

16. Dry saturated steam enters a turbine nozzle at a pressure of 125 lb/in^2 abs. and the pressures at the throat and exit are 72 and 2 lb/in^2 abs. respectively. The heat drop in the steam from entrance to throat is 46 Btu/lb, and from entrance to exit it is 272 Btu/lb. Assuming 8% of the heat drop is lost to friction in the divergent part of the nozzle, calculate the area of the nozzle at the throat and at exit to pass 50 lb weight of steam per minute.

17. An engine develops 1500 i.h.p. The steam consumption is 15 lb per i.h.p. per hour. The condenser is leaking and the hotwell density is 0·08 of the sea density. The boilers are kept at 3·5 times the sea density by blowing. How many tons are blown out per day, how many tons leak into the condenser per day, and

how many tons flow into the bilges? Assume that 10 tons of 'make-up' feed are required per day, this being supplied from the leakage that is taking place in the condenser.

18. A pump pumps oil of specific gravity 0·91 from a tank wherein the oil level is 4 ft 3 in. above the suction. The discharge is 23 ft 6 in. above the suction and the oil is delivered through a 6 in. diameter pipe at 190 feet per minute. Find the horse power to drive the pump if its efficiency is 59%.

19. The consumption of fuel in a diesel engine is 12 tons per day when developing 2500 b.h.p., the calorific value of the fuel being 18000 Btu/lb. Compare the overall efficiency of this engine with that of a steam engine of the same b.h.p. where the boiler efficiency is 75% and the engine efficiency, based on the b.h.p., is 15%.

Find, also, the coal consumption per day for the steam engine if the calorific value of the coal is 13500 Btu/lb.

20. The stroke of a steam engine is 51 inches and the clearance volume is equal to 8·9% of the stroke volume. The pressure at cut-off is 63 lb/in^2 by gauge. The valve opens to exhaust when the piston is at the end of its stroke and the release pressure is 33 lb/in^2 by gauge. If the index for the expansion curve is 1·12, find the cut-off in inches of the stroke. Assume atmospheric pressure is 15 lb/in^2.

21. A surface feed heater is supplied with steam direct from the main steam line at 180 lb/in^2 abs., 0·98 dry, the drain being led direct to the hotwell. The condensate from the main condenser is discharged into the hotwell at 110°F. If the temperature of the feed water is 210°F, calculate the percentage of the main steam used by the heater.

22. Fluids x, y and z have temperatures of 120°F, 145°F and 177°F respectively, and the weights are in the ratio 2 : 3 : 5. When x and y are mixed, the resulting temperature is 130°F. When y and z are mixed the resulting temperature is 157°F. Determine the resulting temperature when x, y and z are mixed.

23. A turbine installation was supplied with superheated steam at 320 lb/in^2 abs. and at 700°F and exhausted to the condenser at 0·5 lb/in^2 abs. 0·85 dry. After the fitting of a re-heater, steam is supplied as before and expanded to 30 lb/in^2 abs. dry and saturated before being passed through the reheater where it is

heated to 700°F; the exhaust to the condenser is now 0·5 lb/in² abs. 0·94 dry. Taking the mean specific heat of superheated steam as 0·61 at 320 lb/in² and as 0·5 at 30 lb/in², calculate the theoretical thermal efficiency of the plant with and without reheat.

24. A slide valve has a steam lap of 1¾ inches, lead ¼ inch, and the steam port opening when the crank is 90° past centre is 1·125 inches. Find the travel of the valve and the angle of advance.

25. A single acting air pump works in conjunction with a rotary bilge pump. The swept volume of the air pump is 0·004 of the volume of the main bilge pump. If the pressure in the main bilge pump is 14·7 lb/in² abs. with the suction valves closed, find the number of suction strokes required to reduce the pressure from 14·7 lb/in² to 4·9 lb/in².

26. Find (a) the total heat per lb, and (b) the total heat per cubic foot, of (i) dry saturated steam at 240 lb/in² abs. and (ii) superheated steam at 240 lb/in² abs. at 700°F. Take the specific heat of superheated steam as 0·56 and use the following formula for the specific volume of superheated steam.
$$v = 1·248\,(h - 835·2)/P + 0·0123 \text{ ft}^3/\text{lb}.$$

27. A slide valve has 6 inches travel, 1 5/16 inches steam lap and 3/16 inch lead. How far is the valve from the bottom of its travel when the crank has turned through 35° from the top centre?

28. Wet steam at 160 lb/in² abs. 0·88 dry, is mixed with steam at the same pressure having 80 F° of superheat, the ratio of the weights mixed, wet steam to superheated steam, being 3 to 2. The mixture is now passed through a reducing valve and reduced in pressure to 50 lb/in² abs. Assuming no heat losses, find the state of the steam (a) after mixing, (b) after throttling.

29. A calorimeter weighing 223 grams and of specific heat 0·1, contains 600 grams of water at 15°C. A piece of brass and a piece of steel, each of the same volume and temperature of 250°C are placed in the water. The resulting temperature was 19·5°C. If no heat is lost by radiation, or other causes, find the weights of brass and steel.

30. If it requires 9 per cent. more heat units to completely dry a sample of wet steam having a temperature of 393·7°F and which has been formed from water at 200°F, what was the dryness fraction of the sample?

31. A piece of copper weighs 2 lb and has a temperature of 350°F. It is dropped into a quantity of water which has four times the

number of heat units that the copper has, both being measured from 32°F. If the resulting temperature is 84°F, find the weight and the temperature of the water.

32. Steam at 250 lb/in² abs. with 240 F degrees of superheat is expanded in turbine nozzles to 25 lb/in² abs. 0·97 dry. The blade velocity of the turbine is 1100 ft/sec and the angle of the nozzles is 20° to the direction of movement of the blades. Find the absolute velocity of the steam at entrance to the turbine blades and the inlet angle of the blades.

33. In an ideal diesel cycle, the compression ratio is 13·5 to 1 and the cut off ratio is 2·2 to 1. At the beginning of compression the pressure is 14·7 lb/in² abs. and the temperature is 80°F. Taking the specific heats at constant pressure and constant volume as 0·238 and 0·17 respectively, calculate the temperatures at the remaining cardinal points of the cycle and the ideal thermal efficiency.

34. In a two-stage compressor, one lb of air is compressed in the 1st stage from 14 lb/in² abs. and 70°F to 32·2 lb/in² abs. It is then cooled at constant pressure in the intercooler to its original temperature. In the 2nd stage, the air is compressed to 64·4 lb/in² abs. and again cooled to its original temperature in the aftercooler. Sketch the PV diagram and, assuming adiabatic compression, calculate the percentage decrease in volume due to cooling at the end of each stage. Take $R = 53·3$ ft lb per lb per °F, and $\gamma = 1·4$.

35. Prove the following relationship for a perfect gas between the specific heat at constant pressure, represented by C_P, and the specific heat at constant volume, represented by C_V; R being the gas constant for one lb and J being Joule's mechanical equivalent of heat,

$$C_P - C_V = \frac{R}{J}$$

36. A sample of the fuel burned in the boilers contained 81% carbon, 5·6% hydrogen, 5·7% oxygen, remainder ash, etc., and the air supplied to the furnaces was 50% in excess of the theoretical weight required. The temperature of the air was 80°F on entering the air heater and 230°F on leaving. The flue gas temperature was 800°F before entering the heater. Taking the specific heat of air as 0·24, specific heat of the gas as 0·25, and assuming perfect heat transfer, estimate the temperature of the flue gases leaving the heater.

37. Define Latent Heat. Steam at 220 lb/in² abs. and 0·93 dry was blown into a tank of water and caused the weight of water in the tank to be increased by one twenty-fifth. The temperature of the condensate and water mixed was 180°F. Find the initial temperature of the water in °C.

38. An engine piston valve with outside steam admission has a steam lap and lead of 2 and 0·25 inches respectively. When the crank is 80° past centre, the port opening to steam is 1·5 inches. Find the angle of advance and the valve travel.

39. A triple expansion engine has an H.P. cylinder 26 inches diameter. The stroke is 4 feet, the connecting rod length is 100 inches and the shaft 13 inches diameter. The boiler pressure is 180 lb/in² by gauge, and cut-off takes place in the H.P. cylinder at 0·6 of the stroke. Find the number of times the steam is expanded. Diameter of the shaft for a triple expansion engine with three cranks at 120° is given by:

$$\sqrt[3]{\frac{C \times P \times D^2}{f(2 + \frac{D^2}{d^2})}}$$

Where C = length of crank in inches.
P = boiler pressure (absolute).
D = diameter of L.P. cylinder.
d = diameter of H.P. cylinder.
f = 1110

40. When the crank of a reciprocating engine has passed through 45° from the bottom centre the effective load on the piston was 47 tons. The length of the connecting rod is 112 inches and the stroke is 4 feet 4 inches. If the shaft is 13 inches diameter, find the torsional stress.

41. In an engine working on the diesel cycle, the compression ratio is 13·5 to 1. Fuel is admitted for $\frac{1}{16}$ of the stroke and exhaust commences at $\frac{7}{8}$ of the stroke. The temperature of the air at the beginning of compression is 100°F and compression and expansion follows the law $PV^n = k$, where $n = 1·35$. Calculate the temperatures at the following points:
  (i) at the end of compression,
  (ii) at the end of fuel supply,
  (iii) at opening to exhaust.

FIRST CLASS EXAMINATION QUESTIONS 507

42. Water is heated in a sealed vessel to 350·2°F. An escape valve is now opened and the temperature falls to 212°F. Some of the water evaporates and is allowed to escape. If the experiment is repeated without adding more water to the vessel, to what temperature must the water be heated so that the same weight of water evaporates when the temperature again falls to 212°F?

43. A slide valve has $\frac{1}{8}$ inch lead, $1\frac{7}{16}$ inches steam lap and 2 inches maximum port opening to steam. If the exhaust lap is $-\frac{1}{8}$ inch, find the angle of advance, and the fraction of the stroke when release occurs. The connecting rod is $4\frac{1}{2}$ cranks long.

44. Find the indicated horse power and the pounds of fuel per i.h.p. per hour, of a 4-stroke single-acting diesel engine which uses 3·6 tons of fuel per day. It has 8 cylinders 508 mm bore and 974 mm stroke. The revolutions are 85 per minute and the mean effective pressure from the cards is 6·35 kilograms per square centimetre.

45. The following readings were taken from the intermediate cylinder of a steam reciprocating engine when the barometer stood at 736 mm; Exhaust pressure 14·5 lb/in² gauge, admission pressure 59 lb/in² gauge. The top and bottom clearances are 9% of the stroke volume. Assuming the admission of the steam to be at top dead centre and the compression to raise the pressure from exhaust up to initial pressure, find the fraction of the exhaust stroke where the slide valve closes to exhaust.

46. Owing to a leaky condenser scumming is resorted to, and the boiler density is kept at 3·7 times the sea density. The coal consumption is found to have increased by 4·3 per cent. since the scumming began. The boiler pressure is 260 lb/in² abs., and the feed temperature is 133°F. What is the feed water density?

47. Prove that the relationship between temperature and pressure during adiabatic operations is given by $\dfrac{T_1}{T_2} = \left(\dfrac{P_1}{P_2}\right)^{\frac{n-1}{n}}$ where $PV^n =$ constant.

If the temperature and pressure at the beginning of compression of the air in an air compressor were 90°F and 14·7 lb/in² abs. respectively, and at the end of compression 700°F and 250 lb/in² abs., find the value of n.

48. One wall of a cold room is 15 ft long and 7 ft 6 in. high and is constructed of $4\frac{1}{2}$ inch brick at the outside, 3 inches of cork, and 1 inch wood on the inside. Taking the coefficients of thermal conductivity of brick, cork and wood respectively as 6·5, 0·3 and

1·2 Btu per hour, per sq ft of area, per inch of thickness, per degree Fahrenheit, estimate the heat leakage per 24 hours through this wall and also the interface temperatures, when the outside temperature is 65°F and inside temperature 28°F.

49. A refrigerating machine is required to produce 300 lb of ice per hour at 22°F from water at 67°F. The refrigerant enters the evaporator 0·5 dry and leaves 0·95 dry, the pressure being 218 lb/in² and temperature — 5°F. The compressor, which is single-acting, runs at 215 r.p.m. and the stroke/bore ratio is 2 to 1. Determine the dimensions of the compressor cylinder.

    Note, Latent heat of the refrigerant at — 5°F is 130 Btu/lb and the specific volume of the gas at 218 lb/in² is 0·35 ft³/lb.

50. A mixture of air and gas is compressed in the cylinder of a gas engine whose stroke is 21 inches. The initial pressure is 15 lb/in² abs., and the pressure at the end of the stroke is 150 lb/in² abs. If the compression follows the law $PV^{1·35}$ = constant, express the clearance in inches of the stroke.

51. A diesel engine is using 0·43 lb of oil per b.h.p. per hour. The calorific value of the oil is 20,000 Btu/lb and the mechanical efficiency of the engine is 80 per cent. If the heat units passing away in the exhaust gases represent 30 per cent. of those in the fuel consumed, what is the percentage loss in the cooling water?

52. An engine working on the constant volume cycle takes in air at 90°F; the compression ratio is 6 to 1 and the temperature at the end of the heat supply at constant volume is 3000°F. Assuming compression and expansion to be adiabatic, i.e., the index of the law is 1·4, find (a) the temperature at the end of compression, (b) the temperature at the end of expansion, (c) the theoretical thermal efficiency of the cycle.

53. A boiler, working at 200 lb/in² has water gauges connected to shut off cocks near the top and bottom of the boiler. The temperature of the water in the boiler is 383·5°F, and in the gauge pipe 212°F. If the water level in the gauge is 7 feet above the bottom cock, what is the difference between the level of the water in the boiler and in the gauge?

    The relationship between temperature and volume of water is expressed by $V = 1 + 0·0000119\ (t° — 4)^{1·8}$ where V is the volume of that water, whose volume at 4°C is unity.

54. The clearance of a gas engine cylinder is 6000 cc. The diameter of the cylinder is 265 mm and the stroke 530 mm. If gas and air is drawn in at atmospheric pressure (15 lb/in² abs.),

FIRST CLASS EXAMINATION QUESTIONS 509

find the pressure at the end of the compression stroke. The law of the compression is $PV^{1\cdot 38} = $ constant.

55. The indicated horse power of a diesel engine is 5660 and the brake horse power is 4550. The engine uses 21·5 tons of fuel per day. If 30 per cent. of the heat in the fuel is lost in the exhaust gases and 32 per cent. is lost in the cooling water, calculate the mechancial efficiency of the engine, its overall efficiency and the calorific value of the fuel.

56. An engine of 1500 i.h.p. uses 14 lb of steam per i.h.p. per hour. The leakage of steam at the glands, etc., is 2 per cent. of the steam supplied to the engine. The condenser is defective and some sea water enters the hotwell causing the feed density to be 0·15 of the sea density. Find how much water overflows from the hotwell per day, upon the assumption that the boiler is not blown down and that a constant water level is maintained.

57. The throw of an eccentric is $3\frac{9}{16}$ inches and its angle of advance is 44° 44'. The valve has $\frac{7}{16}$ inch lead. Calculate the steam lap of the valve, its maximum port opening to steam, and the fraction of the stroke at which cut off takes place.

58. A vertical boiler has an internal diameter of 10 feet and the diameter of the internal uptake is 3 feet 3 inches. It contains 9 tons of fresh water and the water level is 3 inches above the bottom of the gauge glass when the temperature is 55°F. What will be the change in water level by the time steam has been raised and the water temperature is 375°F? Make use of the formula:

$$\text{Volume of water} = \tfrac{1}{2}\left[\frac{T + 460}{500} + \frac{500}{T + 460}\right]$$

where the vol. of water at its max. density is unity.

59. The stroke of an engine is $4\frac{1}{2}$ feet and the length of the connecting rod is 2·2 times the stroke. Cut off of the steam occurs at 0·6 stroke. If the maximum twisting moment is 800000 lb ft, find the load on the piston and the position of the piston when maximum twisting moment occurs.

60. A slide valve has $\frac{1}{8}$ inch lead, $1\frac{7}{16}$ inch steam lap and 2 inches maximum port opening to steam. Find the travel, and if the valve has $\frac{3}{16}$ inch negative exhaust lap, at what fractions of the stroke does the valve open and close to exhaust?

61. An air compressor takes in air at atmospheric pressure and compresses it to 25 atmospheres. Find the volume of 5 lb of this compressed air, and also its temperature, if the engine room temperature is 70°F. Take the law of compression to be $PV^{1\cdot4}$ = constant, and that 1 cu ft of air at 32°F and at 14·7 lb/in² abs. pressure weighs 0·0807 lb.

62. A shaft has 8 bearings and the load on each is 29·4 tons. Its diameter is 345 mm. When running at 85 r.p.m. the coefficient of friction is 0·014. Find the horse power expended in overcoming friction, and also express this in Chu per minute.

63. Before wet steam at 240 lb/in² abs. is passed to the superheaters, a sample is drawn off and tested in a throttling calorimeter. The temperature of the steam in the calorimeter was 260°F and pressure 6·8 inches of mercury gauge when the atmospheric pressure was 29·9 inches of mercury. Find (a) the dryness fraction of the steam entering the superheaters, (b) the heat required per lb of steam in superheating if the wet steam at 240 lb/in² is superheated to 600°F, and (c) the percentage increase in volume of the steam after superheating.

Take specific heat of superheated steam as 0·55 and assume the volume of superheated steam to vary as its absolute temperature.

64. A two-stroke engine of 3,600 i.h.p. uses 0·37 lb of fuel per i.h.p. per hour, and was fitted with an exhaust gas boiler. The data obtained from the boiler were: Exhaust gas inlet 620°F. Exhaust gas outlet 420°F. Steam temperature 389·9°F, and dryness fraction 0·96. Feed temperature 120°F. Thermal efficiency of boiler 0·8. The engine used 18 lb of air per lb of fuel burnt in the cylinders. Calculate the weight of steam formed by the boiler per hour. Assume the mean specific heat of the gases to be 0·203.

65. The output of a turbo-generator is 7500 kilowatts. The electrical efficiency is 95 per cent. neglecting all other losses. 35000 cubic feet of air at 70°F per minute are blown through the windings at a pressure of 6 inches of water. The windings are kept at 55°C. What is the difference in temperature between the windings and the final temperature of the air? The weight of 1 cubic foot of air at 32°F and at a pressure equal to a head of 34 feet of water is 0·0807 lb. The specific heat of air at constant pressure is 0·24.

FIRST CLASS EXAMINATION QUESTIONS 511

66. A six cylinder two-stroke diesel engine developing 3600 i.h.p. at 85 revolutions per minute, burns 0·35 lb of fuel per i.h.p. per hour. At each stroke 10 cubic feet of air at 2 lb/in^2 gauge and at 85°F enter the cylinder. An analysis of the fuel gave 84 per cent. carbon, 15 per cent. hydrogen and 1 per cent. incombustible matter. The atomic weight of carbon is 12, oxygen 16, and hydrogen 1, and the air contains 23 per cent. by weight of oxygen. One cubic foot of air at atmospheric pressure and at 32°F weighs 0·0807 lb. Find: (a) The theoretical weight of air required per cycle. (b) The actual weight of air supplied per cycle. (c) The weight of excess air per cycle.

67. A three stage single-acting tandem air compressor has cylinders 3 inches, 13$\frac{1}{2}$ inches and 15 inches diameter. It is connected to 3 air storage bottles of equal size which have internal diameters of 12 inches and are 8 feet long overall with hemispherical ends. How long will it take to lift the relief valves on the bottles, which are set to lift at 1200 lb/in^2 gauge, starting from atmospheric pressure? The stroke of the compressor is 12 inches, the revolutions 150 per minute, and the volumetric efficiency is 0·91. The volume of the pipes, etc., connecting the bottles to the compressor is equal to 10 per cent. of the volume of the bottles.

68. A twin screw diesel driven ship has 4 air reservoirs having a total capacity of 1000 cubic feet. The compressor is motor driven and has cylinders 3 inches, 6$\frac{1}{2}$ inches and 15 inches diameter, a stroke of 15 inches, and the volumetric efficiency is 0·9. The compressor runs at 180 r.p.m. The compressor supplies air to blast bottles at 900 lb/in^2 abs. and a leak off from this line supplies the air reservoirs at 350 lb/in^2 abs. While manoeuvring, the pressure in the air reservoirs falls to 250 lb/in^2 abs. The compressor is working continuously, and the engine takes 50 per cent. of the air delivered. How long will it take to bring the pressure in the air reservoirs back to 350 lb/in^2 abs?

69. A slide valve has a travel of 7 inches. The bottom lead is $\frac{1}{4}$ inch and the angle of advance of the eccentric sheave is 37° 22′. Assuming the valve to have outside steam admission, find the angle turned through by the crank from its bottom centre when the steam port is open one inch.

70. A steam turbine was originally supplied with steam at 235 lb/in^2 abs. and having a dryness fraction of 0·94. The steam was expanded down to 2·4 lb/in^2 abs. when its dryness fraction was 0·755. A vacuum augmentor is now fitted and the terminal

pressure is reduced to 1 lb/in² abs. and also the initial steam is superheated 200 F°. Taking the final dryness fraction of the steam as 0·828, find the percentage increase in power due to these improvements, if the horse power is proportional to the heat drop of the steam during its passage through the turbine.

71. 91·5 kilograms of water at 54°C and 97·5 kilograms of water at 0°C are mixed with 31·5 kilograms of ice at — 25°C. If the temperature of the mixture becomes uniform, find the resultant temperature in degrees Centigrade.

72. 3 cubic feet of air at atmospheric pressure are compressed to a gauge pressure of 450 lb/in². Find the final volume in cubic inches (a) if compressed isothermally, (b) if compressed adiabatically, following the law $PV^{1 \cdot 4}$ = constant.

73. Define boiler efficiency, and equivalent evaporation from and at 212°F. A test was carried out on a boiler with a low grade fuel and the following results were obtained:
Weight of coal burnt per hour = 160 lb.
Calorific value of coal = 10110 Btu/lb.
Weight of feed water entering boiler per hour = 816 lb.
Temperature of feed water = 145°F.
Pressure of steam = 195 lb/in² abs.
Saturation temperature of steam = 379·7°F.
Superheat temperature of steam = 720°F.
Find the boiler efficiency and the equivalent evaporation from and at 212°F.

74. A sample of coal weighing 1·34 grams is burnt in a bomb calorimeter and the water, which weighs 1634 grams, is raised in temperature from 14·3°C to 20·4°C. The water equivalent of the calorimeter was 223 grams. Find the calorific value of the coal (a) in gram calories per gram (b) in Btu /lb. Make a sketch of the calorimeter.

75. The total load on a single collar thrust block is 12 tons, the effective radius of the pads is 7·5 inches, and the speed of the shaft is 77 r.p.m. Find the quantity of oil required to pass through the block, in gallons per hour, to limit the rise in temperature of the oil to 10 F°. Take the coefficient of friction as 0·05, specific heat of oil 0·48, specific gravity of oil 0·92.

76. Calculate suitable cylinder diameters for a compound steam engine of 1200 horse power, taking the cut off in the H.P. cylinder at 0·5 stroke, cylinder area ratio 1 : 3·5, diagram factor 0·68, initial steam pressure 140 lb/in² abs., back press. 3 lb/in² abs., piston speed 600 feet per minute.

FIRST CLASS EXAMINATION QUESTIONS 513

77. 50 kilograms of dry steam at 115 lb/in² abs. pressure and at a temperature of 170°C are blown into 4000 litres of water at 45°C. Find the resultant temperature.

78. An engine of 7800 i.h.p. uses 15·3 lb of steam per i.h.p. per hour. The temperature of the exhaust is 162·3°F and its dryness is 0·81. The hotwell is 120°F. If the circulating water inlet is 55°F and outlet 84°F, and the work done in pumping the circulating water through the condenser is equivalent to pumping it to a height of 30 feet, find the horse power developed by the circulating pump.

79. The stroke of a steam engine is 3 feet 6 in. and cut off takes place when the crank has passed through 130° from its top centre. In this position the load on the piston is 28·9 tons and the load on the crank pin is 29·37 tons. Find (a) the total pressure on the guide, (b) the length of the connecting rod, and (c) the distance that the piston is from the bottom of its stroke.

80. An engine is coupled to a rope brake, and when running at 300 r.p.m. the load on the brake is 370 lb on one end and 10 lb on the other. The effective diameter of the brake is 5 feet. 85% of the heat generated by friction at the brake is carried away by a stream of cooling water which has a rise in temperature of 10 C°. Find the gallons of water supplied per hour, and the brake horse power of the engine.

81. The cylinder of a vertical steam engine is 12 in. diam., the stroke is 18 in., and the length of the connecting rod is 45 in. The weight of the moving parts is 275 lb. When the crank is horizontal the effective steam pressure on the piston is 53·2 lb/in² and the de-acceleration of the piston is 161 ft/sec². Find the effective crank effort in this position, and also the turning moment.

82. Steam at 220 lb/in² abs. is blown into water whose initial temperature is 60°F and causes the absolute temperature and weight of the water to be increased by 10% and 5% respectively. Find the dryness fraction of the steam.

83. Two copper pipes of a telemotor system are each 0·5 inch internal diameter and 250 feet long when filled with oil at a temperature of 70°F. If the temperature of the oil rises to 95°F, how much oil will be released into the replenishing tank? Neglect the amount of oil in the cylinders.

   Volumetric expansion of oil = 0·00043 per F°.
   Linear expansion of copper = 0·00001 per F°.

84. In the manufacture of lead pipes, the hot ingot is forced through a die by a ram having a force of 1400 kilograms per square centimetre. Find the rise in temperature of the lead in degrees Centigrade, due to it being forced through the die.

85. An oil engine uses 0·38 lb of fuel per b.h.p. per hour. The mechanical efficiency is 80%. Find the thermal efficiency based on the indicated horse power if the fuel is composed of 86% carbon, 12·1% hydrogen, 1·5% oxygen, and the remainder impurities.

86. State what is meant by the terms 'Lower Calorific Value' and 'Higher Calorific Value'. 0·015 lb of coal is burnt in a bomb calorimeter whose water equivalent was 3 lb. The weight of water in the calorimeter was 12 lb and its rise in temperature was 6·8 C°. Find the calorific value of this fuel and state whether your answer is the higher or lower value.

87. Find suitable cylinder diameters for a triple expansion engine to develop 2500 i.h.p. with a piston speed of 600 feet per min. The initial pressure of the steam is 200 lb/in^2 abs., and the back pressure is 3 lb/in^2 abs. Take the diagram factor as 0·65, number of expansions 12, and ratio of cylinder volumes 1 : 2·7 : 7·2.

88. A throttling calorimeter was fitted on to a steam pipe wherein the steam pressure was 120 lb/in^2 abs. The pressure of steam in the calorimeter was 15·5 lb/in^2 abs. and the temperature of the steam was 259°F. Taking the specific heat of superheated steam as 0·5, find the dryness fraction of the steam in the main steam pipe. Sketch a throttling calorimeter and explain how the dryness fraction of steam may be determined by the instrument.

89. Steam from a nozzle enters an impulse turbine at a velocity of 3200 feet per second, the angle of the nozzle to the direction of movement of the blades is 18°, and the linear velocity of the blades is 750 feet per second. Calculate the entrance angle of the blades and, if the exit angle is the same as that at entrance, find, neglecting friction, the absolute velocity of the exit steam.

90. A cold storage room is 15 feet long, 12 feet wide, and 7 feet high, and every wall is covered with 6 inches thickness of insulating material. The coefficient of thermal conductivity of the insulation is 0·0003 gram-calorie per centimetre cube per second, per °C. Find the quantity of heat to be taken away from the room every minute, in Btu to maintain the temperature inside at − 5°C, when the outside temperature is 25°C.

FIRST CLASS EXAMINATION QUESTIONS 515

91. The consumption of oil in a boiler is 1400 lb per hour. The constituents of the oil are 85% carbon, 13% hydrogen and 2% oxygen. Find (a) the theoretical amount of air required per lb of fuel for perfect combustion, and (b) if the actual amount of air supplied is 70% in excess of the theoretical quantity, find the weight of flue gases passing up the chimney every hour.

92. 8025 lb of coal having a calorific value of 12950 Btu/lb, are burnt in the boilers of a steam engine plant every hour, and 80800 lb of dry steam are generated. The pressure of the steam is 235 lb/in^2 abs. and the temperature of the hotwell water 169°F. Find the efficiency of the boilers, and, if the overall efficiency of the plant is 11%, find the thermal efficiency based on the i.h.p. if the mechanical efficiency of the engine is 80%.

93. Steam at 195 lb/in^2 gauge is supplied to the H.P. cylinder of an engine, and is cut off at 0·42 of the stroke. The clearance volume is equal to 8% of the piston swept volume, the back pressure is 58 lb/in^2 gauge and the compression pressure at the end of the exhaust stroke is 190 lb/in^2 gauge. Find (a) the fraction of the exhaust stroke where the valve closes to exhaust, (b) mean gross pressure, and (c) mean effective pressure. Assume $PV = $ constant for both curves.

94. Steam leaves the nozzles of a De Laval turbine at a velocity of 3600 feet per sec. The steam jet is at an angle of 20° to the direction of motion of the blades, and the linear velocity of the blades is 600 feet per second. If the turbine uses 1800 lb of steam per hour, find the force exerted on the blades, (a) neglecting friction, (b) if the friction loss is 12%, and (c) the horse power developed when the friction loss is 12%. Assume the blades are symmetrical, that is the entrance and exit angles are equal.

95. A refrigerating machine makes ½ ton of ice per hour at 30°F, from water at 48°F. The brine in the evaporator coils is at 16°F, and the latent heat of CO_2, at this temperature, is 105·5 Btu/lb. On entering the evaporator coils the CO_2 has a dryness fraction of 0·34 and on leaving its dryness fraction is 0·92. Estimate the pounds of CO_2 that pass through the coils per hour.

96. In a vapour compression refrigerating machine the ammonia employed is liquefied at 70°F. After passing through the regulating valve its temperature is 14°F. Find the dryness fraction of the ammonia immediately after passing the valve, upon the assumption that it was entirely liquefied at the higher temperature.   (*continued on next page*)

Latent heat of ammonia = 566 − 0·8t Btu/lb where t is the temperature in F°.
Specific heat of ammonia = 1·1.

97. Define Willans' Law. A steam engine uses 37000 lb of steam per hour when developing 2500 i.h.p., and 67000 lb per hour when the i.h.p. is 5000. Estimate the consumption of steam per hour, and per i.h.p. per hour when developing 4000 i.h.p.

98. The values of the net steam force acting on the piston of a horizontal steam engine for one stroke, obtained by measurement from the indicator diagram, and the accelerating forces found by calculation, are given in the table for various crank angles. Plot the piston effort on a base of crank angle turned through, and from the curve find the crank angle when the piston effort is zero. If the stroke is 12 in. and the connecting rod is 30 in., find the twisting moment when the connecting rod and crank are at 90°.

Crank Angle in Degrees	0	30°	60°	90°	120°	150°	180°
Net Steam Load in lb	13100	15400	14200	9800	5600	1300	−5000
Accelerating Force in lb	4600	3330	1450	−860	−2200	−2300	−3000

99. The fuel valve of a diesel engine is open for 45° total movement of the crank, and it opens at 7° before the crank reaches the top centre. If the piston moves through a total linear distance of 134 millimetres while the fuel valve is open, find the stroke of the engine in millimetres and in inches.

100. The mean speed of a flywheel of a reciprocating engine is 750 revolutions per minute. If the speed varies from 0·5% above this mean speed to 0·5% below, find the angular acceleration of the flywheel.
    Note. At top and bottom of the stroke the acceleration is nil. Assume the acceleration and the retardation to be equal and opposite.

101. 9 kilograms of ice at − 7°C and 11·5 kilograms of water at 79° C are put into a copper tank which weighs 10 kilograms and has the initial temperature of 60°C. Find the resulting temperature in Fahrenheit degrees.

102. A ship's engines indicate 9000 horse power when the speed is 12 knots, and the load on the thrust is 51 tons. The efficiency of the propeller is 90% of the mechanical efficiency of the engine. Find the propeller efficiency and the mechanical efficiency.

103. Ten cubic feet of steam at 500 lb/in² abs. and superheated 320 F° is cooled at constant volume to 300 lb/in² abs. Find (i) the weight of the steam, (ii) the dryness fraction at the reduced pressure, (iii) the total heat loss.

Use the following equation for the specific volume of superheated steam:
$$v = 1 \cdot 248 \, (h - 835 \cdot 2)/P + 0 \cdot 0123 \text{ ft}^3/\text{lb}$$

104. Find the difference in the level of the water in the gauge glass and in the boiler, when the temperature of the boiler water is 325°F, and the water in the gauge glass is 250°F. The water level in the glass is 4 feet above the bottom cock on the boiler.

Use the formula $V = 1 + \dfrac{(T - 39 \cdot 2)^2}{711 \, (697 + T)}$

V (the volume of water) is unity when the temperature is 39·2°F.

T is the temperature in Fahrenheit degrees.

105. A ship has 3 boilers, each 17 feet diameter and 13 feet long. Before lagging the boilers, the temperature of the boiler plates was 350°F, and the stokehole temperature was 100°F. After lagging the shells and end plates, the temperature of the cleading was 150°F and the stokehole 90°F. If the calorific value of the coal used is 12500 Btu/lb, find the tons of coal saved per day by lagging.

Note. $Q = K \, T^4$, where Q = quantity of heat in Btu radiated per sq foot of surface per hour. T is the absolute temperature in Fahrenheit degrees. K is a constant of value 16×10^{-10}.

106. Steam leaves the guide blades of a reaction turbine at a velocity of 440 feet per second, the exit angle being 20°. The linear velocity of the moving blades is 284 feet per second. Assuming the channel section of fixed and moving blades to be identical, and assuming ideal conditions find (a) the entrance angle of the moving blades, (b) the work done per pound of steam per second in this stage.

107. Wet steam, with 2% moisture, at 215 lb/in² abs., is passed through a reducing valve and reduced in pressure to 115 lb/in² abs. Find the condition of the reduced pressure steam. The specific heat of superheated steam is 0·48.

108. A solid metal sphere at 60°F was placed into 24 gallons of water at 200°F. When the temperature of the sphere and the water had equalised it was found that the diameter of the sphere had increased by 0·15%. The specific gravity of the metal of the sphere was 2·56; its specific heat was 0·22, and the coefficient of linear expansion was 0·0000128 per degree Fahrenheit. Find the original diameter of the sphere.

109. 1·2 tons of fuel are burnt every hour in the boiler furnaces of a ship. The analysis of the fuel is 85% carbon, 12% hydrogen, 1·5% oxygen and 1·5% impurities. Find the theoretical weight of air to burn one pound of the fuel. Actually, 50% excess air is supplied. If the stokehole temperature is 76°F, the funnel temperature 550°F, and the specific heat of the gases of combustion is 0·24, find the quantity of heat passing up the funnel every hour.

110. The exhaust steam from a turbine passes into the condenser at 1·5 lb/in² abs. and it is 13% wet. The hotwell temperature is 104°F. The circulating water enters the condenser at 55°F and leaves at 90°F, its weight per hour being 520 tons. The tubes of the condenser are ¾ inch outside diameter and 12 feet long. If 14 lb weight of steam are condensed per square foot of tube surface per hour, find the number of tubes in the condenser.

111. An internal combustion engine works on the dual combustion cycle, taking in air at 14·7 lb/in² abs. and 80°F. The compression ratio is 11, maximum pressure and maximum temperature in the cycle is 650 lb/in² abs. and 3000°F. Assuming ideal conditions and taking $C_V = 0·17$, $C_P = 0·238$, calculate the remaining pressures and temperatures at the cardinal points of the cycle, and the ideal thermal efficiency.

112. A copper calorimeter of weight 12 grams, contains 70 grams of a liquid at 10°C. When 100 grams of steel at 150°C are placed in the liquid the temperature rises to 44·5°C. If there is no heat lost, determine the specific heat of the liquid.

113. Two cubic feet of air are compressed in a compression cylinder from atmospheric pressure (15 lb/in² abs.) to 330 lb/in² by gauge. The clearance volume of the cylinder is 100 cu inches. Find the volume of air delivered per stroke if it is compressed (a) isothermally, (b) adiabatically. State which is the more economical method of compression, and give reasons.

114. Steam is supplied to an engine at a pressure of 250 lb/in² abs. and temperature 581°F, and expanded isentropically to a pressure of 1 lb/in² abs. Calculate (i) the dryness fraction of the steam at the end of expansion, (ii) the Rankine efficiency. Show by a sketch how the Rankine cycle appears on the temperature-entropy diagram.

115. A rope brake on the flywheel of an engine is kept at a uniform temperature by a stream of water amounting to 530 gallons per hour, the water entering at 20°C and leaving at 45°C. The flywheel rim is 5 feet diameter, and the engine runs at 70 r.p.m. Assuming that the water carries away 85 per cent. of the heat generated, find the effective load on the brake, and the brake horse power.

116. The fuel valve of a 2 stroke diesel engine opens at 9° before the top dead centre, and it closes when the piston has completed 9 per cent. of its downward stroke. If the piston moves through 4 inches whilst the valve is open, find (a) the length of the stroke, (b) for how many degrees of the crank circle the fuel valve is open.

117. A steam reciprocating engine was supplied with dry saturated steam at 200 lb/in² abs., and exhausted at 2 lb/in² abs. dryness fraction 0·786. Superheaters are now fitted, the steam is superheated 240 F degrees at constant pressure and the exhaust is 2 lb/in² abs. dryness fraction 0·866. Calculate the percentage increase in work done per lb of steam, and, if the cut off was at 0·625 stroke initially, at what fraction of the stroke should it now take place in order to develop the same power?

118. The output of a dynamo is 60 kilowatts at 240 r.p.m. The dynamo is driven by a 6 cylinder 4-stroke single-acting oil engine, the mechanical efficiency of which is 0·8. Find the diameter of the cylinders if the mean effective pressure is 8·5 kilograms per sq centimetre, and the stroke is 1·5 times the cylinder diameter. Assume the efficiency of the dynamo is 0·92. Give the answer in millimetres.

119. A six-cylinder 4-stroke single acting internal combustion engine has a mean effective pressure of 85 lb/in² in each cylinder. The cylinders are 650 mm diameter, the stroke is twice the cylinder diameter, and the b.h.p. developed is 1000. If the mechanical efficiency expressed as a percentage is equal to the square root of the product of the revolutions per minute and the mean effective pressure, find the revolutions per minute and the mechanical efficiency.

120. A single screw ship has a quadruple expansion steam engine which uses 1·38 lb of fuel per i.h.p. per hour. The calorific value of the fuel is 13650 Btu/lb and the boiler efficiency is known to be 73%. Find (a) the thermal efficiency of the engine, (b) the i.h.p. when the consumption is 63 tons per day, (c) the total loss of heat per lb of fuel burnt.

121. Steam leaves the nozzles of an impulse turbine at a velocity of 3826 ft/sec, the nozzles being inclined at 18° to the direction of movement of the blades. The blade speed is 347 ft/sec, and the exit angle of the blades is 24°. The steam loses 20% of its velocity due to friction in passing through the blades. Calculate the entrance angle of the blades, and the absolute velocity and direction of the steam at exit.

122. The average dimensions of an engine room are 36 feet by 56 feet and its permeability is 75%. A carbon dioxide system of fire extinguishing is fitted, which is capable of producing a saturation of 25% of CO_2 gas in the engine room up to a height of 20 feet. One cubic foot of air at 14·7 lb/in² abs. and at 32°F weighs 0·0807 lb, the relative density of CO_2 gas to air is 1·518 to 1. Taking the engine room temperature as 60°F, find the weight of CO_2 gas required. If 2 lb of liquid CO_2 have the same volume as 3 lb of water, find the volume of the bottles in which the CO_2 liquid is stored.

123. An evaporator is worked at a constant density of $2\frac{1}{2}$ thirty-seconds. The water supplied to the evaporator is at 70°F, and the vapour pressure is 18 lb/in² abs. The steam to the coils is at a pressure of 205 lb/in² abs., and the water from the drain 233·1°F. If 1 lb of coal burnt in the furnaces of the boiler generates 7·5 lb of steam, what weight of coal is required to make 24 tons of make-up feed if the evaporator is worked continuously?

124. The following data were taken from a 4-stroke compression ignition engine during a test. i.h.p. = 308; b.h.p. = 195; Consumption of fuel per hour = 109·5 lb, calorific value of this fuel = 19200 Btu/lb; weight of circulating water used per minute = 171 lb, inlet temperature 63°F, outlet temperature 118°F. Calculate the indicated and brake thermal efficiencies; find the necessary data for, and draw up, a heat balance.

125. A turning engine is capable of producing a torque of 1600 lb-inches in its crank shaft, and the gear ratio is 1000 to 1. The efficiency of the gear is 12%. A bar has been left across the guide and is held in position by two bolts which are each $\frac{7}{8}$ inch

diameter. The engine is turned and the guide shoe comes on to the bar when the piston is exactly half way down the cylinder. Find the stress set up in the bolts, the stroke of the piston being 42 inches and the connecting rod 7 feet long.

126. An air motor has a stroke of 12 inches and the clearance is equal to 0·5 inch of the stroke. Air is admitted at a pressure of 90 lb/in² gauge for three-eighths of the stroke, and is expanded to 15 lb/in² gauge at the end of the stroke. Find the law connecting pressure and volume and the fraction of the stroke where the pressure is 70 lb/in² gauge.

127. Two ships, A and B, are fitted with turbine machinery. A is supplied with steam at 200 lb/in² abs. and 2% wet, and exhausts at 1·5 lb/in² abs., the dryness fraction of the exhaust steam being 0·88. B is supplied with steam at 400 lb/in² abs. having 200 F° of superheat, and exhausts at 0·5 lb/in² abs., the dryness fraction being 0·8. Calculate the theoretical thermal efficiency of each engine.

128. A diesel engine has a compression ratio of 13 to 1, and 35 lb of air are supplied per lb of fuel burnt. Air is drawn into the cylinder at 14 lb/in² abs. and at 100°F. The law of compression is $PV^{1.4}$ = constant, calorific value of the fuel 18500 Btu/lb, specific heat of air at constant pressure 0·2375. Assume ideal conditions and find the maximum pressure and temperature in the cylinder during the cycle.

129. In a multi-collar thrust block there are 6 shoes and 7 collars, the inside and outside diameters of the rubbing surfaces are 12 and 20 inches respectively, and the pressure on them is 45 lb/in². Assume propeller and engine efficiencies combined to be 69% and calculate the i.h.p. of the engine when the ship is travelling at 12 knots. Make a sketch of the shoe showing the bearing surface.

130. The cylinders of a triple expansion engine are 27, 45 and 74 inches diameter respectively and the mean effective pressure referred to the L.P. is 33 lb/in². If the power developed in the I.P. is 2% more than in the H.P., and the power developed in the L.P. is 5% more than in the H.P., find the mean effective pressure in each cylinder.

131. The difference in pressure between the two sides of the piston in a cylinder 20 inches diameter of a horizontal engine, is 110 lb/in² at a certain part of the stroke, and the acceleration of the piston is then 280 feet per sec per sec. If the weight of the moving parts is 1400 lb, find the load on the crosshead.

132. A heavy oil engine has a crank pin circle radius of 21 inches, the connecting rod is 6 feet long and the centre line of the cylinder is off-set 2 inches. When the crank has travelled through 30° past the top vertical centre the load on the piston is 20 tons, find the pressure in lb/in² on the guide if the area of the guide shoe is 173 sq inches. Find also the distance in inches the piston has moved down its stroke for this position of the crank.

133. An ordinary slide valve has 7·5 inches travel. The lead is 0·24 inch and the positive exhaust lap is 0·17 inch. When the crank is on the top centre the port is 2 inches open to exhaust. Find the steam lap, the angle of advance of the eccentric and the angular position of the crank when compression begins.

134. A marine engine is fitted with a surface heater which takes steam from the main engine supply line. The engine takes 93·65% of the steam generated, and the heater takes the remainder. The engine steam, as water at 126°F, mixes with the drain water from the heater coils in the hotwell. All then passes through the heater, and is delivered to the boilers at 196°F. Assume no loss of heat, find the initial temperature of the steam.

135. Explain what is meant by the 'Coefficient of Performance' of a refrigerating machine. The capacity of a refrigerating machine is calculated from the weight of ice it can make at 32°F from water at 32°F in 24 hours. What weight of ice can be made in 24 hours by a 4 ton machine? The ice is made from water at 67°F, and its temperature is 25°F. If the condenser temperature were 70°F, calculate the temperature of the evaporator. Assume a co-efficient of performance of 5·8.

136. 115 ft³ of wet steam at 40 lb/in² abs., is blown into a tank containing 160 lb of water at 50°F and the final temperature of the mixture is 125°F. Calculate the initial dryness fraction of the steam.

137. The absolute pressure in a condenser is 1·9 lb/in². The temperature of the air and vapour present is 115·7°F and this is the temperature of steam at the absolute pressure of 1·5 lb/in². Find the weight of air present in 100 ft³ space of the condenser. For air the characteristic constant is 53·2 ft lb per lb per F°.

138. Find the ratio of the work done per 1 pound of steam in the following cases. (a) When the cut off occurs at ¼ stroke; (b) when cut off is at ½ stroke; (c) when steam is carried for the full stroke. The initial pressure is 230 lb/in² abs. The back pressure

is 30 lb/in² abs. At 230 lb/in² abs. one pound weight of steam occupies 2 ft³. Assume the expansion follows the law P × V = constant.

139. The load on the crank pin of an engine was 32 tons and the guide load was 5·1 tons when the crank was 127° past the top centre. The connecting rod is 9 ft 3 in. long. Find the piston load, the length of the stroke and the distance between crosshead and shaft centres.

140. A steam reciprocating engine has a stroke of 4 feet and the connecting rod is 8 feet long. The engine runs at 120 r.p.m. It has a slide valve of the ordinary D type, of travel 8 inches and angle of advance 37°. Assume the slide valve moves with simple harmonic motion. The velocity of the piston is given by:

$$v = \omega r \left[\sin \theta + \frac{\sin (2\theta)}{2n} \right] \text{ ft/sec}$$

ω = angular velocity of the crank in radians per sec.
r = length of crank in feet.
θ = angle of crank past centre in degrees.

$$n = \frac{\text{length of connecting rod}}{\text{length of crank}}$$

Find the relative velocity of piston and valve when the crank is 140° past the top centre.

141. 3 ft³ of air at 135 lb/in² abs. are admitted to a cylinder and expand isothermally. The clearance is 5% and the cut off is at 0·3 of the stroke.

   Calculate (a) the final pressure and volume of the air,
   (b) the pressure at 0·8 of the stroke,
   (c) the heat added during expansion.

142. Due to a leaky condenser the feed water density is 0·25 of the sea density, and the fuel consumption increases by 3·8% due to blowing down. The boilers produce dry steam at 250 lb/in² abs. from feed water at 172°F. Find the boiler density being maintained, given that the solids in solution in sea water are:
   3·7 oz per gallon Sodium Chloride
   0·3 ,, ,, ,, Magnesium Chloride
   0·4 ,, ,, ,, Magnesium Sulphate
   0·3 ,, ,, ,, Calcium Sulphate
   0·3 ,, ,, ,, Calcium Carbonate.

143. Steam is supplied to a turbine at 250 lb/in² abs. and 601°F temperature, and exhausts at 0·5 lb/in² abs. and 0·87 dry. At a certain stage in the turbine, where the pressure is 20 lb/in² abs. 0·1388 lb of each 1 lb of steam passing is withdrawn to a feed heater, and this is sufficient to heat the feed water to the saturation temperature of the steam withdrawn. Compare the thermal efficiences without, and with feed heating.

144. Steam at 250 lb/in² abs., superheated 240 F° passes through the H.P. turbine and exhausts at 30 lb/in² abs., being then dry and saturated. It then passes through the L.P. turbine, finally exhausting at 1 lb/in² abs., and 0·87 dry. A certain percentage of the H.P. exhaust is passed to a feed heater.

Calculate (a) the percentage of the total steam passed to the heater if equal power is developed by each turbine.

(b) the feed temperature if the temperature of the condensate is 100°F.

145. A single-acting air compressor cylinder is 12 in. diameter and the stroke is 18 inches. It runs at 300 r.p.m. After overhaul the law of compression was found to be $PV^{1·15} = k$, and after running for some time it was $PV^{1·35} = k$. Find the actual increase in horse power, and the increase per cent. if the speed and output remain unchanged. The initial pressure is 15 lb/in² abs., and the delivery 60 lb/in² abs.

146. A common slide valve has 7·875 inches travel. The lead is 0·375 inch, and the negative exhaust lap is 0·2 inch.

When the crank was 20° before the bottom centre, the opening to exhaust was 1·25 inches. Calculate the steam port opening when the crank had passed through 80° from the top centre.

147. Sketch the ideal diesel cycle, showing combustion at constant pressure and exhaust at constant volume. In such a cycle, the compression ratio is 14 to 1 and the cut off ratio is 2·2 to 1. Taking the ratio of specific heats as 1·4 and the pressure at the beginning of compression as 15 lb/in² abs., calculate the mean effective pressure.

148. On test, a 4 cylinder internal combustion engine gave a b.h.p. of 500, and with one cylinder cut out in turn, 364, 345, 343, 354·8 b.h.p. respectively. Assuming the mechanical efficiency and the speed to remain unchanged, find the mechanical efficiency of the engine.

FIRST CLASS EXAMINATION QUESTIONS

149. Steam is supplied to an engine at 150 lb/in² abs. A combined separating and throttling calorimeter is fitted to ascertain the quality of the steam. The data obtained were:
    Weight of water collected in separating calorimeter = 0·25 lb.
    Weight of condensed steam after throttling = 3·5 lb.
    Pressure in the throttling calorimeter = 15 lb/in² abs., and its temperature 235°F. Determine the dryness fraction of the steam supplied to the engine, taking the spec. ht. of superheated steam as 0·48.

150. In a CO_2 refrigerating machine, the pressure and temperature of the liquid CO_2 leaving the condenser is 950 lb/in² and 25°C, and on entering the evaporator the pressure and temperature of the liquid-vapour mixture is 350 lb/in² and — 10°C. If the specific entropy of the refrigerant after throttling through the reducing valve is 0·022 more than before throttling, calculate the dryness fraction of the gas entering the evaporator. Use the formula given below for specific entropy (s), and take the latent heat of CO_2 at 350 lb/in² as 63 Chu/lb.

$$s = C_f \log_\varepsilon \frac{T_K}{233} + \frac{q\, h_{fg}}{T_K}$$

where, C_f = spec. ht. of liquid CO_2 = 0·66
T_K = saturation temperature in °C abs.
q = dryness fraction
h_{fg} = latent heat in Chu/lb

SOLUTIONS TO FIRST-CLASS EXAMINATION QUESTIONS

1. $$\text{Brake horse power} = \frac{\text{Torque (lb ft)} \times 2\pi \times \text{r.p.m.}}{33000}$$

With one cylinder cut out, b.h.p. developed by three remaining cylinders $= \dfrac{80 \times 2\pi \times 1200}{33000} = 18\cdot 28$

∴ i.h.p. of each cylinder cut out in turn
$= 26\cdot 2 - 18\cdot 28 = 7\cdot 92$
i.h.p. of 4 cyls. $= 4 \times 7\cdot 92 = 31\cdot 68$ i.h.p. Ans. (i)

Fuel consumption
$= 0\cdot 55$ lb/b.h.p. hour
$= \dfrac{0\cdot 55 \times 26\cdot 2}{31\cdot 68}$ lb/i.h.p. hour

Indicated thermal efficiency
$= \dfrac{\text{Heat turned into work per i.h.p. hour}}{\text{Heat supplied per i.h.p. hour}}$

$= \dfrac{2545 \times 31\cdot 68}{0\cdot 55 \times 26\cdot 2 \times 18500}$

$= 0\cdot 3024$ or $30\cdot 24\%$ Ans. (ii)

2. Latent heat of water $= 144 \times \tfrac{5}{9} = 80$ Chu/lb
Let T = temp. of resultant mixture, in °C
Heat gained by ice = Heat lost by water
$18 (0\cdot 5 \times 5 + 80 + T) = 19\cdot 5 (89 - T)$
$45 + 1440 + 18T = 1735\cdot 5 - 19\cdot 5 T$
$37\cdot 5 T = 250\cdot 5$
$T = 6\cdot 68°C$
$6\cdot 68 \times \tfrac{9}{5} + 32 = 44\cdot 02°F$. Ans. (i)

Resultant temp. after second mixing is to be
$$44{\cdot}02 + 10 = 54{\cdot}02°F$$

Let x kg be taken out, heated to 212°F, and poured back.

Heat before mixing = Heat after mixing
$$(37{\cdot}5 - x)(44{\cdot}02 - 32) + x(212 - 32)$$
$$= 37{\cdot}5(54{\cdot}02 - 32)$$
$$450{\cdot}75 - 12{\cdot}02x + 180x = 825{\cdot}75$$
$$167{\cdot}98x = 375$$
$$x = 2{\cdot}232 \text{ kg}$$
$$2{\cdot}232 \times 2{\cdot}2 = 4{\cdot}91 \text{ lb. Ans. (ii)}$$

3.

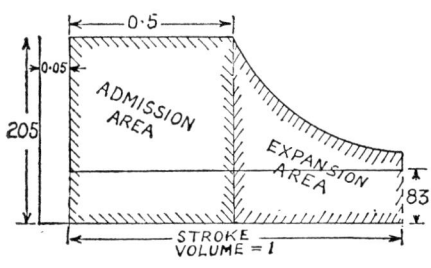

Let stroke = 1
Clearance = 0·05
Ratio of expansion
$$= \frac{1{\cdot}05}{0{\cdot}55} = 1{\cdot}909$$
$\log_e 1{\cdot}909$ from hyperbolic log tables
$$= 0{\cdot}6465$$

Area representing work done during admission of steam
$$= 205 \times 0{\cdot}5 = 102{\cdot}5$$

Area representing work done during expansion of steam
$$= P_1 V_1 \log_e r = 205 \times 0{\cdot}55 \times 0{\cdot}6465 = 72{\cdot}9$$

Gross area = $102{\cdot}5 + 72{\cdot}9 = 175{\cdot}4$

Mean height = area ÷ length
$$= 175{\cdot}4 \div 1 = 175{\cdot}4$$

∴ Mean gross pressure = $175{\cdot}4$ lb/in² abs.
Mean effective pressure = mean gross press.—back press.
∴ Hypothetical m.e.p. = $175{\cdot}4 - 83 = 92{\cdot}4$ lb/in²
Ans. (a)

Actual m.e.p. = $92{\cdot}4 \times 0{\cdot}7 = 64{\cdot}68$ lb/in²
Ans. (b)

$$\text{i.h.p.} = \frac{p\text{ALN}}{33000}$$

$$= \frac{64{\cdot}68 \times 0{\cdot}7854 \times 26^2 \times 4{\cdot}5 \times 75 \times 2}{33000}$$

= 702·5 Ans. (c)

4. Let the law of compression be $pv^n = $ constant

$$p_1 v_1^n = p_2 v_2^n$$
$$15 \times (18 + 3 \cdot 4)^n = (161 + 15) \times (3 \cdot 4)^n$$
$$15 \times 21 \cdot 4^n = 176 \times 3 \cdot 4^n$$
$$\text{Log } 15 + n \log 21 \cdot 4 = \log 176 + n \log 3 \cdot 4$$
$$n \log 21 \cdot 4 - n \log 3 \cdot 4 = \log 176 - \log 15$$

$$\therefore n = \frac{\log 176 - \log 15}{\log 21 \cdot 4 - \log 3 \cdot 4} = \frac{1 \cdot 0694}{0 \cdot 7989} = 1 \cdot 338$$

The law is $pv^{1 \cdot 338} = $ constant. Ans.

When 12 in. of the stroke are completed then 6 in. remain, and the volume is proportional to $6 + 3 \cdot 4 = 9 \cdot 4$

$$p_2 v_2^n = p_1 v_1^n$$
$$p_2 \times 9 \cdot 4^{1 \cdot 338} = 15 \times 21 \cdot 4^{1 \cdot 338}$$
$$\log p_2 + 1 \cdot 338 \log 9 \cdot 4 = \log 15 + 1 \cdot 338 \log 21 \cdot 4$$
$$\log p_2 = \log 15 + 1 \cdot 338 (\log 21 \cdot 4 - \log 9 \cdot 4)$$
$$= 1 \cdot 1761 + 0 \cdot 4781 = 1 \cdot 6542$$
$$1 \cdot 6542 \text{ is the log of } 45 \cdot 1$$

Pressure is $45 \cdot 1$ lb/in² abs. Ans.

5. From steam tables,
$P = 195$ lb/in², $\quad h_f = 353 \cdot 2, \quad h_{fg} = 846, \quad h_g = 1199 \cdot 2$

Heat to form one lb steam before blowing
$\quad = $ Heat to be given to one lb of feed water
$\quad = 1199 \cdot 2 - (160 - 32)$
$\quad = 1071 \cdot 2$ Btu ... ... (i)

With leaking condenser, to maintain constant density in boiler:

Salt put into boiler = Salt blown out
Amount of feed × feed density = Amount blown out × B.O. density

For one lb of steam generated,
$\quad$ let x lb = blow out
$\quad$ then $(1 + x)$ lb = feed
$\quad (1 + x) \times 3 \cdot 5 = x \times 18$
$\quad x = \tfrac{7}{29}$ lb
$\therefore$ feed = $1\tfrac{7}{29}$ lb

SOLUTIONS TO FIRST CLASS EXAMINATION QUESTIONS 529

Heat required to generate one lb of steam

= sensible heat to $1\tfrac{7}{29}$ lb water + latent heat to 1lb
 ($\tfrac{7}{29}$ lb is blown out)   (to make 1 lb of steam)

$= 1\tfrac{7}{29} [353\cdot 2 - (160 - 32)] + 846$
$= 279\cdot 5 + 846 = 1125\cdot 5$ Btu   ...   . (ii)

Consumption increases in the ratio of the heat required,

$$\therefore \text{New consumption} = 20 \times \frac{1125\cdot 5}{1071\cdot 1}$$

$= 21\cdot 01$ tons per day. Ans.

6.   Before alteration, $v_1 = 1200 + 120 = 1320$
$v_2 = 120$
The law is $p\, v^{1\cdot 4} =$ constant
$\therefore p_1 \times 1320^{1\cdot 4} = 528 \times 120^{1\cdot 4}$

$$p_1 = \frac{528 \times 120^{1\cdot 4}}{1320^{1\cdot 4}} \quad \ldots \quad \ldots \quad (1)$$

After alteration, $v_1 = 1200 + 119 = 1319$
$v_2 = 119$
$p_1 \times 1319^{1\cdot 4} = p_2 \times 119^{1\cdot 4}$, where p_2 is the new compression pressure

Substitute value of p_1 from (1)

$$\frac{528 \times 120^{1\cdot 4}}{1320^{1\cdot 4}} \times 1319^{1\cdot 4} = p_2 \times 119^{1\cdot 4}$$

$\therefore \log 528 + 1\cdot 4 \log 120 + 1\cdot 4 \log 1319 - 1\cdot 4 \log 1320 = \log p_2 + 1\cdot 4 \log 119$
$\therefore \log p_2 = \log 528 + 1\cdot 4 (\log 120 + \log 1319 - \log 1320 - \log 119)$

$= 2\cdot 7226 + 0\cdot 00462 = 2\cdot 7272$
$\therefore p_2 = 533\cdot 6$ lb/in² abs.
Increase of press. $= 533\cdot 6 - 528 = 5\cdot 6$ lb/in². Ans.

7.

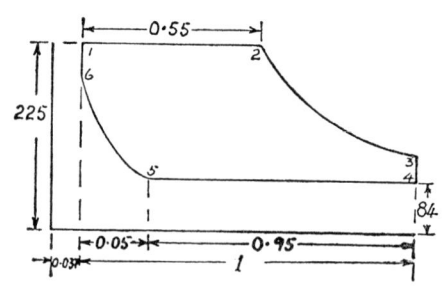

Let stroke = 1
Ratio of expansion
$= \frac{1 \cdot 037}{0 \cdot 587} = 1 \cdot 767$
$\log_\varepsilon 1 \cdot 767$ from tables
$= 0 \cdot 5693$
Ratio of compression
$= \frac{0 \cdot 087}{0 \cdot 037} = 2 \cdot 352$
$\log_\varepsilon 2 \cdot 352$ from tables
$= 0 \cdot 8553$

$P_5 V_5 = P_6 V_6$ $84 \times 0 \cdot 087 = P_6 \times 0 \cdot 037$
$P_6 = 197 \cdot 5$ lb/in^2 abs. Ans. (a)

Admission area $= 225 \times 0 \cdot 55 = 123 \cdot 75$
Expansion area $= P_2 V_2 \log_\varepsilon r$
$= 225 \times 0 \cdot 587 \times 0 \cdot 5693 = 75 \cdot 18$
Gross area $= 123 \cdot 75 + 75 \cdot 18 = 198 \cdot 93$
Exhaust area $= 84 \times 0 \cdot 95 = 79 \cdot 8$
Compression area $= P_5 V_5 \log_\varepsilon r = 84 \times 0 \cdot 087 \times 0 \cdot 8553$
$= 6 \cdot 252$
Back pressure area $= 79 \cdot 8 + 6 \cdot 252 = 86 \cdot 052$
Net effective area $=$ gross area — back pressure area
$= 198 \cdot 93 - 86 \cdot 052$
$= 112 \cdot 878$

Mean effective pressure $=$ effective area $\div$ length
$= 112 \cdot 878 \div 1$
$\therefore$ Mean effective pressure $= 112 \cdot 878$ lb/in^2. Ans. (b)

8.   Cylinder volume ratios $= 1 : 9^{\frac{1}{3}} : 9^{\frac{2}{3}} : 9$
$= 1 : 2 \cdot 08 : 4 \cdot 326 : 9$

$$\text{Total expansion} = \frac{\text{vol. of L.P.}}{\text{vol. of H.P.} \times \text{cut off}}$$

$$14 = \frac{9}{1 \times \text{cut off}}$$

$\therefore$ cut off in H.P. $= \frac{9}{14} = 0 \cdot 643$ stroke. Ans.

Assuming expansion follows the law $pv =$ constant,
$225 \times 0 \cdot 643 =$ Terminal expansion press. in 1st I.P. $\times 2 \cdot 08$
$\therefore$ Terminal press. in 1st I.P. $= \dfrac{225 \times 0 \cdot 643}{2 \cdot 08} = 69 \cdot 55$ lb/in^2

SOLUTIONS TO FIRST CLASS EXAMINATION QUESTIONS 531

Since cut off in 1st I.P. is at 0·62 stroke,

$$\therefore \text{Initial press. in 1st I.P.} = \frac{69 \cdot 55}{0 \cdot 62} = 112 \cdot 2 \text{ lb/in}^2$$

Again, $225 \times 0 \cdot 643 =$ Terminal expansion press. in 2nd I.P. $\times 4 \cdot 326$

$\therefore$ Terminal press. in 2nd I.P. $= 33 \cdot 45$ lb/in^2

Cut off is at 0·63 stroke,

$$\therefore \text{Initial press. in 2nd I.P.} = \frac{33 \cdot 45}{0 \cdot 63} = 53 \cdot 09 \text{ lb/in}^2$$

Also, $225 \times 0 \cdot 643 =$ Terminal expansion press. in L.P. $\times 9$
$\therefore$ Terminal press. in L.P. $= 16 \cdot 07$ lb/in^2

Cut off is at 0·75 stroke,

$$\therefore \text{Initial press. in L.P.} = \frac{16 \cdot 07}{0 \cdot 75} = 21 \cdot 43 \text{ lb/in}^2$$

Initial steam pressures are:
225, 112·2, 53·09, 21·43 lb/in^2 abs. Ans.

9. Refer to Chapter 11 for Equivalent Evaporation.
From steam tables,
$P = 250$ lb/in^2, $h_g = 1202 \cdot 1$ Btu/lb
$P = 215$ lb/in^2, $h_g = 1200 \cdot 4$ Btu/lb

BOILER A:
Heat utilised per lb of fuel
$= 9 \cdot 5 [1202 \cdot 1 - (210 - 32)] = 9729$ Btu

BOILER B:
Heat utilised per lb of fuel
$= 8 \cdot 9 [1200 \cdot 4 - (160 - 32)] = 9544$ Btu

Thermal efficiency of B is 0·72, therefore,

$$\text{Calorific value of fuel used in B} = \frac{9544}{0 \cdot 72} \text{ Btu/lb}$$

and this is 10% more than that of fuel A, hence,

Calorific value of fuel used in A

$$= \frac{9544}{0\cdot72 \times 1\cdot1} = 12050 \text{ Btu/lb. Ans. (i)}$$

$$\text{Therm. Eff. of A} = \frac{9729}{12050} = 0\cdot8073 \text{ or } 80\cdot73\% \text{ Ans. (ii)}$$

Equivalent evaporation

$$\left.\begin{array}{l} \text{of A} = \dfrac{9729}{970\cdot6} = 10\cdot02 \text{ lb} \\[2ex] \text{of B} = \dfrac{9544}{970\cdot6} = 9\cdot83 \text{ lb} \end{array}\right\} \text{Ans. (iii)}$$

10.

$$y = 250 \sin 20° = 85\cdot5$$
$$x = 85\cdot5 \div \tan 38° = 109\cdot4$$
$$V_w = 250 \times \cos 20° = 234\cdot9$$
$$\text{Mean blade speed} = u$$
$$= V_w - x$$

Mean blade speed
$$= 234\cdot9 - 109\cdot4 = 125\cdot5 \text{ ft/sec. Ans. (i)}$$

Let D feet = diameter of rotor, then,

Diameter at mean blade height
$$= D + (2 \times 0\cdot25) = D + 0\cdot5 \text{ ft.}$$

Linear velocity of mean blade height
$$= \pi (D + 0\cdot5) \times \text{rev per sec}$$

$$\therefore \pi \times (D + 0\cdot5) \times \frac{500}{60} = 125\cdot5$$

$$D + 0\cdot5 = \frac{125\cdot5 \times 60}{\pi \times 500} = 4\cdot793 \text{ ft}$$

Diameter of rotor = 4·293 ft = 4 ft 3·5 in. Ans. (ii)

SOLUTIONS TO FIRST CLASS EXAMINATION QUESTIONS 533

11. Mean press. of H.P. referred to L.P. $= \dfrac{78 \cdot 8 \times 1}{6 \cdot 75}$

$= 11 \cdot 67 \text{ lb/in}^2$

Mean press. of I.P. referred to L.P.

$= \dfrac{39 \times 2 \cdot 35}{6 \cdot 75} = 13 \cdot 58 \text{ lb/in}^2$

Mean press. of L.P. $= 13 \cdot 7 \text{ lb/in}^2$

Mean press. all referred to L.P. $= 38 \cdot 95 \text{ lb/in}^2$.

Ans. (a)

Area of L.P. piston $= 0 \cdot 7854 \times 33 \cdot 5^2 \times 6 \cdot 75$

Total i.h.p.

$= \dfrac{38 \cdot 95 \times 0 \cdot 7854 \times 33 \cdot 5^2 \times 6 \cdot 75 \times 4 \cdot 5 \times 75 \times 2}{33000}$

$= 4739.$ Ans. (b)

Powers developed are in the ratio of their mean pressures referred to L.P.

Ratio of powers $=$ H.P. : I.P. : L.P.
$= 11 \cdot 67 : 13 \cdot 58 : 13 \cdot 7$
$= 1 : 1 \cdot 163 : 1 \cdot 174$
$= 100\% : 116 \cdot 3\% : 117 \cdot 4\%$ Ans. (c)

12. Mean pressure of H.P. referred to L.P.

$= \dfrac{26^2}{71^2} \times 73 = 9 \cdot 788 \text{ lb/in}^2$

Mean press. of I.P. referred to L.P.

$= \dfrac{42^2}{71^2} \times 27 = 9 \cdot 444 \text{ lb/in}^2$

Mean pressure in L.P. $= 10 \cdot 5 \text{ lb/in}^2$

Mean pressure all referred to L.P. $= 29 \cdot 732 \text{ lb/in}^2$

When vacuum drops 4 in., reduction in mean pressure
$$= 4 \times 0.491 = 1.964 \text{ lb/in}^2$$
∴ Second mean pressure $= 29.732 - 1.964 = 27.768$ lb/in^2

$$\frac{\text{Mean pressure}}{\text{Revs.}^2} = \text{constant}$$

∴ New revs. $= 72 \sqrt{\dfrac{27.768}{29.732}} = 69.58$ r.p.m. Ans.

13. From steam tables,
P = 220 lb/in^2, $T_R = 849.6$, $h_{fg} = 836.5$
P = 20 lb/in^2, $T_R = 687.7$, $h_{fg} = 960.4$

Entropy of high pressure steam

$$= \log_\varepsilon \frac{849.6}{492} + \frac{836.5}{849.6} + 0.6 \log_\varepsilon \frac{1160}{849.6}$$

$$= 0.5463 + 0.9847 + 0.6 \times 0.3114$$
$$= 1.7178 \quad \ldots \quad \ldots \quad \ldots \quad \ldots \quad \ldots \quad \ldots \quad \text{(i)}$$

Entropy of reduced pressure steam

$$= \log_\varepsilon \frac{687.7}{492} + \frac{q \times 960.4}{687.7}$$

$$= 0.3349 + 1.396\, q \ldots \quad \ldots \quad \ldots \quad \ldots \quad \ldots \quad \text{(ii)}$$

Equating (i) and (ii),
$$0.3349 + 1.396\, q = 1.7178$$
$$1.396\, q = 1.3829$$
$$q = 0.9908 \text{ Ans.}$$

14. $\dfrac{T_2}{T_1} = \left\{\dfrac{v_1}{v_2}\right\}^{n-1}$

∴ $\dfrac{v_1}{v_2} = \left\{\dfrac{T_2}{T_1}\right\}^{\frac{1}{n-1}} = \left\{\dfrac{T_2}{T_1}\right\}^{\frac{1}{1.35-1}} = \left\{\dfrac{T_2}{T_1}\right\}^{\frac{1}{0.35}}$

$T_2 = 520 + 460 = 980°$F abs., $T_1 = 560°$F abs.

Now $\dfrac{v_1}{v_2} = r$, the compression ratio

SOLUTIONS TO FIRST CLASS EXAMINATION QUESTIONS 535

$$\therefore r = \left\{\frac{980}{560}\right\}^{\frac{1}{0\cdot 35}}$$

$$\log r = \frac{1}{0\cdot 35}(\log 980 - \log 560) = 0\cdot 6943$$
$$\therefore r = 4\cdot 947 \quad \text{Ans.}$$

15. Sat. temp. of steam at 600 lb/in² = 486·2°
    Sat. temp. of steam at 550 lb/in² = 476·9°

    Difference for 50 lb/in² = 9·3°
    $\therefore$ Difference for 30 lb/in² = $\frac{3}{5} \times 9\cdot 3 = 5\cdot 58°$
    $\therefore$ sat. temp. of steam at 580 lb/in² = 476·9 + 5·58
    = 482·48, say 482·5°
    $\therefore$ steam at 580 lb/in² and 800°F has (800 − 482·5)
    = 317·5 F° of superheat.
    For steam at 600 lb/in² with 320° superheat, h = 1411·2
    For steam at 600 lb/in² with 280° superheat, h = 1388·6

    Difference for 40° superheat = 22·6
    Difference for (317·5 − 280) = 37·5° superheat
    $$= \frac{37\cdot 5}{40} \times 22\cdot 6 = 21\cdot 18$$
    $\therefore$ steam at 600 lb/in² with 317·5° superheat
    $h = 1388\cdot 6 + 21\cdot 18 = 1409\cdot 78$ ... ... (i)
    For steam at 500 lb/in² with 320° superheat, h = 1405·3
    For steam at 500 lb/in² with 280° superheat, h = 1383·4

    Difference for 40° superheat = 21·9

    Difference for 37·5° superheat
    $$= \frac{37\cdot 5}{40} \times 21\cdot 9 = 20\cdot 52$$
    $\therefore$ steam at 500 lb/in² with 317·5° superheat,
    $h = 1383\cdot 4 + 20\cdot 52 = 1403\cdot 92$ ... ... (ii)

    From (i) and (ii),

    Difference for 100 lb/in²
    = 1409·78 − 1403·92 = 5·86

S

Difference for 80 lb/in²
$= \frac{8}{10}$ of $5 \cdot 86 = 4 \cdot 688$

∴ steam at 580 lb/in² with 317·5° superheat (temp. 800°F.)
$h = 1403 \cdot 92 + 4 \cdot 688 = 1408 \cdot 61$ Btu/lb

Steam at 600 lb/in², 0·98 dry
$= 471 \cdot 8 + 0 \cdot 98 \times 732 \cdot 4 = 1189 \cdot 55$ Btu/lb

∴ heat supplied per lb of steam through superheater
$= 1408 \cdot 61 - 1189 \cdot 55 = 219 \cdot 06$, say 219 Btu. Ans.(a)

By Callendar's equation:
$v = 1 \cdot 248 \, (h - 835 \cdot 2)/P + 0 \cdot 0123$ ft³/lb

$= \dfrac{1 \cdot 248 \, (1408 \cdot 61 - 835 \cdot 2)}{580} + 0 \cdot 0123$

$= 1 \cdot 246$ ft³/lb   Ans. (b).

Alternatively, the volume of the steam may be assumed to vary directly as the absolute temperature, as explained in Chapter 8, thus:

$$v = v_g \times \frac{T}{T_R}$$

16.   From steam tables,
$P = 125$ lb/in², $h_g = 1192 \cdot 1$
$P = 72$ lb/in², $h_f = 274 \cdot 6$, $h_{fg} = 907 \cdot 4$, $v_g = 6 \cdot 044$
$P = 2$ lb/in², $h_f = 94$, $h_{fg} = 1022 \cdot 2$, $v_g = 173 \cdot 7$

$\begin{array}{c}\text{Heat in steam}\\\text{at 72 lb/in}^2\end{array} = \begin{array}{c}\text{Heat in steam}\\\text{at 125 lb/in}^2\end{array} - \text{heat drop}$

$274 \cdot 6 + q \times 907 \cdot 4 = 1192 \cdot 1 - 46$
$q \times 907 \cdot 4 = 871 \cdot 5$
$q = 0 \cdot 9603$

Volume per lb of steam at throat $= q v_g$
$= 0 \cdot 9603 \times 6 \cdot 044 = 5 \cdot 803$ ft³/lb

Velocity through throat $= \sqrt{2gJ \times \text{heat drop}}$
$= \sqrt{2 \times 32 \cdot 2 \times 778 \times 46} = 1518$ ft/sec.

Rate of flow (ft³/sec) = Area (ft²) × Velocity (ft/sec)

$\dfrac{50 \times 5 \cdot 803}{60} = \dfrac{\text{area (in}^2)}{144} \times 1518$

SOLUTIONS TO FIRST CLASS EXAMINATION QUESTIONS 537

$$\therefore \text{Area at throat} = \frac{50 \times 5 \cdot 803 \times 144}{60 \times 1518}$$

$$= 0 \cdot 4586 \text{ in}^2. \quad \text{Ans. (i)}.$$

Effective heat drop from entrance to exit
$= 0 \cdot 92 \times 272 = 250 \cdot 2 \text{ Btu/lb}$

$$\frac{\text{Heat in steam}}{\text{at 2 lb/in}^2} = \frac{\text{Heat in steam}}{\text{at 125 lb/in}^2} - \text{heat drop}$$

$$94 + q \times 1022 \cdot 2 = 1192 \cdot 1 - 250 \cdot 2$$
$$q \times 1022 \cdot 2 = 847 \cdot 9$$
$$q = 0 \cdot 8296$$

Volume per lb of steam at exit $= qv_g$
$= 0 \cdot 8296 \times 173 \cdot 7 = 144 \cdot 1 \text{ ft}^3/\text{lb}$

Velocity at exit
$$= \sqrt{2 \times 32 \cdot 2 \times 778 \times 250 \cdot 2} = 3540 \text{ ft/sec}$$

Rate of flow = Area × Velocity

$$\frac{50 \times 144 \cdot 1}{60} = \frac{\text{area (in}^2)}{144} \times 3540$$

$$\therefore \text{Area at exit} = \frac{50 \times 144 \cdot 1 \times 144}{60 \times 3540}$$

$$= 4 \cdot 885 \text{ in}^2. \quad \text{Ans. (ii)}$$

17.   Steam used per day $= \dfrac{1500 \times 15 \times 24}{2240} = \dfrac{6750}{28}$

$$= 241 \cdot 07 \text{ tons.}$$

Blow out $= \dfrac{\text{Feed density}}{\text{Boiler density}} = \dfrac{0 \cdot 08}{3 \cdot 5} = \tfrac{8}{350}$ of the feed.

$\therefore$ Weight of steam used $= 1 - \tfrac{8}{350} = \tfrac{342}{350}$ of the feed.
Actual blow out per day $= 241 \cdot 07 \times \tfrac{8}{342} = 5 \cdot 64$ tons.
Fresh water in condenser from steam condensed
$= 241 \cdot 07 - 10 = 231 \cdot 07$ tons.

Let x = tons leaking per day
$$(231 \cdot 07 \times 0) + (x \times 1) = (231 \cdot 07 + x)\, 0 \cdot 08$$

$$x - 0 \cdot 08\, x = 231 \cdot 07 \times 0 \cdot 08$$

$$x = \frac{231 \cdot 07 \times 0 \cdot 08}{0 \cdot 92} = 20 \cdot 09 \text{ tons.}$$

Amount available at hotwell
$= 231 \cdot 07 + 20 \cdot 09$ ... ... $= 251 \cdot 16$ tons.
Amount necessary as feed
$= 241 \cdot 07 + 5 \cdot 64$ ... ... $= 246 \cdot 71$ tons.

Amount passing to bilges $=$ $4 \cdot 45$ tons.

Blow out $= 5 \cdot 64$ tons per day. Ans.
Leakage $= 20 \cdot 09$ tons per day. Ans.
To bilges $= 4 \cdot 45$ tons per day. Ans.

18. Hydraulic lift $= 23$ ft. 6 in. $- 4$ ft. 3 in. $= 19 \cdot 25$ feet
Rate of flow $= 0 \cdot 7854 \times 0 \cdot 5^2 \times 190$ cu. ft. per min.
Weight of oil $= 0 \cdot 7854 \times 0 \cdot 5^2 \times 190 \times 0 \cdot 91 \times 62 \cdot 5$ lb per min.

Horse power output
$=$ Wt. lifted per min. $\times$ distance lifted $\div 33000$

$\therefore$ H.P. to drive pump
$$= \frac{0 \cdot 7854 \times 0 \cdot 5^2 \times 190 \times 0 \cdot 91 \times 62 \cdot 5 \times 19 \cdot 25}{33000} \times \frac{100}{59}$$
$= 2 \cdot 097$ Ans.

19.
$$\text{Fuel used per hour} = \frac{12 \times 2240}{24} = 1120 \text{ lb}$$

Heat supplied per hour $= 1120 \times 18000$ Btu

Heat equivalent of one b.h.p. hour
$$= \frac{33000 \times 60}{778} = 2545 \text{ Btu}$$

Heat equivalent of 2500 b.h.p.
$= 2500 \times 2545$ Btu per hour

SOLUTIONS TO FIRST CLASS EXAMINATION QUESTIONS 539

$$\therefore \text{Overall thermal effic.} = \frac{2500 \times 2545}{1120 \times 18000} = 0.3156 \text{ or } 31.56\%$$

Overall effic. of steam engine $= 0.15 \times 0.75 = 0.1125$ or 11.25%.

Comparison is:—
Diesel : Steam engine :: 31·56 : 11·25 or 2·8 : 1.   Ans. (i)

$$\text{Overall effic. of steam engine} = \frac{\text{Heat equivalent of 2500 b.h.p. hour}}{\text{Cal. value of coal} \times \text{Coal per hour}}$$

$$\therefore \text{Coal per hour} = \frac{2500 \times 2545}{0.1125 \times 13500} = 4190 \text{ lb}$$

$$\text{Coal per day} = \frac{4190 \times 24}{2240} = 44.9 \text{ tons.} \qquad \text{Ans. (ii)}$$

20.
$$\text{Clearance} = \frac{8.9}{100} \times 51 = 4.539 \text{ in.}$$

Final volume $= 51 + 4.539 = 55.539$
Final press. $= 48 \text{ lb/in}^2$ abs.
Initial volume $= v$. Initial press. $= 78 \text{ lb/in}^2$ abs.
The law of expansion is $p\,v^{1 \cdot 12} = $ constant
$\therefore 78 \times v^{1 \cdot 12} = 48 \times 55.539^{1 \cdot 12}$
Log $78 + 1.12 \log v = \log 48 + 1.12 \log 55.539$
$$\therefore \text{Log. } v = \log 55.539 - \frac{\log 78 - \log 48}{1.12}$$
$= 1.7446 - 0.1883 = 1.5564$
$v = 36.01$ inches, which includes clearance.
$36.01 - 4.539 = 31.471$ in.

Cut off is at 31·471 inches of stroke.   Ans.

21. Considering heater and hotwell as one combined unit:
Total heat entering heater and hotwell = Total heat leaving
Let weight of main steam supply = 100 lb.
then weight of feed water = 100 lb.
Steam used by heater = x lb.
Steam used by engines = $(100 - x)$ lb.

$$x(346 \cdot 1 + 0 \cdot 98 \times 851 \cdot 9) + (100 - x)(110 - 32) = 100(210 - 32)$$
$$x \times 1181 + 7800 - 78x = 17800$$
$$1103x = 10000$$
$$x = 9 \cdot 065\% \quad \text{Ans.}$$

22.

Fluid	Temperature F°	Weight	Specific heat	Water equivalent
x	120	2	a	$2a$
y	145	3	b	$3b$
z	177	5	c	$5c$

Mixing x and y, $(2a \times 120) + (3b \times 145) = (2a + 3b)\,130$
from which $45b = 20a$, or $9b = 4a$
Mixing y and z, $(3b \times 145) + (5c \times 177) = (3b + 5c)\,177$
from which $100c = 36b$, or $25c = 9b$

$$\therefore 4a = 9b = 25c\,;\, a = \frac{25}{4}c\,;\, b = \frac{25}{9}c$$

Mixing x, y and z, let T = resultant temperature.
$(2a \times 120) + (3b \times 145) + (5c \times 177) = (2a + 3b + 5c) \times \text{T}$
$(240a) + (435b) + (885c) = (2a + 3b + 5c) \times \text{T}$
Substituting for 'a' and 'b'

$$\left[\frac{240 \times 25}{4} c \right] + \left[\frac{435 \times 25}{9} c \right] + 885c$$

$$= \left[\frac{25}{2} c + \frac{25}{3} c + 5c \right] \times \text{T}$$

'c' cancels and $3593\tfrac{1}{3} = 25\tfrac{5}{6} \times \text{T}$
$\text{T} = 139 \cdot 1°\text{F}.$ Ans.

SOLUTIONS TO FIRST CLASS EXAMINATION QUESTIONS 541

23. Total heat (in Btu) per lb of steam and water at cardinal points of the cycle:

(i) *Without reheater,*
At engine stop valve $= 1204\cdot3 + 0\cdot61\,(700 - 423\cdot3)$
$= 1373\cdot09$ Btu/lb
At exhaust $= 47\cdot6 + 0\cdot85 \times 1048\cdot5$
$= 938\cdot83$ Btu/lb
At hotwell $= 47\cdot6$ Btu/lb

(ii) *With reheater,*
At engine stop valve $= 1373\cdot09$ Btu/lb
At reheater inlet $= 1164\cdot6$ Btu/lb
At reheater outlet $= 1164\cdot6 + 0\cdot5\,(700 - 250\cdot3)$
$= 1389\cdot5$ Btu/lb
At exhaust $= 47\cdot6 + 0\cdot94 \times 1048\cdot5$
$= 1033\cdot2$ Btu/lb
At hotwell $= 47\cdot6$ Btu/lb

$$\text{Efficiency} = \frac{\text{Heat used by engine}}{\text{Heat supplied by boiler}}$$

$$= \frac{1373\cdot09 - 938\cdot83}{1373\cdot09 - 47\cdot6}$$

without reheat $= 0\cdot3277$ or $32\cdot77\%$ Ans. (i)

$$= \frac{(1373\cdot09 - 1164\cdot6) + (1389\cdot5 - 1033\cdot2)}{(1373\cdot09 - 47\cdot6) + (1389\cdot5 - 1164\cdot6)}$$

$$= \frac{208\cdot49 + 356\cdot3}{1325\cdot49 + 224\cdot9}$$

with reheat $= 0\cdot3643$ or $36\cdot43\%$ Ans. (ii)

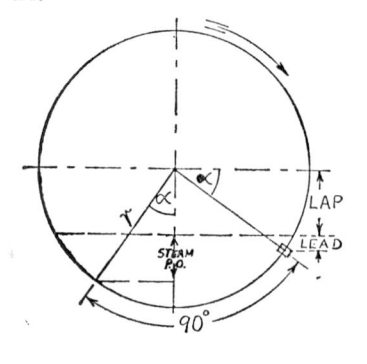

24.

Let r = half travel

α = angle of advance

$r \cos \alpha$ = Lap + port opening

$\qquad\quad$ = 1·75 + 1·125

$r \cos \alpha = 2·875$ ... (i)

$$\frac{\text{Lap + lead}}{r} = \sin \alpha$$

$r \sin \alpha = 2$ ... ... (ii)

Dividing (ii) by (i):

$$\frac{r \sin \alpha}{r \cos \alpha} = \frac{2}{2·875}$$

$\tan \alpha = 0·6955$

$\alpha = 34° 49'$ Ans. (a)

From (ii) $r \sin 34° 49' = 2$

$$r = \frac{2}{0·5709} = 3·502$$

∴ Travel of valve = 7·004 inches. Ans. (b)

25.
$$P_1 V_1 = P_2 V_2$$
$$14·7 \times 1 = P_2 \times 1·004$$

$$P_2 = \frac{14·7}{1·004} = \text{pressure after first stroke}$$

$$P_3 V_3 = P_4 V_4$$

$$\frac{14·7}{1·004} \times 1 = P_4 \times 1·004$$

$$P_4 = \frac{14·7}{1·004^2} = \text{pressure after 2nd stroke}$$

SOLUTIONS TO FIRST CLASS EXAMINATION QUESTIONS 543

Similarly, pressure after 3rd stroke $= \dfrac{14\cdot 7}{1\cdot 004^3}$

and, pressure after nth stroke $= \dfrac{14\cdot 7}{1\cdot 004^n}$

$\therefore \dfrac{14\cdot 7}{1\cdot 004^n} = 4\cdot 9$

$1\cdot 004^n = \dfrac{14\cdot 7}{4\cdot 9} = 3$

$\therefore n = \dfrac{\log 3}{\log 1\cdot 004}$

$= 280\cdot 6$ suction strokes. Ans.

26. From steam tables,
$P = 240$ lb/in^2, $t_F = 397\cdot 4$, $h_g = 1201\cdot 7$, $v_g = 1\cdot 918$

Total heat of dry sat. steam $= 1201\cdot 7$ Btu/lb. Ans. (ai)

Total heat of superheated steam
$= 1201\cdot 7 + 0\cdot 56\,(700 - 397\cdot 4) = 1371\cdot 2$ Btu/lb. Ans. (aii)

Heat per cubic foot of dry sat. steam
$= \dfrac{1201\cdot 7}{1\cdot 918} = 626\cdot 5$ Btu/ft^3. Ans. (bi)

Specific volume of superheated steam
$= \dfrac{1\cdot 248\,(1371\cdot 2 - 835\cdot 2)}{240} + 0\cdot 0123 = 2\cdot 8$ ft^3/lb

Heat per cubic foot of superheated steam
$= \dfrac{1371\cdot 2}{2\cdot 8} = 489\cdot 7$ Btu/ft^3. Ans. (bii)

27. Sine of angle of advance
$$= \frac{\text{Steam lap} + \text{lead}}{\frac{1}{2}\text{ travel}}$$
$$= \frac{1\frac{5}{16} + \frac{3}{16}}{3} = 0\cdot 5$$
∴ Angle of advance = 30°
AB = 3 sin (30° + 35°)
  = 2·7189 in.
Distance from bottom of travel
  = 3 — 2·7189
  = 0·2811 in. Ans.

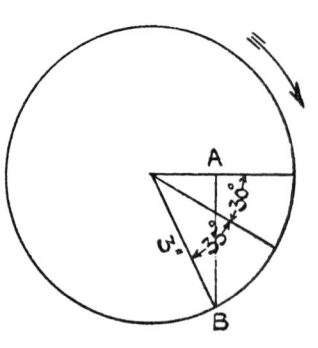

28. Total heat before mixing = Total heat after mixing
3 (336 + 0·88 × 860·1) + 2 × 1242·1
$$= 5 (336 + q \times 860\cdot 1)$$
3279 + 2484 = 1680 + 4301 q
     4301 q = 4083
         q = 0·9493

∴ After mixing, dryness fraction of the steam = 0·9493
Ans. (a)

Total heat before throttling = Total heat after throttling
336 + 0·9493 × 860·1 = 250·2 + q × 924·6
336 + 816·6 = 250·2 + 924·6 q
    924·6 q = 902·4
         q = 0·9759

∴ After throttling, dryness fraction of the steam
  = 0·9759 Ans. (b)

29. Let the volume of each metal be x cc
Weight of brass = 8·4x grams, and weight of steel = 7·86x grams

Heat lost by metals = Heat gained by water and calorimeter
(250—19·5) (8·4x × 0·094 + 7·86x × 0·116) = (19·5—15)(600 + 223 × 0·1)
230·5 (0·7896 x + 0·91176 x) = 4·5 × 622·3
    230·5 × 1·70136 x = 4·5 × 622·3

$$x = \frac{4\cdot 5 \times 622\cdot 3}{230\cdot 5 \times 1\cdot 70136} = 7\cdot 143 \text{ cc}$$

∴ Weight of brass = 8·4 × 7·143 = 60 grams ⎫
                                              ⎬ Ans.
Weight of steel = 7·86 × 7·143 = 56·14 grams ⎭

30. From tables, $t_F = 393·7$; $h_f = 368·3$; $h_{fg} = 833$

Heat required to form 1 lb of steam, dryness q
$= 368·3 - (200 - 32) + 833 q$
$= 200·3 + 833 q$ Btu

Heat to form 1 lb of dry steam $= 200·3 + 833$ Btu

∴ $109 (200·3 + 833 q) = 100 (200·3 + 833)$
$109 \times 833 q = 83300 - 9 \times 200·3$
$= 81497·3$

$$\therefore q = \frac{81497·3}{109 \times 833} = 0·897 \text{ Ans.}$$

31. Let W = weight of the water in lb
Heat before mixing = Heat after mixing
$2 \times 0·095 (350 - 32) + 4 \{2 \times 0·095 (350 - 32)\}$
$\qquad = 2 \times 0·095 (84 - 32) + W (84 - 32)$
$5 \times 2 \times 0·095 (350 - 32)$
$\qquad = 2 \times 0·095 (84 - 32) + W (84 - 32)$
$302·1 = 9·88 + 52W$

$$W = \frac{292·22}{52} = 5·62 \text{ lb}$$

Heat in copper × 4 = Heat in water
$4 \times 2 \times 0·095 (350 - 32) = 5·62 (T - 32)$

$$T = \frac{4 \times 2 \times 0·095 \times 318}{5·62} + 32$$

$= 43 + 32 = 75°F$

The weight of the water is 5·62 lb, and its temperature is 75°F. Ans.

32. From superheat steam tables,
    P = 250 lb/in² supht. 240 F°, $h = 1340$

   From saturation steam tables,
    P = 25 lb/in², $h_f = 208{\cdot}6$, $h_{fg} = 952{\cdot}5$

   Heat drop through nozzle
   $$= 1340 - (208{\cdot}6 + 0{\cdot}97 \times 952{\cdot}5)$$
   $$= 207{\cdot}5 \text{ Btu/lb}$$

   Considering 1 lb of steam:
   K.E. acquired by steam = Equivalent of heat energy given up
   $$\frac{1 \times v^2}{2g} = 1 \times 207{\cdot}5 \times 778$$
   $$v = \sqrt{207{\cdot}5 \times 778 \times 64{\cdot}4}$$
   $$= 3225 \text{ ft/sec. Ans. (i)}$$

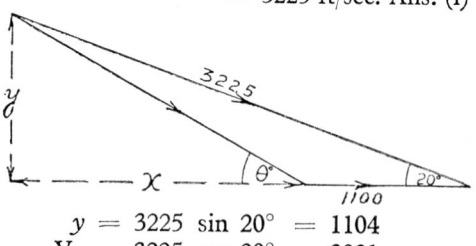

   $y = 3225 \sin 20° = 1104$
   $V_w = 3225 \cos 20° = 3031$
   $x = 3031 - 1100 = 1931$

   $$\operatorname{Tan} \theta = \frac{1104}{1931} = 0{\cdot}5718$$

   $\theta = 29° 45'$ inlet angle of blades. Ans. (ii)

33.

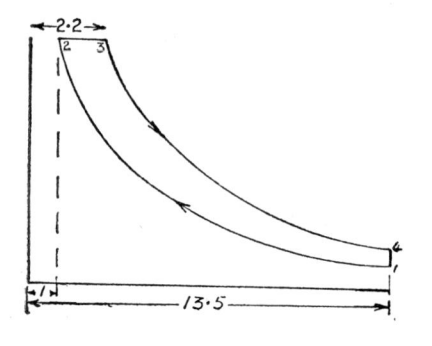

   $P_1 = 14{\cdot}7$
   $V_1 = 13{\cdot}5$
   $T_1 = 540$
   $V_2 = 1$
   $V_3 = 2{\cdot}2$
   $V_4 = 13{\cdot}5$
   $\gamma = \dfrac{0{\cdot}238}{0{\cdot}17}$
   $= 1{\cdot}4$

$$\frac{T_2}{T_1} = \left\{\frac{V_1}{V_2}\right\}^{\gamma-1}$$

$$\therefore T_2 = 540 \times 13.5^{0.4} = 1530°F \text{ abs.}$$

Temperature at end of compression
$$= 1530 - 460 = 1070°F. \text{ Ans. (i)}$$

$$T_3 = T_2 \times \frac{V_3}{V_2} = 1530 \times 2.2 = 3366°F \text{ abs.}$$

Temperature at end of combustion
$$= 3366 - 460 = 2906°F. \text{ Ans. (ii)}$$

$$\frac{T_4}{T_3} = \left\{\frac{V_3}{V_4}\right\}^{\gamma-1}$$

$$\therefore T_4 = 3366 \times \left\{\frac{2.2}{13.5}\right\}^{0.4} = 1629°F \text{ abs.}$$

Temperature at end of expansion
$$= 1629 - 460 = 1169°F. \text{ Ans. (iii)}$$

$$\text{Thermal Effic.} = \frac{\text{Heat supplied} - \text{Heat rejected}}{\text{Heat supplied}}$$

$$= 1 - \frac{\text{Heat rejected}}{\text{Heat supplied}}$$

$$= 1 - \frac{C_V (T_4 - T_1)}{C_P (T_3 - T_2)}$$

$$= 1 - \frac{0.17 \times 1089}{0.238 \times 1836}$$

$$= 1 - 0.4236$$
$$= 0.5764 \text{ or } 57.64\% \text{ Ans. (iv)}$$

34. See Fig. 123 for similar PV diagram and reference to state points 1 to 5.

From $PV = wRT$,

$$V_1 = \frac{wRT_1}{P_1} = \frac{1 \times 53\cdot3 \times 530}{14 \times 144} = 14\cdot02 \text{ ft}^3$$

$$P_2 V_2^{1\cdot4} = P_1 V_1^{1\cdot4}$$
$$32\cdot2 \times V_2^{1\cdot4} = 14 \times 14\cdot02^{1\cdot4}$$
$$V_2 = 14\cdot02 \times \sqrt[1\cdot4]{\frac{14}{32\cdot2}} = 7\cdot729 \text{ ft}^3$$

Volume after perfect intercooling is the same as if compression had been isothermal:

$$P_3 V_3 = P_1 V_1$$

$$V_3 = \frac{14 \times 14\cdot02}{32\cdot2} = 6\cdot091 \text{ ft}^3$$

$$P_4 V_4^{1\cdot4} = P_3 V_3^{1\cdot4}$$
$$64\cdot4 \times V_4^{1\cdot4} = 32\cdot2 \times 6\cdot091^{1\cdot4}$$
$$V_4 = 6\cdot091 \times \sqrt[1\cdot4]{\frac{32\cdot2}{64\cdot4}} = 3\cdot713 \text{ ft}^3$$

Volume after isothermal compression would be:
$$P_5 V_5 = P_1 V_1$$
$$64\cdot4 \times V_5 = 14 \times 14\cdot02$$
$$V_5 = 3\cdot046 \text{ ft}^3$$

% decrease in volume due to cooling after 1st stage

$$= \frac{7\cdot729 - 6\cdot091}{7\cdot729} \times 100 = 21\cdot19\% \text{ Ans.(i)}$$

% decrease in volume due to cooling after 2nd stage

$$= \frac{3\cdot713 - 3\cdot046}{3\cdot713} \times 100 = 17\cdot97\% \text{ Ans. (ii)}$$

SOLUTIONS TO FIRST CLASS EXAMINATION QUESTIONS 549

35. See Chapter 3.

Heat supplied = increase in internal energy + external work done

Considering heating 1 lb of gas from T_1 to T_2 at constant volume,

$$C_V (T_2 - T_1) = \text{increase in I.E.} + 0$$

Considering heating 1 lb of gas from T_1 to T_2 at constant pressure,

$$C_P (T_2 - T_1) = \text{Increase in I.E.} + \frac{P(V_2 - V_1)}{J}$$

$$C_P (T_2 - T_1) = C_V (T_2 - T_1) + \frac{R(T_2 - T_1)}{J}$$

$$C_P = C_V + \frac{R}{J}$$

$$\text{or, } C_P - C_V = \frac{R}{J}$$

36. Theoretical weight of air per lb of fuel
$$= \tfrac{100}{23} [2\tfrac{2}{3} C + 8 \{H - \tfrac{O}{8}\}]$$
$$= \tfrac{100}{23} [2\tfrac{2}{3} \times 0.81 + 8\{0.056 - \tfrac{0.057}{8}\}]$$
$$= \tfrac{100}{23} [2.16 + 0.391]$$
$$= \tfrac{100}{23} \times 2.551 = 11.09 \text{ lb air/lb fuel}$$

Actual weight of air supplied = $1.5 \times 11.09 = 16.635$ lb

Weight of gases passing up funnel for each lb of fuel burned
$$= 16.635 + 0.81 + 0.056 + 0.057$$
$$= 17.558 \text{ lb.}$$

Assuming perfect heat transfer,

Heat given up by gases = Heat taken in by air
$$17.558 \times 0.25 \times (800 - T) = 16.635 \times 0.24 \times (230 - 80)$$

$$800 - T = \frac{16.635 \times 0.24 \times 150}{17.558 \times 0.25}$$

$$800 - T = 136.5$$
$$T = 663.5°F. \text{ Ans.}$$

37. Latent heat is the heat given to or taken from a substance which changes its physical state without change of temperature.

Let t = initial temp. of water in °F
,, initial weight of water in tank = 25,
then weight of steam blown in = 1

Total heat before mixing = Total heat after
$$(364 \cdot 2 + 0 \cdot 93 \times 836 \cdot 5) + 25\,(t - 32) = 26\,(180 - 32)$$
$$1142 + 25\,t - 800 = 3848$$
$$25\,t = 3506$$
$$t = 140 \cdot 24°F$$

$$C = (F - 32) \times \tfrac{5}{9}$$
$$= 60 \cdot 1°C. \text{ Ans.}$$

38.

Let r = half travel
α = angle of advance
$\theta = \alpha - 10°$

$$\frac{2 + 0 \cdot 25}{r} = \sin \alpha$$

$$r = \frac{2 \cdot 25}{\sin \alpha} \quad \ldots \text{ (i)}$$

$$\frac{2 + 1 \cdot 5}{r} = \cos \theta$$

$$\frac{3 \cdot 5}{r} = \sin(100 - \alpha)$$

$$r = \frac{3 \cdot 5}{\sin(100 - \alpha)} \quad \text{(ii)}$$

(i) = (ii) .·. $\dfrac{2 \cdot 25}{\sin \alpha} = \dfrac{3 \cdot 5}{\sin(100 - \alpha)}$

$$\sin(100 - \alpha) = 1 \cdot 555 \sin \alpha$$

From, $\sin(A - B) = \sin A \cos B - \cos A \sin B$:
$$\sin 100 \times \cos \alpha - \cos 100 \times \sin \alpha = 1 \cdot 555 \sin \alpha$$
$$0 \cdot 9848 \cos \alpha + 0 \cdot 1736 \sin \alpha = 1 \cdot 555 \sin \alpha$$

Dividing throughout by cos α:
$$0.9848 + 0.1736 \tan \alpha = 1.555 \tan \alpha$$
$$1.3814 \tan \alpha = 0.9848$$
$$\tan \alpha = 0.7131$$
$$\alpha = 35° \ 30'. \text{ Ans. (a)}$$

From (i) $r = \dfrac{2.25}{\sin 35° \ 30'} = 3.875$

∴ Travel $= 2 \times 3.875 = 7.75$ in. Ans. (b)

39. Assuming atmospheric pressure $= 15 \text{ lb/in}^2$

$$13^3 = \dfrac{24 \times 195 \times D^2}{1110 \left(2 + \dfrac{D^2}{d^2}\right)}$$

$$13^3 \times 1110 \left(2 + \dfrac{D^2}{d^2}\right) = 24 \times 195 \times D^2$$

$$d^2 = 26^2 = 676$$

$$13^3 \times 1110 \left(\dfrac{1352 + D^2}{676}\right) = 24 \times 195 \times D^2$$

$$1352 + D^2 = \dfrac{24 \times 195 \times D^2 \times 676}{13^3 \times 1110}$$

$$1352 + D^2 = 1.297 \, D^2$$

$$0.297 \, D^2 = 1352 \text{ and } D^2 = \dfrac{1352}{0.297} = \dfrac{26^2 \times 2}{0.297}$$

$$\dfrac{LP^2}{HP^2} = \dfrac{D^2}{d^2} = \dfrac{26^2 \times 2}{0.297} \times \dfrac{1}{26^2} = 6.75$$

Number of expansions $= \dfrac{6.75}{0.6} = 11.25$ Ans.

Note that it is not essential to find the actual diameter of the L.P. cylinder.

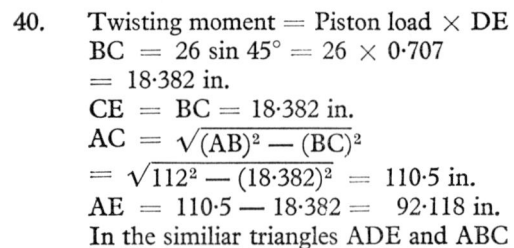

40. Twisting moment = Piston load × DE
BC = 26 sin 45° = 26 × 0·707
= 18·382 in.
CE = BC = 18·382 in.
AC = $\sqrt{(AB)^2 - (BC)^2}$
= $\sqrt{112^2 - (18·382)^2}$ = 110·5 in.
AE = 110·5 − 18·382 = 92·118 in.
In the similiar triangles ADE and ABC

AE : DE : : AC : BC

$$\therefore DE = \frac{AE \times BC}{AC}$$

$$= \frac{92·118 \times 18·382}{110·5} = 15·33 \text{ in.}$$

Twisting moment (T) = 47 × 2240 × 15·33 lb/in.
From T = $\frac{\pi}{16} D^3 q$

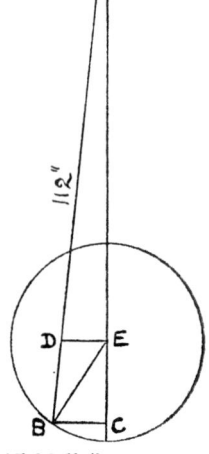

$$\text{Stress} = \frac{16 \times 47 \times 2240 \times 15·33}{\pi \times 13^3}$$

= 3742 lb/in². Ans.

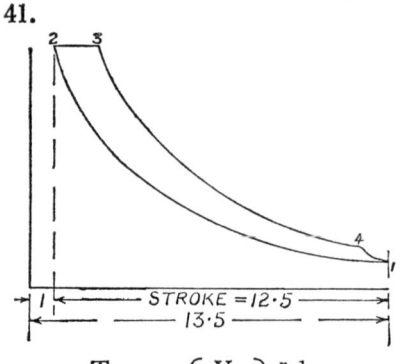

41.

Let stroke = 13·5—1
= 12·5
$\frac{1}{16}$ stroke = 0·78125
$\frac{7}{8}$ stroke = 10·9375
V_1 = 13·5
V_2 = 1
V_3 = 1·78125
V_4 = 11·9375

$$\frac{T_2}{T_1} = \left\{ \frac{V_1}{V_2} \right\}^{n-1} \therefore T_2 = 560 \times 13·5^{0·35} = 1393°\text{F abs.}$$

∴ Temperature at end of compression = 1393 − 460
= 933°F. Ans. (i)

SOLUTIONS TO FIRST CLASS EXAMINATION QUESTIONS 553

$$\frac{T_3}{T_2} = \frac{V_3}{V_2} \quad \therefore T_3 = 1393 \times 1\cdot 78125 = 2480°F \text{ abs.}$$

∴ Temperature at end of fuel supply = 2480 — 460
= 2020°F. Ans. (ii)

$$\frac{T_4}{T_3} = \left\{\frac{V_3}{V_4}\right\}^{n-1}$$

$$\therefore T_4 = 2480 \times \left\{\frac{1\cdot 78125}{11\cdot 9375}\right\}^{0\cdot 35} = 1274°F \text{ abs.}$$

∴ Temperature at opening to exhaust = 1274 — 460
= 814°F. Ans. (iii)

42. From steam tables,
when $t_F = 350\cdot 2$, $h_f = 321\cdot 9$, and when $t_F = 212$, $h_f = 180\cdot 1$

∴ heat given up by 1 lb of water when the pressure falls on the escape valve being opened is 321·9 — 180·1
= 141·8 Btu

Weight of water evaporated $= \dfrac{141\cdot 8}{970\cdot 6} = 0\cdot 1461$ lb since it is atmospheric pressure steam which will be formed.

If the vessel contained 1 lb of water at first, then 1 — 0·1461 = 0·8539 lb remains at 212°F, and this must be heated to a certain temperature in order that the sensible heat given up when the pressure is again released shall be 141·8 Btu.

If t_F is that temperature, then h_f is its sensible heat above 32°F and h_f — 180·1 Btu will be given up per 1 lb.

But there is only 0·8539 lb left

∴ 0·8539 $(h_f - 180\cdot 1) = 141\cdot 8$
  0·8539 $h_f - 153\cdot 8 = 141\cdot 8$
  0·8539 $h_f = 295\cdot 6$
    $h_f = 346\cdot 1$

From tables, when $h_f = 346\cdot 1$, $t_F = 373\cdot 1$, $p = 180$

The water must be heated to 373·1°F, which corresponds to a pressure of 180 lb/in² abs. Ans.

43.

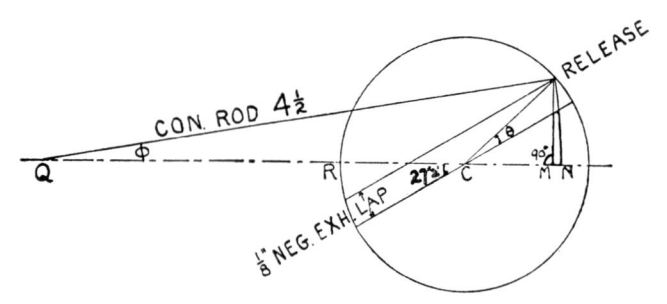

[Figure: Circle diagram showing CON. ROD 4½, Q, R, C, M, N, RELEASE, 27°2′, ⅛″ NEG. EXH. LAP]

$$\text{Sine of angle of advance} = \frac{1\tfrac{7}{16} + \tfrac{1}{8}}{1\tfrac{7}{16} + 2} = 0.4545$$

Angle of advance = 27° 2′. Ans.

If the valve had no exhaust lap then release would occur when crank is 27° 2′ from the line of stroke.

$$\text{Sine } \theta = \frac{\tfrac{1}{8}}{3\tfrac{7}{16}}, \quad \theta = 2° 5′$$

Due to negative exhaust lap being given the valve opens to exhaust 2° 5′ earlier.

∴ Crank is at 27° 2′ + 2° 5′ = 29° 7′ to line of stroke, at release.

Let crank = 1 and con. rod = 4½

$$\frac{1}{\text{Sin } \phi} = \frac{4\tfrac{1}{2}}{\text{Sin }(180° - 29° 7′)}, \text{ Sine } \phi = 0.1081, \phi = 6° 12′$$

The angularity of the con. rod is 6° 12′ at the point of release.

MN is the effect of the angularity of the con. rod.

$$\text{MN} = 4\tfrac{1}{2} - 4\tfrac{1}{2} \cos 6° 12′ = 4\tfrac{1}{2}(1 - \cos 6° 12′)$$
$$= 4\tfrac{1}{2} \times 0.0059 = 0.02655$$

This effect is positive on the 'out' or 'down' stroke, and negative on the 'in' or 'up' stroke.

$$\text{RM} = 1 + \cos 29° 7′ = 1.8736$$

SOLUTIONS TO FIRST CLASS EXAMINATION QUESTIONS 555

If the data given refer to the top end of the valve, then
release is at $\dfrac{1\cdot 8736 + 0\cdot 02655}{2} = 0\cdot 95$, or 95% of the stroke. Ans.

If the data refer to the bottom end of the valve, then
release is at $\dfrac{1\cdot 8736 - 0\cdot 02655}{2} = 0\cdot 9235$, or 92·35% of the stroke. Ans.

44. i.h.p. $= \dfrac{pALN}{4560} \times 8$ (for 8 cylinders)

$= \dfrac{6\cdot 35 \times 0\cdot 7854 \times 50\cdot 8^2 \times 0\cdot 974 \times 42\cdot 5}{4560} \times 8$

$= 935$ i.h.p. Ans. (i)

Specific fuel consumption
$= \dfrac{3\cdot 6 \times 2240}{24 \times 935} = 0\cdot 3593$ lb/i.h.p. hour. Ans. (ii)

45. Atmospheric pressure $= 736$ mm of mercury
$736 \div 25\cdot 4 = 28\cdot 98$ in. of mercury
$28\cdot 98 \times 0\cdot 491 = 14\cdot 23$ lb/in^2
Exhaust pressure $= 14\cdot 5 + 14\cdot 23 = 28\cdot 73$ lb/in^2 abs.
Admission pressure $= 59 + 14\cdot 23 = 73\cdot 23$ lb/in^2 abs.

Let stroke $= 1$, clearance $= 0\cdot 09$,
$x =$ fraction of exhaust stroke when compression begins.
$P_1 V_1 = P_2 V_2$
$28\cdot 73 \times V_1 = 73\cdot 23 \times 0\cdot 09$
$V_1 = 0\cdot 2294$
$x = 1\cdot 09 - 0\cdot 2294 = 0\cdot 8606$ of stroke. Ans.

46. From steam tables,
$P = 260$ lb/in^2, $h_f = 379\cdot 9$, $h_g = 1202\cdot 5$ Btu/lb

Heat to form one lb of steam before scumming
$= 1202\cdot 5 - (133 - 32) = 1101\cdot 5$ Btu

Since the coal consumption increases by 4·3% when scumming, then the heat to form 1 lb of steam is now $1\cdot 043 \times 1101\cdot 5$ Btu or, the extra heat required per lb is $1101\cdot 5 (1\cdot 043 - 1)$
$= 47\cdot 36$ Btu

This represents the sensible heat in the water scummed out.
Sensible heat per 1 lb = 379·9 — (133 — 32)
= 278·9 Btu

∴ 47·36 Btu is the sensible heat of $\dfrac{47 \cdot 36}{278 \cdot 9}$

= 0·1698 lb

∴ when scumming, to form 1 lb of steam 1·1698 lb of feed water must be supplied, and 0·1698 will be scummed out.

Feed × Feed density = Amount scummed × Boiler density
1·1698 × Feed density = 0·1698 × 3·7

$$\text{Feed density} = \dfrac{0 \cdot 1698 \times 3 \cdot 7}{1 \cdot 1698} = 0 \cdot 537$$

The feed density is 0·537 of the sea density. Ans.

47. See Chapter 4 for proof.

$$\dfrac{T_1}{T_2} = \left\{ \dfrac{P_1}{P_2} \right\}^{\frac{n-1}{n}} \quad \therefore \dfrac{1160}{550} = \left\{ \dfrac{250}{14 \cdot 7} \right\}^{\frac{n-1}{n}}$$

$$2 \cdot 109 = 17^{\frac{n-1}{n}}$$

$$\log \text{ of } 2 \cdot 109 = \tfrac{n-1}{n} \times \log \text{ of } 17$$

$$0 \cdot 3241 = \tfrac{n-1}{n} \times 1 \cdot 2306$$

$$0 \cdot 3241 \, n = 1 \cdot 2306 \, n - 1 \cdot 2306$$

$$1 \cdot 2306 = 0 \cdot 9065 \, n$$

$$n = 1 \cdot 358$$

48. $\quad Q = \dfrac{Cat\theta}{d} \quad \therefore \theta = \dfrac{Qd}{Cat}$

$$\theta_1 + \theta_2 + \theta_3 = \dfrac{Q}{at} \left\{ \dfrac{d_1}{C_1} + \dfrac{d_2}{C_2} + \dfrac{d_3}{C_3} \right\}$$

$$65 - 28 = \dfrac{Q}{15 \times 7 \cdot 5 \times 24} \left\{ \dfrac{4 \cdot 5}{6 \cdot 5} + \dfrac{3}{0 \cdot 3} + \dfrac{1}{1 \cdot 2} \right\}$$

$$Q = \frac{37 \times 15 \times 7\cdot 5 \times 24}{11\cdot 526}$$

$Q = 8670$ Btu/24 hours. Ans. (a)

Through brick: $\theta_1 = \dfrac{8670 \times 4\cdot 5}{6\cdot 5 \times 15 \times 7\cdot 5 \times 24} = 2\cdot 22$ F°

∴ Temperature at interface of brick and cork
$= 65 - 2\cdot 22 = 62\cdot 78°$F. Ans. (bi)

Through cork: $\theta_2 = \dfrac{8670 \times 3}{0\cdot 3 \times 15 \times 7\cdot 5 \times 24} = 32\cdot 1$ F°

∴ Temperature at interface of cork and wood
$= 62\cdot 78 - 32\cdot 1 = 30\cdot 68°$F. Ans. (bii)

49. Heat to be taken from water to make ice
$= 300\,[(67 - 32) + 144 + 0\cdot 5\,(32 - 22)]$
$= 300 \times 184$ Btu/hour

Heat absorbed by refrigerant in evaporator
$= (0\cdot 95 - 0\cdot 5) \times 130 = 0\cdot 45 \times 130$ Btu/lb

Volume of refrigerant leaving evaporator
$$= \frac{300 \times 184 \times 0\cdot 35 \times 0\cdot 95}{0\cdot 45 \times 130} \text{ cu ft per hour}$$

Let $d =$ diameter of cylinder, in feet
$2d =$ stroke, in ft

Assuming 100% volumetric efficiency:

$$0\cdot 7854 \times d^2 \times 2d = \frac{300 \times 184 \times 0\cdot 35 \times 0\cdot 95}{0\cdot 45 \times 130 \times 60 \times 215}$$

$$d = \sqrt[3]{\frac{300 \times 184 \times 0\cdot 35 \times 0\cdot 95}{0\cdot 45 \times 130 \times 60 \times 215 \times 0\cdot 7854 \times 2}}$$

$= 0\cdot 2492$ ft

$0\cdot 2492 \times 12 = 2\cdot 99$ in., say 3 in.

∴ Diameter of cylinder $= 3$ in. ⎫
⠀⠀⠀⠀⠀⠀⠀⠀Stroke $= 6$ in. ⎬ Ans.
⠀⠀⠀⠀⠀⠀⠀⠀⠀⠀⠀⠀⠀⠀⠀⠀⠀⠀⠀⎭

50. $15 \times v_1^{1\cdot35} = 150 \times v_2^{1\cdot35}$

$$\left\{\frac{v_1}{v_2}\right\}^{1\cdot35} = \frac{150}{15} = 10$$

$$\therefore \frac{v_1}{v_2} = 10^{\frac{1}{1\cdot35}} = 5\cdot504$$

$$\therefore \frac{21 + C}{C} = 5\cdot504$$

$\therefore 4\cdot504\ C = 21$
$\therefore C = 4\cdot66$ in. Ans.

51.   Oil used per i.h.p. per hour $= 0\cdot43 \times 0\cdot8 = 0\cdot344$ lb

$$\text{Thermal efficiency} = \frac{2545}{0\cdot344 \times 20000} \times 100 = 37\%$$

$100 - 37 = 63\%$ loss, and this is due to heat in exhaust gases and cooling water.

$\therefore$ Loss in water $= 63 - 30 = 33\%$ Ans.

52. $T_1 = 90 + 460 = 550,\qquad V_1 = V_4 = 6$
$T_3 = 3000 + 460 = 3460,\qquad V_2 = V_3 = 1$

$$\frac{T_2}{T_1} = \left\{\frac{V_1}{V_2}\right\}^{n-1} \quad \therefore T_2 = 550 \times 6^{0\cdot4} = 1127°\text{F abs.}$$

$\therefore$ Temperature at end of compression $= 1127 - 460$
    $= 667°$F. Ans. (a)

$$\frac{T_4}{T_3} = \left\{\frac{V_3}{V_4}\right\}^{n-1} \quad \therefore T_4 = 3460 \times (\tfrac{1}{6})^{0\cdot4} = 1690°\text{F abs.}$$

$\therefore$ Temperature at end of expansion $= 1690 - 460$
    $= 1230°$F. Ans. (b)

$$\text{Efficiency} = \frac{T_3 - T_4}{T_3} = \frac{3460 - 1690}{3460}$$

$$= 0\cdot5116 = 51\cdot16\% \text{ Ans. (c)}$$

SOLUTIONS TO FIRST CLASS EXAMINATION QUESTIONS 559

53. $383 \cdot 5°F = 195 \cdot 3°C.$ $212°F = 100°C$

Height of water varies as the volume

$$\therefore \text{Height} = 84 \times \frac{1 + 0 \cdot 0000119 \times 191 \cdot 3^{1 \cdot 8}}{1 + 0 \cdot 0000119 \times 96^{1 \cdot 8}}$$

$$= 84 \times \frac{1 + 0 \cdot 1522}{1 + 0 \cdot 0441}$$

$$= 92 \cdot 69 \text{ in.}$$

$\therefore$ Difference $= 92 \cdot 69 - 84 = 8 \cdot 69$ in. Ans.

54. Clearance volume expressed in cm of stroke

$$= \frac{6000}{26 \cdot 5^2 \times 0 \cdot 7854} = 10 \cdot 88 \text{ cm}$$

$$15 \times (53 + 10 \cdot 88)^{1 \cdot 38} = p_2 \times 10 \cdot 88^{1 \cdot 38}$$

$$\therefore p_2 = 15 \times \left\{ \frac{63 \cdot 88}{10 \cdot 88} \right\}^{1 \cdot 38} = 172 \cdot 5 \text{ lb/in}^2 \text{ abs.}$$

$$= 157 \cdot 5 \text{ lb/in}^2 \text{ gauge. Ans.}$$

55. Mechanical efficiency $= \frac{4550}{5660} = 0 \cdot 804$ Ans.
Thermal efficiency $= 1 - 0 \cdot 3 - 0 \cdot 32 = 0 \cdot 38$
$\therefore$ Overall efficiency $= 0 \cdot 804 \times 0 \cdot 38 = 0 \cdot 3055$ Ans.

$$\text{Oil per i.h.p. per hour} = \frac{21 \cdot 5 \times 2240}{24 \times 5660} = 0 \cdot 3545$$

$$\text{Thermal efficiency} = \frac{2545}{0 \cdot 3545 \times \text{Calorific value}}$$

$$\therefore \text{Calorific value} = \frac{2545}{0 \cdot 3545 \times 0 \cdot 38}$$

$$= 18900 \text{ Btu/lb. Ans.}$$

56.
$$\text{Steam used per day} = \frac{1500 \times 14 \times 24}{2240} = 225 \text{ tons}$$

Fresh water returned to hotwell $= 0.98 \times 225 = 220.5$ tons
Let $x =$ tons leaking per day
$220.5 \times 0 + x \times 1 = (220.5 + x)\, 0.15$
$0.85x = 220.5 \times 0.15$
$x = 39$ tons
   Total water in hotwell $= 220.5 + 39 = 259.5$ tons
   Water necessary as feed $= 225$ tons
$\therefore$ Overflow to bilges $= 259.5 - 225 = 34.5$ tons. Ans.

57. Lap $+$ lead $= 3\tfrac{9}{16}$. Sin $44° 44' = 2.508$ in.
$\therefore$ lap $= 2.508 - 0.4375 = 2.0705$ in. Ans. (i)
Max. opening to steam $=$ throw $-$ lap
$= 3.5625 - 2.0705 = 1.492$ in. Ans. (ii)

$$\text{Sine of angle of advance due to lap} = \frac{2.0705}{3.5625}$$

$= \text{Sin } 35° 32'$
$\therefore$ Steaming angle $= 180° - 2\,(35° 32') = 108° 56'$
Angle of preadmission $= 44° 44' - 35° 32' = 9° 12'$
At cut off, angle of crank from dead centre
$= 108° 56' - 9° 12' = 99° 44'$ and $180° - 99° 44' = 80° 16'$

$$\therefore \text{Cut off occurs at } \frac{1 + \text{Cos } 80° 16'}{2} = 0.5845 \text{ stroke.}$$
Ans. (iii)

58.
$$\text{Vol. at } 55°F = \tfrac{1}{2}\left[\frac{55 + 460}{500} + \frac{500}{55 + 460}\right]$$
$= 1.00045$
$$\text{Vol. at } 375°F = \tfrac{1}{2}\left[\frac{375 + 460}{500} + \frac{500}{375 + 460}\right]$$
$= 1.1344$
Increase per unit vol. $= 1.1344 - 1.00045 = 0.13395$
$$\text{Actual increase vol.} = \frac{0.13395 \times 9 \times 2240}{62.5}$$
$$\text{Difference in water level} = \frac{0.13395 \times 9 \times 2240}{62.5\,(10^2 - 3\tfrac{1}{4}{}^2)\,0.7854}$$
$= 0.615$ ft $= 7.38$ in. Ans.

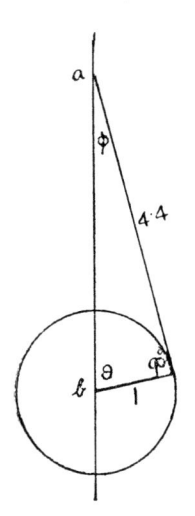

Assuming down stroke.

Tan θ = 4·4 $\therefore \theta$ = 77° 12′

$$a\,b = \frac{4\cdot 4}{\text{Sin } 77° \, 12′} = 4\cdot 512$$

Distance piston is from top

$$= \frac{4\cdot 4 + 1 - 4\cdot 512}{2}$$

= 0·445 of stroke.

= 2·0025 feet from top.   Ans.

$$\text{Twisting moment} = \frac{\text{Piston load}}{\text{Cos } \phi} \times 2\cdot 25$$

$$\therefore \text{Piston load} = \frac{800000 \times \text{Cos } 12° \, 48′}{2\cdot 25}$$

= 346700 lb = 154·8 tons.   Ans.

60.   Valve travel = $(1\frac{7}{16} + 2)\,2 = 6\frac{7}{8}$ in.   Ans.

Sine of angle of advance = $\dfrac{1\frac{9}{16}}{3\frac{7}{16}}$ = 0·4545

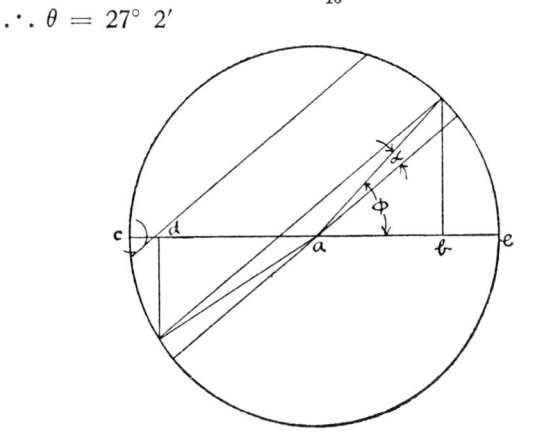

$\therefore \theta$ = 27° 2′

$$\text{Sin } \alpha = \frac{\frac{3}{16}}{3\frac{7}{16}} = 0{\cdot}0545 \quad \therefore \alpha = 3°\,8'.$$

$\phi = 27°\,2' + 3°\,8' = 30°\,10'$
$a\,b = 3\frac{7}{16} \text{ Cos } 30°\,10' = 2{\cdot}972 \text{ in.}$
$c\,b = 3{\cdot}4375 + 2{\cdot}972 = 6{\cdot}4095 \text{ in.}$

$$\text{Fraction of stroke at release} = \frac{6{\cdot}4095}{6{\cdot}875} = 0{\cdot}9323 \text{ Ans.}$$

$27°\,2' - 3°\,8' = 23°\,54'$
$d\,a = 3\frac{7}{16} \text{ Cos } 23°\,54' = 3{\cdot}143$
$e\,d = 3{\cdot}4375 + 3{\cdot}143 = 6{\cdot}5805$

$$\text{Fraction of stroke at compression} = \frac{6{\cdot}5805}{6{\cdot}875} = 0{\cdot}957 \text{ Ans.}$$

61. Volume of one lb of air at 32°F and 14·7 lb/in²

$$= \frac{1}{0{\cdot}0807} = 12{\cdot}39 \text{ ft}^3$$

$\therefore$ Volume of 5 lb of air at 70°F and 14·7 lb/in²

$$= 5 \times 12{\cdot}39 \times \frac{530}{492} = 66{\cdot}75 \text{ ft}^3$$

$$P_1V_1^{1{\cdot}4} = P_2V_2^{1{\cdot}4}$$
$$1 \times 66{\cdot}75^{1{\cdot}4} = 25 \times V_2^{1{\cdot}4}$$
$$V_2 = 6{\cdot}698 \text{ ft}^3 \quad \ldots \quad \ldots \quad \ldots \text{ Ans. (i)}$$

$$\frac{P_1V_1}{T_1} = \frac{P_2V_2}{T_2}$$

$$\frac{1 \times 66{\cdot}75}{530} = \frac{25 \times 6{\cdot}698}{T_2}$$

$T_2 = 1330°\text{F abs.} = 870°\text{F.} \quad \ldots \quad \ldots \quad \ldots \text{ Ans. (ii)}$

SOLUTIONS TO FIRST CLASS EXAMINATION QUESTIONS 563

62.
$$\text{Diam. of shaft} = \frac{345}{25\cdot 4 \times 12} \text{ feet.}$$

Horse power expended on friction per minute

$$= \frac{29\cdot 4 \times 2240 \times 8 \times 0\cdot 014 \times 345 \times \pi \times 85}{25\cdot 4 \times 12 \times 33000} = 67\cdot 53 \quad \text{Ans.}$$

$$\text{Chu/min.} = \frac{67\cdot 53 \times 42\cdot 42 \times 5}{9} = 1592 \quad \text{Ans.}$$

63. Absolute pressure of steam in calorimeter
$$= 6\cdot 8 + 29\cdot 9 = 36\cdot 7 \text{ inches of mercury.}$$
$$36\cdot 7 \times 0\cdot 491 = 18 \text{ lb/in}^2$$

Total heat before throttling = Total heat after throttling
$$372\cdot 3 + 829\cdot 4\, q = 1154\cdot 6 + 0\cdot 55\,(260 - 222\cdot 4)$$
$$829\cdot 4\, q = 802\cdot 98$$
$$q = 0\cdot 9681 \quad \text{Ans. (a)}$$

Additional heat to dry and superheat 1 lb of steam at 240 lb/in²

$$= (1 - 0\cdot 9681) \times 829\cdot 4 + 0\cdot 55(600 - 397\cdot 4)$$

$$= 0\cdot 0319 \times 829\cdot 4 + 0\cdot 55 \times 202\cdot 6$$

$$= 26\cdot 46 + 111\cdot 43$$

$$= 137\cdot 9 \text{ Btu/lb} \quad \text{Ans. (b)}$$

Volume per lb of wet steam $= 0\cdot 9681 \times 1\cdot 918$
$$= 1\cdot 857 \text{ ft}^3$$

Volume per lb of superheated steam
$$= 1\cdot 918 \times \frac{600 + 460}{397\cdot 4 + 460}$$

$$= 2\cdot 371 \text{ ft}^3$$

$$\% \text{ increase in volume} = \frac{2\cdot 371 - 1\cdot 857}{1\cdot 857} \times 100$$

$$= 27\cdot 69\% \quad \text{Ans. (c)}$$

64. Fuel per hour = 3600 × 0·37 lb
Gases per hour = 3600 × 0·37 × 19 lb
(Note, 1 lb of fuel + 18 lb of air = 19 lb of gases)

Available heat
= 3600 × 0·37 × 19 × (620 − 420) × 0·203
= 1,028,000 Btu per hour.

Heat utilised in forming steam = 1,028,000 × 0·8
= 822,400 Btu

From tables, t_F = 389·9 ; h_f = 364·2 ; h_{fg} = 836·5

Heat to form 1 lb of steam
= 364·2 − (120 − 32) + 0·96 × 836·5 = 1079·24 Btu

∴ Steam formed = $\dfrac{822400}{1079\cdot24}$ = 762 lb per hour. **Ans.**

65. Since losses are 5% of input to generator,
then, Output = 95% of input,
∴ Losses are $\tfrac{5}{95}$ of output = $\tfrac{5}{95}$ × 7500 kilowatts
Heat equivalent of 1 kilowatt hour = 3412 Btu

Heat equivalent of losses = $\tfrac{5}{95}$ × 7500 × $\dfrac{3412}{60}$ Btu/min.

Weight of air blown through
= 35000 × 0·0807 × $\dfrac{492}{530}$ × $\dfrac{34\cdot5}{34}$ lb

Heat gained by air = Heat equivalent of losses

35000 × 0·0807 × $\dfrac{492}{530}$ × $\dfrac{34\cdot5}{34}$ × 0·24 × temp. rise

= $\tfrac{5}{95}$ × 7500 × $\dfrac{3412}{60}$

From which, temp. rise = 35·18 F°
Temperature of windings = 55°C = 131°F
∴ Difference in temperature between windings and **final** temperature of the air = 131 − (70 + 35·18) = 25·82 F°. **Ans.**

SOLUTIONS TO FIRST CLASS EXAMINATION QUESTIONS 565

66. Weight of fuel burned per working stroke

$$= \frac{0.35 \times 3600}{60 \times 85 \times 6} = 0.04119 \text{ lb}$$

Weight of air required per lb of fuel
$$= \tfrac{100}{23} \{2\tfrac{2}{3} \text{ C} + 8 (\text{H} - \tfrac{\text{O}}{8})\}$$
$$= \tfrac{100}{23} \{2\tfrac{2}{3} \times 0.84 + 8 \times 0.15\}$$
$$= 14.96 \text{ lb}$$

Weight of air required per working stroke
$$= 0.04119 \times 14.96 = 0.6161 \text{ lb. Ans. (a)}$$

Weight of air taken in per stroke
$$= 0.0807 \times 10 \times \tfrac{17}{15} \times \tfrac{492}{545} = 0.8255 \text{ lb. Ans. (b)}$$

Excess air $= 0.8255 - 0.6161 = 0.2094$ lb. Ans. (c)

67. Volume of each bottle
$$= \tfrac{\pi}{6} \times 1^3 + \tfrac{\pi}{4} \times 1^2 \times 7 = 6.0214 \text{ ft}^3$$

Volume of 3 bottles and pipes
$$= 3 \times 6.0214 \times 1.1 = 19.87 \text{ ft}^3$$

Volume of free air to be pumped
$$= 19.87 \times \frac{1215}{15} - 19.87 = 1590 \text{ ft}^3$$

Volume of free air taken in per stroke
$$= \frac{(15^2 - 3^2) \times 0.7854 \times 1 \times 0.91}{144} = 1.072 \text{ ft}^3$$

$$\therefore \text{Time} = \frac{1590}{1.072 \times 150} = 9.883 \text{ minutes. Ans.}$$

566 REED'S HEAT AND HEAT ENGINES FOR MARINE ENGINEERS

68. The diameter of the I.P. cylinder in relation to the diameters of the H.P. and L.P. indicates that this compressor is an in-line type, and not a tandem as in the previous (No. 67) problem. The full volume of the L.P. cylinder is therefore available for drawing air in from the atmosphere.

Volume of free air (15 lb/in^2 abs.) required to fill reservoirs of 1000 ft^3 at 350 lb/in^2 = $1000 \times \frac{350}{15}$ ft^3

Equivalent volume of free air initially in reservoirs when the pressure is 250 lb/in^2 = $1000 \times \frac{250}{15}$ ft^3

∴ Volume of free air to be taken into compressor

$$= \frac{1000 \times 350}{15} - \frac{1000 \times 250}{15} = \frac{1000 \times 100}{15} \text{ ft}^3$$

Volume of free air taken into compressor per stroke

$$= \frac{0.7854 \times 15^2 \times 15 \times 0.9}{1728} = 1.38 \text{ ft}^3$$

Half of this supplies reservoirs = 0·69 ft^3/stroke

At 180 effective strokes per minute:

$$\text{Time required} = \frac{1000 \times 100}{15 \times 0.69 \times 180} = 53.68 \text{ minutes. Ans.}$$

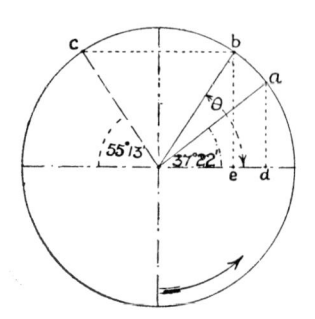

69. When the engine is on the centre the displacement of the valve from its mid-travel = lap + lead.

This is represented by ad
ad = lap + lead = 3·5 sin 37° 22′

∴ lap = 3·5 sin 37° 22′ − 0·25
= 1·8745 in.

When the port is 1 in. open to steam the displacement of the valve from its mid-travel
= 1 + 1·8745 = 2·8745 in. = be

SOLUTIONS TO FIRST CLASS EXAMINATION QUESTIONS 567

The eccentric centre may be at b or c, in other words, there are two positions of the eccentric, and of the crank, when the port is 1 inch open to steam.

$$\sin \theta = \frac{2 \cdot 8745}{3 \cdot 5}, \theta = 55° \; 13'$$

$$55° \; 13' - 37° \; 22' = 17° \; 51'$$
$$180° - 55° \; 13' - 37° \; 22' = 87° \; 25'$$

The steam port is 1 in. open when the crank has moved through 17° 51', or 87° 25' from the centre. Ans.

70. From steam tables, 1st case

$P = 235,\qquad h_f = 370 \cdot 3,\qquad h_{fg} = 831 \cdot 2$
$P = 2 \cdot 4,\qquad h_f = 100 \cdot 8,\qquad h_{fg} = 1018 \cdot 3$
Heat per lb of high press. steam $= 370 \cdot 3 + 0 \cdot 94 \times 831 \cdot 2$
$\qquad = 1151 \cdot 628$ Btu
Heat per lb of exhaust press. steam $= 100 \cdot 8 + 0 \cdot 755 \times 1018 \cdot 3$
$\qquad = 869 \cdot 6165$ Btu
$\therefore$ heat drop through turbine $= 1151 \cdot 628 - 869 \cdot 6165$
$\qquad = 282 \cdot 0115,$
$\qquad\qquad$ say 282 Btu/lb

From tables, 2nd case
h of steam at 250 lb/in², superheated 200 F°
$\qquad = 1318 \cdot 5$ Btu

h of steam at 200 lb/in² superheated 200 F°
$\qquad = 1312 \cdot 1$ Btu

$\qquad$ Difference for 50 lb/in² $= \quad 6 \cdot 4$ Btu

$\therefore$ difference for 35 lb/in² $= \dfrac{35}{50} \times 6 \cdot 4 = 4 \cdot 48$ Btu

$\therefore$ Heat of steam at 235 lb/in² and superheated 200 F°
$= 1312 \cdot 1 + 4 \cdot 48 = 1316 \cdot 58$ Btu/lb

Also from tables, $P = 1$, $h_f = 69 \cdot 7$, $h_{fg} = 1036 \cdot 1$
Heat per lb of exhaust steam $= 69 \cdot 7 + 0 \cdot 828 \times 1036 \cdot 1$
$\qquad = 927 \cdot 5908$ Btu

T

∴ heat drop through turbine = 1316·58 — 927·5908
= 388·9892, say 389 Btu/lb

Additional heat drop = 389 — 282 = 107 Btu/lb

$$\text{Increase in power} = \frac{107}{282} \times 100 = 37\cdot94\% \quad \text{Ans.}$$

71. Latent heat of water = 144 Btu/lb = 80 cal/gram
Working in kilocalories, let $t°C$ = final temperature

Heat gained by ice and cold water = Heat lost by hot water

$31\cdot5\,(0\cdot5 \times 25 + 80 + t) + 97\cdot5 \times t = 91\cdot5\,(54 - t)$
$393\cdot75 + 2520 + 31\cdot5\,t + 97\cdot5\,t = 4941 - 91\cdot5\,t$
$220\cdot5\,t = 2027\cdot25$
$t = 9\cdot194°C$. Ans.

72. Assuming atmospheric pressure is 15 lb/in²
Final pressure = 450 + 15 = 465 lb/in² abs.

Isothermal compression,
pv = constant
$15 \times 3 \times 1728 = 465 \times v_2$

$$v_2 = \frac{15 \times 3 \times 1728}{465} = 167\cdot23 \text{ cu in. Ans.}$$

Adiabatic compression:
$pv^{1\cdot4}$ = constant
$15 \times (3 \times 1728)^{1\cdot4} = 465 \times v_2^{1\cdot4}$
$(3 \times 1728)^{1\cdot4} = 31 \times v_2^{1\cdot4}$
$1\cdot4 \log 5184 = \log 31 + 1\cdot4 \log v_2$
∴ $1\cdot4 \log v_2 = 1\cdot4 \log 5184 - \log 31$

$$\log v_2 = \log 5184 - \frac{\log 31}{1\cdot4}$$

$= 3\cdot7146 - 1\cdot0653 = 2\cdot6493$
$v_2 = 446$ cu in. Ans.

SOLUTIONS TO FIRST CLASS EXAMINATION QUESTIONS 569

73. From steam tables,

h of steam at 200 lb/in^2
superheated 360 F° = 1394·9 Btu

h of steam at 180 lb/in^2
superheated 360 F° = 1391·1 Btu

Difference for 20 lb/in^2 = 3·8 Btu
∴ Difference for 5 lb/in^2 = 0·95 Btu

∴ h of steam at 195 lb/in^2
superheated 360 F° = 1394·9 — 0·95
= 1393·95 Btu

Again, h of steam at 200 lb/in^2
superheated 320 F° = 1374·3 Btu

And, h of steam at 180 lb/in^2
superheated 320 F° = 1370·7 Btu

Difference for 20 lb/in^2 = 3·6 Btu
∴ Difference for 5 lb/in^2 = 0·9 Btu

h of steam at 195 lb/in^2
superheated 320 F° = 1374·3 — 0·9
= 1373·4 Btu

h of steam at 195 lb/in^2
superheated 360 F° = 1393·95 Btu

h of steam at 195 lb/in^2
superheated 320 F° = 1373·4 Btu

Difference for 40°F = 20·55 Btu
∴ Difference for 20°F = 10·28 Btu

and h of steam at 195 lb/in^2
superheated 340 F° = 1373·4 + 10·28
= 1383·68 Btu

Since the feed temperature is 145°F
∴ heat to form 1 lb of steam
= 1383·68 — (145 — 32) = 1270·68 Btu

Heat in steam formed per hour
= 1270·68 × 816 Btu

Heat in coal burnt per hour
= 10110 × 160

$$\therefore \text{Boiler effic.} = \frac{1270 \cdot 68 \times 816}{10110 \times 160} = 0 \cdot 641 \text{ or } 64 \cdot 1\% \quad \text{Ans. (a)}$$

$$\text{Equivalent evaporation} = \frac{1270 \cdot 68 \times 816}{970 \cdot 6 \times 160} = 6 \cdot 67 \text{ lb. Ans. (b)}$$

74. $1 \cdot 34 \times \text{C.V.} = (1634 + 223) \times (20 \cdot 4 - 14 \cdot 3)$

$$\text{C.V.} = \frac{1857 \times 6 \cdot 1}{1 \cdot 34} = 8452 \text{ gram calories per gram. Ans.}$$

8452 gram calories per gram

$$= \frac{8452}{252} \text{ Btu per gram}$$

$$= \frac{8452}{252} \times \frac{1000}{2 \cdot 2} = 15250 \text{ Btu/lb. Ans.}$$

See Chapter 13 for diagram and description of **Bomb Calorimeter**.

75. Friction horse power at thrust,

$$= \frac{12 \times 2240 \times 0 \cdot 05 \times 2\pi \times 7 \cdot 5 \times 77}{33000 \times 12} = 12 \cdot 32$$

Heat equivalent of friction h.p. = $12 \cdot 32 \times 2545$ Btu per hour.

Let x = gallons of oil required per hour

1 gallon weighs $10 \times 0 \cdot 92 = 9 \cdot 2$ lb

Heat gained by oil = Heat given out at bearings
$x \times 9 \cdot 2 \times 10 \times 0 \cdot 48 = 12 \cdot 32 \times 2545$

$$x = \frac{12 \cdot 32 \times 2545}{9 \cdot 2 \times 10 \times 0 \cdot 48} = 710 \text{ gallons. Ans.}$$

76. $\text{Ratio of expansion} = \dfrac{\text{Full volume of L.P. cyl.}}{\text{Vol. of H.P. up to cut off}}$

$$R = \frac{3 \cdot 5}{0 \cdot 5 \times 1} = 7$$

SOLUTIONS TO FIRST CLASS EXAMINATION QUESTIONS 571

$$\text{Referred mean press.} = \left\{ \frac{P}{R}(1 + \log_\varepsilon R) - P_b \right\} \times f_o$$

$$= \left\{ \frac{140}{7}(1 + \log_\varepsilon 7) - 3 \right\} \times 0.68$$

$$= 38 \text{ lb/in}^2$$

$$\text{i.h.p.} = \frac{pALN}{33000}$$

$$1200 = \frac{38 \times 0.7854 \times D^2 \times 600}{33000}$$

$$D^2 = \frac{1200 \times 33000}{38 \times 0.7854 \times 600} = 2211$$

$$D = \sqrt{2211} = 47.01 \text{ in. (say 47 in.)}$$

$$d^2 = \frac{2211}{3.5} = 631.6$$

$$d = \sqrt{631.6} = 25.13 \text{ in. (say 25 in.)}$$

∴ Suitable cylinder diameters are:
$$\left. \begin{array}{l} \text{H.P.} = 25 \text{ in.} \\ \text{L.P.} = 47 \text{ in.} \end{array} \right\} \text{Ans.}$$

77. Assuming only Btu version of steam tables are available:

From steam tables,
P = 115 lb/in², h_g = 1190·7 Btu/lb

45°C × 9/5 + 32 = 113°F

4000 litres of water weighs 4000 kilograms

Let t = final temperature in °F

Heat lost by steam = Heat gained by water
$$50[1190.7 - (t - 32)] = 4000(t - 113)$$
$$1222.7 - t = 80t - 9040$$
$$81t = 10262.7$$
$$\left. \begin{array}{l} t = 126.7°F \\ \text{or } 52.6°C \end{array} \right\} \text{Ans.}$$

78. From steam tables,
$t_F = 162\cdot3$, $P = 5$ lb/in^2, $h_f = 130\cdot2$, $h_{fg} = 1001\cdot6$

Heat given up by 1 lb of steam when condensed
$= 130\cdot2 - (120 - 32) + 0\cdot81 \times 1001\cdot6 = 853\cdot5$ Btu

$$\text{Steam condensed per minute} = \frac{7800 \times 15\cdot3}{60} \text{ lb}$$

Heat taken up by circulating water $= 84 - 55 = 29$ Btu/lb

∴ Circulating water required per minute

$$= \frac{7800 \times 15\cdot3 \times 853\cdot5}{60 \times 29} = 58530 \text{ lb}$$

$$\text{Horse power of pump} = \frac{58530 \times 30}{33000} = 53\cdot21 \text{ Ans.}$$

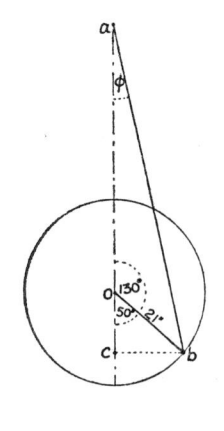

79.

$cb = 21 \sin 50° = 16\cdot086$ in.
$oc = 21 \cos 50° = 13\cdot499$ in.
Load on guide $= \sqrt{(29\cdot37)^2 - (28\cdot9)^2}$
$= 5\cdot233$ tons. Ans. (a)
cb represents 5·233 tons to the same scale that ab represents 29·37 tons
$$ab = 16\cdot086 \times \frac{29\cdot37}{5\cdot233} = 90\cdot3 \text{ in.}$$
Connecting rod length $= 90\cdot3$ in. Ans. (b)

$$\frac{21}{\sin \phi} = \frac{90\cdot3}{\sin 130°}$$

$$\sin \phi = \frac{21}{90\cdot3} \times \sin 130°, \phi = 10° 16'$$

$ac = 90\cdot3 \cos 10° 16' = 88\cdot85$ in.

Effect of angularity of con rod $= 90\cdot3 - 88\cdot85 = 1\cdot45$ in.
Piston is $13\cdot499 + 1\cdot45 = 14\cdot949$ in. below half stroke.
$21 - 14\cdot949 = 6\cdot051$ in. from bottom of stroke. Ans. (c)

SOLUTIONS TO FIRST CLASS EXAMINATION QUESTIONS 573

80. $$\text{b.h.p.} = \frac{(370 - 10) \times \pi \times 5 \times 300}{33000} = 51.41. \text{ Ans. (b)}$$

Heat generated at brake = 51·41 × 2545 Btu per hour
Heat carried away = 51·41 × 2545 × 0·85 Btu per hour

10 C° rise in temperature = $10 \times \frac{9}{5}$ = 18 F° rise in temperature

Let x = gallons of water per hour. 1 gall. weighs 10 lb

$$10 \times x \times 18 = 51.41 \times 2.545 \times 0.85$$

$$x = \frac{51.41 \times 2545 \times 0.85}{10 \times 18} = 618 \text{ galls. per hour. Ans. (a)}$$

81. Weight of moving parts per sq. in. of piston area

$$= \frac{275}{12^2 \times 0.7854} = 2.431 \text{ lb}$$

Force to accelerate moving parts = mass × acceleration

$$= \frac{2.431 \times 161}{32.2} = 12.155 \text{ lb}$$

The engine is vertical, and the weight of the moving parts acts downwards.

∴ effective pressure on piston = 53·2 + 12·155 + 2·431
= 67·786 lb/in²

$$\text{Crank effort} = \frac{\text{effective piston load} \times \text{OT}}{\text{crank length}} \quad \text{(see Chapter 13, Vol. 2)}$$

When the crank is horizontal OT = crank length

∴ Crank effort = 67·786 × 12² × 0·7854 = 7668 lb.
Ans.
Turning moment = 7668 × 0·75 = 5751 ft lb. **Ans.**

82. From steam tables,
   $P = 220 \text{ lb/in}^2$, $h_f = 364 \cdot 2$, $h_{fg} = 836 \cdot 5$

   Initial temp. of water $= 60 + 460 = 520°$F abs.
   Final temp. of water $= 1 \cdot 1 \times 520 = 572°$F abs.
   $572 - 460 = 112°$F

   Let initial weight of water $= 100$ lb
   then, final weight of water $= 105$ lb
   and, wt. of steam blown in $= 5$ lb

   or, in the same ratio, let 1 lb of steam be blown into 20 lb of water.

   Heat lost by steam $=$ Heat gained by water
   $364 \cdot 2 - (112 - 32) + 836 \cdot 5 \, q = 20 \, (112 - 60)$
   $284 \cdot 2 + 836 \cdot 5 \, q = 1040.$
   $836 \cdot 5 \, q = 755 \cdot 8$
   $q = 0 \cdot 9038$ Ans.

83. Initial volume of pipes and oil $= 0 \cdot 7854 \times (0 \cdot 5)^2 \times 500 \times 12$
   $= 1178$ cu in.

   Change of volume of the oil $= 1178 \, (0 \cdot 00043 \times 25)$ cu in.
   Co-efficient of linear expansion of copper $= 0 \cdot 00001$
   Co-efficient of cubical expansion of copper $= 0 \cdot 00001 \times 3$
   $= 0 \cdot 00003$

   Change of volume of pipes $= 1178 \, (0 \cdot 00003 \times 25)$
   ∴ volume of oil released
   $= 1178 \, (0 \cdot 00043 \times 25) - 1178 \, (0 \cdot 00003 \times 25)$
   $= 1178 \times 25 \, (0 \cdot 00043 - 0 \cdot 00003)$
   $= 1178 \times 25 \times 0 \cdot 0004 = 11 \cdot 78$ cu in. Ans.

84. 1400 kilograms per sq. centimetre
   $= 1400 \times 2 \cdot 2 \times 2 \cdot 54^2 \times 144 \text{ lb/ft}^2$
   Work done $=$ Pressure (lb/ft^2) $\times$ Volume (ft^3)

   ∴ Work done on 1 cubic foot
   $= 1400 \times 2 \cdot 2 \times 2 \cdot 54^2 \times 144 \times 1$ ft lb

   1 Chu $= 1400$ ft lb, hence,

   Heat equivalent of work done

   $= \dfrac{1400 \times 2 \cdot 2 \times 2 \cdot 54^2 \times 144 \times 1}{1400}$

   $= 2 \cdot 2 \times 2 \cdot 54^2 \times 144$ Chu

SOLUTIONS TO FIRST CLASS EXAMINATION QUESTIONS 575

Weight of lead = 712 lb/ft^3
Spec. ht. of lead = 0·029

Heat given to lead = Wt. × spec. ht. × temp. rise
2·2 × 2·54^2 × 144 = 712 × 0·029 × temp. rise

$$\therefore \text{Temp. rise} = \frac{2·2 \times 2·54^2 \times 144}{712 \times 0·029}$$

= 98·97 C degrees. Ans.

85. Available hydrogen = $12·1\% - \dfrac{1·5\%}{8} = 11·9125\%$

C.V. = 14500 × 0·86 + 62000 × 0·119125
= 19856 Btu/lb

Oil used per i.h.p. hour = 0·8 × 0·38 = 0·304 lb

Thermal efficiency, based on i.h.p. = $\dfrac{2545}{0·304 \times 19856}$

= 0·4216, or 42·16% Ans.

86. See Chapter 13 for lower and higher calorific values
6·8 C° rise of temperature = 6·8 × $\frac{9}{5}$ = 12·24 F°
0·015 × C.V. = (3 + 12) × 12·24

$$\text{C.V.} = \frac{15 \times 12·24}{0·015} = 12240 \text{ Btu/lb. Ans.}$$

This is the higher calorific value of the coal.

87. Referred mean press. = $\left\{\dfrac{P}{R}(1 + \log_e R) - P_b\right\} \times f_o$

$= \left\{\dfrac{200}{12}(1 + \log_e 12) - 3\right\} \times 0·65$

= 35·77 lb/in^2

Let dia. of H.P., I.P. and L.P. be represented by d, x and D

$$\text{i.h.p.} = \frac{pALN}{33000}$$

$$2500 = \frac{35 \cdot 77 \times 0 \cdot 7854 \times D^2 \times 600}{33000}$$

$D^2 = 4893, \qquad D = \sqrt{4893} \qquad = 70 \text{ in.}$
(practically)

$d^2 = \dfrac{4893}{7 \cdot 2} = 680, \qquad d = \sqrt{680} \qquad = 26 \text{ in.}$
(practically)

$x^2 = 680 \times 2 \cdot 7, \qquad x = \sqrt{680 \times 2 \cdot 7} = 43 \text{ in.}$
(practically)

Practical cylinder diameters are

26 in., 43 in., and 70 in. Ans.

88. From steam tables,
$P = 120 \text{ lb/in}^2, \qquad h_f = 312 \cdot 5, \qquad h_{fg} = 878 \cdot 9 \text{ Btu/lb}$

Abridged steam tables do not include values for the pressure of $15 \cdot 5$ lb/in², therefore, taking the average values between 16 and 15 lb/in²:

$P = 15 \cdot 5 \text{ lb/in}^2, \qquad t_F = \frac{1}{2}(216 \cdot 3 + 213) = 214 \cdot 65°F$
$hg = \frac{1}{2}(1152 \cdot 4 + 1151 \cdot 2) = 1151 \cdot 8 \text{ Btu/lb}$

Heat before throttling = Heat after throttling
$312 \cdot 5 + 878 \cdot 9\, q = 1151 \cdot 8 + 0 \cdot 5\,(259 - 214 \cdot 65)$
$878 \cdot 9\, q = 861 \cdot 475$
$q = 0 \cdot 9804$ Ans.

See Chapter 8 for sketch and description of throttling calorimeter.

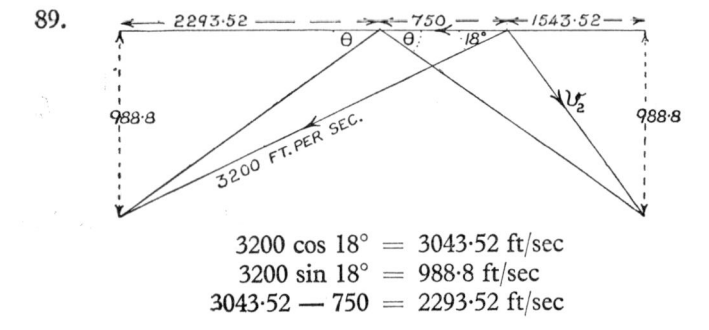

89.

3200 cos 18° = 3043·52 ft/sec
3200 sin 18° = 988·8 ft/sec
3043·52 — 750 = 2293·52 ft/sec

SOLUTIONS TO FIRST CLASS EXAMINATION QUESTIONS 577

$$\text{Tan } \theta = \frac{988 \cdot 8}{2293 \cdot 52} = 0 \cdot 4311, \quad \theta = 23° \, 19'$$

Entrance and exit angles are 23° 19' Ans.

$2293 \cdot 52 - 70 = 1543 \cdot 52$ ft/sec

$$v_2 = \sqrt{(988 \cdot 8)^2 + (1543 \cdot 52)^2}$$
$$= 1833 \text{ ft/sec (absolute vel. at exit). Ans.}$$

90. Total area $= \{(15 \times 7) + (12 \times 7) + (15 \times 12)\} \times 2$
$= 738$ sq ft

Total area $= 738 \times 144 \times (2 \cdot 54)^2$ sq cm $= a$
Depth of insulation $= 6 \times 2 \cdot 54$ cm $= d$

Temperature difference $= 25 - (-5) = 30$ C° $= \theta$
Time $= 60$ seconds $= t$

$$Q = \frac{kat\theta}{d}$$

$$= \frac{0 \cdot 0003 \times 738 \times 144 \times (2 \cdot 54)^2 \times 60 \times 30}{6 \times 2 \cdot 54} \text{ gram calories}$$

$$Q = \frac{0 \cdot 0003 \times 738 \times 144 \times (2 \cdot 54)^2 \times 60 \times 30}{6 \times 2 \cdot 54 \times 252} = 96 \cdot 4 \text{ Btu}$$
Ans.

91. Available hydrogen $= 13 - \frac{2}{8} = 12 \cdot 75\%$

Oxygen to burn the carbon in 1 lb of fuel
$= 2\frac{2}{3} \times 0 \cdot 85 \quad = 2 \cdot 267$ lb

Oxygen to burn the hydrogen in 1 lb of fuel
$= 8 \times 0 \cdot 1275 \quad = 1 \cdot 02$ lb

Oxygen to burn 1 lb of fuel $= 3 \cdot 287$ lb
Theoretical air per lb of fuel $= 3 \cdot 287 \times \frac{100}{23} = 14 \cdot 29$ lb
Ans. (a)

Actual air per lb of fuel $= 14 \cdot 29 \times 1 \cdot 7 = 24 \cdot 293$ lb
Weight of gases formed per lb of fuel burnt $= 24 \cdot 293 + 1$
$= 25 \cdot 293$ lb

∴ Weight of gases per hour $= 1400 \times 25 \cdot 293 = 35410$ lb
Ans. (b)

92. From steam tables, $P = 235$ lb/in^2, $h_g = 1201\cdot5$ Btu/lb

Heat required to form 1 lb of steam
$= 1201\cdot5 - (169 - 32) = 1064\cdot5$ Btu

$$\text{Thermal Eff.} = \frac{\text{Heat given to steam}}{\text{Heat supplied}}$$

$$= \frac{80800 \times 1064\cdot5}{8025 \times 12950} = 0\cdot8275 \text{ or } 82\cdot75\% \text{ Ans.}$$

Overall eff. = Boiler eff. × therm. eff. of eng. × mech. eff.
$0\cdot11 = 0\cdot8275 \times$ therm. eff. of eng. $\times 0\cdot8$
∴ Therm. eff. of engine = $0\cdot1662$ or $16\cdot62\%$ Ans.

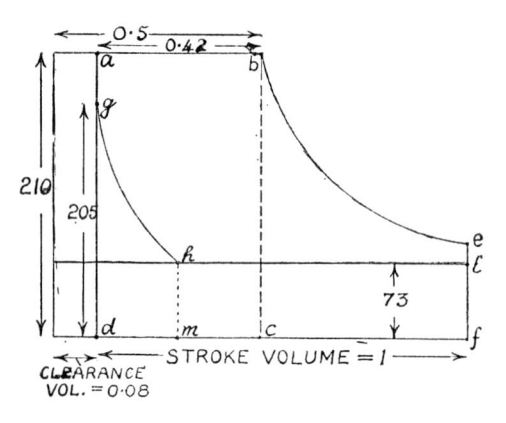

93.

Assuming atmospheric pressure = 15 lb/in^2

Absolute pressures are,
Initial steam = $195 + 15 = 210$ lb/in^2
Back pressure = $58 + 15 = 73$ lb/in^2
At end of compression = $190 + 15 = 205$ lb/in^2

Let the stroke volume be 1

Then the clearance volume is $0\cdot08$, and the volume at cut-off is $0\cdot42 + 0\cdot08 = 0\cdot5$

Ratio of expansion
$$= \frac{1\cdot08}{0\cdot5} = 2\cdot16, \text{ and } \log_e 2\cdot16 = 0\cdot76935$$

SOLUTIONS TO FIRST CLASS EXAMINATION QUESTIONS 579

Ratio of compression

$$= \frac{205}{73} = 2·809, \text{ and } \log_e 2·809 = 1·032$$

Volume at h = vol. at $g \times 2·809$
,, = $0·08 \times 2·809 = 0·2247$

$\therefore hl = 1·08 - 0·2247 = 0·8553$

Valve closes to exhaust at 0·8553, or 85·53 % of return stroke.
Ans. (a)

Area $abcd$ = 210 × 0·42 = 88·2 units
Area $befc$ = 210 × 0·5 × $\log_e$ of ratio of expansion
= 210 × 0·5 × 0·76935 = 80·782 units

$\therefore$ area $abefd$ = 88·2 + 80·782 = 168·982 units

$$\text{Gross mean pressure} = \frac{\text{area } abefd}{\text{stroke volume}} = \frac{168·982}{1}$$

= 168·982 lb/in². Ans. (b)

Area $ghmd$ = 205 × 0·08 × $\log_e$ of ratio of compression
= 205 × 0·08 × 1·032 = 16·92 units

Area $hlfm$ = 0·8553 × 73 = 62·45 units
16·92 + 62·45 = 79·37 units

Net area = $abelhg$ = 168·982 − 79·37 = 89·612 units

$$\text{Theoretical m.e.p.} = \frac{\text{Net area}}{\text{stroke volume}}$$

$$= \frac{89·612}{1} = 89·612 \text{ lb/in}^2. \text{ Ans. (c)}$$

The actual m.e.p. will be some fraction of the theoretical m.e.p., and it may be about 0·75 × 89·612, or about 67 lb/in². Note that the diagram factor is not assumed to be 0·75. It is something less than this, because the area under the compression curve has already been subtracted.

94.

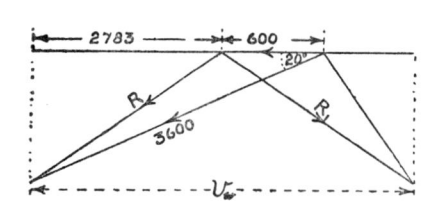

$3600 \cos 20° = 3383$ ft/sec.

$3383 - 600 = 2783$ ft/sec.

If there is no friction loss, then R (relative velocity at entrance) must equal R_1 (relative velocity at exit).

Change of velocity of the steam $= v_w = 2 \times 2783$ ft/sec.

Steam supplied per sec. $= \frac{1800}{3600} = 0.5$ lb

Force = change of momentum per second

Force $= \dfrac{0.5}{32.2} \times 2 \times 2783 = 86.42$ lb. Ans. (a)

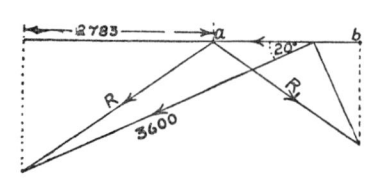

If $R_1 = 0.88$ R due to friction, then it follows that $ab = 0.88 \times 2783$

Change of velocity of the steam

$= 2783 + 0.88 \times 2783$
$= 5232$ ft/sec

Force $= \dfrac{0.5}{32.2} \times 5232 = 81.24$ lb. Ans. (b)

Work done per sec. $= 81.24 \times 600$ ft lb
One horse power $= 550$ ft lb/sec

∴ Horse power $= \dfrac{81.24 \times 600}{550} = 88.65$ Ans. (c)

95. Heat to be taken from 1 lb of water at 48°F to change to ice at 30°F

$= (48 - 32) + 144 + (32 - 30) \times 0.5 = 161$ Btu

Total heat to be extracted per hour $= 1120 \times 161$ Btu

Heat taken up by 1 lb of CO_2 when its dryness fraction increases from 0.34 to $0.92 = (0.92 - 0.34) \times 105.5$
$$= 61.19 \text{ Btu}$$

$\therefore$ Pounds of CO_2 required $= \dfrac{1120 \times 161}{61.19}$

$= 2947$ lb. Ans.

96. The passage of the liquid ammonia through the regulating valve is a throttling, or wire-drawing process. A wire-drawing process is a constant heat operation, that is the total heat after wire-drawing is the same as the total heat before. Measuring the total heat above $0°F$:

Let x = final dryness

Heat before = Heat after

$$70 \times 1.1 = 14 \times 1.1 + x\{566 - 0.8 \times 14\}$$
$$61.6 = 554.8\,x$$

$$x = \dfrac{61.6}{554.8} = 0.111 \text{ Ans.}$$

97. Willans' Law: The steam consumption in a steam engine is a linear function of the indicated horse power, if the ratio of expansion is not changed.

Therefore the steam consumption can be expressed by the straight line law:

$$W = a + bP$$

where W = weight of steam used per hour,
P = indicated horse power.

Inserting values of W and P from the data given to produce a simultaneous equation:

$$37000 = a + b \times 2500 \quad \ldots \text{ (i)}$$
$$67000 = a + b \times 5000 \quad \ldots \text{ (ii)}$$

Multiplying (i) by 2 ... ...
$$74000 = 2a + 5000b$$
$$67000 = a + 5000b$$

Subtracting ... ...
$$7000 = a$$

Substituting $a = 7000$ into (ii)
$$67000 = 7000 + 5000b$$
$$b = 12$$

$\therefore$ Willans' Law for this engine is $W = 7000 + 12\,P$

When P = 4000:
W = 7000 + 12 × 4000
= 55000 lb per hour. Ans. (a)
Specific consumption = 55000 ÷ 4000
= 13·75 lb/i.h.p. hour. Ans. (b)

98.

Crank angle in degrees	0°	30°	60°	90°	120°	150°	180°
Net steam load in lb	13100	15400	14200	9800	5600	1300	—5000
Accelerating force in lb	4600	3330	1450	—860	—2200	—2300	—3000
Piston effort in lb	8500	12070	12750	10660	7800	3600	—2000

At 0°, the net steam load is 13100 lb and of this 4600 lb is required to accelerate the moving parts. Therefore the piston effort is 13100 — 4600 = 8500 lb.

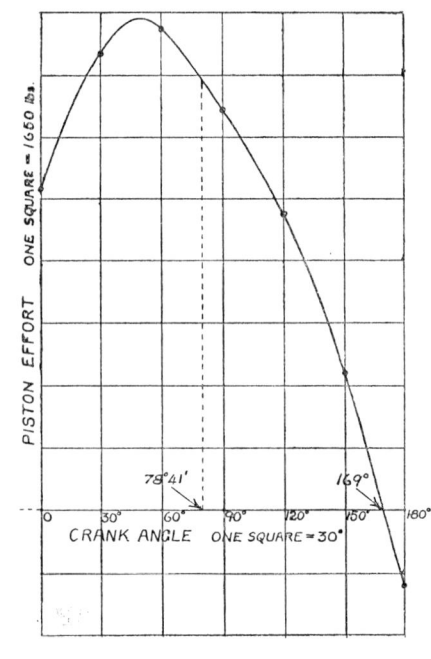

At 90°, the net steam load is 9800, whilst the accelerating force is — 860 lb. The acceleration here is negative, or the moving parts are being retarded, and the piston effort is 9800 — (— 860) = 10660 lb.

At 180° the piston effort is — 5000 — (— 3000) = — 2000 lb.

The diagram shows piston effort plotted on a base line of crank angle turned through. The curve crosses the base line at 169°, and here the piston effort is zero.

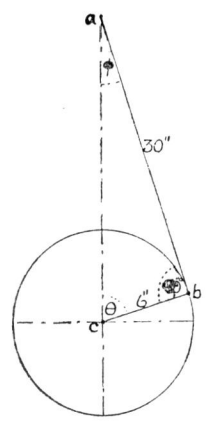

SOLUTIONS TO FIRST CLASS EXAMINATION QUESTIONS 583

Tan $\phi = \frac{3}{60} = 0.2$, ϕ (angularity of con. rod) $= 11°\ 19'$

$\theta = 90° - 11°\ 19' = 78°\ 41'$ and this is the crank angle when the crank and connecting rod are at 90°

From the graph, the piston effort at 78° 41' measures 11470 lb.

$ac = \sqrt{6^2 + 30^2} = 30.6$ in.

ab represents the piston effort, and to the same scale ac represents the force acting in the connecting rod.

Force in connecting rod $= 11470 \times \dfrac{30.6}{30}$ lb

Twisting moment $= 11470 \times \dfrac{30.6}{30} \times 6 = 70200$ lb in. Ans.

99.

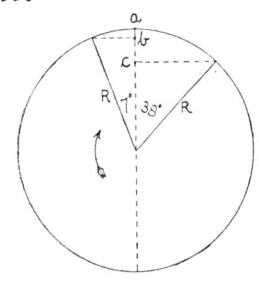

Let the radius of the crank circle be R millimetres.

The valve closes at $45° - 7° = 38°$ past the top centre.

Referring to the diagram,
$ab = R - R \cos 7°$
$\quad = R(1 - \cos 7°)$
$ac = R - R \cos 38°$
$\quad = R(1 - \cos 38°)$
$ab + ac = 134$ mm

$\therefore R(1 - \cos 7° + 1 - \cos 38°) = 134$
$R(2 - 0.9925 - 0.7880) = 134$
$0.2195\ R = 134$

$R = \dfrac{134}{0.2195} = 610.5$ mm

Stroke $= 2 \times 610.5 = 1221$ mm

$= \dfrac{1221}{25.4} = 48.08$ in.

$\Bigg\}$ Ans.

100. Maximum speed

$$= 750 + \frac{0.5}{100} \text{ of } 750 = 753.75 \text{ r.p.m.}$$

Minimum speed

$$= 750 - \frac{0.5}{100} \text{ of } 750 = 746.25 \text{ r.p.m.}$$

Variation of speed $= 753.75 - 746.25$
$= 7.5$ r.p.m.

The flywheel is accelerated during the first 90° movement of the crank from the top centre, and retarded during the next 90°, when the crank has reached the bottom centre. On the up stroke it is accelerated for the first 90°, and retarded during the next 90° movement. It follows, therefore, that the flywheel attains its maximum speed at 90° past top, or bottom centres, because acceleration has just ceased and retardation has not yet begun. When on the centres, the flywheel attains its lowest speed, because retardation has just finished and acceleration has not yet begun. At each of these four points the acceleration is zero.

Time for crank to move through 90°, or ¼ revolution

$$= \frac{60}{750 \times 4} \text{ second}$$

Change of speed in this time $= 7.5$ r.p.m.

$$= \frac{7.5 \times 2\pi}{60} \text{ rad/sec}$$

Angular acceleration

$$= \frac{\text{Change of angular speed}}{\text{Time to change}}$$

$$= \frac{7.5 \times 2\pi}{60} \div \frac{60}{750 \times 4}$$

$$= \frac{7.5 \times 2\pi \times 750 \times 4}{60 \times 60} = 39.27 \text{ rad/sec}^2$$

SOLUTIONS TO FIRST CLASS EXAMINATION QUESTIONS

101. If we take the weights to be pounds instead of kilograms we have the same proportions and the final temperature will therefore be the same.

Latent heat of water $= 144 \times \frac{5}{9} = 80$ Chu/lb

Let t be the final temperature of the mixture in °C

Chu gained by the ice = Chu lost by the water and tank

$$9 [0.5 \times 7 + 80 + t] = 11.5 (79-t) + 10 \times 0.095 (60-t)$$
$$751.5 + 9 t = 908.5 - 11.5 t + 57 - 0.95 t$$
$$21.45 t = 214$$
$$t = 9.977°C$$

$9.977 \times \frac{9}{5} + 32 = 49.96°F.$ Ans.

102. $12 \text{ knots} = \dfrac{12 \times 6080}{60} = 1216$ ft per minute

Thrust horse power $= \dfrac{51 \times 2240 \times 1216}{33000} = 4210$

Combined efficiency of engine and propeller

$$= \frac{4210}{9000}$$

Let x be the mechanical efficiency of the engine
Then $0.9 x$ is the propeller efficiency

$$\therefore x \times 0.9 x = \frac{4210}{9000}$$

$x^2 = \dfrac{4210}{8100}$, and $x = 0.721$

$0.9 x = 0.9 \times 0.721 = 0.6489$

Mechanical efficiency $= 0.721$, or 72.1% ⎫
⎬ Ans.
Propeller efficiency $= 0.6489$, or 64.89% ⎭

103. Specific volume of the superheated steam

$$= \frac{1\cdot 248\ (1405\cdot 3 - 835\cdot 2)}{500} + 0\cdot 0123 = 1\cdot 435\ \text{ft}^3/\text{lb}$$

$$\therefore \text{Wt. of steam} = \frac{10}{1\cdot 435} = 6\cdot 969\ \text{lb. Ans. (i)}$$

Spec. vol. of dry sat. steam at 300 lb/in² = $1\cdot 543\ \text{ft}^3/\text{lb}$

∴ Dryness fraction at 300 lb/in²

$$= \frac{1\cdot 435}{1\cdot 543} = 0\cdot 9298\ \text{Ans. (ii)}$$

Heat/lb at reduced pressure = $h_f + q\ h_{fg}$
 = 394 + 0·9298 × 809·8 = 1146·8 Btu/lb

$$\text{Total heat loss} = 6\cdot 969\ (1405\cdot 3 - 1146\cdot 8)$$
$$= 1802\ \text{Btu. Ans. (iii)}$$

104. At 325°F,
$$V = 1 + \frac{(325 - 39\cdot 2)^2}{711\ (697 + 325)} = 1 + 0\cdot 1124 = 1\cdot 1124$$

At 250°F,
$$V = 1 + \frac{(250 - 39\cdot 2)^2}{711\ (697 + 250)} = 1 + 0\cdot 066 = 1\cdot 066$$

Height of water varies directly as the volume

$$\therefore \frac{H_1}{V_1} = \frac{H_2}{V_2},\quad H_2 = \frac{H_1 \times V_2}{V_1}$$

$$H_2 = \frac{48 \times 1\cdot 1124}{1\cdot 066} = 50\cdot 07\ \text{in.}$$

Difference in levels = 50·07 − 48 = 2·07 in.
 ∴ level in boiler is 2·07 inches above level shown in glass.
Ans.

105. Assuming one end plate only of each boiler is lagged,

$$\text{Surface lagged} = 3\left\{\pi \times 17 \times 13 + 17^2 \times \frac{\pi}{4}\right\}$$

$$= 2763 \text{ ft}^2$$

Heat radiated per hour before lagging
$= \text{Area} \times K[810^4 - 560^4]$ Btu

Heat radiated per hour after lagging
$= \text{Area} \times K[610^4 - 550^4]$ Btu

Saving per hour
$= \text{Area} \times K[(810^4 - 560^4) - (610^4 - 550^4)]$ Btu
$= \text{Area} \times K \times 28517 \times 10^7$ Btu

Saving of coal per day

$$= \frac{2763 \times 16 \times 28517 \times 10^7 \times 24}{10^{10} \times 12500 \times 2240}$$

$= 1{\cdot}08$ tons. Ans.

106.

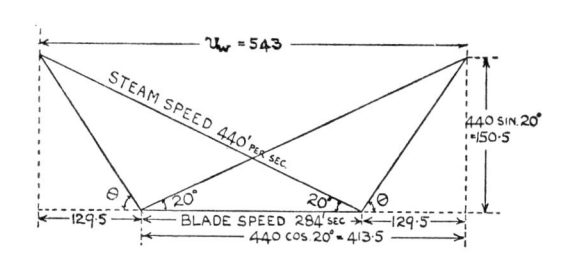

$440 \cos 20° = 413{\cdot}5$, and $413{\cdot}5 - 284 = 129{\cdot}5$ ft/sec

$440 \sin 20° = 150{\cdot}5$ ft/sec

$\text{Tan } \theta = \dfrac{150{\cdot}5}{129{\cdot}5}$, $\theta = 49° 17'$, entrance angle to moving blades. Ans. (a)

$v_w = 2 \times 440 \cos 20° - 284 = 543$ ft/sec

or $v_w = 129{\cdot}5 + 284 + 129{\cdot}5$ ft/sec

Force on blades = Change of momentum per second.

$$= \frac{1 \times 543}{32 \cdot 2} \text{ lb}$$

Work done per lb of steam per second

$$= \frac{1 \times 543}{32 \cdot 2} \times 284$$

$$= 4789 \text{ ft lb. Ans. (b)}$$

107. From steam tables,
$\text{P} = 215 \text{ lb/in}^2$, $h_f = 362 \cdot 1$, $h_{fg} = 838 \cdot 3 \text{ Btu/lb}$
$\text{P} = 115 \text{ lb/in}^2$, $t_F = 338 \cdot 1°\text{F}$, $h_f = 309 \cdot 2$,
$h_{fg} = 881 \cdot 5$, $h_g = 1190 \cdot 7$

Reducing the pressure of steam in a reducing valve is a constant total heat operation.

$$\text{Heat in high pressure steam} = 362 \cdot 1 + 0 \cdot 98 \times 838 \cdot 3$$
$$= 1183 \cdot 6 \text{ Btu/lb}$$

Heat in dry saturated steam at 115 lb/in² is 1190·7 Btu/lb, therefore the steam must still be wet, and its temperature is 338·1°F.

$$\text{Heat/lb @ 115 lb/in}^2 = \text{Heat/lb at 215 lb/in}^2$$

$$309 \cdot 2 + q \times 881 \cdot 5 = 1183 \cdot 6$$

$$q \times 881 \cdot 5 = 874 \cdot 4$$

$$q = 0 \cdot 9919$$

The condition of the reduced pressure steam is WET, its dryness fraction is 0·9919, and its temperature is 338·1°F. Ans.

108. Let d = original diameter of sphere in inches,

then, increase in diameter $= \dfrac{0 \cdot 15}{100} d$

Original dia. × Change of temp. × Co-efficient = Increase in dia.

$$d \times \text{Change of temp.} \times 0 \cdot 0000128 = \frac{0 \cdot 15}{100} d$$

SOLUTIONS TO FIRST CLASS EXAMINATION QUESTIONS 589

$$\therefore \text{Change of temp.} = \frac{0 \cdot 15}{100 \times 0 \cdot 0000128} = 117 \cdot 19°F$$

Final temp. of sphere and water
= 60 + 117·19 = 177·19°F

Fall in temp. of the water
= 200 — 177·19 = 22·81 F°

Weight of sphere × Change of temp. × Sp. heat
= Weight of water × Change of temp.

$$\frac{\pi}{6} \times d^3 \times 2 \cdot 56 \times \frac{62 \cdot 5}{1728} \times 117 \cdot 19 \times 0 \cdot 22 = 24 \times 10 \times 22 \cdot 81$$

$$\therefore d = \sqrt[3]{\frac{24 \times 10 \times 22 \cdot 81 \times 6 \times 1728}{\pi \times 2 \cdot 56 \times 62 \cdot 5 \times 117 \cdot 19 \times 0 \cdot 22}} = 16 \cdot 36 \text{ in.}$$

Ans.

109.
$$\text{Available hydrogen} = 12 - \frac{1 \cdot 5}{8} = 11\tfrac{13}{16}\%$$

Oxygen required for combustion

$$= 2\tfrac{2}{3} \times 0 \cdot 85 + 8 \times \frac{11\tfrac{13}{16}}{100}$$

= 3·212 lb per pound of fuel.

Theoretical weight of air
$$= 3 \cdot 212 \times \frac{100}{23} = 13 \cdot 965 \text{ lb.} \quad \text{Ans.}$$

Actual weight of air = 13·965 × 1·5 = 20·948 lb.

Weight of gases resulting from the combustion of 1 pound of fuel = 20·948 + 0·985 = 21·933 lb.

$\therefore$ heat carried away per hour
= 21·933 × 1·2 × 2240 × (550 — 76) × 0·24

= 6704000 Btu.   Ans.

110. From steam tables,
$P = 1\cdot5$ lb/in^2, $h_f = 83\cdot7$, $h_{fg} = 1028\cdot1$ Btu/lb
Dryness fraction of exhaust steam $= 1 - 0\cdot13 = 0\cdot87$
Let W = weight of steam condensed in lb per hour

Heat lost by steam = Heat gained by water

$$W[0\cdot87 \times 1028\cdot1 + 83\cdot7 - (104 - 32)] = \frac{520 \times 2240}{(90 - 55)}$$

$$W = 45000 \text{ lb/hour}$$

$$\text{Area of tube surface required} = \frac{45000}{14} = 3214 \text{ ft}^2$$

$$\text{Surface area of one tube} = \frac{\pi \times 0\cdot75}{12} \times 12$$

$$= 2\cdot3562 \text{ ft}^2$$

$$\therefore \text{Number of tubes required} = \frac{3214}{2\cdot3562} = 1364 \quad \text{Ans.}$$

111. Referring to Fig. 38 for state points 1 to 5.

$$\gamma = \frac{C_P}{C_V} = \frac{0\cdot238}{0\cdot17} = 1\cdot4$$

$$P_1 V_1^\gamma = P_2 V_2^\gamma$$

$$P_2 = \frac{14\cdot7 \times 11^{1\cdot4}}{1^{1\cdot4}} = 422 \text{ lb/in}^2 \text{ abs.} \quad \text{(i)}$$

$$\frac{P_1 V_1}{T_1} = \frac{P_2 V_2}{T_2}$$

$$T_2 = \frac{540 \times 422 \times 1}{14\cdot7 \times 11} = 1409°\text{F abs.}$$

$$1409 - 460 = 949°\text{F} \quad \ldots \quad \text{(ii)}$$

$$T_3 = T_2 \times \frac{P_3}{P_2}$$

$$= \frac{1409 \times 650}{422} = 2171°F \text{ abs.}$$

$$2171 - 460 = 1711°F \qquad \ldots \quad \text{(iii)}$$

$$V_4 = V_3 \times \frac{T_4}{T_3}$$

$$= \frac{1 \times 3460}{2171} = 1\cdot594$$

$$P_5 V_5^{1\cdot4} = P_4 V_4^{1\cdot4}$$

$$P_5 = \frac{650 \times 1\cdot594^{1\cdot4}}{11^{1\cdot4}} = 43\cdot49 \text{ lb/in}^2 \text{ abs. (iv)}$$

$$T_5 = T_1 \times \frac{P_5}{P_1}$$

$$= \frac{540 \times 43\cdot49}{14\cdot7} = 1598°F \text{ abs.}$$

$$1598 - 460 = 1138°F \qquad \ldots \quad \text{(v)}$$

$$\text{Ideal Efficiency} = 1 - \frac{\text{Heat rejected}}{\text{Heat supplied}}$$

$$= 1 - \frac{C_V(T_5 - T_1)}{C_V(T_3 - T_2) + C_P(T_4 - T_3)}$$

$$= 1 - \frac{0\cdot17 \times 1058}{0\cdot17 \times 762 + 0\cdot238 \times 1289}$$

$$= 1 - \frac{1058}{762 + 1\cdot4 \times 1289}$$

$$= 1 - 0\cdot4121$$
$$= 0\cdot5879 \text{ or } 58\cdot79\% \qquad \ldots \quad \ldots \quad \text{(vi)}$$

Pressure @ end of compression = 422 lb/in² abs.
Temperature @ end of compression = 949°F
Temp. @ end of combustion @ const. vol. = 1711°F
Pressure @ end of expansion = 43·49 lb/in² abs.
Temperature @ end of expansion = 1138°F
Ideal thermal efficiency = 58·79%

112. Specific heat of Copper = 0·095
Specific heat of Steel = 0·116

Heat lost by steel = Heat gained by liquid + heat gained by calorimeter.

100 (150 — 44·5) × 0·116
= 70(44·5 — 10) × Sp. Ht. + 12(44·5 — 10) × 0·095

11·6 × 105·5 = 70 × 34·5 × Sp. Ht. + 1·14 × 34·5

1223·8 = 2415 Sp. Ht. + 39·33

∴ 2415 Sp. Ht. = 1184·47

$$\text{Specific Heat} = \frac{1184·47}{2415} = 0·4904 \quad \text{Ans.}$$

113.

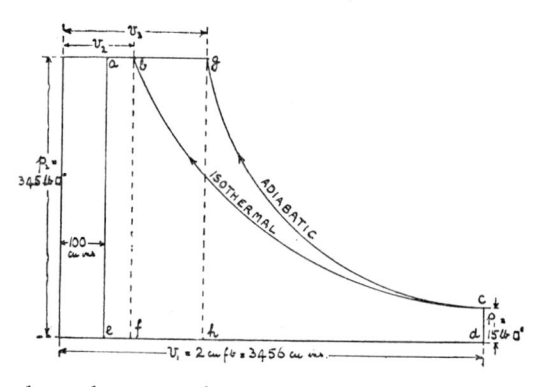

Isothermal compression, pv = constant.

$p_1 \times v_1 = p_2 \times v_2$

15 × 3456 = 345 × v_2

$$v_2 = \frac{15 \times 3456}{345} = 150·3 \text{ in}^3 = 0·08698 \text{ ft}^3$$

SOLUTIONS TO FIRST CLASS EXAMINATION QUESTIONS 593

Volume delivered $= 150·3 - 100 = 50·3$ in^3.   Ans. (a)

Adiabatic compression, $p\,v^{1·4} =$ constant

$$p_1 \times v_1^{1·4} = p_3 \times v_3^{1·4}$$

$$15 \times (3456)^{1·4} = 345 \times v_3^{1·4}$$

By logs, $\log 15 + 1·4 \log 3456 = \log 345 + 1·4 \log v_3$

Divide each term by $1·4$

$$\frac{\log 15}{1·4} + \log 3456 = \frac{\log 345}{1·4} + \log v_3$$

$$\log v_3 = 0·8401 + 3·5386 - 1·8127$$

$$= 2·5660$$

$$v_3 = 368·2 \text{ in}^3 = 0·213 \text{ ft}^3$$

Volume delivered $= 368·2 - 100 = 268·2$ in^3.   Ans. (b)

Referring to the diagram, the area $b\,c\,d\,f$ represents the work done in raising the pressure from p_1 to p_2 under isothermal conditions, and the area $a\,b\,f\,e$ is the work done in delivering the air.

Again, the area $g\,c\,d\,h$ represents the work done in raising the pressure from p_1 to p_2 under adiabatic conditions, the area $a\,g\,h\,e$ is the work done in delivering the air.

Since the area of a $p\,v$ diagram represents work done, and since the same *weight* of air is delivered in each case, then Isothermal Compression is the more economical method.

114.   From steam tables, saturation temperature (t_F) of steam at 250 lb/in^2 = 401°F, therefore the steam has (581 — 401) = 180 F degrees of superheat.

Reading heat and entropy values from the steam tables,

$P = 250$, supht. 180, $h = 1307·7$, $s = 1·6384$
$P = 1$, $h_f = 69·7$, $h_{fg} = 1036·1$, $s_f = 0·1326$,
$\qquad\qquad\qquad\qquad\qquad\qquad\qquad s_g = 1·9783$

Entropy after expansion = Entropy before expansion

$$0\cdot1326 + q(1\cdot9783 - 0\cdot1326) = 1\cdot6384$$

$$0\cdot1326 + 1\cdot8457q = 1\cdot6384$$
$$1\cdot8457q = 1\cdot5058$$
$$q = 0\cdot8158 \quad \text{Ans. (i)}$$

$$\text{Rankine effic.} = \frac{h_{g1} - h_{g2}}{h_{g1} - h_{f2}}$$

$$= \frac{1307\cdot7 - (69\cdot7 + 0\cdot8158 \times 1036\cdot1)}{1307\cdot7 - 69\cdot7}$$

$$= 0\cdot3173 \text{ or } 31\cdot73\% \quad \text{Ans. (ii)}$$

See Chapter 11 for Rankine cycle on the T-s diagram.

115. Rise in temperature of cooling water

$$= 45 - 20 = 25 \text{ C}°$$

$$= 25 \times \tfrac{9}{5} = 45 \text{ F}°$$

Heat carried away per hour $= 530 \times 10 \times 45$ Btu

Heat generated per hour $= \dfrac{530 \times 10 \times 45}{0\cdot85}$ Btu

b.h.p. $= \dfrac{530 \times 10 \times 45}{0\cdot85 \times 2545} = 110\cdot3$ Ans. (ii)

$$\frac{\text{Effective load on brake} \times \pi \times 5 \times 70}{33000} = 110\cdot3$$

Effective load on brake
$$= \frac{110\cdot3 \times 33000}{\pi \times 5 \times 70} = 3309 \text{ lb.} \quad \text{Ans. (i)}$$

116. Let r = length of the crank.
Then $2r$ = length of the stroke.

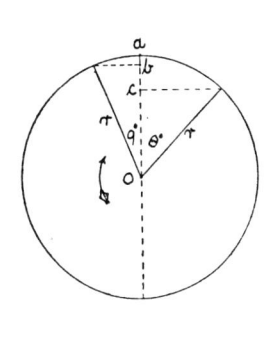

$$ac = \frac{9}{100} \text{ of } 2r = 0.18\,r$$

$$co = r - 0.18\,r = 0.82\,r$$

$$\text{Cos } \theta = \frac{0.82\,r}{r} = 0.82,\ \theta = 34°\,55'$$

∴ valve is open for $9° + 34°\,55'$
$= 43°\,55'$ Ans. (ii)

$ab = r - r \cos 9° = 0.0123\,r$

$ac = r - r \cos \theta = 0.18\,r$

∴ $ab + ac = 0.0123\,r + 0.18\,r = 0.1923\,r$
$0.1923\,r = 4$ in.

∴ $r = \dfrac{4}{0.1923} = 20.8$ in.

Stroke of engine $= 20.8 \times 2 = 41.6$ in.   Ans. (i)

117. From steam tables,
$P = 200$ lb/in², supht. 240 F°, $h = 1332.9$

$P = 200$ lb/in² sat, $h_g = 1199.5$, $v_g = 2.29$

$P = 2$ lb/in² sat, $h_f = 94$, $h_{fg} = 1022.2$

Without superheat:
Heat drop $= 1199.5 - (94 + 0.786 \times 1022.2)$
$= 302.05$ Btu/lb

With superheat;
Heat drop $= 1332.9 - (94 + 0.866 \times 1022.2)$
$= 353.67$ Btu/lb

% increase in heat drop = % increase in work done
$$= \frac{353.67 - 302.05}{302.05} \times 100 = 17.07\%$$ Ans. (i)

With superheat, more work is done per lb of steam in the ratio 353·67/302·05, therefore to do an equal amount of work as saturated steam, less superheated steam is required in the ratio 302·05 ÷ 353·67 = 0·8544. That is, 0·8544 lb of superheated steam will do the work of 1 lb of saturated steam.

Spec. vol. of superheated steam, using equation given in Callendars steam tables:

$$v = \frac{1 \cdot 248 \, (1332 \cdot 9 - 835 \cdot 2)}{200} + 0 \cdot 0123 = 3 \cdot 118 \text{ ft}^3/\text{lb}$$

Volume of 0·8544 lb = 0·8544 × 3·118 = 2·665 ft³

Cut off to accommodate 1 lb of saturated steam at 2·29 ft³/lb
= 0·625 stroke

Cut off to accommodate 0·8544 lb of superheated steam at 2·665 ft³ = $0 \cdot 625 \times \dfrac{2 \cdot 665}{2 \cdot 29}$

= 0·7272 stroke. Ans. (ii)

118.   746 watts = one horse power

$$\therefore \text{i.h.p. of engine} = \frac{60 \times 1000}{0 \cdot 8 \times 0 \cdot 92} = 109 \cdot 3$$

also, $\text{i.h.p.} = \dfrac{p\text{ALN}}{4560} \times 6$ for six cylinders

Let d = dia of cyl. in centimetres

then, $1 \cdot 5d$ = stroke in centimetres

and, $\dfrac{1 \cdot 5d}{100}$ = stroke in metres

$$\therefore 109 \cdot 3 = \frac{8 \cdot 5 \times 0 \cdot 7854 d^2 \times 1 \cdot 5d \times 120}{4560 \times 100} \times 6$$

$$d = \sqrt[3]{\frac{109 \cdot 3 \times 4560 \times 100}{8 \cdot 5 \times 0 \cdot 7854 \times 1 \cdot 5 \times 120 \times 6}}$$

= 19·05 cm = 190·5 mm. Ans.

SOLUTIONS TO FIRST CLASS EXAMINATION QUESTIONS 597

119. Let N = revolutions per minute

$$\% \text{ Mechanical efficiency} = \frac{\text{b.h.p.}}{\text{i.h.p.}} \times 100$$

$$\frac{1000 \times 100}{\text{i.h.p.}} = \sqrt{N \times 85}$$

$$\text{i.h.p.} = \frac{100000}{\sqrt{N} \times \sqrt{85}} \quad \ldots \quad \ldots \quad \ldots \quad \ldots \quad (i)$$

Also:—

$$\text{i.h.p.} = \frac{85 \times 0.7854 \times 650^2 \times 2 \times 650 \times N \times 6}{(25.4)^2 \times 25.4 \times 12 \times 2 \times 33000}$$

$$= 16.95 \, N \quad \ldots \quad \ldots \quad \ldots \quad \ldots \quad \ldots \quad (ii)$$

$$16.95 \, N = \frac{100000}{N^{\frac{1}{2}} \times \sqrt{85}} \qquad N^{\frac{3}{2}} = \frac{100000}{16.95 \times \sqrt{85}}$$

$$N = \sqrt[\frac{3}{2}]{\frac{100000}{16.95 \times \sqrt{85}}} = 74.25 \text{ r.p.m. Ans.}$$

Mechanical efficiency $= \sqrt{74.25 \times 85} = 79.43\%$ Ans.

120. Heat supplied to cylinders, per i.h.p. per hour
$= 0.73 \times 1.38 \times 13650$ Btu

Heat equivalent of one i.h.p. $= 2545$ Btu

∴ Thermal efficiency of engine

$$= \frac{2545}{0.73 \times 1.38 \times 13650}$$

$$= 0.1851 \text{ or } 18.51\% \text{ Ans. (a)}$$

598 REED'S HEAT AND HEAT ENGINES FOR MARINE ENGINEERS

$$\text{i.h.p. for } 1\cdot38 \text{ lb of fuel per hour} = 1$$

$$\text{,, ,, 1 lb ,, ,,} = \frac{1}{1\cdot38}$$

$$\text{,, ,, } \tfrac{63}{24} \times 2240 \text{ lb ,,} = \frac{1 \times 63 \times 2240}{1\cdot38 \times 24}$$

$$= 4260 \text{ i.h.p. Ans. (b)}$$

Overall thermal efficiency of engines and boilers

$$= 0\cdot73 \times 0\cdot1851 = 0\cdot1351 = 13\cdot51\%$$

Percentage of total heat lost $= 100 - 13\cdot51 = 86\cdot49\%$

For each lb of fuel, this loss is $0\cdot8649 \times 13650$

$$= 11805 \text{ Btu/lb Ans. (c)}$$

121.

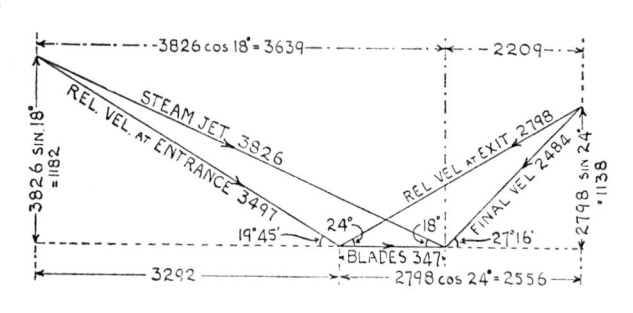

$3826 \cos 18° = 3639$, and $3639 - 347 = 3292$ ft/sec

$3826 \sin 18° = 1182$ ft/sec

$$\text{Tan } \theta = \frac{1182}{3292}, \theta = 19° 45' \text{ (entrance angle of moving blades)}.$$

Velocity of steam relative to moving blades at entrance

$$= \frac{1182}{\sin 19° 45'}$$

$$= 3497 \text{ ft/sec}$$

SOLUTIONS TO FIRST CLASS EXAMINATION QUESTIONS 599

Velocity of steam relative to moving blades at exit

$= 3497 \times 0{\cdot}8 = 2798$ ft/sec

$2798 \cos 24° = 2556$ and $2556 - 347 = 2209$ ft/sec

$2798 \sin 24° = 1138$ ft/sec

$$\tan \phi = \frac{2209}{1138}, \quad \phi = 62° 44'$$

Final absolute velocity of steam

$$= \frac{2209}{\sin 62° 44'} = 2484 \text{ ft/sec}$$

Entrance angle of moving blades $= 19° 45'$
Final abs. velocity of steam $= 2484$ ft/sec
Direction is at $62° 44'$ to turbine axis, or at
$90° - 62° 44' = 27° 16'$ to direction of blade motion } Ans.

122. 1 cu ft CO_2 gas at $14{\cdot}7$ lb/in^2 and $32°$F weighs $0{\cdot}0807 \times 1{\cdot}518$ lb

1 cu ft CO_2 gas at $14{\cdot}7$ lb/in^2 and $60°$F

weighs $\dfrac{0{\cdot}0807 \times 1{\cdot}518 \times 492}{520} = 0{\cdot}1159$ lb.

Permeable space $= 36 \times 56 \times 20 \times 0{\cdot}75$ ft^3

Volume of CO_2 gas required $= 36 \times 56 \times 20 \times 0{\cdot}75 \times 0{\cdot}25$ ft^3

Weight of $CO_2 = 36 \times 56 \times 20 \times 0{\cdot}75 \times 0{\cdot}25 \times 0{\cdot}1159$
$= 876{\cdot}2$ lb. Ans.

Volume occupied by 3 lb of water $= \dfrac{3}{62{\cdot}5}$ ft^3 and this is the volume of 2 lb of CO_2 in the liquid form.

$\therefore$ Volume of bottles $= \dfrac{3 \times 876{\cdot}2}{62{\cdot}5 \times 2} = 21{\cdot}03$ ft^3 Ans.

123. From steam tables,
$P = 205$ lb/in^2, $h_g = 1199 \cdot 8$ Btu/lb

$P = 18$ lb/in^2, $h_f = 190 \cdot 6$, $h_{fg} = 964$ Btu/lb

For 1 lb of vapour formed, let x lb of water be blown out, then the feed is $(1 + x)$ lb.

For constant density of water in evaporator:

$$\text{Salt put in} = \text{Salt blown out}$$
$$\text{Amount of feed} \times \text{feed density} = \text{Amount blown out} \times \text{Evaporator density}$$
$$(1 + x) \times 1 = x \times 2\tfrac{1}{2}$$
$$x = \tfrac{2}{3} \text{ lb}$$

Therefore for every lb of vapour formed, $1\tfrac{2}{3}$ lb of water are fed into the evaporator and $\tfrac{2}{3}$ lb of this water is blown out. Thus the whole $1\tfrac{2}{3}$ lb of water receives sensible heat but only 1 lb receives latent heat.

Heat to produce 1 lb of vapour
$$= 1\tfrac{2}{3}[190 \cdot 6 - (70 - 32)] + 964 = 1218 \cdot 3 \text{ Btu}$$

Heat given up by 1 lb of heating steam
$$= 1199 \cdot 8 - (233 \cdot 1 - 32) = 998 \cdot 7 \text{ Btu}$$

Weight of heating steam required
$$= \frac{24 \times 2240 \times 1218 \cdot 3}{998 \cdot 7} \text{ lb}$$

Weight of coal required $= \dfrac{24 \times 2240 \times 1218 \cdot 3}{998 \cdot 7 \times 7 \cdot 5 \times 2240}$

$$= 3 \cdot 902 \text{ tons.} \quad \text{Ans.}$$

124.

Fuel used per i.h.p. per hour $= \dfrac{109 \cdot 5}{308} = 0 \cdot 3556$ lb

Fuel used per b.h.p. per hour $= \dfrac{109 \cdot 5}{195} = 0 \cdot 5616$ lb

Indicated thermal efficiency
$$= \frac{2545}{0 \cdot 3556 \times 19200} \times 100 = 37 \cdot 29\% \quad \text{Ans.}$$

SOLUTIONS TO FIRST CLASS EXAMINATION QUESTIONS 601

Brake thermal efficiency

$$= \frac{2545}{0{\cdot}5616 \times 19200} \times 100 = 23{\cdot}6\% \quad \text{Ans.}$$

Friction and pumping losses $= 37{\cdot}29 - 23{\cdot}6 = 13{\cdot}69\%$

Heat carried away by cooling water per hour
$= 171 \times 60 \times (118 - 63)$
$= 564300$ Btu

As a percentage of the total heat supplied this is

$$= \frac{564300}{109{\cdot}5 \times 19200} \times 100 = 26{\cdot}83\%$$

The remainder, i.e., $100 - (37{\cdot}29 + 26{\cdot}83) = 35{\cdot}88\%$, is lost in exhaust gases.

Heat balance diagram:—

Heat in Fuel $= 100\%$

i.h.p. $= 37{\cdot}29\%$ Cooling water $= 26{\cdot}83\%$ Exhaust gases $= 35{\cdot}88\%$

b.h.p. $= 23{\cdot}6\%$ Friction, etc. $= 13{\cdot}69\%$

125.

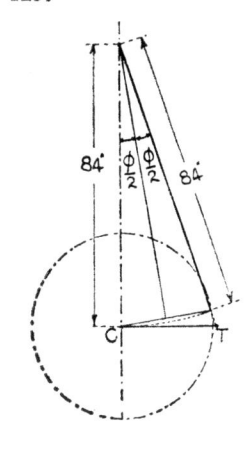

Torque produced in main engine shaft
$= 1600 \times 1000 \times 0{\cdot}12$ inch lb

$$\sin \frac{\phi}{2} = \frac{10{\cdot}5}{84} = 0{\cdot}125, \quad \frac{\phi}{2} = 7°\,11'$$

$$\phi = 14°\,22'$$

$CT = 84 \tan 14°\,22' = 21{\cdot}51$ in.
(note the isosceles triangle).

$$\therefore \text{Vertical pull on crosshead} = \frac{\text{Torque}}{CT}$$

$$= \frac{1600 \times 1000 \times 0{\cdot}12}{21{\cdot}51}\ \text{lb}$$

Cross sectional area of each bolt $= 0.7854 \times (0.875)^2$

∴ Shear stress in bolts

$$= \frac{1600 \times 1000 \times 0.12}{21.51 \times 0.7854 \times (0.875)^2 \times 2} = 7424 \text{ lb/in}^2 \text{ Ans.}$$

126. Assuming atmospheric pressure $= 15 \text{ lb/in}^2$
$\tfrac{3}{8}$ of $12 = 4.5$ in.

$$P_1 V_1{}^n = P_2 V_2{}^n$$
∴ $105 \times (4.5 + 0.5)^n = 30 \times (12 + 0.5)^n$
Log $105 + n \log 5 = \log 30 + n \log 12.5$
Log $105 - \log 30 = n (\log 12.5 - \log 5)$

$$\therefore n = \frac{\log 105 - \log 30}{\log 12.5 - \log 5} = \frac{2.0212 - 1.4771}{1.0969 - 0.6990}$$

$$= \frac{0.5441}{0.3979} = 1.368$$

$n = 1.368$

The law is $PV^{1.368} = $ constant. Ans. (i)

$$105 \times 5^{1.368} = (70 + 15) \times v^{1.368}$$
Log $105 + 1.368 \log 5 = \log 85 + 1.368 \log v$
$2.0212 + 1.368 \times 0.6990 = 1.9294 + 1.368 \log v$

$$\text{Log } v = \frac{1.0478}{1.368} = 0.7663$$

$v = 5.838$

Position of piston $= 5.838 - 0.5 = 5.338$ in. of stroke

$$\text{Fraction of stroke} = \frac{5.338}{12} = 0.4448 \text{ Ans. (ii)}$$

SOLUTIONS TO FIRST CLASS EXAMINATION QUESTIONS 603

127. Values from steam tables, for engine A:
$P = 200 \text{ lb/in}^2$, $h_f = 355 \cdot 5$, $h_{fg} = 844$ Btu/lb
$P = 1 \cdot 5 \text{ lb/in}^2$, $h_f = 83 \cdot 7$, $h_{fg} = 1028 \cdot 1$ Btu/lb

Total heat in supply steam
$= 355 \cdot 5 + 0 \cdot 98 \times 844 = 1182 \cdot 6$ Btu/lb

Total heat in exhaust steam
$= 83 \cdot 7 + 0 \cdot 88 \times 1028 \cdot 1 = 988 \cdot 4$ Btu/lb

Heat drop through turbine
$= 1182 \cdot 6 - 988 \cdot 4 = 194 \cdot 2$ Btu/lb

For the theoretical efficiency, it is assumed (a) that the heat drop in the steam through the turbine is usefully converted into work by the turbine, and (b) that latent heat only is given up by the exhaust steam in the condenser, and the condensate, at the temperature of the exhaust steam, is pumped directly into the boilers as feed water.

Heat supplied to steam in boilers
$= 1182 \cdot 6 - 83 \cdot 7 = 1098 \cdot 9$ Btu/lb

Theoretical thermal efficiency
$= \dfrac{194 \cdot 2}{1098 \cdot 9} = 0 \cdot 1767 = 17 \cdot 67\%$ Ans. (i)

Values from steam tables, for engine B:

$P = 400 \text{ lb/in}^2$ supht. 200 F°, $h = 1331 \cdot 6$ Btu/lb
$P = 0 \cdot 5 \text{ lb/in}^2$, $h_f = 47 \cdot 6$, $h_{fg} = 1048 \cdot 5$ Btu/lb

Total heat in exhaust steam
$= 47 \cdot 6 + 0 \cdot 8 \times 1048 \cdot 5 = 886 \cdot 4$ Btu/lb

Heat drop through turbine
$= 1331 \cdot 6 - 886 \cdot 4 = 445 \cdot 2$ Btu/lb

Heat supplied in boilers
$= 1331 \cdot 6 - 47 \cdot 6 = 1284$ Btu/lb

Theoretical thermal efficiency
$= \dfrac{445 \cdot 2}{1284} = 0 \cdot 3467 = 34 \cdot 67\%$ Ans. (ii)

128. The highest pressure will be attained at the end of compression. During the combustion period there is theoretically no rise in pressure, and the highest temperature is reached at the end of combustion.

$$P_1 V_1^{1.4} = P_2 V_2^{1.4}$$
$$14 \times 13^{1.4} = P_2 \times 1^{1.4}$$
$$P_2 = 507 \cdot 7 \text{ lb/in}^2 \text{ abs. (highest pressure). Ans.}$$

$$\frac{PV}{T} = \text{constant for a gas}$$

$$\frac{14 \times 13}{560} = \frac{507 \cdot 7 \times 1}{T_2}, \; T_2 = 1562°F \text{ abs.}$$

∴ Temp. at end of compression = 1562 — 460 = 1102°F

Assume 35 lb of air is compressed and 1 lb of oil is used, the weight of the gases formed is 36 lb.

Heat given up by fuel = Heat gained by gases

$$18500 = 36 \times \text{Change of temp.} \times 0 \cdot 2375$$

∴ Change of temp. $= \dfrac{18500}{36 \times 0 \cdot 2375} = 2164 \, F°$

∴ Final temp. = 1102 + 2164 = 3266°F (highest temp). Ans.

As the specific heat of the products of combustion was not stated, it has been assumed to be the same as that of air.

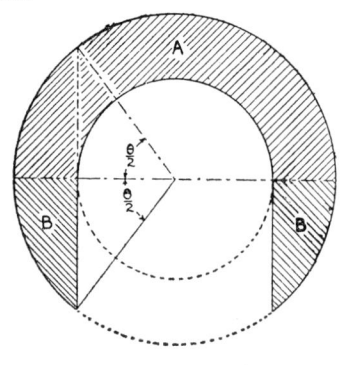

129. The rubbing surface of each face of the shoe consists of the area of a semi-annulus A, and the area of two half segments B and B.

Area of semi-annulus

$$= 0 \cdot 7854 \, [20^2 - 12^2] \times \tfrac{1}{2}$$

$$= 100 \cdot 53 \text{ sq in.}$$

SOLUTIONS TO FIRST CLASS EXAMINATION QUESTIONS 605

$$\operatorname{Cos} \frac{\theta}{2} = \frac{6}{10} = 0\cdot 6, \quad \frac{\theta}{2} = 53° \ 8', \text{ and } \theta = 106° \ 16'$$

$$\text{Area of whole segment} = \frac{r^2}{2} [\theta - \sin \theta]$$

$$= \frac{10^2}{2} \left\{ \frac{106\cdot 26}{57\cdot 3} - \sin 106° \ 16' \right\}$$

$$= 44\cdot 7 \text{ sq in.}$$

∴ Area of rubbing surface $= 100\cdot 53 + 44\cdot 7 = 145\cdot 23$ sq in.

Total load on thrust $= 45 \times 145\cdot 23 \times 6$ lb

$$\text{i.h.p. of engine} = \frac{45 \times 145\cdot 23 \times 6 \times 12 \times 6080}{33000 \times 60} \times \frac{100}{69}$$

$$= 2095 \text{ Ans.}$$

130. L.P. power $= 1\cdot 05 \times$ H.P. power

∴ H.P. power $= \dfrac{\text{L.P.}}{1\cdot 05} = 0\cdot 9523 \times$ L.P.

I.P. power $= 1\cdot 02 \times$ H.P. power
$= 1\cdot 02 \times 0\cdot 9523 \times$ L.P. $= 0\cdot 9714 \times$ L.P.

Ratio of powers developed $=$ H.P. : I.P. : L.P.
$= 0\cdot 9523 : 0\cdot 9714 : 1$

$0\cdot 9523 + 0\cdot 9714 + 1 = 2\cdot 9237$

∴ L.P. develops $\dfrac{1}{2\cdot 9237}$ of the total power

M.E.P. in L.P. $= \dfrac{1}{2\cdot 9237} \times 33 = 11\cdot 28$ lb/in² ⎫

M.E.P. in I.P. $= \dfrac{11\cdot 28 \times 74^2}{45^2} \times 0\cdot 9714 = 29\cdot 63$ lb/in² ⎬ Ans.

M.E.P. in H.P. $= \dfrac{11\cdot 28 \times 74^2}{27^2} \times 0\cdot 9523 = 80\cdot 67$ lb/in² ⎭

131. Steam load on piston
= 0·7854 × 20² × 110 = 34560 lb

Accelerating force = Mass × acceleration

$$\text{,, ,,} = \frac{1400 \times 280}{32\cdot 2} = 12170 \text{ lb}$$

Load on crosshead
= 34560 − 12170 = 22390 lb or 9·995 tons. Ans.

132.

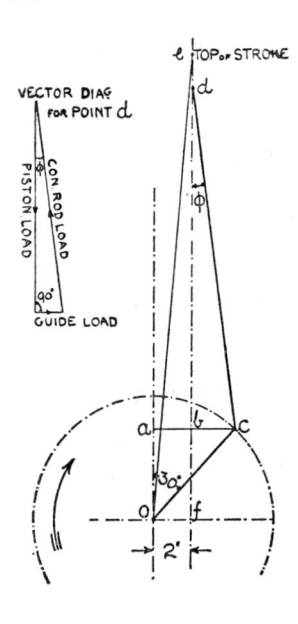

$ac = 21 \times \sin 30° = 10\cdot 5$ in.
$bc = 10\cdot 5 - 2 = 8\cdot 5$ in.

$$\sin \phi = \frac{8\cdot 5}{72}, \phi = 6° 47'$$

$$\frac{\text{Guide load}}{\text{Piston load}} = \tan \phi$$

∴ Guide load
= 20 × 2240 × 0·119 lb
Guide pressure
$$= \frac{20 \times 2240 \times 0\cdot 119}{173}$$

= 30·81 lb/in². Ans.
$db = 72 \cos 6° 47' = 71\cdot 48$ in.
$ao = 21 \cos 30° = 18\cdot 186$ in.

∴ Vertical distance from crosshead to shaft centre when the crank is 30° past the vertical = 71·48 + 18·186 = 89·666 in.

Vertical distance from crosshead to shaft centre when the piston is at its highest position (this is when crank and con. rod are in line) = ef
$ef = \sqrt{93^2 - 2^2} = 92\cdot 977$ in.

∴ Distance piston has moved down its stroke = ed
= 92·977 − 89·666 = 3·311 in. Ans.

133. Lap + lead — exhaust lap = opening to exhaust
Lap + 0·24 — 0·17 = 2

∴ Steam lap = 2 — 0·24 + 0·17 = 1·93 in. Ans. (i)

Sine angle of advance

$$= \frac{\text{Lap + lead}}{\tfrac{1}{2}\text{ travel}} = \frac{1·93 + 0·24}{3·75}$$

Angle of advance (θ) = 35° 22′. Ans. (ii)

$$\sin \phi = \frac{\text{Exhaust lap}}{\tfrac{1}{2}\text{ travel}} = \frac{0·17}{3·75}$$

$\phi = 2° 36′$

∴ At start of compression the crank is at 35° 22′ + 2° 36′
= 37° 58′ to the line of stroke. Ans. (iii)

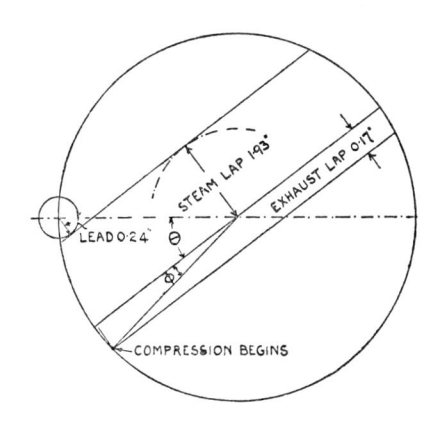

134. If P = the boiler pressure in lb/in² abs.

then t_F = its temperature, and h_g is its total heat per 1 lb above 32°F.

Assume 100 lb of steam are generated by the boiler, and 100 lb of feed water are returned at 196°F.

93·65 lb are used by the engine and recovered as water at 126°F, whilst 6·35 lb are used by the heater.

Since no heat is lost, then the heat in the condensed water from the engine, plus the heat in the steam used by the heater, must be equal to the heat in the feed water.

All heat must be measured above 32°F, since h_g is above 32°F.

$$93 \cdot 65\ (126 - 32) + 6 \cdot 35\ h_g = 100\ (196 - 32)$$
$$8803 \cdot 1 + 6 \cdot 35\ h_g = 16400$$
$$6 \cdot 35\ h_g = 7596 \cdot 9$$
$$h_g = 1196 \cdot 4\ \text{Btu}$$

From tables, $P = 165$, $t_F = 366$, $h_g = 1196 \cdot 6$, and this latter is practically the value calculated.

$\therefore$ temperature of the steam $= 366°F$. Ans.

135. Refer to chapter 14 for co-efficient of performance

The capacity of the machine to extract heat
$= 4 \times 2240 \times 144$ Btu per 24 hours

Heat to take from 1 lb of water at 67°F to turn it into ice at 25°F $= (67 - 32) + 144 + 0 \cdot 5\ (32 - 25) = 182 \cdot 5$ Btu

Ice made in 24 hours

$$= \frac{4 \times 2240 \times 144}{182 \cdot 5 \times 2240} = 3 \cdot 156 \text{ tons, or } 7068 \text{ lb. Ans.}$$

$$\text{Co-eff. of Performance} = \frac{T_2}{T_1 - T_2}$$

$T_2 =$ evaporator temp., $T_1 =$ condenser temp.

$$5 \cdot 8 = \frac{T_2}{(70 + 460) - T_2}$$

$5 \cdot 8 \times 530 - 5 \cdot 8 T_2 = T_2$
$6 \cdot 8\ T_2 = 5 \cdot 8 \times 530$
$T_2 = 452 \cdot 1°F$ abs., or $- 7 \cdot 9°F$. Ans.

136. Spec. vol. of wet steam at 40 lb/in² of dryness q
$= 10 \cdot 5\ q$ ft³/lb

$\therefore$ Weight of steam blown in $= \dfrac{115}{10 \cdot 5\ q}$ lb

SOLUTIONS TO FIRST CLASS EXAMINATION QUESTIONS 609

Total heat before mixing = Total heat after mixing
$$\frac{115}{10\cdot 5q}(236\cdot 1 + 934\cdot 4\,q) + 160\,(50-32) = \left(\frac{115}{10\cdot 5q}+160\right)(125-32)$$

Multiplying throughout by $\frac{10\cdot 5q}{115}$

$$236\cdot 1 + 934\cdot 4\,q + \frac{160 \times 18 \times 10\cdot 5\,q}{115} = 93 + \frac{160 \times 93 \times 10\cdot 5q}{115}$$

$$236\cdot 1 + 934\cdot 4\,q + 262\cdot 9\,q = 93 + 1358\,q$$
$$143\cdot 1 = 160\cdot 7\,q$$
$$q = 0\cdot 8906 \text{ Ans.}$$

137. Dalton's law of partial pressures.

The pressure of a mixture of gases and vapours, on the walls of a vessel containing it, is the sum of the pressures which each would exert if each separately occupied the whole volume of the vessel.

Partial pressure due to air present = $1\cdot 9 - 1\cdot 5 = 0\cdot 4$ lb/in^2 abs.

The question therefore resolves itself into finding the weight of 100 cu ft of air at $0\cdot 4$ lb/in^2 abs. and $115\cdot 7°$F, or $115\cdot 7 + 460 = 575\cdot 7°$F abs.

$PV = wRT$ where V is the total volume in cu ft, w is the weight in pounds.

$$w = \frac{PV}{RT} = \frac{0\cdot 4 \times 144 \times 100}{53\cdot 2 \times 575\cdot 7} = 0\cdot 1881 \text{ lb. Ans.}$$

138. Assume a piston area of 1 sq foot, and that 1 lb weight of steam is admitted when cut off occurs at ¼ stroke.

For ¼ stroke:
Initial volume = 2 cu ft. Final vol. = 8 cu ft
Ratio of expansion = 4, and $\log_e 4 = 1\cdot 385$

Work done = $230 \times 144 \times 2 + 230 \times 144 \times 2 \times 1\cdot 385 - 30 \times 144 \times 8$
= $144 \times 2[230 + 230 \times 1\cdot 385 - 120]$ft lb per 1 lb of steam

For ½ stroke:
Initial volume = 4 cu ft. Final vol. = 8 cu ft
2 lb weight of steam are used
Ratio of expansion = 2, and $\log_e 2 = 0\cdot 6923$

Work done = 230 × 144 × 4 + 230 × 144 × 4 × 0·6923 — 30 × 144 × 8
   ,,   = 144 × 4[230 + 230 × 0·6923 — 60]ft lb for 2 lb steam
∴ Work done = 144 × 2 [230 + 230 × 0·6923 — 60] ft lb per lb steam.

For full stroke:
4 lb weight of steam are used
Work done = 230 × 144 × 8 — 30 × 144 × 8
          = 144 × 8 (230 — 30) ft lb for 4 lb of steam
∴ Work done = 144 × 2 (230 — 30) ft lb for 1 lb of steam

∴ Ratio of work done per 1 lb of steam =
(230+230×1·385—120) : (230+230×0·6923—60) : (230—30)
   = 428·55 : 329·229 : 200
   = 2·1428 : 1·6461 : 1   Ans.

139.

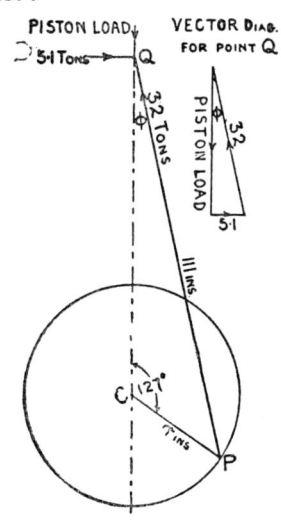

Piston load = $\sqrt{32^2 - (5\cdot1)^2}$
            = 31·6 tons. Ans.

$\sin \phi = \dfrac{5\cdot1}{32}$, $\phi = 9°\ 10'$

In triangle QCP, and by the sine rule,

$$\dfrac{r}{\sin \phi} = \dfrac{111}{\sin 127°}$$

$$r = \dfrac{111 \times 5\cdot1}{32 \times \sin 127°}$$

   = 22·15 in.
∴ Stroke = 22·15 × 2 = 44·3 in. Ans.

Angle QPC = 180° — 127° — 9° 10' = 43° 50'

In triangle QCP, and by the sine rule,

$$\dfrac{QC}{\sin 43°\ 50'} = \dfrac{111}{\sin 127°}, \quad QC = 96\cdot26 \text{ in.}$$

Distance from crosshead to shaft = 96·26 in. Ans.

SOLUTIONS TO FIRST CLASS EXAMINATION QUESTIONS 611

140.
$$\omega = \frac{120}{60} \times 2\pi = 4\pi \text{ rads per sec}$$
$$\omega r = 4\pi \times 2 = 8\pi \text{ ft/sec}$$
$$\text{Sin } 140° = \text{Sin }(180° - 140°) = \text{Sin } 40°$$
$$\text{Sin } 280° = - \text{Sin } 80° \qquad n = \tfrac{8}{2} = 4$$

$$\text{Piston velocity} = 8\pi \left\{ \text{Sin } 140° + \frac{\text{Sin } 280°}{8} \right\}$$

$$\text{,,} \qquad \text{,,} \quad = 13\cdot07 \text{ ft/sec moving down}$$

When the crank is 140° past the top centre, the eccentric driving the slide valve is 140° — (90° — 37°) = 87° past the bottom centre, and the valve is moving up

Velocity of the valve $= \omega r$ Sin 87°

$\omega = 4\pi$ rads. per sec
$r = 4$ in. $= \tfrac{1}{3}$ ft

$$\text{Velocity of the valve} = 4\pi \times \tfrac{1}{3} \times \text{Sin } 87° = 4\cdot183$$
$$\text{ft/sec moving up.}$$

Since piston and valve are moving in opposite **directions**, then their relative velocity is the sum of their velocities.

$$\text{Relative velocity} = 13\cdot07 + 4\cdot183$$
$$= 17\cdot253 \text{ ft/sec. Ans.}$$

141.

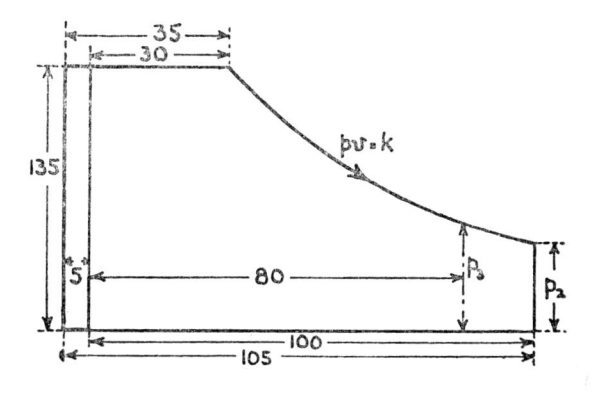

It is assumed that the clearance space is completely empty before the admission of the air.

$$\text{Let stroke} = 100$$
$$\text{then, cut off} = 0.3 \times 100 = 30$$
$$\text{and, clearance} = 5\% \text{ of } 100 = 5$$

$$\text{Ratio of expansion} = \frac{105}{35} = 3,$$
$$\therefore \text{ final vol.} = 3 \times 3 = 9 \text{ ft}^3$$
$$P_1 \times V_1 = P_2 \times V_2$$
$$135 \times 35 = P_2 \times 105$$
$$P_2 = 45 \text{ lb/in}^2$$

Final pressure = 45 lb/in² abs. Volume = 9 ft³. Ans. (a)

$$135 \times 35 = P_3 \times 85$$
$$P_3 = 55.6 \text{ lb/in}^2 \text{ abs.}$$

Pressure at 0·8 stroke = 55·6 lb/in² abs. Ans. (b)

$$\text{Work done during expansion} = P_1 V_1 \log_e r$$
$$= 144 \times 135 \times 3 \times \log_e 3$$
$$= 64000 \text{ ft lb}$$

Since the temperature is constant during isothermal expansion the internal energy must also be constant. The heat given must be equal to the external work done.

$$\text{Heat given} = \frac{64000}{778} = 82.25 \text{ Btu. Ans. (c)}$$

142.

3·7 oz per gal. Sodium Chloride ⎫ 4 oz per gall. These remain
0·3 oz per gal. Magnesium Chloride ⎬ in solution and cause the boiler density to increase.

0·4 oz per gal. Magnesium Sulphate ⎫ 1 oz per gall. These are
0·3 oz per gal. Calcium Sulphate ⎬ precipitated and do not affect the boiler density.
0·3 oz per gal. Calcium Carbonate ⎭

5·0 oz per gal. Total

The feed water density is given as 0·25 of the sea density, but as regards permanently soluble solids its density is 0·25 of 4 = 1 oz per gal.

SOLUTIONS TO FIRST CLASS EXAMINATION QUESTIONS 613

From tables, $P = 250$; $h_f = 376 \cdot 1$, $h_{fg} = 826$ Btu/lb

Heat to form 1 lb of steam normally
$$= 376 \cdot 1 - (172 - 32) + 826$$
$$= 236 \cdot 1 + 826 = 1062 \cdot 1 \text{ Btu}$$

Since the fuel consumption increases by $3 \cdot 8\%$ when blowing down, then the extra heat required to form 1 lb = $0 \cdot 038 \times 1062 \cdot 1 = 40 \cdot 36$ Btu.

This is due to sensible heat given to the water blown out,

and its weight is $\dfrac{40 \cdot 36}{236 \cdot 1} = 0 \cdot 171$ lb per lb of steam formed.

∴ when blowing, in order to form 1 lb of steam, $1 \cdot 171$ lb of feed water are required and $0 \cdot 171$ lb will be blown out.

Feed × Feed density due to permanently soluble solids = Amount blown out × Boiler density

$$1 \cdot 171 \times 1 = 0 \cdot 171 \times \text{Boiler density}$$
from which, Boiler density = $6 \cdot 85$ oz per gallon. Ans.

143. From steam tables, saturation temperature of steam at 250 lb/in² abs. is 401°F, therefore the supply steam is superheated $(601 - 401) = 200$ F°

$P = 250$ lb/in² supht. 200F°, $\quad h = 1318 \cdot 5$ Btu/lb
$P = 20$ lb/in², $\quad t_F = 228$, $\quad h_f = 196 \cdot 3$, $\quad h_{fg} = 960 \cdot 4$
$P = 0 \cdot 5$ lb/in², $\quad h_f = 47 \cdot 6$, $\quad h_{fg} = 1048 \cdot 5$

Without feed heating,
Heat per lb at $0 \cdot 5$ lb/in² and $0 \cdot 87$ dry

$$= 47 \cdot 6 + 0 \cdot 87 \times 1048 \cdot 5 = 959 \cdot 8 \text{ Btu}$$

Heat drop through turbine = $1318 \cdot 5 - 959 \cdot 8 = 358 \cdot 7$ Btu/lb

Heat supplied is that to give to 1 lb of water, having $h_f = 47 \cdot 6$ to convert to superheated steam
$$= 1318 \cdot 5 - 47 \cdot 6 = 1270 \cdot 9 \text{ Btu}$$

∴ thermal efficiency $= \dfrac{358 \cdot 7}{1270 \cdot 9} = 0 \cdot 2823$,

or $28 \cdot 23\%$ Ans. (a)

With feed heating,

0·1388 lb of steam at 20 lb/in² abs. and q dry is sufficient to raise the temperature of $1 - 0·1388 = 0·8612$ lb of water which contains 47·6 Btu/lb so that the final result is 1 lb of water at 228°F, which contains 196·3 Btu/lb

$0·1388 [196·3 + 960·4 q] + 0·8612 \times 47·6 = 196·3$
from which $q = 0·9602$ (dryness at 20 lb/in²)

Heat per lb of this wet steam $= 196·3 + 0·9602 \times 960·4$
$= 1118·8$ Btu

Total heat drop through turbine per lb of steam supplied
$= (1318·5 - 1118·8) + 0·8612 (1118·8 - 959·8)$
$= 336·63$ Btu

Heat to form 1 lb of steam $= 1318·5 - 196·3 = 1122·2$ Btu

$\therefore$ thermal efficiency $= \dfrac{336·63}{1122·2} = 0·3$, or 30%   Ans. (b)

144. From steam tables,
P = 250 lb/in² supht. 240 F°, $h = 1340$ Btu/lb
P = 30 lb/in², $h_g = 1164·6$ Btu/lb
P = 1 lb/in², $h_f = 69·7$, $h_{fg} = 1036·1$ Btu/lb

Heat drop in 1st stage $= 1340 - 1164·6 = 175·4$ Btu/lb

Heat per lb of steam at 1 lb/in² and 0·87 dry
$= 69·7 + 0·87 \times 1036·1 = 971·1$ Btu

Heat drop in 2nd stage $= 1164·6 - 971·1 = 193·5$ Btu/lb

For equal powers the heat drops in each stage must be equal.

$\therefore$ for 1 lb used in 1st stage, only $\dfrac{175·4}{193·5} = 0·9065$ lb is used in 2nd.

$1 - 0·9065 = 0·0935$ lb of each 1 lb which passes through the 1st stage is withdrawn to heat the feed water, and this is 9·35%
Ans. (a)

Heat in 0·0935 lb of dry saturated steam at 30 lb/in²
$= 0·0935 \times 1164·6 = 108·89$ Btu and this goes to heat 0·9065 lb of water at 100°F resulting in 1 lb of feed water at t°F.

… 108·89 + 0·9065 (100 — 32) = 1 (t — 32)
from which t = 202·53°F. Ans. (b)

Note that since 108·89 Btu are measured above 32°F, then the heat in the 0·9065 lb of condensate at 100°F, and in the 1 lb of feed water at 't' must also be taken above 32°F.

145. $V_1 = 0·7854 \times 1^2 \times 1·5 = 1·178$ ft^3

When $n = 1·15$:

$$P_1 V_1^{1·15} = P_2 V_2^{1·15}$$
$$15 \times 1·178^{1·15} = 60 \times V_2^{1·15}$$
from which, $V_2 = 0·3529$ ft^3

Work done per cycle $= \dfrac{n}{n-1} \{P_2 V_2 - P_1 V_1\}$

$$= \dfrac{1·15 \times 144}{1·15 - 1} \{60 \times 0·3529 - 15 \times 1·178\}$$

$= 3868$ ft lb

When $n = 1·35$:

$$P_1 V_1^{1·35} = P_2 V_2^{1·35}$$
$$15 \times 1·178^{1·35} = 60 \times V_2^{1·35}$$
from which, $V_2 = 0·4219$ ft^3

Work done per cycle

$$= \dfrac{1·35 \times 144}{1·35 - 1} \{60 \times 0·4219 - 15 \times 1·178\}$$

$= 4246$ ft lb

Increase in work done per cycle
 = 4246 — 3868 = 378 ft lb

Increase in horse power
$$= \dfrac{378 \times 300}{33000} = 3·436 \text{ Ans. (i)}$$

% increase
$$= \dfrac{378}{3868} \times 100 = 9·772\% \text{ Ans. (ii)}$$

146.

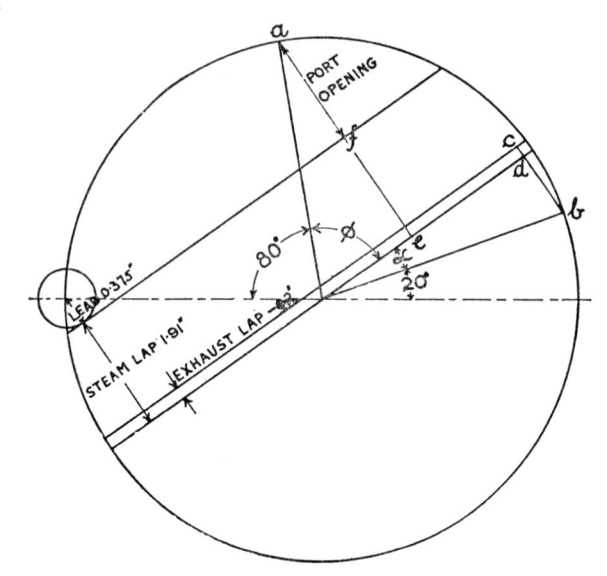

[Figure: Valve diagram showing crank positions 'a' at 80° from centre and 'b' at 20° before bottom centre, with port opening, steam lap 1·91″, exhaust lap, lead 0·375″, and angle φ marked.]

'a' is position of crank at 80° from centre. 'b' is position of crank at 20° before bottom centre

$bc = 1{\cdot}25$ in. (port opening to exhaust at position 'b')
$bd = 1{\cdot}25 - 0{\cdot}2 = 1{\cdot}05$ in.

$$\sin \alpha = \frac{1{\cdot}05}{3{\cdot}9375}, \quad \alpha = 15° \, 28'$$

Angle of advance of eccentric $= 15° \, 28' + 20° = 35° \, 28'$

$$\frac{\text{Lap} + \text{lead}}{\tfrac{1}{2} \text{travel}} = \sin 35° \, 28'$$

$$\text{Lap} + 0{\cdot}375 = 3{\cdot}9375 \sin 35° \, 28'$$
$$\text{Lap} = 1{\cdot}91 \text{ in.}$$

$\phi = 180° - 80° - 35° \, 28' = 64° \, 32'$
$ae = 3{\cdot}9375 \sin 64° \, 32' = 3{\cdot}555$ in.

ae is the displacement of the valve from mid travel when crank is at 'a', i.e., at 80° from top centre.

$ae = af + fe =$ port opening at 80° + steam lap

∴ port opening at 80° $= 3{\cdot}555 - 1{\cdot}91 = 1{\cdot}645$ in. Ans.

147.

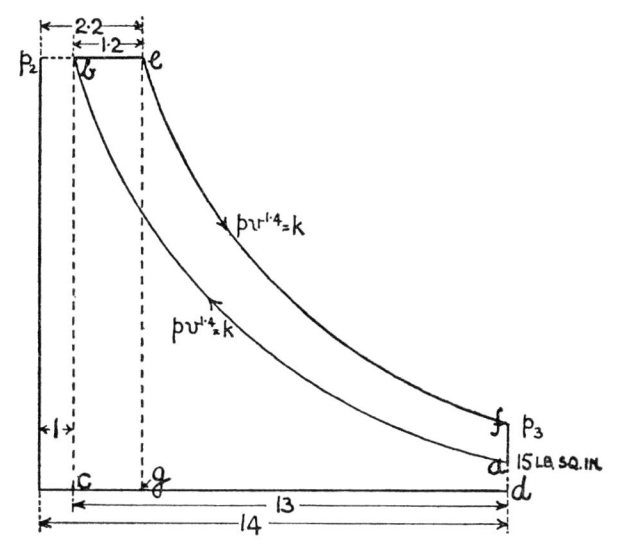

$P_1 V_1^{1.4} = P_2 V_2^{1.4}$
$15 \times 14^{1.4} = P_2 \times 1^{1.4}$, $P_2 = 603.4$ lb/in ^{2}abs.

Work done in compression = area *abcd*

$$= \frac{P_1 V_1 - P_2 V_2}{\gamma - 1}$$

$$= \frac{144}{1.4 - 1} \{15 \times 14 - 603.4 \times 1\}$$

$= 144 \times -983.5$ ft lb

(minus sign indicating work done on the air)

$603.4 \times 2.2^{1.4} = P_3 \times 14^{1.4}$, $P_3 = 45.24$ lb/in^2 abs.

Work done during expansion
= area *efdg*

$$= \frac{144}{1.4 - 1} \{603.4 \times 2.2 - 45.24 \times 14\}$$

$= 144 \times 1735.3$ ft lb

Net work done during cycle = area $befa$

$$= \text{area } begc + \text{area } efdg - \text{area } abcd$$
$$= 144\,(603\cdot4 \times 1\cdot2 + 1735\cdot3 - 983\cdot5)$$
$$= 144 \times 1475\cdot9 \text{ ft lb}$$

Mean effective pressure

$$= \text{mean height} = \frac{\text{area}}{\text{length}}$$

$$= \frac{144 \times 1475\cdot9}{13} \text{ lb/ft}^2$$

$$= \frac{1475\cdot9}{13} \text{ lb/in}^2 = 113\cdot5 \text{ lb/in}^2. \text{ Ans.}$$

148. i.h.p. of No. 1 cylinder = 500 — 364 = 136
,, ,, 2 ,, = 500 — 345 = 155
,, ,, 3 ,, = 500 — 343 = 157
,, ,, 4 ,, = 500 — 354·8 = 145·2

Total i.h.p. = 593·2

Mechanical effic. $= \dfrac{\text{b.h.p.}}{\text{i.h.p.}} = \dfrac{500}{593\cdot2}$

$$= 0\cdot8429 \text{ or } 84\cdot29\%. \text{ Ans.}$$

149. From steam tables,
$P = 150 \text{ lb/in}^2$, $h_f = 330\cdot6$, $h_{fg} = 864\cdot5$ Btu/lb
$P = 15 \text{ lb/in}^2$, $t_F = 213°F$, $h_g = 1151\cdot2$ Btu/lb

Dryness fraction from separating calorimeter:

$$q_1 = \frac{W}{W + w} = \frac{3\cdot5}{3\cdot5 + 0\cdot25} = \frac{14}{15}$$

Dryness fraction from throttling calorimeter:

$$\text{Superheat of throttled steam} = 235 - 213 = 22\ F°$$
$$\text{Heat before throttling} = \text{Heat after throttling}$$
$$330\cdot6 + q_2 \times 864\cdot5 = 1151\cdot2 + 0\cdot48 \times 22$$
$$q_2 \times 864\cdot5 = 831\cdot16$$
$$q_2 = 0\cdot9614$$

Dryness fraction of steam in steam pipe:

$$q = q_1 \times q_2 = \frac{14}{15} \times 0\cdot9614 = \mathbf{0\cdot8973}\ \text{Ans.}$$

150. Entropy before throttling $+ 0\cdot022 =$ Entropy after throttling
      (liquid only)   (liquid and gas)

$$0\cdot66\ \log_\varepsilon \frac{298}{233} + 0\cdot022 = 0\cdot66\ \log_\varepsilon \frac{263}{233} + \frac{q \times 63}{263}$$

$$0\cdot66\ \log_\varepsilon \frac{298}{263} + 0\cdot022 = \frac{q \times 63}{263}$$

$$0\cdot08243 + 0\cdot022 = 0\cdot2395\ q$$
$$0\cdot10443 = 0\cdot2395\ q$$
$$q = \mathbf{0\cdot436}\ \text{Ans.}$$